Microscopic Aspects of Nonlinearity in Condensed Matter

NATO ASI Series

Advanced Science Institutes Series

A series presenting the results of activities sponsored by the NATO Science Committee, which aims at the dissemination of advanced scientific and technological knowledge, with a view to strengthening links between scientific communities.

The series is published by an international board of publishers in conjunction with the NATO Scientific Affairs Division

A	**Life Sciences**	Plenum Publishing Corporation
B	**Physics**	New York and London
C	**Mathematical and Physical Sciences**	Kluwer Academic Publishers
D	**Behavioral and Social Sciences**	Dordrecht, Boston, and London
E	**Applied Sciences**	
F	**Computer and Systems Sciences**	Springer-Verlag
G	**Ecological Sciences**	Berlin, Heidelberg, New York, London,
H	**Cell Biology**	Paris, Tokyo, Hong Kong, and Barcelona
I	**Global Environmental Change**	

Recent Volumes in this Series

Volume 261—Z° Physics: *Cargèse 1990*
edited by Maurice Lévy, Jean-Louis Basdevant, Maurice Jacob, David Speiser, Jacques Weyers, and Raymond Gastmans

Volume 262—Random Surfaces and Quantum Gravity
edited by Orlando Alvarez, Enzo Marinari, and Paul Windey

Volume 263—Biologically Inspired Physics
edited by L. Peliti

Volume 264—Microscopic Aspects of Nonlinearity in Condensed Matter
edited by A. R. Bishop, V. L. Pokrovsky, and V. Tognetti

Volume 265—Fundamental Aspects of Heterogeneous Catalysis Studied by Particle Beams
edited by H. H. Brongersma and R. A. van Santen

Volume 266—Diamond and Diamond-Like Films and Coatings
edited by Robert E. Clausing, Linda L. Horton, John C. Angus, and Peter Koidl

Volume 267—Phase Transitions in Surface Films 2
edited by H. Taub, G. Torzo, H. J. Lauter, and S. C. Fain, Jr.

Series B: Physics

Microscopic Aspects of Nonlinearity in Condensed Matter

Edited by

A. R. Bishop

Los Alamos National Laboratory
Los Alamos, New Mexico

V. L. Pokrovsky

Landau Institute for Theoretical Physics
Academy of Sciences of the USSR
Moscow, USSR

and

V. Tognetti

University of Florence
Florence, Italy

Plenum Press
New York and London
Published in cooperation with NATO Scientific Affairs Division

Proceedings of a NATO Advanced Research Workshop on
Microscopic Aspects of Nonlinearity in Condensed Matter,
held June 7-13, 1990,
in Florence, Italy

Library of Congress Cataloging-in-Publication Data

NATO Advanced Research Workshop on Microscopic Aspects of Nonlinearity in Condensed Matter (1990 : Florence, Italy)
Microscopic aspects of nonlinearity in condensed matter / edited by A.R. Bishop, V.L. Pokrovsky, and V. Tognetti.
p. cm. -- (NATO ASI series. Series B, Physics ; vol. 264)
"Published in cooperation with NATO Scientific Affairs Division."
"Proceedings of a NATO Advanced Research Workshop on Microscopic Aspects of Nonlinearity in Condensed Matter, held June 7-13, 1990, in Florence, Italy"--T.p. verso.
Includes bibliographical references and index.

1. Condensed matter--Congresses. 2. Statistical mechanics--Congresses. 3. Nonlinear theories--Congresses. 4. Mathematical physics--Congresses. I. Bishop, A. R. (Alan R.) II. Pokrovsky, V. L. III. Tognetti, V. IV. North Atlantic Treaty Organization. Scientific Affairs Division. V. Title. VI. Series: NATO ASI series. Series B, Physics ; v. 264.
QC173.4.C65N3716 1990
530.4'1--dc20 91-20200
CIP

ISBN-13: 978-1-4684-5963-0 e-ISBN-13: 978-1-4684-5961-6
DOI: 10.1007/978-1-4684-5961-6

A Division of Plenum Publishing Corporation
233 Spring Street, New York, N.Y. 10013
Softcover reprint of the hardcover 1st edition 1991

SPECIAL PROGRAM ON CHAOS, ORDER, AND PATTERNS

This book contains the proceedings of a NATO Advanced Research Workshop held within the program of activities of the NATO Special Program on Chaos, Order, and Patterns.

Volume 208—MEASURES OF COMPLEXITY AND CHAOS
edited by Neal B. Abraham, Alfonso M. Albano, Anthony Passamante, and Paul E. Rapp

Volume 225—NONLINEAR EVOLUTION OF SPATIO-TEMPORAL STRUCTURES IN DISSIPATIVE CONTINUOUS SYSTEMS
edited by F. H. Busse and L. Kramer

Volume 235—DISORDER AND FRACTURE
edited by J. C. Charmet, S. Roux, and E. Guyon

Volume 236—MICROSCOPIC SIMULATIONS OF COMPLEX FLOWS
edited by Michel Mareschal

Volume 240—GLOBAL CLIMATE AND ECOSYSTEM CHANGE
edited by Gordon J. MacDonald and Luigi Sertorio

Volume 243—DAVYDOV'S SOLITON REVISITED: Self-Trapping of Vibrational Energy in Protein
edited by Peter L. Christiansen and Alwyn C. Scott

Volume 244—NONLINEAR WAVE PROCESSES IN EXCITABLE MEDIA
edited by Arun V. Holden, Mario Markus, and Hans G. Othmer

Volume 245—DIFFERENTIAL GEOMETRIC METHODS IN THEORETICAL PHYSICS: Physics and Geometry
edited by Ling-Lie Chau and Werner Nahm

Volume 256—INFORMATION DYNAMICS
edited by Harald Atmanspacher and Herbert Scheingraber

Volume 260—SELF-ORGANIZATION, EMERGING PROPERTIES, AND LEARNING
edited by Agnessa Babloyantz

Volume 263—BIOLOGICALLY INSPIRED PHYSICS
edited by L. Peliti

Volume 264—MICROSCOPIC ASPECTS OF NONLINEARITY IN CONDENSED MATTER
edited by A. R. Bishop, V. L. Pokrovsky, and V. Tognetti

PREFACE

The collection of articles in this volume represents the proceedings of the inaugural Conference of the Institute of Theoretical Matter Physics at the University of Florence, Italy, June 7-13, 1990. The aim of this Institute (sponsored by the Consorzio Interuniversitario di Fisica della Materia, INFM) is to foster interdisciplinary approaches to theoretical problems in Solid State, Electronics and Optics. Correspondingly, the Conference surveyed a broad scope of modern analytic techniques in nonlinear science and their application to novel liquids, solids and optical environments. The Conference was held at Centro Studi CISL, a magnificent Villa up to Florence.

The major topics addressed through survey and specialized lectures, discussions and posters were: *Complexity and Coherence* in spin glasses, associative distributive memory, low dimensional electronic materials, magnets, Josephson junction arrays, optics, semiconductors, convection cells, material science, high-temperature superconductivity; *Physics of Quantum Devices* such as nonlinear feedback oscillators, macroscopic quantum tunneling, hetero structures, quantum ballistic devices, tunneling in atomic devices; *Field Theory and Statistical Mechanics*, especially for soliton-bearing and quantum chaotic systems.

Many exciting ideas and results were presented and debated in lively fashion. Perhaps two major themes were evident throughout the workshop.

> *Firstly*, "nonlinear science" is now emerging as a combination of two complentary paradigms: (a) The possibility of chaotic or irregular dynamics even in small degree-of-freedom systems (as, e.g., in period-doubling routes to chaos in one-dimensional maps) or chaotic spatial patterns in certain static, discrete lattices (as, e.g., in commensurate-incommensurate phase transition models). The common ingredient here is the presence of competing frequency and/or length scales. Indeed synergistic mappings frequently exist between long-time attractors of time-dependent problems and ground states of time-independent problems in a higher spatial dimension; and (b) The appearance of coherent collective structures in nonlinear space-time systems - structures such as solitons, dislocations, vortices. This duality of chaos and coherence in extended nonlinear systems results in the many fascinating current examples of pattern formation and competition, low-dimensional nonlinear mode reductions and low-dimensional chaos, and so forth. Of course the collective-coordinate description of such coherent structures, of their dynamics and interactions, and of their self-organized mesoscale patterns, is essential to establishing a bridge between microscopic and macroscopic properties of materials.

Secondly, the interface between controlled condensed matter experiments and realistic phenomena in material science, optics, etc., is proving to be extremely fertile. Imaginative, laboratory scale condensed matter experiments are able to isolate essential ingredients of material science leading to complex patterns, dynamics and response. This is where interdisciplinary interactions are, and surely will continue to be, a true driving force for applications of the paradigms of nonlinear science.

Sponsorship of the Conference was generously provided by the NATO (ARW 900372) through the special program Chaos Order and Patterns (COP), the Consorzio di Fisica della Materia (INFM), the Italian National Science Council (CNR) and the Department of Physics of the University of Florence. We are also deeply appreciative of the expert secretarial support from Genny Ferasin, both during the Conference and in the preparation of manuscripts.

Firenze, January 1991

A.R. Bishop
V.L. Pokrovskii
V. Tognetti

CONTENTS

I. Complexity and Chaos

II. Coherent Structures

III. Physics of Quantum Devices

IV. Field Theory and Statistical Mechanics

SPACE TIME COMPLEXITY IN QUANTUM OPTICS

F.T. Arecchi

Dept. of Physics of the University and
Istituto Nazionale di Ottica, Florence - Italy

Abstract

Two recent experiments in quantum optics, namely, i) a waveguide laser supporting many transverse modes, and ii) an optical cavity with a photorefractive gain medium and a variable aperture have displayed controllable routes to space complexity. This fills the gap between the single mode dynamics and the many domain turbulent-like behavior, which so far was unreachable for a radiation field. Due to the high accuracy of optical measurements, we foresee a precise way to test many conjectures formulated for fluids or other nonlinear field problems. Thus we have called this new research area "dry hydrodynamics".

In particular I show the first experimental evidence of two phenomena, recently described theoretically, namely:

i) chaotic itinerancy = self induced switching among different slow manifolds

ii) Space-Time Chaos (STC) = high dimensional chaos, with strongly non Gaussian statistics in real space, but with a Gaussian spectral statistics, up to a critical wave number given by the reciprocal of the correlation length.

1- Introduction

Self organization evolved from a qualitative concept used by bio-mathematicians in the therties [1] to a quantitative appraisal of a physical phenomenon as soon as physicists were able to pick up a low-dimensional laboratory system displaying a satisfactory agreement between its experimental behavior and the corresponding theoretical model. While theoretical models were worked out for non-equilibrium transitions in fluids and chemical reactions, the first clear experimental evidence was provided in the middle sixties by the single mode laser. This system displayed an outstanding agreement between theory and experimental tests carried by photon counting statistics [2].

Later on, similar threshold measurements were performed at the onset of instability phenomena in fluid convection or in chemical reactions, thus opening the field of the so called "non equilibrium phase transitions" [3]. However in fluids it is straightforward to scale up the system size from small to large cells, thus making it possible to explore in many ways the passage from systems coherent (fully correlated) in space to systems made up of many uncorrelated, or weakly correlated, domains.

Crucial questions such as: i) the passage from order to chaos within a single domain and ii) possible synchronizations of time behaviors at different space domains, have been addressed in the past years, with the general idea in mind that space-time organization is what makes a large scale object complex, that

Microscopic Aspects of Nonlinearity in Condensed Matter
Edited by A.R. Bishop *et al.*, Plenum Press, New York, 1991

is, richer in information than the sum of the elementary constituents, due to an additional non trivial cross- information [4].

Thus far, such an investigation was not possible in the optical field, because all coherent optics is based on Schawlow-Townes original idea of a drastic mode selection [5]. The opposite, thermodynamic limit of optics was explored early in this century [6]. A confrontation of the two limits was shown experimentally in terms of photon statistics [7] in 1965, however what is in between has not been adequately explored, at variance with hydrodynamic instabilities or unstirred chemical reactions.

Here I show very recent evidence of space-time complexity in optics. The experimental configurations which have made possible to fulfill this twenty-year long search appear so promising that we can foresee an extensive investigation of space phenomena in optics along the coming decade. Let me call this area of investigation "dry hydrodynamics".

The paper is organized as follows. In Sec. 2 I present a qualitative description of the phenomena which characterize low dimensional chaos, that is, chaotic attractors with space semplicity, then space-time chaos (STC), as well as the transition between the two regimes. In order to stimulate the theoretical interest, we try to connect these phenomena with the present theoretical understanding, even though the experimental findings were not motivated by preliminary theories. On the contrary, they have been based on powerful heuristic conjectures [8], and search for satisfactory theories is still in progress. Secs. 3 and 4 are devoted to the presentation of the experimental systems investigated by my Group and of the preliminary results available.

2 - Periodical alternance, chaotic itinerancy and STC

In this Section I anticipate and explain what we are going to see in the experiments reported in the later Sections. To appreciate the role of space coupling let me summarize the present status of affairs in quantum optics. Since all coherent phenomena take place in a cavity mainly extended in a z-direction (as e.g. the Fabry-Perot cavity), we expand the field $e(r,t)$, which obeys the wave equation

$$\Box^2 e = -\mu \ddot{p} \tag{1}$$

(where $p(r,t)$ is the induced polarization), as

$$e(\vec{r},t) = E(x,y,z,t)e^{-i(\omega t - kz)} \tag{2}$$

If the longitudinal variations are mainly accounted for by the plane wave, then we can take the envelope E as slowly varying in t and z with respect to the variation rates ω and k in the plane wave exponential. Furthermore we call P the projection of p on the plane wave. By neglecting second order envelope derivatives it is easy to approximate the operator on E as

$$\Box^2 \rightarrow 2ik\left(\partial_z + \frac{1}{c}\partial_t\right) + \partial_x^2 + \partial_y^2 \tag{3}$$

as is usually done in the eikonal approximation of wave optics. This further suggests three relevant physical situations.

2.1 (1+0) - dimensional Optics

In such a case there is only a time dependence and no space derivatives, that is, $\Box^2 \rightarrow 2i\omega d_t$. Assuming that the laser cavity is a cylinder of length L, with two mirrors of radius a at the two ends, the cavity resonance spectrum is

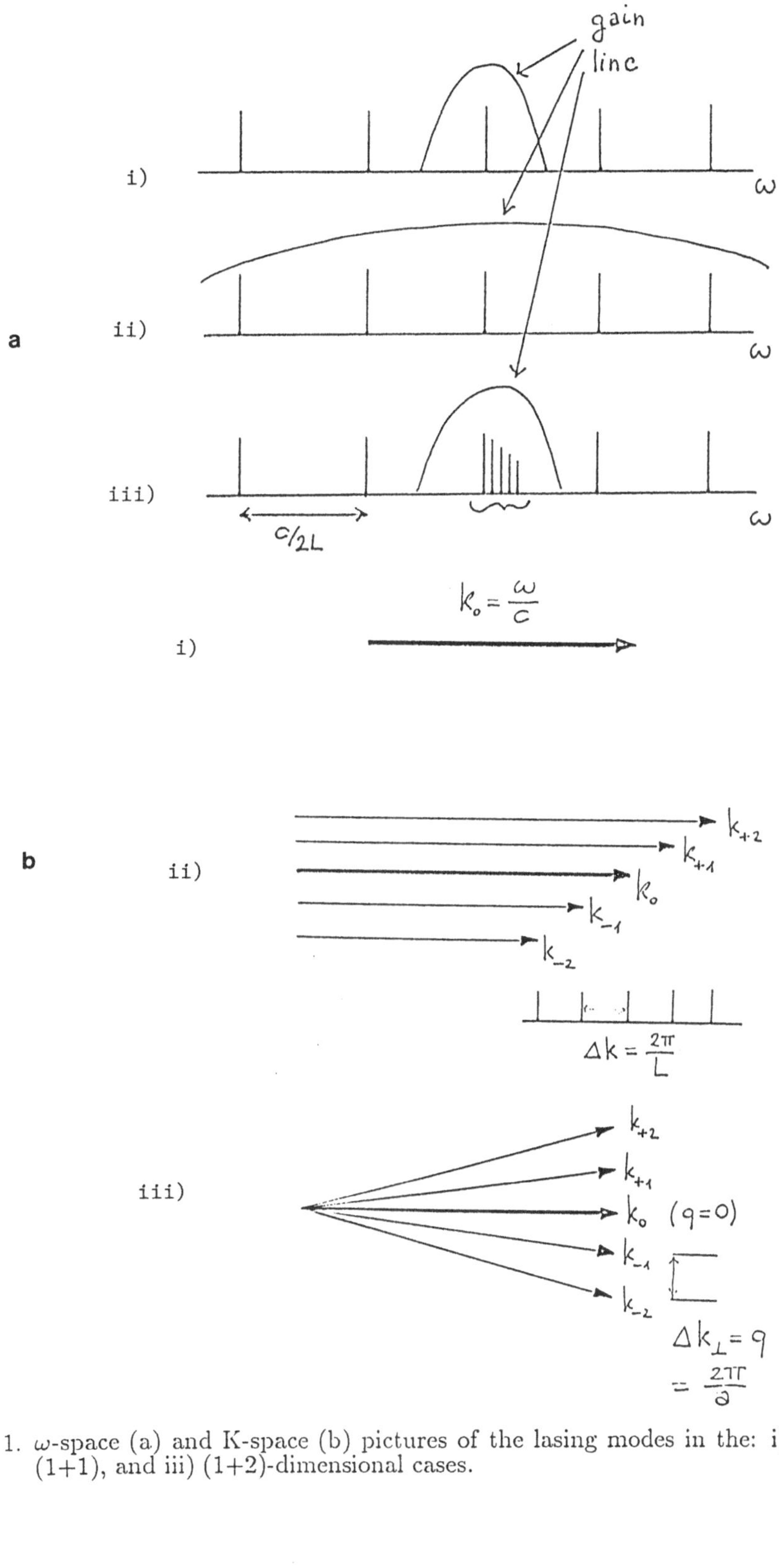

Fig. 1. ω-space (a) and K-space (b) pictures of the lasing modes in the: i) (1+0), ii) (1+1), and iii) (1+2)-dimensional cases.

made of discrete lines separated by $c/2L$ in frequency, each one corresponding to an integer number of half wavelengths contained in L, plus a crown of quasi-degenerate transverse modes at the same longitudinal wavenumber, with their propagation vectors separated from each-other by a diffraction angle λ/a (Fig. 1b).

This case corresponds to a gain line narrower than the longitudinal frequency separation (so called free spectral range) and to a Fresnel number

$$F \equiv \frac{a^2}{\lambda L} \tag{4}$$

of the order of unity, so that the first off axis mode already escapes out of the mirror. Intuitively F is the ratio between the geometric angle a/L of view of one mirror from the other and the diffractive angle λ/a.

In such a case, the resulting ODE replacing the wave PDE has to be coupled to the matter equations giving the evolution of P. In the simple case of a cavity mode resonant with the atomic line, we obtain the so called Maxwell-Bloch equations [9].

$$\begin{aligned} \dot{E} &= -\gamma_E E + gP \\ \dot{P} &= -\gamma_\perp P + gEN \\ \dot{N} &= \gamma_\| N - 2gEP + A \end{aligned} \tag{5}$$

where N is the population inversion (we have modeled the gain medium as a collection of two level atoms), γ_E, $\gamma_\perp$ and $\gamma_\|$ are the loss rates of E, P and N respectively, g is the field-matter coupling constant and A is the pump rate.

Eqs. (5) are isomorphic to Lorenz equations for a model of convective fluid instability. Being three nonlinear equations, they provide the minimal conditions for deterministic chaos. However time scale considerations can rule out some of the three dynamical variables, yielding a dissipative dynamics with only one variable (fixed point attractor) or two variables (limit cycle attractor). Only when the three damping rates γ_E, $\gamma_\perp$ and $\gamma_\|$ are of the same order of magnitude, we do have a three-equation dynamics and hence possibility of a chaotic motion (strange attractor). The above three cases have been classified as class A,B, and C lasers respectively.

A comprehensive review of experiment and theory for these single-domain, (1+0)-dimensional systems is given in the book cited in Ref. 9, covering the period 1982-87 over which these space invariant instabilities have been studied.

2.2 - (1+1) - dimensional Optics

Here, we have a cavity thin enough to reject off axis modes, but fed by a gain line wide enough to overlap many longitudinal modes. The superposition of many longitudinal modes means that one must retain the z gradient. Thus the wave equation reduces to

$$(\partial_t + c\partial_z)E = GP \tag{6}$$

where G is a scaled coupling constant.

Having a PDE, any mode expansion with reasonable wavenumber cut offs provides a large number of coupled ODE's, thus it is immaterial whether P and N are adiabatically eliminated, as in class A and B lasers, or whether they keep their dynamical character as in class C laser. Anyway, we have enough equations to see space time chaos.

Equations as (6) have been solved numerically in the sixties [10], to explain space variations on a length scale much smaller than L, as seen in regular or

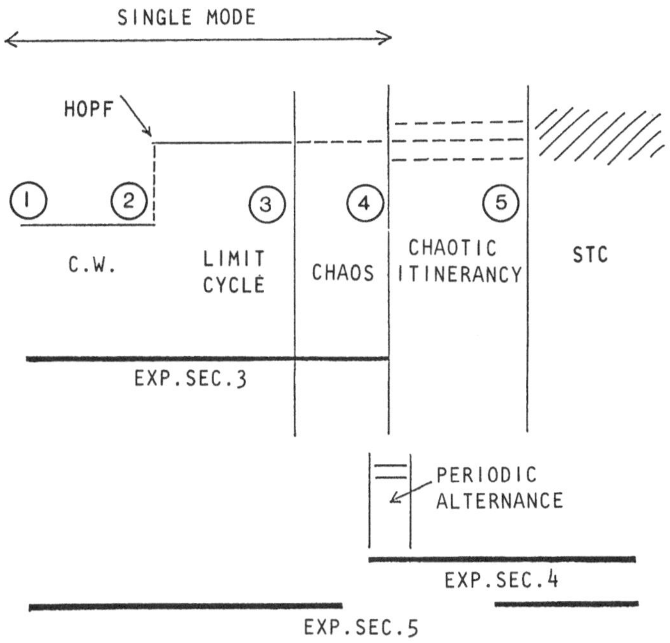

Fig. 2. Space-time complexity in optics. Qualitative plots of different behaviors observed in laboratory experiments and in numerical solutions of model equations. In Sec. 4 we report periodic alternance of modes instead of chaos in a single mode, between thresholds 3 and 4.

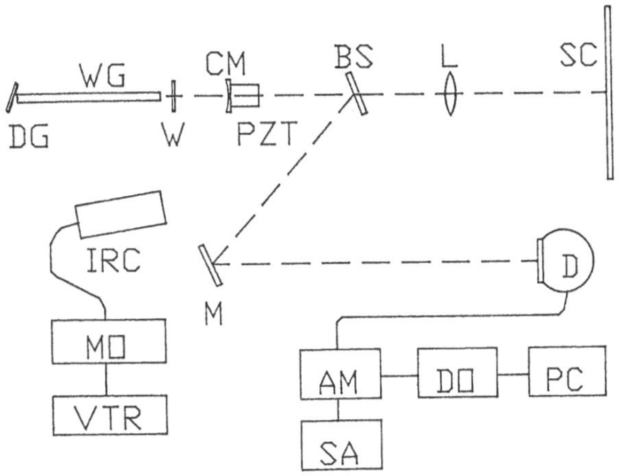

Fig. 3. Experimental set up: DG, diffraction grating: WG, Pyrex hollow cylindrical guide; W, ZnSe window; CM, coupling mirror; PZT, piezotranslator; BS, beam splitter (10%); S, screen; IRC, infrared camera; Mo, monitor; VTR, videotape recorder; D, detector; A, video-band amplifier; DO, digital oscilloscope; PC, personal computer; SA, spectrum analyzer (20-40 MHz).

erratic mode locking [11]. Fig. 2 collects a sequence of possible behaviors as one increases an intensive control parameter (the pump to loss ratio) for a cavity long enough to provide a high ratio of gain linewidth to free spectral range, or alternatively, as one increases an extensive parameter, that is, the latter ratio for a fixed pump- to-loss ratio. Since the free spectral range is given by $c/2L$, increasing the extensive parameter amounts to increasing the cavity length L.

The circled numbers 1 to 5 in Fig. 2 denote the transition points, and the solid lines denote the range of experimental observations reported in Secs. 3 to 5.

Threshold n. 1 is the usual laser threshold, whereby uncorrelated spontaneous emission selforganizes into single mode coherent laser action. Mathematically Eqs.(5) provide a pitchfork bifurcation, with critical divergence of the fluctuation amplitude and correlation time (critical slowing down). These transition phenomena have been experimentally demonstrated in the middle sixties in a series of experiments reported in Ref. 2.

An analogy with the Landau model of a single domain phase transition (no space features) stems from the trivial properties of such a bifurcation [12]. In order to consider space variations, one must couple Eq.(6) with the matter equations (second and third Eq.(5)). This was done by a mode expansion of Eq.(6), and a second threshold, n.2, which is a Hopf bifurcation toward an oscillatory regime, was introduced [13]. A more appropriate analogy with thermodynamic phase transitions, including space correlations, was developed on such a new set of equations [14].

A third threshold marks the onset of deterministic chaos in a single mode laser. In fact it is a cascade of bifurcations depending on the specific route to chaos, which is influenced by possible laboratory perturbations, as modulations or feedback [9]. The isomorphism of Eqs.(5) with Lorenz equations was first pointed out by Haken [15]. After that, a large amount of experimental and theoretical investigation has been devoted to chaos in a single mode laser.

Recent consideration of a space extended optical system [16] by a model made up of Eq.(6) and the last two Eqs.(5) has shown evidence of a further behavior, called "chaotic itinerancy". It consists in the jump from one slow manifold to another, i.e., from one quasi-attractor to another. At any time, a single mode with a chaotic behavior is present, but after a while it is replaced by another mode, and so on. Alternatively, in Sec. 4 we will show experimental evidence of a non chaotic, but periodic alternance of modes, schematized in the lower part of Fig. 2. As stressed in Sec. 4, the main indicator of chaotic itinerancy is that, while a local measurement provides a chaotic signal, measurement of the space correlations provides a highly correlated signal.

Above the threshold n. 5 we enter a new regime, called spatio- temporal chaos (STC) where a large number of modes coexist. This regime has been characterized on very general grounds by Hohenberg and Shraiman [8]. Suppose that we have a generic field $u(r,t)$ ruled by a PDE including nonlinear and gradient terms. Such is indeed the situation of our (1+1)- dimensional optical system. Let us take the field of deviations away from the local time average

$$\delta u(r,t) = u(r,t) - < u(r,t) >$$

where $< ... >$ denote time average. Under very broad assumptions, we can take the leading part of the correlation function as an exponential, that is,

$$C(r,r') = < \delta u(r,t) \delta u(r',t) > \simeq e^{-|r-r'|/\xi} \qquad (7)$$

Whenever the correlation length ξ is larger than the system size $L(\xi > L)$ we have low dimensional chaos, that is, even though the system can be chaotic in time, it is coherent in space (single mode, in a suitable mode expansion). The corresponding chaotic attractor is low dimensional. In the opposite limit of $\xi << L$, a local chaotic signal is not confined in a low dimensional space. However a new outstanding feature appears. If we collect a local time series of data $\delta u(r,t)$ at a given point r, the corresponding statistical distribution $P(\delta u(r,t))$ is strongly non-Gaussian. No wander about that: after all δu stems

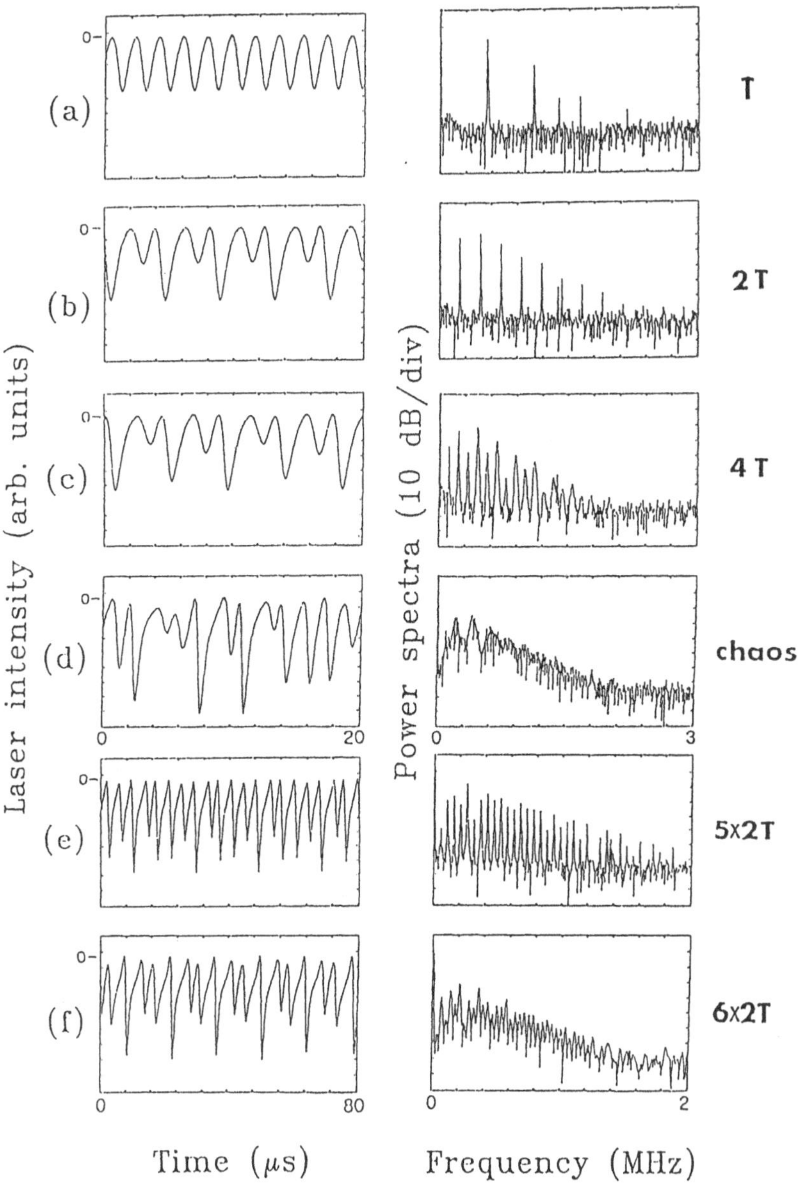

Fig. 4. Temporal behavior of the laser intensity (left) and related power spectra (right) for increasing values of the PZT voltage. The sequence shows a subharmonic route to chaos, plus two periodic windows beyond the onset of chaos. T indicates the period. Left-hand side: (a)-(d) 2 us/div; (e) and (f) 8 us/div. Right-hand side: (a)-(d) 300 kHz/div; (e) and (f) 200 kHz/div.

from a strongly nonlinear dynamics. But if we now Fourier-transform $\delta u(r,t)$, the corresponding dynamical variable$\delta u(q,t)$ in wave number space displays a Gaussian statistics. This is somewhat surprising, because a Fourier transform is a linear operation and can not introduce a Gaussian property where that was absent.

Considering however the Fourier transform

$$\delta u(q,t) = \int dr e^{-iqr} \delta u(r,t) \tag{8}$$

we realize that, up to a cut off wave number $q_c = 1/\xi$, the phase factor qr changes very little if r is confined within a segment smaller than ξ. Thus, in the phased sum (8) we can replace δu with its coarse grained approximation

$$\delta \bar{u}(r,t) \equiv \frac{1}{\xi} \int_r^{r+\xi} \delta u(r',t) dr' \tag{9}$$

where $\delta \bar{u}$ is an average over a correlation length.

With this replacement, (8) becomes a sum of uncorrelated objects and hence it has a Gaussian distribution, by the central limit theorem. We can say that the coarse graining operation has been the nonlinear device distorting the statistics. The first experimental evidence of such an STC feature was given by our Group (see Secs. 4 and 5).

2.3 - (1+2) - dimensional Optics

As shown in Fig. 1 iii), let us consider a gain line allowing for a single longitudinal mode, but take a Fresnel number high enough to allow for many transverse modes.

We rescale the transverse coordinates x, y with respect to the cross cavity size a, and the time t to the longitudinal photon lifetime $L/(cT)$, where T is the mirror transmittivity. The new variables are

$$\begin{aligned} x' &= x/a \quad , \quad y' = y/a \\ t' &= \frac{t}{L/cT} \end{aligned} \tag{10}$$

Furthermore we neglect the longitudinal gradient. Then, as shown by Eq.(3), the wave equation reduces to

$$\left(\partial_{t'} - i\alpha \nabla_\perp^2\right) E = GP \tag{11}$$

where $\nabla_\perp^2$ is the transverse Laplacian and

$$\alpha = \frac{1}{4\pi F T} \tag{12}$$

As in the (1+1) case, Eq.(11) must be coupled with the material equation. If P has a fast relaxation toward a local equilibrium with the field, and if we expand its dependence to the lowest orders, we have a relation as

$$P = aE - b \mid E \mid^2 E \tag{13}$$

Introducing this into Eq.(11), one has a nonlinear Schrodinger equation (NLS) which has been recently considered in many theoretical investigations [17,18].

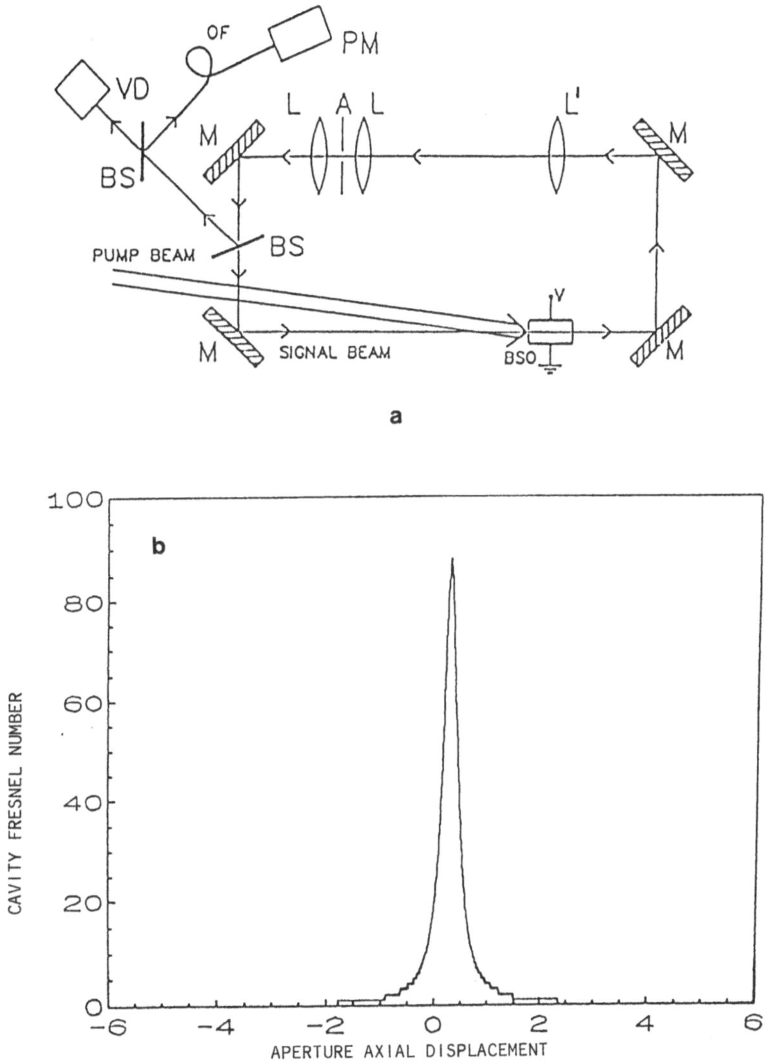

Fig. 5. a) Experimental set up. Video camera VD records the wavefront pattern; photomultiplier PM measures the time evolution at a point selected by fiber OF.
b) Effect of the variable pupil on the cavity Fresnel number F. The horizontal axis reports the displacement of aperture A away from the focus of lens L. The slight asymmetry is due to the presence of lens L'.

On the other hand, important considerations have been developed for the complex Ginzburg-Landau equation (CGL) [19]. This can be written as

$$\partial_t u = (\alpha_i + i\alpha_2)\nabla^2 u - \mu u - (1 - i\beta) \mid u \mid^2 u \tag{14}$$

The CGL includes NLS (for $\alpha_1 = 0$) the chemical reaction-diffusion equation (for $\alpha_2 = 0$) and the single mode laser equation (for $\alpha_1 = \alpha_2 = 0$).

Writing the complex field as $u = \mid u \mid e^{i\phi}$, under broad assumptions one may localize regimes where the modulus has small variations, whereas the phase has crucial changes described by the Kuramoto-Sivashinski equation [20].

The CGL and the corresponding phase equation have played a fundamental role in understanding many features of hydrodynamic turbulence [21], including the formation of cellular patterns [22] and of STC [8]. The local equilibrium assumption (13) may appear too naive. In fact, an improved center manifold treatment shows that in this case the laser equations are well approximated by a full CGL [23].

We are still far from a full knowledge of all the potentialities of CGL. Coullet and coworkers [19,24] exploring a parameter range around $\mu = 1, \ \alpha_1 = 1, \ \alpha_2 = -1 \div +1, \ \beta = 0.5 \div 3$ have described the onset of vortices as topological defects as well as defect induced turbulence. Vortex formation has also been explored by Lugiato et al. [23,25] by a modal expansion of Eq.(14). While the vortices of Coullet result from a space-time integration of Eq.(14) with a grid of a very fine resolution, the phase singularities of Lugiato et al. emerge from the interaction among a few modes, and thus results from a simple analitical treatment. Notice that the evidence of defect mediated turbulence is only numerical [24] and the phase singularity theory [25] was developed to explain the experimental evidence of steady vortices [26].

In order to describe phenomena such as chaotic itinerancy and STC, the CGL must be considered in a completely different parameter regime. This has been done by Otsuka and Ikeda [27]. They used a discretized version of CGL on $N = 5$ sites with a space coupling which mimics the second derivative i.e.:

$$\nabla^2 u \rightarrow u_{i+1} + u_{i-1} - 2u_i \quad (i = 1, ...N)$$

and give solutions for $\mu = 1, \alpha_1 = \alpha_2 = 0.1$, and β increasing for 10 to 30.

For increasing β the system displays a variety of dynamical behaviors as shown in Fig. 2. It is crucial to notice that in Ref. 27 the transverse space coordinate is 1-dimensional (1D), whereas Coullet vortices refer to a 2-dimensional case (2D).

The dynamical difference between 1D and 2D CGL equations is discussed in Ref. 28, where a new feature is theoretically presented, that is, a regime of strong turbulence, which manifests mathematically as follows: steady state (i.e. time asymptotic) pointwise estimates of $\mid u \mid$ may be many orders of magnitude greater than the estimates of spatial and temporal averages. When this occurs, then these fluctuations must be narrow in both space and time (intermittent spikes). In the coming Sections we offer experimental evidence of the phenomena collected in Fig. 3 and that we have seen as associated with 1D CGL equation [(1+1- dimensional dynamics)]. Since some of our physical systems are (1+2), we should expect future evidence of intermittent spikes, not yet observed.

3 - The CO-2 waveguide laser [29]

We have devised a gas laser with a waveguide gain tube, so that many off-axis modes will be confined with almost equal losses. The mode selection is then provided by the angular acceptance of an external curved mirror (CM in Fig. 3). When the CM is far away from the right tube end, we have a standard

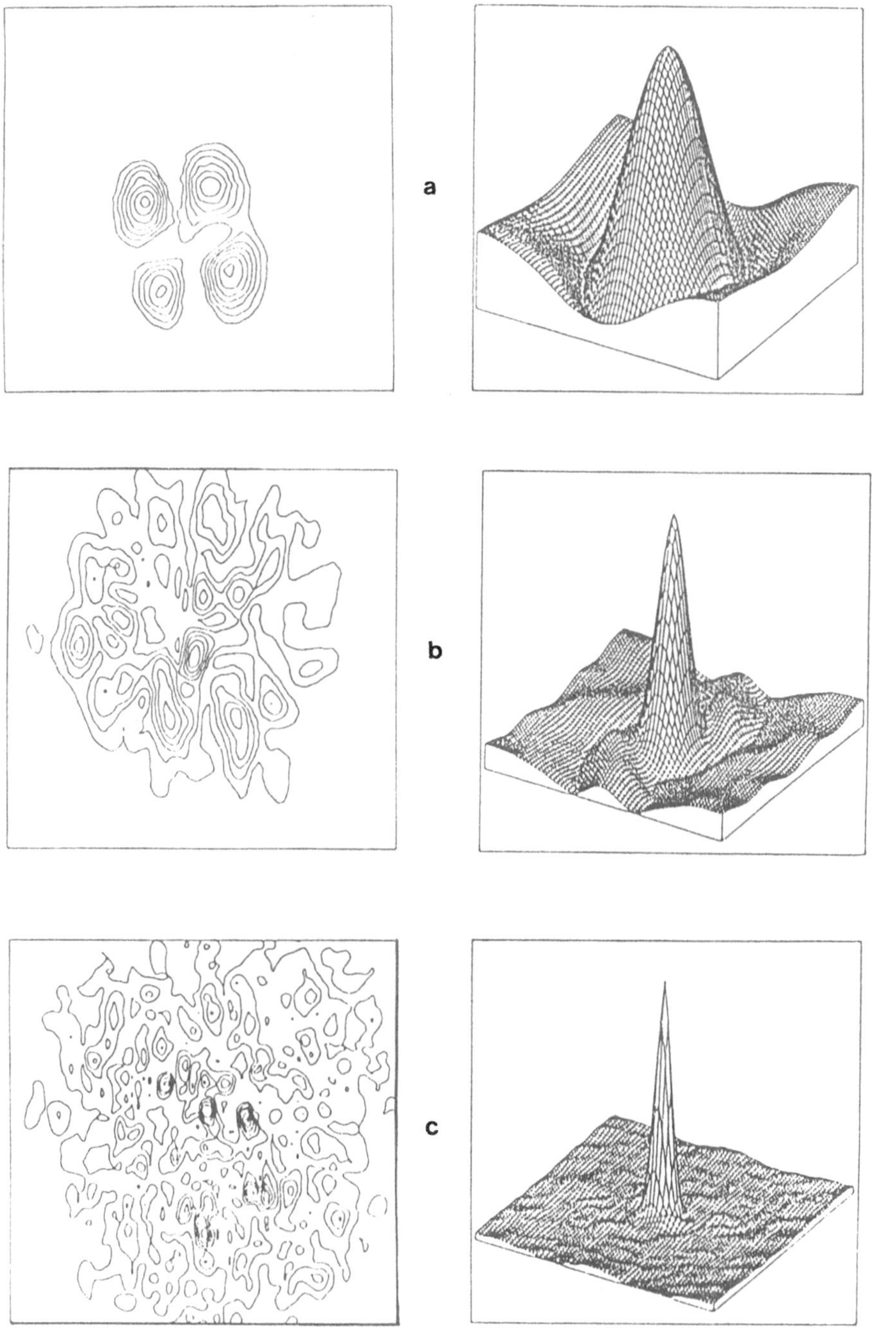

Fig. 6. Intensity distribution on the wavefront (left) and space autocorrelation function (right) for increasing Fresnel number.

a) $F = 5$, one single mode at a time is present, ratio between coherence length ξ and frame size D is $\xi/D \simeq 1$.

b) $F = 20$, $\xi/D \simeq 0.25$

c) $F = 70$, $\xi/D \simeq 0.1$

single-mode operation, as CM is put close to the tube, the laser oscillates over a large number of modes.

Here we show the spontaneous onset of chaos due to the nonlinear interaction of two transverse modes when the cavity length is driven slightly away from the stable operation. Our system, shown in Fig. 3, consists of a CO_2 waveguide laser resonator that contains a hollow Pyrex guide of circular section with $4mm$ inner diameter and $88cm$ length, which is terminated by a window W (anti-reflection coated on both surfaces). The optical cavity consists of the waveguide, coupled directly on the left-hand side to a diffraction grating, selecting the P(20) rotational line at $10.6\mu m$, and on the right-hand side, to an external coupling mirror CM (80% reflectivity). This spherical mirror with 2 m of radius of curvature can be placed at various distances from the end of the guide. The coupling mirror is mounted on a piezoelectric translator (PZT) which allows micrometric control of the cavity length. The free spectral range of the optical cavity is large enough to avoid simultaneous multilongitudinal mode emission. The gas flows at a pressure of 40 mbar, providing a homogeneously broadened linewidth for the laser transition. The excitation of the medium is obtained by means of a dc discharge inside the guide, current stabilized to 0.1%.

The spatial profile of the modes has been visualized with an infrared camera by projecting on a wide screen the laser beam expanded by a germanium lens. The size of the screen has been chosen in order to avoid distortions of the image due to the speckle pattern (typical of laser-diffused radiation). We follow the temporal behavior of the intensity by means of a photodiode. The signal is then amplified and sent to a digital oscilloscope, controlled by a PC.

By careful setting of the coupling mirror, we can obtain simultaneous oscillation of two or more transverse modes, characterized by a complication in the spatial profile of a laser emission and by the presence of beat notes in the output intensity, in the range 1-40 MHz. In particular, for a mirror distance of $25cm$, the pattern corresponds to a stationary emission. Starting from this pattern we find, by a slight increase of the cavity length with the PZT, the onset of a single note beat at about 2 MHz (two-mode operation). Then this frequency slows down to 400 kHz, loosing in this process the *dc* component typical of beating phenomena. Thus, at the end of this slowing down the beating has disappeared and we have instead a nonlinear competition between two modes. On increasing our control parameter, the system undergoes a subharmonic bifurcation cascade leading to a chaotic regime, within which we detected the odd- periodic windows predicted by the theory. The situation is shown in Fig. 4, where in the left-hand side we report the temporal behavior for increasing values of the control parameter, amd on the right-hand side the associated power spectra.

A further increase of the cavity length drives the system through the inverse bifurcation sequence, until a new stable situation is reached. The whole instability window covers about 1/40 of the free spectral range, that corresponds to a cavity length variation of $0.13\mu s$ (or to a detuning of 3.3 MHz). The low dimensionality of this chaos has been confirmed by measuring the correlation dimension of the attractor with the Grassberger and Procaccia algorithm using 32000 points and finding a value $D_2 = 2.6 \pm 0.2$.

In conclusion, we have provided experimental evidence of transverse-mode interaction inducing a spontaneous chaos in the output intensity of a $C0_2$ waveguide laser for critical values of the cavity length. So far the experimental investigation has been limited to part of the diagram of Fig. 3. We plan to explore a wide domain in the future.

From a theoretical point of view we expect that our waveguide laser is asymptotically ruled by a CGL.

Indeed, if the waveguide walls were perfectly reflecting we would have a gain medium (the waveguide) separated from the empty diffractive region (the space between the right end of the guide and the mirror CM).

All tilted modes propagate within the guide, with multiple reflections on the wall. Outside the guide, they fan out and are ruled by a diffractive Kernel Eq. (11). However, the reflection losses on a glassy wall are small only for grazing incidence (on axis mode $q = 0$). As the grazing angle (proportional to

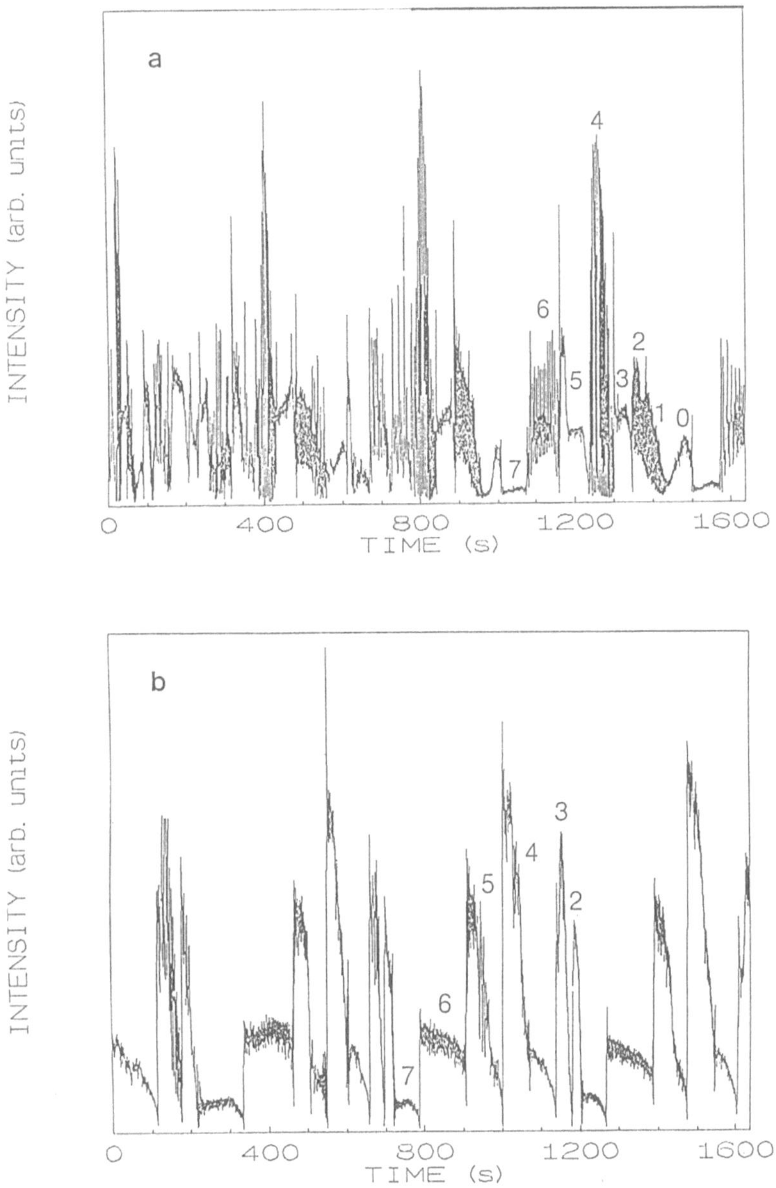

Fig. 7. Time records of the local intensity (samples collected at 10 Hz rate) at F=8.
a) with the low pass filter
b) with the band pass filter

q) increases, the corresponding losses increase [30]. We expect a loss factor more than linear, approximately quadratic $k(q) = k_0 + \alpha_1 q^2$.

Now, it is an easy matter to realize that, the Fourier transform of Eq.(14) is

$$d_t u_q = -(\alpha_1 + i\alpha_2)q^2 u_q + \mu u_q - (1 + i\beta)u_q \star u_q \star u_q \tag{15}$$

where the star denotes a convolution.

Thus, we see that the presence of wall losses provides the transformation of the diffractive NLS to a CGL.

4 - Chaotic itinerancy and Spatio-Temporal Chaos in Optics [31]

In this Section I report the first experimental evidence of (1+2)-dimensional physics in an extended optical medium. Precisely, we seed a ring cavity with a photorefractive gain medium pumped by an Argon laser and study the time and space features of the generated field. By varying the size of the cavity pupil, we control the number of transverse modes which can oscillate. We report two different regimes, namely one of a low dimensional chaos, where a single mode at a time is oscillating, and a small number of modes alternates in a fashion which displays close similarities and one of STC where many modes oscillate simultaneously yielding a very small transverse correlation length and spectral fluctuations with Gaussian statistics.

The experimental set up, shown in Fig. 5a) consists of a ring cavity with photorefractive gain. The gain medium is a BSO (Bismuth Silicon Oxide) crystal to which a dc electric field is applied. The crystal is pumped by a CW Argon laser with an intensity around $1mW/cm^2$. The basic cavity configuration consists of four high reflectivity dielectric mirrors and a lens L which provides a near confocal configuration.

The Fresnel number of the cavity is controlled by a variable aperture. A pinhole of 300 um diameter is inserted on the optical path between two confocal lenses L of short focal length (50 mm). Small displacements of the pinhole along the optical axis yield a continuous change of the aperture to spot size ratio, and consequently inhibit a different number of transverse modes. The effective Fresnel number F is the ratio of the area of the diffracting aperture that limits the system (pupil) to the spot size of the fundamental Gaussian mode spot, evaluated in the plane where the aperture is placed. F can be varied in the range from 0 to 100 approximately (Fig. 5b). This corresponds roughly to the variation of the number of transverse modes that can oscillate. The mechanical and thermal stability are ensured on time intervals longer those of the measurements (half an hour).

Fig. 6 shows the transverse (x, y) intensity pattern recorded by the video camera (left) and its spatial autocorrelation function (right). For low F ($F = 5$) one single mode at a time oscillates and the wavefront is wholly correlated, indeed the correlation length ξ is of the same order as the cross size D of the beam (Fig. 6a). For high F ($F = 70$) many modes oscillate simultaneously, yielding a speckle-like pattern (Fig. 6c) whose correlation length is very small ($\xi/D < 0.1$). The correlation test is crucial, otherwise one might suppose that the intensity pattern at left refers to a pure mode with a large mode number. Between these two asymptotic limits, we have a smooth variation of the ratio ξ/D, with intermediate situations as shown in Fig. 6b.

The low F limit corresponds to a periodic alternance of a few modes of the diffraction limited propagation followed by a dark period. The radial quantum number is always 0 and the azimuthal quantum number changes from $q = 0$ to q around 10. (From now on, we identify the modes by their azimuthal quantum number).

To study the time behavior, the input of an optical fiber picks up the intensity at a generic point on the wavefront (the signal level is a suitable code of each mode). The time plot shows fine details on a time scale of seconds, correspond-

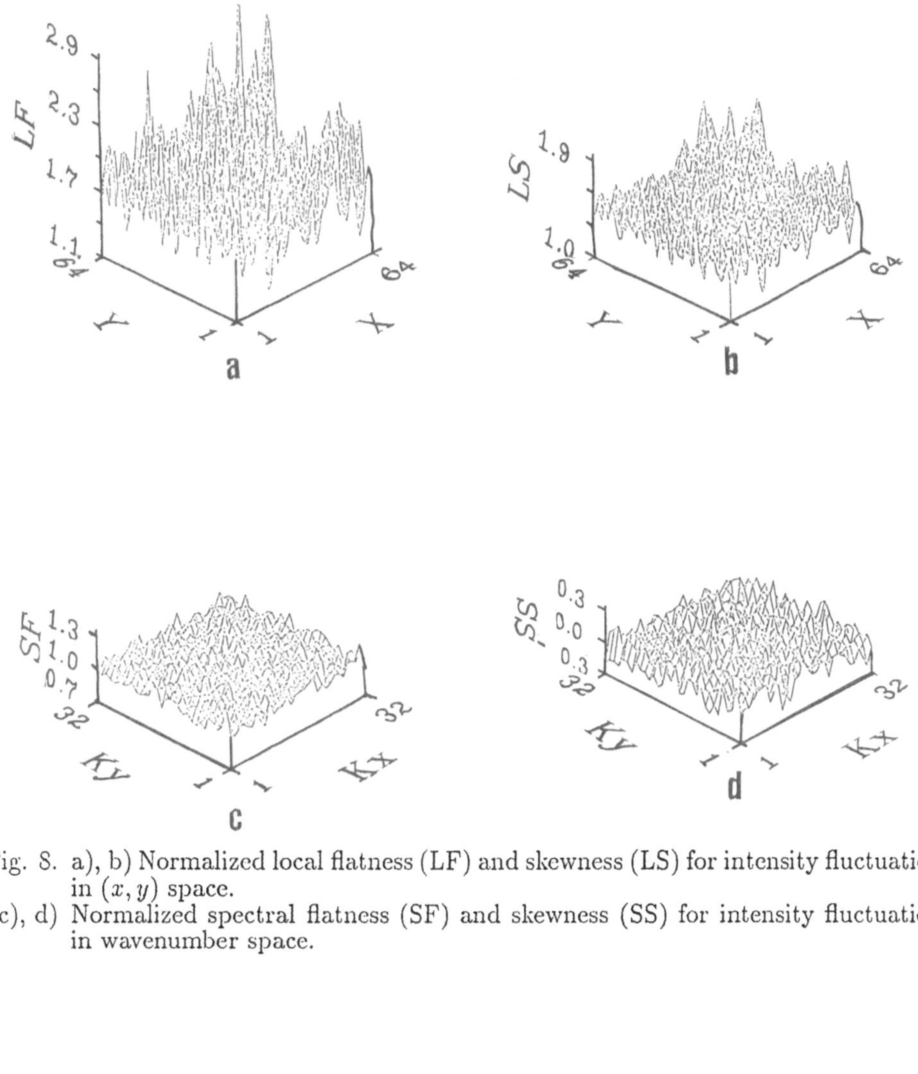

Fig. 8. a), b) Normalized local flatness (LF) and skewness (LS) for intensity fluctuations in (x, y) space.
c), d) Normalized spectral flatness (SF) and skewness (SS) for intensity fluctuations in wavenumber space.

ing to the dielectric relaxation time of BSO. This time scale is typical of the fluctuations in a pure mode and of the intermode switches. Each mode persist for a time of the order of a few minutes. The mode pattern (e.g. 7,6,5,4,3,2 in Fig. 7b) repeats almost periodically. To improve the selectivity we commute from the pinhole (low pass filter) to the pinhole plus an axial stop (band pass filter). For the same aperture size, introduction of the axial stop cuts off the lowest modes (1 and 0) and produces the regularization shown in going from Fig 7a) to b).

At the minimum Fresnel number for which some signal is observed (F around 2) still 4 different modes oscillate one at a time, followed by a dark interval, in a very regular periodic sequence. We call such a behavior "periodic alternance". Increasing the pump intensity, the frequency of the alternance increases but it remains regular. For a slight increase of F above 5 the regularity is lost, that is, the duration of each mode is no longer repeatable. This is an experimental evidence of "chaotic itinerancy".

In the high F limit, when $\xi/D << 1$, we expect spatio-temporal chaos (STC). Considering the local intensity fluctuations at each point (x, y) of the wavefront, a distinctive feature of STC is that, even though a collection of local fluctuations samples has a non Gaussian statistics, its Fourier transform, has a Gaussian distribution [8]. STC can therefore be seen as a coarse grained set of uncorrelated "pixels", each one of size ξ. To prove such a conjecture, we have measured (Fig. 8) the normalized skewness $M_3/M_2^{3/2}$ and flatness $M_4/(3M_2^2)$ for local and spectral intensity fluctuations (M_i) being the i-order moment of the statistical distributions). It is evident that local flatness (LF) and skewness (LS) have deviations from the Gaussian values (1 and 0 respectively) which are much larger than the residual fluctuations. On the contrary, in k-space, both spectral flatness (SF) and skewness (SS) are centered at 1 and 0 respectively, showing evidence of STC.

In conclusion, we have reported experimental evidence of periodic alternance and STC as two asymptotic limits for very small and large Fresnel numbers in a (1+2)-dimensional optical system. At the lower edge of the intermediate region we have observed chaotic itinerancy. For still larger F we should expect transition phenomena which are not simply a mathematical bifurcation as the usual laser threshold but which display the scaling properties of phase transitions in extended media.

Acknowledgement

I am grateful to the colleagues and students with whom I have shared excitement and enthusiasm in doing the works reported in Secs. 3, 4 and 5.

References

1. J. Von Neumann, **Theory of self-reproducing automata**, ed. by A. W. Burks, University of Illinois Press, 1966.

2.a F.T. Arecchi, in **Quantum Optics** ed. by R. J. Glauber, (Academic Press, New York, 1969), p. 57.

b F.T. Arecchi in **Order and fluctuations in equilibrium and non equilibrium statistical mechanics**, Proc. XVII Int. Solvay Conference on Physics edited by G. Nicolis, G. Dewel and J.V. Turner, (J. Wiley, 1981), p. 107.

3. For a review of these phenomena, as well as for a critical discussion on such a term, see the Proc. of the XVII Solvay Conference on Physics, referred to in Ref. 2b).

4. P.W. Anderson, Science, **177**, 393 (1972).

5. A.L. Schawlow and C.H. Townes, Phys Rev. **112**, 1940 (1958).

6. See the reference to the work of M. Von Laue in: A. Sommerfeld, **Optics**, (Academic Press, New York 1954), p. 194 .
7. F.T. Arecchi, Phys. Rev. Lett. **24**, 912 (1965).
8. P.C. Hohenberg and B.I. Shraiman, Physica **D37**, 109 (1989)
9. For a review, see F.T. Arecchi in F.T. Arecchi and R.G. Harrison, eds., **Instabilitis and Chaos in Quantum Optics**, (Springer 1987), p. 9.
10. J. Fleck, Phys. Rev. Lett. **21**, 131 (1968).
11. In solid state lasers, reports on random spiking go back to the first ruby laser, cfr: T. Maiman, Nature **187**, 463 (1967) and Phys. Rev. **123**, 1145 (1961). In gas lasers, evidence of random mode locking was given by: A.G. Fox and P.W. Smith, Phys. Rev. Lett. **18**, 826 (1967).
12. V. Degiorgio and M.O. Scully, Phys. Rev. **A2**, 1170 (1970).
13. H. Risken and K. Nummedal, J. Appl. Phys. **39**, 4662 (1968).
14. R. Graham and H. Haken, Z. Phys. **237**, 31 (1970).
15. H. Haken, Phys. Lett. **53A**, 77 (1975).
16. K. Ikeda, K. Otsuka and K. Matsumoto, Progr. Theor. Phys. Supplement N. 99, 295 (1989); K. Otsuka, Phys. Rev. Lett. **65**, 329 (1990).
17. J.V. Moloney, F.A. Hopf and H.M. Gibbs, Phys. Rev. **A25**, 3442 (1982); D.W. Laughlin, J.V. Moloney and A.C. Newell, Phys. Rev. Lett. **51**, 75 (1983).
18. W. J. Firth and E.M. Wright, Phys. Lett. **92A**, 211 (1982).
19. L.D. Landau and V.L. Ginzburg, **On the theory of Superconductivity**, in **Collected papers of L.D. Landau** ed. D. Terhaar, (Pergamon Press 1965), p. 217.
20. C.R. Doering, J.D. Gibbon, D. Holm and B. Nicolaenko, Nonlinearity **1**, 279 (1988).
21. P. Constantin, C. Foias and R. Temam, Physica **D30**, 284 (1988).
22. B.I. Shraiman, Phys. Rev. Lett. **57**, 325 (1987).
23. L.A. Lugiato, G.L. Oppo, J.R. Tredicce, L.M. Narducci, and M.A. Pernigo, Journ. Opt. Soc. Am. **B 7**, 1019 (1990); G.L. Oppo (unpublished).
24. P. Coullet, L. Gil and F. Rocca, Opt. Comm., **73**, 403 (1985).
25. M. Brambilla, F. Battipede, L.A. Lugiato, V. Penna, F. Prati, C. Tamm and C.O. Weiss, "Transverse Laser Patterns I. Phase Singularity Crystals" and M. Brambilla, L.A. Lugiato, V. Penna, F. Prati, C. Tamm and C.O. Weiss, "Transverse Laser Patterns II. Variational Principle for Pattern-Selection, Spatial Multistability, Laser Hydrodynamics" submitted to Phys. Rev. A.
26. C. Tamm, Phys. Rev. **A38**, 5960 (1988).
27. K. Otsuka and K. Ikeda, **Global chaos in a discrete time-dependent CGL Equation**, Preprint 1990.
28. M. Bartuccelli, P. Constantin, C.R. Doering, J.D. Gibbs and M. Gisselfalt, Physica **D44**, 421 (1990).
29. F.T. Arecchi, R. Meucci and L. Pezzati, Phys. Rev. **A42**, 5791 (1990).
30. M. Born and E. Wolf, **Principles of Optics**, Pergamon, Oxford 1980.
31. F.T. Arecchi, G. Giacomelli, P.L. Ramazza and S. Residori, Phys. Rev. Lett. **65**, 2531 (1990).

CRITICAL PHENOMENA IN HAMILTONIAN CHAOS*

Boris V. Chirikov

Institute of Nuclear Physics
630090 Novosibirsk, USSR

ABSTRACT

An overview of critical phenomena in Hamiltonian dynamics is presented, including the renormalization chaos, based upon a fairly simple resonant theory. First estimates for the critical structure and related statistical anomalies in arbitrary dimensions are given. The results of some numerical experiments with two-dimensional maps are discussed.

The main idea I would like to expose here is the inexhaustible diversity and richness of dynamical chaos whatever description you choose: trajectories, statistics or, recently, renormalization. The importance of this relatively new phenomenon - the dynamical chaos - is in that it presents, even in very simple models to be discussed below, the surprising complexity of the structures and evolution characteristic for a broad range of processes in nature including the highest levels of its organization. Moreover, dynamical chaos is the only stationary source of any new information and, hence, a necessary part of any creative activity, science included. This is a direct implication of the Alekseev-Brudno theorem and Kolmogorov's development in the information theory (see, e.g., Refs. 1 and 2). Chaos is not always that bad!

Below I restrict myself to classical mechanics only. So-called "quantum chaos" is another story (see, e.g., Ref. 3 and 4). Let me just mention that apart from very exotic examples there is no "true" chaos in quantum mechanics contrary to a common belief. On the other hand, the inavoidable statistical element of quantum mechanics related to measurement is very likely associated with the same classical chaos in the measuring device.

With a bit of imagination and fantasy one may even conjecture that any macroscopic event in this World, which formally is a result of some quantum "measurement," would be impossible without chaos.

Also, I am not going to consider any dissipative models (very important in practical applications) because they are not as fundamental as Hamiltonian systems. Besides, strictly speaking, the dissipative systems are not purely dynamical as the dissipation is inevitably related to some noise.

In what follows I take a physicist's approach to the problem, that is my presentation will be based on a simple (sometimes even qualitative) theory combined with the results of extensive numerical (computer) experiments. For a good physical overview of nonlinear dynamics and chaos see books.[5,6]

The principal concept of such a theory is nonlinear resonance whose quite familiar by now phase space picture is depicted, e.g., in Fig. 4 below. The essential part of this resonance structure is a pair of periodic orbits, the most important being the unstable one as it gives rise to the separatrix and, under almost any perturbation, to a chaotic surrounding layer. This is precisely the place where chaos is dawning.

Again, I have to restrict myself to a simpler case of strong nonlinearity which does not vanish with perturbation. A very interesting weakly nonlinear resonance will be briefly mentioned in Section I.2 below.

The paper is organized as follows. In the next Section I simple models are described which are currently extensively used in the studies of nonlinear phenomena and chaos. They well represent the whole spectrum of complexity classified in Section II. The main Sections III and IV are devoted to a detailed description of the so-called critical phenomena in dynamics which reveal the most complicated behavior presently known.

I. SIMPLE MODELS

First, let us consider a number of simple models which are currently very popular in the studies of dynamical chaos. Most of them are specified by some mappings, or maps, rather than by differential equations. This considerably simplifies both theoretical analysis and, especially, computer experiments.

In conservative Hamiltonian systems the chaos requires, at least, two freedoms. Then, the corresponding so-called Poincaré map is two-dimensional.

1. Strong nonlinearity.[7] Below we shall consider 2D maps of the following form:

$$\bar{y} = y + f(x)\ ; \qquad \bar{x} = x + g(\bar{y})\ . \tag{1.1}$$

This map is area-preserving, or canonical, which corresponds to the Hamiltonian nature of the model. Function $f(x)$, periodic in x, describes a perturbation usually assumed to be small. Hence, y is the unperturbed motion integral. Function $g(y)$, even when linear (see Eq. (1.6) below), represents the nonlinearity of x oscillation.

The simplest example of an analytic perturbation is given by

$$f(x) = K \cdot \sin x\ . \tag{1.2}$$

We shall consider also a smooth perturbation specified by the Fourier series

$$f(x) = \sum_m f_m\, e^{imx}\ ; \qquad f_m \sim K|m|^{-\beta}\ , \tag{1.3}$$

where β is the smoothness parameter. The term 'smooth' means actually 'not smooth enough.' For $\beta = 2$, for example, function $f(x)$ is continuous but the first derivative is discontinuous.

I mention two particular forms of nonlinearity. The first one,

$$g(y) = \lambda \cdot \ln|y|\ , \tag{1.4a}$$

models the motion near the separatrix of a nonlinear resonance, so that map (1.1) with this nonlinearity and perturbation (1.2) describes, particularly, a separatrix chaotic layer.[7]

Another form of nonlinearity,

$$g(E) = 2\pi\, \omega(-2E)^{-3/2}\ , \tag{1.4b}$$

corresponds to the Coulomb interaction, and it is actually the Kepler law. Here it is convenient to use the unperturbed energy E<0 as a dynamical variable, and ω is the perturbation frequency (see Ref. 4).

The map (1.1) with nonlinearity (1.4b) and perturbation (1.2) is called the Kepler map, and it was applied in both celestial mechanics and atomic physics. In the former case the motion of Halley's comet driven by Jupiter and Saturn was proved to be chaotic.[8] In atomic physics the Kepler map is a simple model to describe, in particular, a new type of photoelectric effect, the so-called diffusive ionization of Rydberg (highly excited) atoms.[9]

The two latter examples show that map (1.1) can be considered also as a model for time-dependent dynamical systems that is for those driven by a periodic perturbation. This is, of course, simply a very convenient approximation in which the feedback from the perturbed freedom is completely neglected. Then, the model (1.1) can be described by the Hamiltonian

$$H(x,y,t) \;=\; G(y) \;+\; F(x)\,\delta_1(t) \rightarrow G(y) \;+\; K \sum_m \cos(x - 2\pi m t)\;, \tag{1.5}$$

where $\delta_1(t)$ is δ-function of period 1 (one map's iteration), $G'(y) = g(y), F'(x) = -f(x)$, and the last series represents perturbation (1.2).

A fairly simple map (1.1) can be simplified still further by linearizing the second equation. In this way we arrive, upon appropriate change of the action y, at the so-called standard map

$$\bar{y} \;=\; y \;+\; K \cdot \sin x\;; \qquad \bar{x} \;=\; x \;+\; \bar{y}\;, \tag{1.6}$$

which describes the original model (1.1) locally in y, and which is also very popular now in studies of nonlinear phenomena in Hamiltonian systems. Model (1.6) is completely characterized by a single parameter K. In Hamiltonian representation (1.5) the 'kinetic energy' for the standard map is $G(y) = y^2/2$. Since x is an angle (phase) variable and y is the angular momentum, the model (1.6) is also called the 'kicked rotator.'

Each term in series (1.5) describes a particular first-order (primary) nonlinear resonance with the 'pendulum' Hamiltonian (for the standard map)

$$H_m \;=\; \frac{y^2}{2} \;+\; K \cdot \cos(x - 2\pi m t)\;. \tag{1.7}$$

The resonant value of the momentum is $y_m = \dot{x}_m \;=\; 2\pi m$. In variables $\tilde{x} = x - 2\pi m t$ and $\tilde{y} = y - y_m$ any single resonance is a conservative system. Its motion is strictly bounded in y by the resonance width $\Delta y_m = 4\sqrt{K}$ owing to the nonlinearity, that is the dependence of frequency $\dot{x} = y$ on momentum y.

<u>2. Weak nonlinearity.</u>[10] The structure of resonance drastically changes if we add to Hamiltonian (1.7) the term $\omega_0^2 x^2/2$:

$$H_m \;=\; \frac{y^s}{2} \;+\; \frac{\omega_0^2 x^2}{2} \;+\; K \cdot \cos(x - 2\pi m t)\;, \tag{1.8}$$

which breaks down the integrability of the system for any $\omega_0 \neq 0$.

Actually, the model (1.8) is quite different from model (1.7) as now variable x is no longer confined to the interval $(0, 2\pi)$, and y is not the angular momentum. One may interpret Hamiltonian (1.8) as describing a particle-wave interaction. Such models have been studied by many authors in plasma physics (see, e.g., Ref. 5), yet the true understanding has been achieved only recently (see, e.g., Ref. 10). The peculiarity of model (1.8) is the weak nonlinearity, that is the unperturbed (K = 0) oscillation is linear (isochronous) which turns out to be a much more difficult problem as compared with strong (unperturbed) nonlinearity (1.4). The resonance is now determined not by initial conditions but by the parameters of the model: $2\pi m = n\omega_0$ with any integer $n \neq 0$. In the action-angle variables (I, φ) of the harmonic oscillator a single resonance is approximately described by the Hamiltonian

$$H_m \;\approx\; K \cdot \mathcal{J}_m(a)\, \cos\!\left(m\varphi \;+\; \frac{\pi m}{2}\right)\,, \tag{1.9}$$

where $a = (2I/\omega_0)^{1/2}$ is the oscillation amplitude, and $\mathcal{J}_m$ the Bessel function. There are now infinitely many stable and unstable periodic orbits (instead of two for strong nonlinearity) while the separatrices, connecting unstable points, form an unbounded network on the phase plane (I, φ). As a result, even a single weakly nonlinear resonance can make the motion completely unstable and unbounded.

II. LEVELS OF DISORDER

In this Section I attempt to 'organize' the great variety of chaos into a series of levels with increasing disorder and complexity.

1. Complete integrability.[15] This, zero, level of the maximal order is characterized by a stable and dynamically predictable motion in terms of individual trajectories. The motion is quasiperiodic that is of a purely discrete spectrum. One may call it simple dynamics. Yet, in the general theory of dynamical systems this 'simple' motion includes the whole quantum chaos, typically on a finite time scale (see, e.g., Ref. 3). The latter is dynamically equivalent to a many-dimensional linear oscillator which is apparently the simplest model of the quantum chaos.[11] On the other hand, in the formal thermodynamic limit of infinitely many freedoms this model provided the foundations of the traditional statistical mechanics, both classical and quantal, for macroscopic systems (for a rigorous theory see, e.g., Ref. 12).

The standard map, as the simplest model, is completely integrable for $K = 0$ only, that is in the unperturbed limit. In this case $y =$ const is the motion integral, and $x = 2\pi rt$, where

$$r = \frac{\omega}{2\pi} = \frac{\Delta x}{2\pi\Delta t} \tag{2.1}$$

is called the rotation number. This very important parameter of a trajectory is the ratio of the motion frequency (ω) to that of the perturbation (2π). Particularly, this ratio determines resonances (with zero perturbation in this limit!) which correspond to ratio $r = p/q$. Any resonant trajectory is just q separate points on the phase plane (x, y). For irrational r the trajectory is a continuous straight line $y = 2\pi r$ which is called an invariant curve.

In spite of a great recent success in constructing the whole families of completely integrable systems (see, e.g., Ref. 13) they all are exceptional, or non-generic, in the sense that almost any perturbation destroys the integrability.

2. KAM integrability.[14] Is the generic property of a completely integrable system under sufficiently weak perturbation. The theory of such systems had been initiated by Kolmogorov and was essentially developed by Arnold and Moser (see, e.g., Ref. 15); hence the abbreviation KAM.

For the standard map this first level of disorder corresponds to a non-zero $K \to 0$. Most invariant (KAM) curves survive weak perturbation that is they are only slightly deformed but remain continuous and, hence, inpenetrable for other trajectories. For this reason the KAM curve is called an absolute barrier (for the motion). This property depends on the rotation number r of the curve which must be sufficiently irrational for the stability against perturbation. Hence, the importance of parameter r which is used as the label for identification of a given KAM curve at different perturbations.

Curves with resonant $r = p/q$ are all destroyed by any perturbations to form a different structure of the nonlinear resonance (Fig. 4). However, the nonintegrable part of this structure is confined to an exponentially narrow chaotic layer only. From a physical point of view such a motion can be well considered in most cases as integrable to a very high accuracy. This is reminiscent of the adiabatic invariance which is very important in physics even though it is not exact. Actually, there is a deep relation between the two, and we call KAM integrability the inverse adiabaticity.[14,16]

Approximately, the dynamics on this level is as simple as on the previous one. Yet, the chaotic component of motion, being of an exponentially small measure, is everywhere dense. As a result, the whole motion structure becomes very complicated. For more than two freedoms the phase space is cut through by a connected network of channels which support a global diffusion.[7] Even though the rate of this Arnold diffusion is also exponentially small it may be important in some special cases. For a weakly nonlinear system Arnold diffusion is possible even in two freedoms as well, for instance, in model (1.8).[10]

3. Complete chaos.[20] Now we turn to the opposite limiting case when the motion is fully chaotic. In the standard map, as $K \to \infty$, there is a single chaotic component of motion stretched over the whole phase space (cylinder) of the model. The motion spectrum is purely continuous while a typical individual trajectory is most complicated. The latter means that Kolmogorov's complexity, which is equal to the information associated with the trajectory, is finite, per unit time, and equal to the rate of local exponential instability of motion.[1] Hence, the dynamics on this level is most complicated to the extent that the trajectory actually loses its physical meaning.

Nevertheless, the dynamical equations, e.g., map (1.6), can still be applied to completely derive the statistical properties of the unstable motion. Moreover, on this level the statistics turns out to very simple and already well-known from the traditional statistical mechanics. For example, in the standard map it is simply a homogeneous diffusion in y with the rate

$$D_y \equiv \frac{\langle(\Delta y)^2\rangle}{t} = \frac{K^2}{2}\,C(K) \to \frac{K^2}{2}\,, \tag{2.2}$$

where function $C(K)$ accounts for the dynamical correlation of phase x, and $C(K) \to 1$ as $K \to \infty$.[17] For this reason the complexity of motion on this level is still not the highest one.

4. Critical phenomena: scale invariance.[21] For a typical (generic) perturbation, neither very weak nor very strong, the whole structure of motion is most complicated because the phase space is generally divided in many separate domains with both regular and chaotic motions. In the standard map, for example, such an intricate behavior corresponds to $K \sim 1$ (see Fig. 1) that is around the global critical perturbation $K = K_G \approx 1$. The latter is the border between strictly bounded motion for any initial conditions ($K \le K_G$) and unbounded motion for some initial conditions ($K > K_G$).

In the unbounded chaotic component (for $K > K_G$) the motion is still diffusive with the rate[18]

$$D_y \approx 0.3\,(K - K_G)^3\,, \tag{2.3}$$

vanishing toward the critical perturbation (cf. Eq. (2.2) where the correlation $C(K) \approx 0.6\,(K - K_G)^3/K^2$. The main difficulty here is a hierarchical (fractal) structure of the chaotic component. The invariant measure-phase area, known beforehand, doesn't help in this case. The ultimate origin of that complexity is the chaos border in the phase space between chaotic and regular components of motion which also results in very peculiar statistical properties of the chaotic motion (see Section IV).

Thus, the chaos border makes both individual (chaotic) trajectories as well as the statistical properties of the motion very complicated. Is there any way to simplify the description of such a motion? Or: would it be possible to find any order in that mess? Surprisingly it is indeed possible in some cases if one compares the critical structure at different scales in the phase plane (Section III.3). Asymptotically, as you enlarge the structure more and more it exactly repeats itself with all the dynamical and statistical complexity (see also Fig. 5 below)! This peculiar property is called the scale invariance, and it is described by the so-called renormalization group, or in brief, renormgroup.

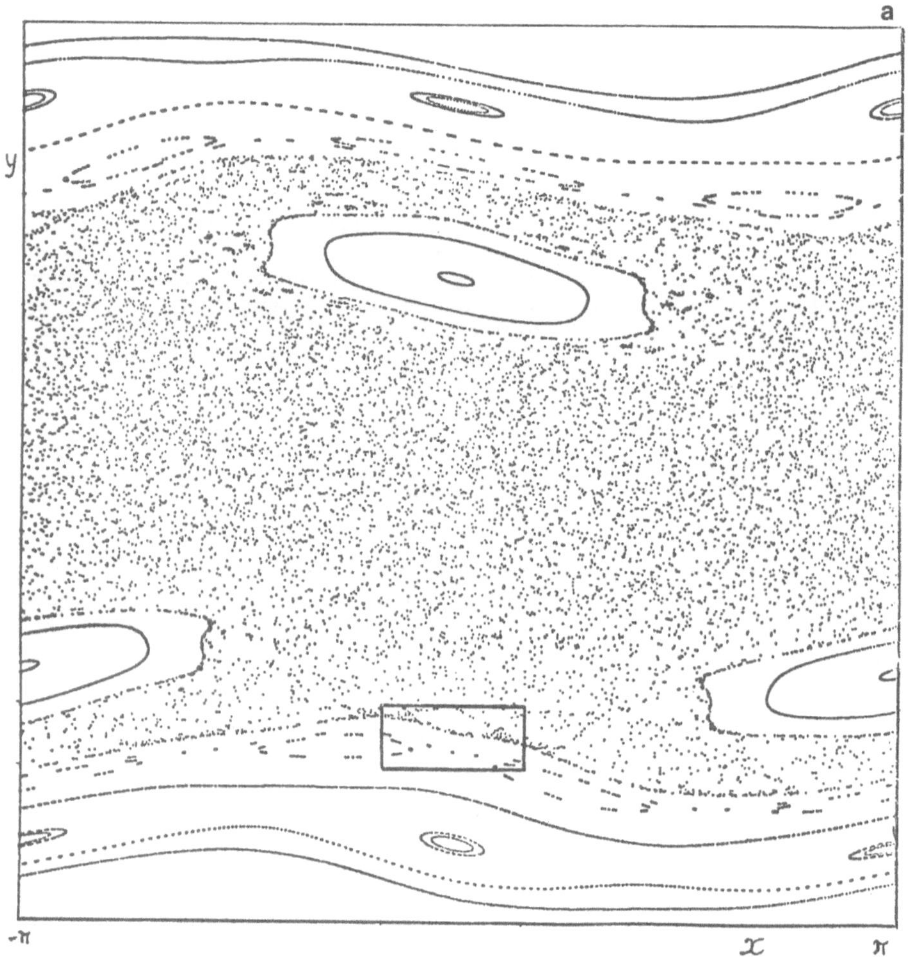

Fig. 1. An example of critical structure in the map (3.1) with $\lambda = 5$; scattered points belong to a single chaotic trajectory: (a) the whole chaotic layer; (b) enlarged part near the chaos border $y \approx -\lambda$ where the motion is described locally by the standard map (1.6) with $K \approx 1$.[19]

Fig. 1b

5. Critical phenomena: renormchaos.[22] The variation of the motion structure with the scale in phase space can be considered as a certain abstract dynamics (see Section III.4) which we termed the renormalization dynamics, or renormdynamics.[22] Here the scale plays a role of 'time' and we call it renormtime. The simplest case of any dynamics is a fixed point (for maps) which in renormdynamics corresponds to the scale invariance described above (see also Section III.3). But typically the dynamics is chaotic, and so there must be a sort of renormalization chaos (renormchaos) as well. Guided by this analogy we have indeed found[22] such a chaos.

In this case the renormalization is as complicated as an individual chaotic trajectory of the original dynamical system. Yet, some remnants of order still persist, namely, the universality of renormalization. It means that asymptotically for a big renormtime, that is for small spatial scales, the critical structure is a universal functional of a single irrational number - the rotation number r_c of the critical curve, e.g., the border curve, in almost any 2D map.[21]

Moreover, one can introduce the renormstatistics, that is statistical description of the renormalization. Then, for almost any r_c the renormstatistics is the same, i.e. universal, and it is fairly simple.

6. Critical phenomena: the breakdown of universality.[23] Recently, the first example of still more complicated behavior has been found in Ref. 23, where some quasiperiodic driving perturbation was studied. Specifically, the standard map (1.6) was used in numerical experiments with the periodically time-dependent parameter $K(t) = k_1 + k_2\cos(2\pi r_2 t)$ incommensurable with the map's time step.

To some extent such a model represents also a higher-dimensional behavior. A critical curve is now characterized by the two irrational rotation numbers r_1, r_2. For a particular choice of irrationals r_1, r_2 it was found that the renormdynamics is different in dependence of parameters k_1, k_2. It is thus not clear whether such a breakdown is typical. If so, one would expect also a more complicated renormstatistics.

Here we have come to the frontier of the unknown. Currently, there is no idea what would be still higher levels of disorder, if any.

III. CRITICAL DYNAMICS

In this Section I consider in some detail the two levels of disorder briefly described in Section II above (levels 4 and 5). This work was done in collaboration with D. L. Shepelyansky.

1. Statistical 'anomalies' in dynamical chaos.[24] We encountered the critical phenomena in studying some statistical properties of motion in a simple map

$$\bar{y} = y + \sin x\,; \qquad \bar{x} = x + \lambda \cdot \ln|\bar{y}| \tag{3.1}$$

of the type described in Section I.1 above. These our studies were stimulated by paper[25] with an intriguing title 'Numerical Experiments in Stochasticity and Heteroclinic Oscillation.' Actually, the motion in a chaotic separatrix layer had been studied, and we went on with a much simpler model (3.1) (see Ref. 7).

We studied the statistics of times t_n when a trajectory crosses the symmetry line $y = 0$. We call differences $\tau_n = t_{n+1} - t_n$ the times of Poincare's recurrences (to line $y = 0$). The same was implicitly done in Ref. 25. Our results are shown in Fig. 2 where $P(\tau)$ is (integral) probability for $\tau_n > \tau$. The initial part of the distribution is very close to

$$P_f = \frac{1}{\sqrt{\tau}}\,; \qquad \tau \geq 1 \tag{3.2}$$

and it is explained by a free homogeneous diffusion within the chaotic layer before the trajectory reaches the layer border ($y_b \approx \lambda$). It takes time

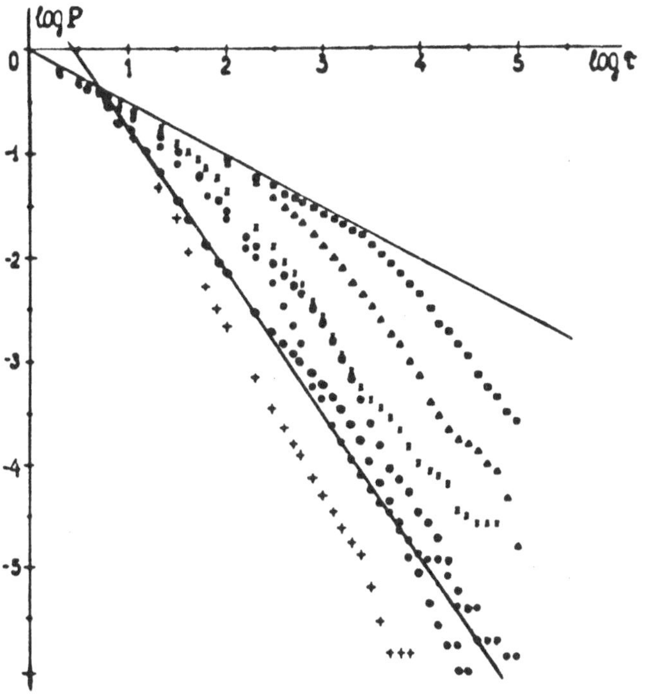

Fig. 2. Poincare's recurrences in the chaotic layer of the map (3.1) for various $\lambda = 1$ (lower points) through 100 (upper points). The two straight lines are power laws with exponents - 0.5 and 1.37, respectively (after Ref. 24).

$$\tau_f \approx 0.3 \lambda^2 , \tag{3.3}$$

where the coefficient was obtained from the numerical data.

Curiously, in Ref. 25 only this (trivial) part of distribution $P(\tau)$ was observed. This was the cost for the authors' great concern about the exponential error growth at a chaotic trajectory. To overcome the instability the computation was performed with the record accuracy of 358 decimal places! As a result the chaotic trajectory can be followed during a rather short time interval.

Error growth is a serious problem, indeed, as the structural stability of Hamiltonian motion is almost unknown rigorously. Yet, all the numerical experience up to now strongly suggests such stability and, hence, the stability of statistical properties which are of the primary interest for a chaotic motion. Besides, only structural stability justifies the use of various simple models and approximations.

In our studies of model (3.1) we directly checked that distribution $P(\tau)$, which is a statistical property of the motion, does not depend on a particular trajectory within expected statistical fluctuations. The latter noticeably influence the lowest part of distribution $P(\tau)$ where the number of events per bin is ~ 1 (see Fig. 2).

Most interesting is the asymptotics of $P(\tau)$ for $\tau \gg \tau_f$ (3.3). This part characterizes the motion of the structure of the chaos border at $|y| = y_b \approx \lambda$, or the critical structure as we call it.

The following features of $P(\tau)$ asymptotics seem to be of importance. First, the distribution is a power law and not an exponential:

$$P(\tau) \approx \frac{\tau_f^{p-1/2}}{\tau^p} \;;\quad \tau \gtrsim \tau_f \;;\quad \langle p \rangle \approx 1.5 < 2 \,. \tag{3.4}$$

This suggests a heirarchical (fractal) structure of the border. The accuracy of the numerical value for p is not very good, yet we are sure that the important inequality (3.4) always holds.

On the other hand the distribution $P(\tau)$ is a power law only approximately, on the average. Irregular oscillations of the local exponent $p(\tau) \equiv d \ln P/d \ln \tau$ clearly show up in Fig. 2. These do not depend on the trajectory and, hence, characterize not statistical fluctuations but, again, the structure of the chaos border. Such a structure with variable exponent $p(\tau)$ is now called multifractality (see, e.g., Ref. 26).

Statistics of Poincare's recurrences $P(\tau)$ proved to be the most convenient and reliable numerical data to study (cf. Ref. 25). On the other hand, it is directly related to the most important statistical property of motion—the time correlation,[27] e.g.,

$$C_y(\tau) \;=\; \frac{\overline{y(t)y(t+\tau)}}{\overline{y^2(t)}} \;, \tag{3.5}$$

which characterizes the 'sticking' of trajectory near the border. Notice that $\overline{y(t)} = 0$ for map (3.1).

Indeed, the correlation is proportional to the sticking time that is (cf. Ref. 27)

$$C_y \;\sim\; \frac{\tau(P(\tau)}{\langle\tau\rangle} \;\sim\; \tau^{-p_c} \;;\quad p_c \;=\; p - 1 < 1 \;, \tag{3.6}$$

for $\tau \gtrsim \tau_f$. Here $\langle\tau\rangle \approx 3\lambda$ is the mean recurrence time, and the latter important inequality follows from Eq. (3.4) (see also Fig. 3b). For chaotic motion $C_y \to 0$ as $\tau \to \infty$ (mixing property), hence, $p_c > 0$, and, for bounded motion, $p > 1$. Also, notice that due to ergodicity of the motion $C_y \sim \mu(\tau)$, the measure of the sticking domain (a strip) which is $\sim |y - y_b|/y_b$ where $y_b \approx \lambda$ is a half-width of the chaotic layer for map (3.1).

Slow correlation decay due to the sticking of a chaotic trajectory near the chaos border, and especially the inequality (3.6), are responsible for all other statistical 'anomalies' of the motion with a chaos border to be discussed below. A power law decay (3.6) is especially remarkable in view of the strong exponential instability of the motion which is characterized by a positive Lyapunov's exponent Λ_+ and the KS-entropy (per map's iteration): $h = \Lambda_+ \approx 0.7$ (see Section 6.3 in Ref. 7). The apparent contradiction is explained as follows. The instability rate h is mainly determined by the central part of the chaotic layer while the sticking is a peripheral effect which has a negligible impact on the mean local instability. In other words, KS-entropy does not discern such statistical anomalies. It can be done using the so-called Renyi entropy K_q which is a generalization of $h = K_1$ (see, e.g., Ref. 28), and which drops to zero for all values of the parameter $q > 1$ in the presence of the chaos border.[29]

The critical phenomena at the chaos border and related statistical anomalies are 'universal' (a very popular word in this field of research!) in that they are approximately the same in any 2D map. In Fig. 3a, for example, our results are compared with those in Ref. 27 for a different map on a torus

$$\bar{y} \;=\; y \;+\; 2\,(x^2 - a^2)\;,\quad \bar{x} \;=\; x \;+\; \bar{y}\;, \tag{3.7}$$

with a closed chaos border surrounding the domain of regular motion around the stable fixed point at $y = 0, x = -a$ $(0 < a < 1)$. Notice that the two distributions $P(\tau)$ are not identical but rather similar (see below).

2. The resonant theory.[30] To understand the statistical anomalies described above we have developed a resonant theory of critical phenomena in dynamics.[30,31] Let us begin with a simpler problem of an isolated critical KAM curve whose rotation number is some irrational r. According to the KAM theory most invariant curves are preserved under a sufficiently weak perturbation in the sense that they remain

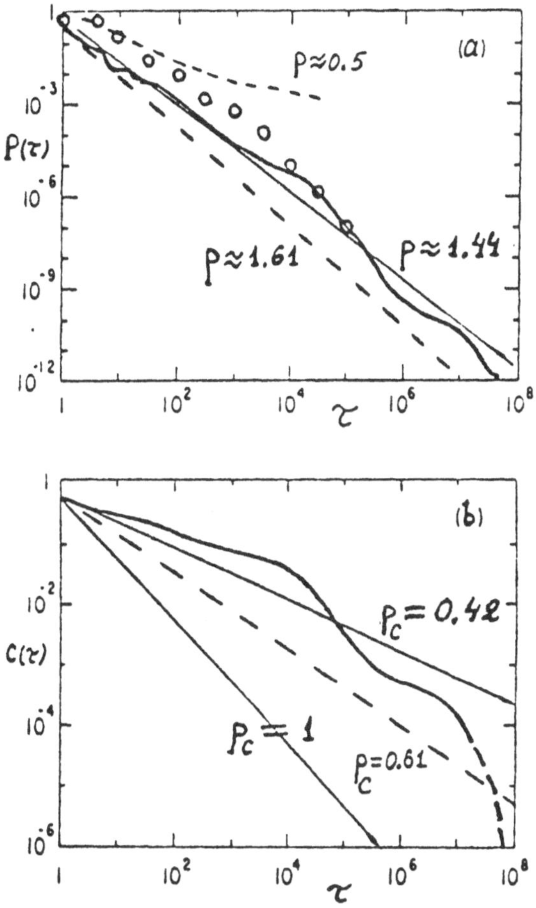

Fig. 3. Statistical properties of motion with chaos border: (a) Poincare's recurrences; (b) correlation decay. Solid curves are for map (3.7)[27] while circles are our data for $\lambda = 3$. Straight lines indicate power law with the exponent shown. Dashed curve is the effect of noise.[22]

continuous and are only slightly deformed by the perturbation. The theory of critical phenomena follows the transformation of a KAM curve up to the critical perturbation which destroys the curve.

The critical perturbation, e.g., $K_c(r)$ for the standard map, crucially depends on the arithmetic of r. Remember that for the everywhere dense set of rationals $r = p/q$ the critical $K_c(p/q) = 0$ (Section II.1). The whole dependence $K_c(r)$ is a fractal function.[32]

The physical explanation of this behavior lies in resonances. Their profound impact on the critical structure is clearly seen in all numerical data (see, e.g., Fig. 1). For irrational r the principal resonances correspond to the best rational approximations of r which are known to be the so-called convergents $r_n = p_n/q_n$ of the infinite continued fraction:

$$r = \frac{1}{m_1 + \frac{1}{m_2 + \ldots}} \equiv (m_1, m_2, \ldots) \; ; \quad r_n = (m_1, \ldots, m_n) \to r \; , \quad n \to \infty \; . \qquad (3.8)$$

The arithmetic of continued fractions gives for almost any r

$$|\rho_n| \equiv |r_n - r| \sim \frac{1}{q_n^2} \sim |r_{n+1} - r_n| \; . \qquad (3.9)$$

From a physical viewpoint ρ_n is the detuning of the n-th principal resonance with respect to the critical motion. Then, from the resonance overlap criterion[7] the main critical scaling, or the criticality condition, is

$$\Delta\rho_n \sim |\rho_n| \sim \frac{1}{q_n^2} \; , \qquad (3.10)$$

where $\Delta\rho_n$ is the resonance width. These resonances determine the principal scales of the critical structure whose scheme is outlined in Fig. 4 (cf. Fig. 5 below).

To estimate $\Delta\rho_n$ we need the critical Hamiltonian which describes all resonances $r_{pq} = p/q$ (p, q any integers), and not the primary ones $r_m = m$ only from the original Hamiltonian of the type (1.5). Integer resonances $r_m - m$ are obtained in the first approximation $x_c(t) \approx 2\pi t_c t \equiv \xi_c$, the mean motion on the critical KAM curve.

Extrapolating the KAM theory (see, e.g., Ref. 15) the following expression can be assumed for the critical motion:

$$x_c(t) + \xi_c + \sum_q a_q \sin(q\xi_c) \; . \qquad (3.11)$$

Locally in y, in the standard-map approximation the critical Hamiltonian H_c, which describes some vicinity of the critical KAM curve $r = r_c$, can be written as a natural generalization of the original Hamiltonian (1.5) (with $G(y) = y^2/2$) in the following form

$$H_c(\theta, \rho, t) = \frac{\rho^2}{2} + \sum_{p,q} \frac{v_{pq}}{(2\pi)^2} \cos 2\pi(q\theta - \nu_{pq} t) \; . \qquad (3.12)$$

Here $\rho = r - r_c$; $2\pi\theta = x - \xi = x - 2\pi\, rt$, and $\nu_{pq} = p - q\, r_c$ are the driving frequencies. Resonances $\rho_{pq} = p/w - r_c$ are characterized by perturbation amplitudes v_{pq} to be found below from the criticality condition. The factor $(2\pi)^2$ is introduced to recover the original Hamiltonian for which $v_{p1} = K$ (in variables θ, ρ).

For principal resonances ($p = p_n$, $q = q_n$) the frequencies $\nu_n \sim q^{-1}$ are minimal, and they determine the time scales $t_n \sim \nu_n^{-1} \sim q_n$ which are the motion periods at the resonances. Instead we can introduce the scaled variables, e.g.,

$$T = \frac{t_n}{q_n} \sim 1 \; , \qquad (3.13)$$

which remain of the same order of magnitude on all scales n.

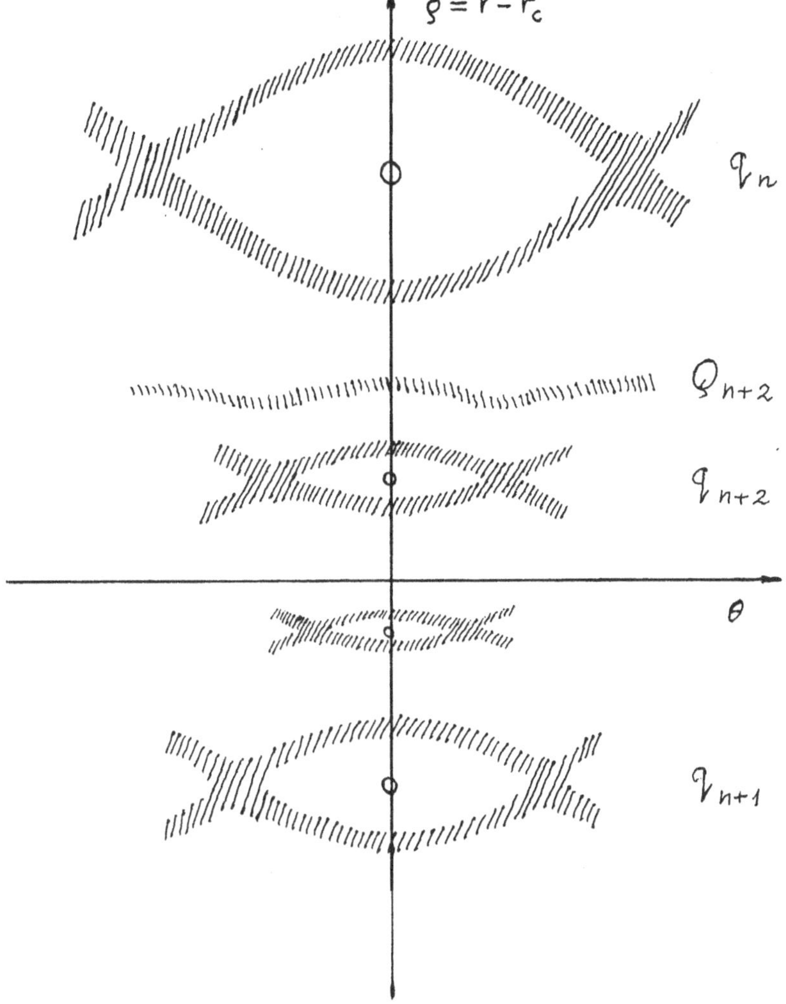

Fig. 4. Outline of the critical structure with a few principal resonances, represented by the separatrix chaotic layers (hatched) and stable periodic orbits (circles), and the corresponding scales q_n. Another chaotic layer Q_n is a bottleneck between the scales (Section IV.3).

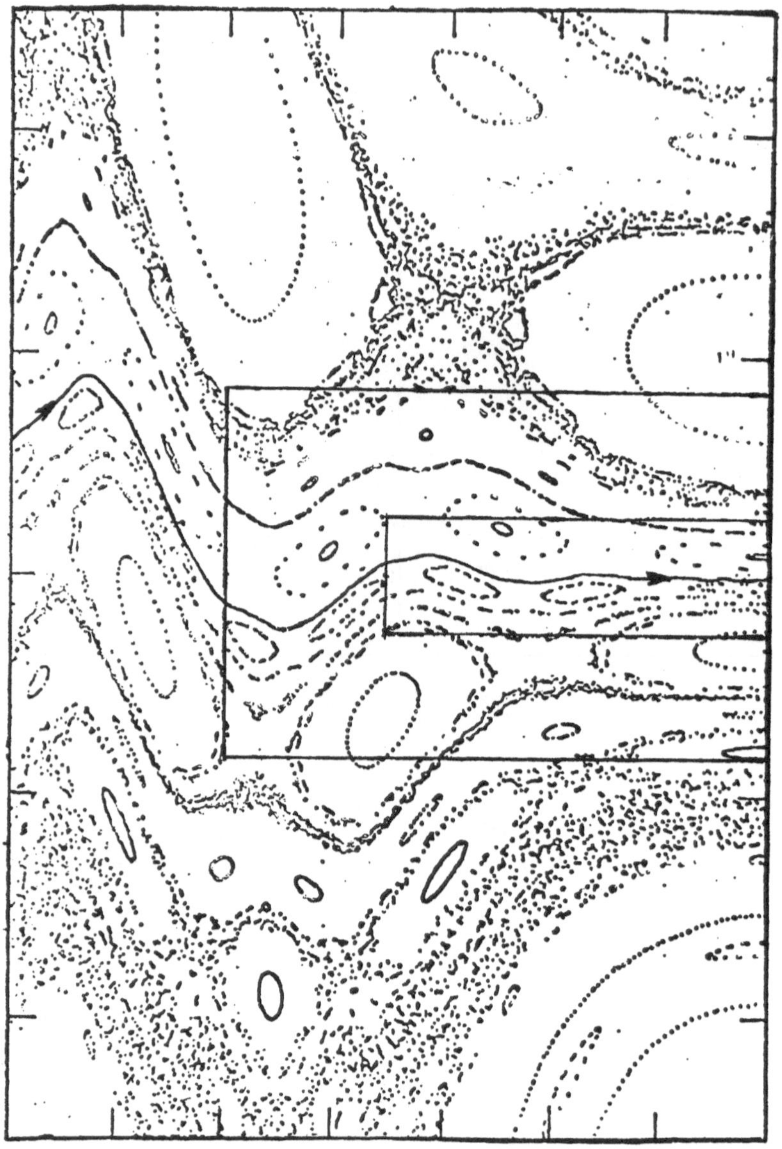

Fig. 5. A small part of the critical structure with 3 successive scales shown by rectangles (including the whole picture). Critical curve is indicated by 2 arrows (after Ref. 21).

The width of a principal resonance is $\Delta\rho_n \sim v_n^{1/2} \sim q_n^{-2}$ (see Eq. (3.10)). Hence $v_n \sim q_n^{-4}$, and another variable scales

$$V = v_n\, q_n^2 \sim 1\,. \tag{3.14}$$

Now we can approximately solve the equation

$$\ddot{\theta} = \dot{\rho} = -\frac{\partial H_c}{\partial \theta} \approx -\frac{\partial H_c}{\partial \theta}\bigg|_{\theta=0}\,.$$

The latter approximation means that we substitute mean motion ξ for $x_c(t)$. As our original model (3.1) is a map, time t is integer, and we can drop term pt in the solution $\theta(t)$ which then takes the form (3.11) with

$$a_q \approx \frac{q}{(2\pi)^3}\sum_p \frac{v_{pq}}{(p-qr)^2}\,; \qquad a_n \sim v_n q_n^3 \sim q_n^{-1}\,.$$

Hence, the longitudinal scaled amplitude

$$A = a_n q_n \sim 1 \tag{3.15}$$

and x scale is q_n^{-1} which is simply one cell of the resonance chain (see Fig. 4).

In the standard-map approximation

$$y_c(t) \approx \dot{x}_c(t) = 2\pi r_c + \sum_q B_q \cos(q\xi_c - 2\pi pt)$$

with $B_n = 2\pi\, a_n\, \nu_n$, and the transverse scaled amplitude

$$B = b_n\, q_n^2 \sim 1\,. \tag{3.16}$$

Hence, the y scale is q_n^{-2} which is the resonance width.

Consider now the periodic orbit at a resonance's center (Fig. 4). Its stability is determined by the so-called Greene residue[33] (see also Ref. 5)

$$R = \sin^2\left(\frac{t_n\,\omega_n}{2}\right) \sim 1\,, \tag{3.17}$$

where $t_n = q_n$ is the period, and $\omega_n \approx q_n\, v_n^{1/2}$ is the small oscillation frequency (see Eq. (3.12)). Obviously, R is a scaled variable.

Finally, the scaled rotation number, or rather the scaled detuning

$$D = \rho_n\, q_n^2 \sim 1 \tag{3.18}$$

determines actually all the other scaled variables.

So far we considered exactly critical conditions that is $K = K_c(r)$. What would be the impact of any deviation $\Delta K = K - K_c(r) \neq 0$? It can be evaluated as follows. Perturbation amplitudes v_n in Eq. (3.12) appear in q_n-th order of the perturbation theory and are proportional to $(K/K_c)^q = \exp(q \ln \frac{K}{K_c})$. Hence, for a small deviation from criticality ($\Delta K \to 0$) the amplitude $v_n \sim \exp(Cq_n\,\Delta K)$ with some $C \sim 1$. At $\Delta K = 0$ the exponential dependence cancels, and only a power low (3.14) remains. Generally,

$$v_n \sim \frac{1}{q_n^4}\exp(Cq_n\,\Delta K)\,. \tag{3.19}$$

For $\Delta K > 0$ all scales $g_n \gtrsim (\Delta K)^{-1}$ are destroyed and a chaotic layer of width $\Delta y \sim (\Delta K)^2$ is formed. From Eq. (3.19) the scaled perturbation can be introduced

$$P = q_n \Delta K_n \sim 1\,, \tag{3.20}$$

which describes approaching the renormalization limit for a fixed V, for example.

If the original perturbation is non-analytic that is with some power law spectrum $v_q^0 \sim q^{-\beta-1}$ (cf. model (1.3) where $f_m \sim mv_m^0$) the critical conditions are only possible for $\beta > 3$, otherwise $K_c = 0$. Thus, $\beta_c = 3$ is the critical smoothness of the perturbation. I shall come back to this point in Section IV.1 below.

3. The renormalization group.[21] This powerful method, well known and widely applied in hydrodynamical turbulence, phase transitions and quantum field theory, was first used in nonlinear dynamics and chaos theory in Ref. 33. Later on, the exact renormalization equations were formulated and studied in Ref. 34 for (dissipative) 1D maps, and in Ref. 21 for 2D area-preserving (Hamiltonian) maps. The renormgroup equations are an abstract map acting in the space of dynamical maps, and it is based on the arithmetical map for successive convergents $r_n = p_n/q_n$ of the critical rotation number $r_c(m_n)$:

$$\bar{p} = \bar{m}p + \underline{p} \; ; \qquad \bar{q} = \bar{m}q = \underline{q} \; , \tag{3.21}$$

where $\bar{p} \equiv p_{n+1}$; $p \equiv p_n$; $\underline{p} \equiv p_{n-1}$ etc. Besides a qualitative understanding of critical phenomena (particularly their universality) this approach provides very efficient numerical algorithms for computing all the parameters of critical structure. Unlike this, our resonant theory, being inherently approximate, allows some analytical estimates.

The resonance overlap criterion, on which the theory is essentially based, can directly provide order-of-magnitude estimates only as, for example, for scaled variables (3.14-3.17). However, there exists another group of critical parameters which can be evaluated to a surprising accuracy. Those are the scaling factors, that is the ratios of particular quantities on the neighboring scales. For example,

$$s_a = \frac{a_n}{a_{n+1}} \; ; \quad s_b = \frac{B_n}{B_{n+1}} \; ; \quad s_K = \frac{\Delta K_n}{\Delta K_{n+1}}$$

are the renormalization factors for x, y, and perturbation K, respectively.

The structure of scaled variables shows that all scaling factors are some powers of the main arithmetical factor

$$s_q = \frac{q_n}{q_{n-1}} \; . \tag{3.22}$$

To compare both approaches let us consider the simplest case of a homogeneous continued fraction $r = (m, m, \ldots, m, \ldots) \equiv (m^\infty)$. In this case all the scaled variables become asymptotically, as $n \to \infty$, exact invariants of the renormgroup. This is called the scale invariance. For example, $D \to (4 + m^2)^{-1/2}$ (see. Eq. (3.21)) which is a simple arithmetical property. The other invariants are not yet shown except the case of $r = r_G = (1^\infty) = (\sqrt{5} - 1)/2 = 0.618\ldots$ which is called the golden tail (because for asymptotic properties only the tail of the continued fraction matters).

In this particular case, studied in great detail, the normalization invariants are: $T = 1$ (if, by definition, $t_n = q_n$, see Eq. (3.13)); $R = 0.2500888\ldots; V \approx (2 \text{ arc} \sin\sqrt{R})^2 = 1.097052\ldots; A \approx 0.167; B \approx 2\pi A/\sqrt{5} \approx 0.470$. Notice that from the above relation $R \sim V \sim \Delta\rho_n/\rho_n$ the Greene résidue also characterizes the resonance overlap.

Now consider the scaling factor for the area $c_n \sim a_n b_n$ of a resonance cell (the corresponding scaled variable $C = c_n q_n^3 \approx AB \approx 0.0787$):

$$s_c = c_n/c_{n+1} = s_q^3 = 4.236\ldots \tag{3.23}$$

while the exact numerical value via the renormgroup is 4.339... The two numbers are not equal but very close which was a puzzle for the formal renormgroup approach.

A similar situation occurs for the perturbation factor: $s_K = s_q = 1.618\ldots$ (resonant theory), and $s_K = 1.627\ldots$ (numerically).

The differences in scaling factors of the two theories can be interpreted as small changes of the exponents of q in scaled variables. For the two above examples we can write:

$$c = c_n q_n^{\alpha} \; ; \; \alpha = 3.049960\ldots$$

$$P = \Delta K_n q_n^{\beta} \; ; \; \beta = 1.0126966\ldots \qquad (3.24)$$

Other examples will be given below.

The behavior of asymptotic renormalization invariants A and R is shown in Fig. 6 below. Remarkably, the invariant critical structure, which repeats itself on finer and finer scales with rapidly increasing precision, is itself of the highest complexity as it contains both chaotic trajectories and an intricate admixture of regular and chaotic components of motion. An example of a tiny part ($\sim 0.01 \times 0.01$) of that structure is shown in Fig. 5.[21] The scale invariance is clearly seen within 3 successively scaled areas indicated by rectangles.

Notice that the scale invariance holds on a particular discrete set of scales, infinite though, because the renormgroup equations are based on the arithmetical map (3.21).

4. Renormalization chaos.[22] Variation of the critical structure from scale to scale can be viewed as some abstract dynamics. The corresponding dynamical space is infinitely dimensional but we may consider various few-dimensional projections of that are described by a set of scaled variables such as A_n, R_n, V_n, etc. (see, e.g., Fig. 6). The serial scale number n plays a role of 'time,' and we call it the renormtime. It is proportional to the logarithm of spatial and temporal scales:

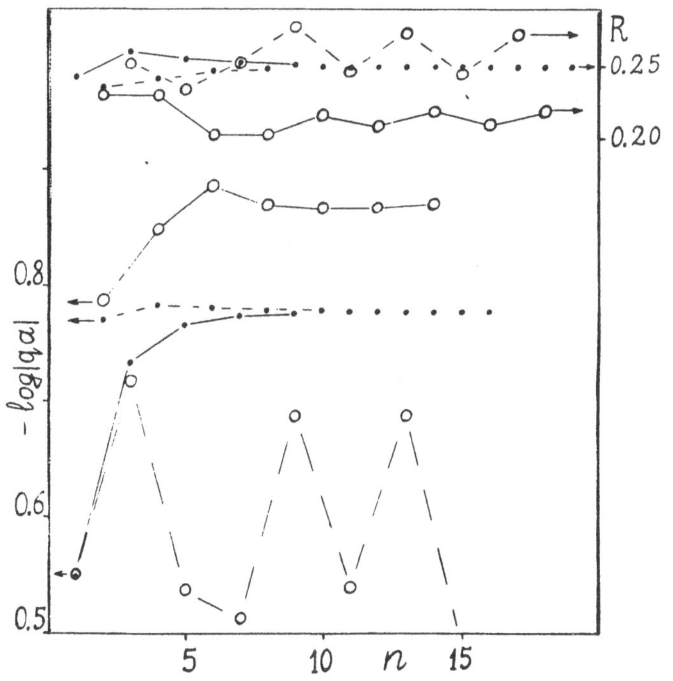

Fig. 6. An example of renormchaos for a random r_c (circles). Arrows indicate the corresponding scales for $A = aq$ (lower part), and for R (upper part); n is renormtime (the number of a principal scale). For comparison the same data are given for $r_c = r_G$ (dots) which illustrate the scale invariance (after Ref. 30).

$$n \sim |\ln a_n| \sim |\ln b_n| \sim \ln t_n \sim \ln q_n \ . \qquad (3.25)$$

The renormtime is discrete, as is the renormdynamics based on the arithmetical map (3.21).

The scale invariance described in the previous Section is the simplest type of renormdynamics, namely, a fixed point of the renormmap. The dynamical interpretation of renormalization suggests other, more complicated, scalings up to a chaotic one which would be the opposite limiting case. Guided by this heuristic approach we conjectured a new type of chaotic behavior - the renormchaos,[22] and presented an example of the latter in Ref. 30. A similar possibility was also considered in Ref. 35 for dissipative systems as modelled by a 1D map.

Our basic idea was to achieve the most complicated renormalization by using a random critical rotation number that is one with a random sequence of the elements $\{m_n\}$. As is known from modern ergodic theory (see, e.g., Ref. 20) this is just the case for almost any irrational r. Indeed, we may introduce a sequence of rotation numbers via the Gauss map

$$r = \frac{1}{m+\bar{r}} \ ; \qquad \bar{r} = \frac{1}{r} \bmod 1 \ , \qquad (3.26)$$

which is known to be chaotic.[20] Moreover, the basic arithmetical factor in renormalization (3.22) also obeys the same map

$$\omega = \frac{1}{\omega} \bmod 1 \ ; \qquad \omega = \frac{1}{s_q} < 1 \qquad (3.27)$$

backwards in renormtime, and with the 'initial' $\omega_\infty = \tilde{r}$ where $\tilde{r}$ is irrational, with reversed sequence of elements in respect to r. Clearly, the variation of critical structure in this case would be as random and unpredictable as a chaotic trajectory. An example of renormchaos is presented in Fig. 6 as described by A and R scaled variables. Irregular character of this renormdynamics is clear from Fig. 6 but the proof of its randomness is related to Gauss's map (3.27).

A chaotic trajectory is completely determined, in principle, by the initial conditions via the formal equations of motion. By analogy, we can conjecture that the chaotic variation of critical structure is related to the rotation number r. This would imply that the scaled variables are some universal functions of r. Then, asymptotically, as $n \to \infty$, the renormdynamics is described by an infinitely dimensional map

$$\bar{A}(r) = A(\bar{r}) \ ; \quad \bar{R}(r) = R(\bar{r}) \ \ etc \ \ \bar{r} = \frac{1}{r} \bmod 1 \ . \qquad (3.28)$$

Some numerical confirmation of this conjecture was presented in Ref. 30.

Thus, particular critical structure essentially depends on r, and in this sense is not universal. Nevertheless, the statistical properties of chaotic renormalization are the same for almost any r. Particularly, the average arithmetical factor (3.22)

$$\langle s_q \rangle = e^{h/2} \approx 3.28 \ ; \qquad h = \frac{\pi^2}{6\ln 2} \approx 2.37 \ , \qquad (3.29)$$

where h is KS-entropy of the Gauss map (3.27). This may be compared with a non-generic case of the scale invariance for $r(m^\infty)$:

$$s_q = \frac{m+\sqrt{4+m^2}}{2} \to 1.618\ldots \qquad (3.30)$$

The numerical value is given for $m = 1$ (golden tail).

A grandiose example of renormchaos is the oscillation of the whole Universe near a singularity in homogeneous but anistropic cosmological models.[36] So far there is no sign of such oscillations in our early Universe. Yet, the equations of general relativity allow that type of solution. Remarkably, the very complicated relativistic equations are approximately reduced to the trivial Gauss map.

5. Higher dimensions. A general picture of overlapping resonances, which destroy KAM tori, holds for an arbitrary number of freedoms.[7] This allows us to extend our resonant theory of critical phenomena to higher dimensions. There are two generally different cases of the latter: (i) $N > 2$ freedoms, and (ii) a driving quasiperiodic perturbation of one freedom. In the resonant theory both are similar, the principal parameter being the number of frequencies N.[37]

First of all, for arbitrary N the number or all resonances with $\sim q$ harmonics of each basic frequency is $\sim q^N$, hence the detuning $\rho \sim q^{-N}$ (cf. Eq. (3.9)), and

$$D = \rho\, q^N \sim 1 . \tag{3.31}$$

Now the main rotation number r is defined with respect to one of perturbation frequencies. The remaining N-2 rotation numbers enter driving frequencies $\nu_n \sim q\rho \sim q^{1-N}$ in the critical Hamiltonian (3.12). The resonance width $\Delta\rho \sim v_q^{1/2}$, and from the overlap criterion (3.10) the criticality condition is $v_q \sim q^{-2N}$, or (cf. Eq. (3.14))

$$R \sim V = v_q\, q^{2N} \sim 1 . \tag{3.32}$$

Hence, the critical perturbation smoothness $\beta_c = 2N - 1$ increases with N (cf. Ref. 37).

Longitudinal amplitudes $a_q \sim q\ v_q/\nu_q^2 \sim q^{-1}$ of the critical motion $x_c(t)$ do not depend on N, and

$$A = q\, a_q \sim 1 , \tag{3.33}$$

as before. The transverse amplitudes $b_q \sim a_q \nu_q \sim q^{-N} \sim \Delta\rho$ decrease with N but remain of the order of the resonance's width. Finally, the perturbation scaling does not, approximately, depend on N also:

$$P = q\, \Delta K_q \sim 1 . \tag{3.34}$$

However, in a many-dimensional case ($N > 2$) there is no simple procedure to single out the principal resonances like for $N = 2$.

The renormalization group in higher dimensions was generally discussed already for dissipative systems (see, e.g., Ref. 35). Yet, I am not aware of any particular results concerning the scaling properties in such systems.

To the best of my knowledge the only numerical data for $N = 3$ (standard map with a time-periodic parameter $K(t)$ were presented recently in Ref. 23. They seem to confirm the scalings related to R, P and A.

On the other hand, the authors did not find scale invariance in this model, and it seems to not exist at all. What is even more important, they discovered a breakdown of the renormalization universality in the sense that irregular oscillations of the critical structure depend, generally, not only on the two rotation numbers but also on model's parameters. Thus, many-dimensional renormdynamics appears to be even more complicated (chaotic?) as compared to the simplest case $N = 2$.

IV. CRITICAL STATISTICS

The most difficult, and as yet unsolved, problem is the impact of the critical structure at chaos border on the statistical properties of motion.

1. Smooth Perturbation: $\beta < \beta_c$. To begin with let us consider a simpler problem of a smooth perturbation (1.3) with $\beta < \beta_c = 3$. First, we can evaluate β_c directly from the resonance overlap criterion as applied to the original perturbation (1.3). The simplest estimate is as follows. The total width of all primary resonances $r_{pq} = p/q$ on the unit r interval is

$$\sim \sum_q q\, v_q^{1/2} \sim K^{1/2} \sum_q q^{\frac{1-\beta}{2}} \sim 1 . \tag{4.1}$$

This sum diverges for $\beta \leq 3$, hence $\beta_c = 3$ in agreement with the previous estimate in Section III.2. The latter estimate in Eq. (4.1) determines those resonances which provide the overlapping for $\beta < 3$. The critical $q_c \sim K^{(\beta-3)^{-1}}$. The corresponding resonance width $\Delta\rho_c \sim K^{1/2} q_c^{-(\beta+1)/2}$, and the frequency (cf. Eq. (3.17)) $\omega_c \sim q_c \Delta\rho_c$. Hence, the diffusion rate in r (or in y) is

$$D \sim \omega_c (\Delta\rho_c)^2 \sim K^{3/2} q_c^{-\frac{1+3\beta}{2}} \sim K^{\frac{5}{3-\beta}} . \tag{4.2}$$

The border case $\beta = 3$ requires more accurate estimates.

Estimate (4.2) agrees with numerical results in Ref. 38 for $\beta = 1$ (discontinuous $f(x)$).

2. Critical Perturbation.[31] One peculiarity of the standard map (1.6) is the periodicity not only in X but also in y with the same period 2π. As a result there is exact critical perturbation $K = K_G \approx 1$,[33] such that for $K > K_G$ the motion is unbounded in y and diffusive for some initial conditions (Section II.3). The problem I am going to discuss now is to explain scaling (2.3) for the diffusion rate as $K \to K_G$.

For $K > K_G$ the last (most robust) KAM curve is destroyed and transformed into a chaotic layer comprising all critical scales $q_n \gtrsim q_\varepsilon$ where $q_\varepsilon \sim \varepsilon^{-1}$, and $\varepsilon = K - K_G \to 0$ (see Eq. (3.19) and around). This chaotic layer is just the critical 'bottleneck' which controls the transition time between integer resonances $r = m$, and hence, global diffusion. The time scale in the layer is $\sim q_\varepsilon$, and the same is for the exit time (t) from the layer. However, entering into this thin layer ($\Delta r_\varepsilon \sim q_\varepsilon^{-2}$, Eq. (3.16)) from a big region ($\Delta r \sim 1$) takes much more time:

$$t_+ = t_- \frac{\Delta r}{\Delta r_\varepsilon} \sim q_\varepsilon^3 \sim \varepsilon^{-3} \sim D^{-1} . \tag{4.3}$$

This determines the transition time which is inversely proportional to the diffusion rate D in agreement with recent numerical results (see Eq. (2.3) and Ref. 18). Notice that the first value for the exponent ≈ 2.6[7] was not very accurate. The above estimate $\Delta r_\varepsilon / t_- \sim \Delta r / t_+$ (4.3) is simply the flux balance in statistical equilibrium.

It is interesting to mention that the renormgroup theory[39] gives the value $\ln s_c / \ln s_K = 3.011\ldots$ This is another example of surprising accuracy in the apparently primitive resonant theory.

3. The Chaos Border.[30] The impact of the critical structure at the chaos border in phase space on the statistical properties of the chaotic motion is the most difficult, and as yet unsolved, problem. The straightforward approach would be as follows. The transition time τ_n between adjacent scales is proportional to the time scale $t_n \sim q_n$ which, in turn, scales like $\mu_n^{-1/2} \sim c_y^{-1/2}$, where $\mu_n \sim \rho_n \sim q_n^{-2}$ is the sticking measure, and where c_y is the correlation (Sections III.1 and III.2). Hence, $c_y \sim \tau^{-2}$, and $p_c = 2$; $p = 3$. In a more sophisticated way the same result was obtained in Ref. 40. But this is a sheer contradiction with numerical data: $p \approx 1.5 < 2$.

The only way I see to avoid this contradiction is the conjecture that at exact criticality all transition times $\tau_n = \infty$ that is all scales are dynamically disconnected. Then, why does a connected chaotic component near the chaos border exists? The natural answer is in that the exact criticality is achieved on the border only, while inside a chaotic region the motion is supercritical. Consider, for example, model (3.1). Locally it is described by the standard map with $K \approx \lambda / y$. In a small vicinity of the border $y = y_b \approx \lambda$ the perturbation K indeed increases like $\Delta K \sim \Delta y \sim \rho \sim q^{-2}$. However, this is not enough to destroy the corresponding scale q_n, as $q_n \Delta K_n \sim q_n^{-1} \ll 1$ (see Eq. (3.19)). Only resonances with $q \gtrsim Q_{\tilde{n}} q_n^2$ would be destroyed and form a very narrow ($\sim Q_n^{-2} \sim q_n^{-4}$) chaotic layer which could play a role of the bottleneck controlling the transition time τ_n. Similarly to the derivation of Eq. (4.3) we obtain

$$\tau_n \sim Q_n \frac{Q_n^2}{q_n^2} \sim q_n^4 \sim \mu_n^{-2} \sim c_y^{-2} . \tag{4.4}$$

Hence

$$c_y(\tau) \sim \tau^{-1/2} ; \quad P(\tau) \sim \tau^{-3/2} \tag{4.5}$$

now in agreement with numerical data.

The same result can be obtained in a different, more formal, way. Namely, we can rescale dependence (4.3) for transition between integer resonances ($q_n = 1$) to arbitrary scale q_n. To this end we rewrite Eq. (4.3) in scaled variables

$$\frac{\tau_n}{t_n} \sim (q_n \Delta K)^{-3} . \tag{4.6}$$

With $t_n \sim q_n$ and $\Delta K \sim q_n^{-2}$ we arrive at Eq. (4.4). Notice that a different relation $\tau_n(q_n)$ in Ref. 22 was due to a mistake in scaling.

A weak point of the latter approach (4.6) is that the scaling (4.3) is asymptotic ($q_n \to \infty$) while integer resonances ($q_n = 1$) are not. In any event, further studies of the mechanism of critical statistics are certainly required.

In higher dimensions (Section III.5) the supercriticality $\Delta K \sim \rho \sim q^{-N}$; the bottleneck harmonic $Q \sim (\Delta K)^{-1} \sim q^N$; and transition time (cf. Eq. (4.4))

$$\tau \sim Q^{N-1} \frac{Q^N}{q^N} \sim Q^{2N-2} \sim \mu^{2-2N} . \tag{4.7}$$

Hence

$$c_y \sim \mu \sim \tau^{\frac{-1}{2N-2}} ; \quad p_c = \frac{1}{2N-2} . \tag{4.8}$$

As $N \to \infty, p_c \to 0$, and correlations do not decay at all. I am going to come back to this interesting case in Section IV.5 below.

Another difficult problem is the arithmetic of rotation numbers r_b of the critical border curves. In Ref. 22 it was conjectured that the set of r_b consists of all combinations of only two elements $m = 1$ and 2 in the continued fraction representation. This is sufficient for r_b to be random, and hence to explain irregular oscillation of the local exponent in the distribution of Poincare's recurrences (Section III.1). This conjecture was partially confirmed numerically in Ref. 45. Our recent refined conjecture is that r_b are the so-called Markov numbers.[30]

4. Internal Borders. Typically, the central part of principal critical resonances is not destroyed (see, e.g., Fig. 5). Hence, in any neighborhood of the main chaos border there is an infinite set of internal chaos borders, each one with its own critical structure. Assuming universality of critical phenomena at any chaos border we arrive at the following estimate in scaled variables:

$$(\mu_q q^2) \sim \left(\frac{\tau}{q}\right)^{-p_c} , \tag{4.9}$$

for a principal resonance q where μ_q is the sticking measure at the internal border.

The main difficulty here is in that the internal border exists not only inside the principal resonances but also in many others, near the critical border curve, which are not destroyed by the local supercritical perturbation. To estimate the total number of such resonances we can make use of Eq. (3.19) which determines the stability zone $\Delta K_s \sim \rho_s \sim q^{-1}$ for any q (as a crude approximation, of course). Then, for a given q only$M_q/q \sim 1$ resonances fall into this zone, where $M_q \sim q$ is the total number of resonances p/q for a fixed q. Again, as a crude approximation we can extend the estimates, particularly Eq. (4.9), on all undestroyed resonances. As a result, the total internal border contribution to the correlation is

$$\tilde{c}_y \sim \sum_q \mu_q \sim \tau^{-p_c} \sum_q^{\tau} q^{p_c-2} , \tag{4.10}$$

where the sum is taken over all q up to τ. This contribution is essential if $p_c \geq 1$. But for $p_c > 1$ the above estimate is not self-consistent as $\tilde{c}_y \sim \tau^{-1}$, contrary to the assumed universality. However, the latter holds for $p_c = 1$ (to logarithmic accuracy). This was the preliminary conclusion in Ref. 41 which was confirmed in Ref. 42.

It would be a nice solution in the spirit of universality of critical phenomena. Yet, first, the value $p_c = 1$ seems still to be incompatible with numerical data (Section III.1), and second, there is another possibility missed in Ref. 41, namely, $p_c < 1$ as is suggested by numerical data. Then, the effect of internal borders is not decisive, at least, for exponent p_c whose value is determined by another mechanism, for example the one described in the previous Section.

In higher dimensions we have instead of Eq. (4.9) (see Section III.5)

$$(\mu_q \, q^N) \sim \left(\frac{\tau}{q^{N-1}}\right)^{-p_c} . \tag{4.11}$$

In calculating the total contribution of all internal borders we need to take into account that now there are as many as $\sim q^{N-2}$ undestroyed resonances, for a given q, within the stability zone. Hence, the total 'internal' correlation is

$$\tilde{c}_y \sim \sum_q \mu_q \, q^{N-2} \sim \tau^{-p_c} \sum_q^{\tau^{1/(N-1)}} q^{p_c(N-1)-2} . \tag{4.12}$$

The critical value p_c^* of the critical exponent is $p_c^* = (N-1)^{-1}$. Only this value preserves universality based entirely on the internal borders. And, again, there is another possibility that $p_c < p_c^*$ so that internal borders are irrelevant. This is just the case if the above estimate (4.8) is true: $p_c = p_c^*/2$.

Preliminary numerical results obtained in collaboration with V. V. Vecheslavov ($p_c \approx 0.26$ and 0.19 for $N = 3$ and 4, respectively) seem to confirm (or, at least, do not contradict) prediction (4.8).

5. Superfast Diffusion.[22] Slow correlation decay with $p_c < 1$ (4.5) may result in a superfast diffusion. Indeed, if this correlation determines the diffusion, the rate

$$D_z \sim \int c_y(\tau) d\tau$$

formally diverges. Here the diffusion goes in a new variable z, and $\dot{z} = y$. The divergence means that the dispersion (the second moment of the distribution function)

$$\sigma^2 \sim \int D \, d\tau \sim t^{2-p_c} \tag{4.13}$$

grows faster than time t. Hence the term 'superfast diffusion' which we use. This phenomenon was studied from different points of view in many papers (see, e.g., Refs. 43, 44).

The simplest example is again the standard map for special values of the parameter $K \approx 2\pi m$ with any integer $m \neq 0$. At these K the so-called accelerator modes exist,[7] that is relatively small areas of regular motion with linearly increasing momentum: $y \sim \pm\, t$ while phase X is fixed. A chaotic trajectory cannot penetrate into these domains but it sticks at their borders. As a result, a superfast diffusion in y occurs which was first observed numerically in Ref. 46. Notice that in the above notation $z = y$ now while y plays the role of a new coordinate normal to the chaos border surrounding the regular regions. According to Eq. (4.13)

$$\sigma^2 \approx \alpha \mu_s \frac{K^2}{2} t^{3/2} \sim t^{3/2} , \tag{4.14}$$

where $\alpha \approx 0.5$ from numerical data[46] for $K \approx 2\pi$, and a relative stable area $\mu_s \approx 0.02$. As $\mu_s \sim K^{-2}$,[7] the rate of this anomalous diffusion ($\sigma^2/t^{3/2}$) does not depend either on $K \to \infty$ or on $\mu_s \to 0$.

In Refs. 44, 47 more complicated accelerator modes were shown to produce a superfast diffusion also corresponding to $p_c \approx 2/3$ in reasonable agreement with our numerical data. A simple expression for the growth of all moments of the distribution function was also given in Ref. 44 namely:

$$\sigma^k \sim t^{k-p_c} \tag{4.15}$$

for k even. In higher dimensions when $N \to \infty$, and $p_c \to 0$ this relation becomes especially simple but somewhat puzzling. It appears to describe almost a free motion but in both directions of the Z variable! The limiting case $p_c = 0$ corresponds to the fastest homogeneous diffusion possible.

Further insight into the nature of superfast diffusion can be obtained from the motion power spectrum which is the Fourier transform of the correlation.[30] For $\omega \to 0$ we have from Eq. (4.8)

$$s_y(\omega) \sim \omega^{p_c-1} \sim \omega^{-\frac{2N-3}{2N-2}} . \tag{4.16}$$

As $N \to \infty$ it approachs the famous $1/\omega$ spectrum which, thus, produces the fastest diffusion. If $\dot{z} = y$, the spectrum of the Z-motion is

$$S_z = \frac{S_y}{\omega^2} \sim \omega^{p_c-3} . \tag{4.17}$$

From normalization (Parseval's theorem)

$$\overline{z^2} = \int s_z(\omega)d\omega \sim w^{p_c-2} \sim t^{2-p_c} . \tag{4.18}$$

If $p_c < 1$ the integral diverges as $\omega \to 0$. For a finite time interval the minimal $\omega \sim t^{-1}$, and the diffusion law (4.13) is recovered, including the limiting $p_c = 0$. However, in the latter case the velocity dispersion $\overline{y^2} \sim \ln\omega$ diverges (see Eq. (4.16)). In our models with a chaos border this is impossible, hence, always $p_c > 0$ (4.8).

The theory of superfast diffusion can be applied to a broad variety of different problems. A nice example is the tangle of a long polymeric molecule in a certain environment. Approximately, such a molecule can be considered as a trajectory of the self-avoiding random walk. The constraint imposes a long-term correlation which can be estimated as follows. Suppose that the molecule length l and the tangle size σ are related by

$$\sigma^2 \sim l^{2\nu} \tag{4.19}$$

with some so far unknown parameter ν. Then, the correlation due to avoided crossings of the molecule line is roughly proportional to the probability of self-crossing:

$$c \sim \frac{l}{\sigma^d} \sim l^{1-\nu d} , \tag{4.20}$$

where the integer d is the space dimension. Hence, we have power-law correlation with the exponent $p_c = \nu d - 1$. Using Eq. (4.13) with $t = l$ and Eq. (4.19) we arrive at the relation

$$2\nu = 3 - \nu d ; \quad \nu = \frac{3}{2+d} , \tag{4.21}$$

which is known at Flory's formula (see, e.g., Ref. 48). It was derived in a completely different way (from the thermodynamics of a polymeric molecule), and holds for $d \leq 4$, otherwise $\nu = 1/2$. In our dynamical approach the latter limitation follows from the condition $p_c < 1$ for anomalous diffusion. In the border case $\sigma^2 \sim l \ln l$ (see, e.g., T. Geisel et al. in Ref. 43) which slightly differs from Flory's formula.

6. Fractal properties.[49] The critical structure in Hamiltonian systems is also called 'random fractals' (R. Voss, Ref. 43) because of the renormchaos (Section III.4) or 'fat fractals'[49] for their finite measure unlike dissipative systems. Some fractal properties were studied numerically in Ref. 49.

Here I am going to explain one property — the fractal dimension d_L of the set of all chaos borders (mainly internal ones, of course). It is inferred from the dependence of the measure μ_{ch} of a chaotic component on spatial linear resolution $\varepsilon \to 0$:

$$\mu_{ch}(\varepsilon) = \mu_{ch}(0) + \alpha\varepsilon^{\beta} \; . \tag{4.22}$$

Here $\mu_{ch}(0) > 0$ is the measure of the whole chaotic component, hence, its dimension $d_S = 2$ is topological. The second term represents the borders whose total length and dimension are

$$L(e) \sim \varepsilon^{\beta-1} \; ; \quad d_L = 2 - \beta \; . \tag{4.23}$$

The simplest evaluation of this scaling can be made as follows (see Section IV.4). Each undestroyed resonance has internal borders of the total length $l_q \sim 1$. This estimate follows from the fact that an individual border is a non-fractal curve whose dimension $d_l = 1$ is topological. It is because the ratio of transverse to longitudinal scaling factors of the critical structure $s_b/s_a = q_n \to \infty$, as $n \to \infty$ (see Sections III.2 and III.3). Thus, the border curve $y_b(x)$ is very smooth (see Fig. 5). The number of undestroyed resonances is of the order of maximal $q = q_{max}$ which is determined by the resolution: $\varepsilon \sim q_{max}^{-2}$. Hence

$$L \sim \varepsilon^{-1/2} \; ; \quad \beta = \frac{1}{2} \; ; \quad d_L = \frac{3}{2} \; , \tag{4.24}$$

in a reasonable agreement with the numerical $\beta = 0.3$ - 0.7.[49]

Notice also that the total number of resolved borders scales, in this approximation, as

$$N_b \sim q_{max}^2 \sim \varepsilon^{-1} \; . \tag{4.25}$$

In higher dimensions with N frequencies we need to consider an N-dimensional map with non-fractal border surfaces of N-1 dimensions. Now there are $\sim q^{N-1}$ undestroyed resonances up to $q_{max} \sim \varepsilon^{-1/N}$. The border surface in each such resonance $s_1 \sim 1$, and total border surface and its dimension are

$$s_b \sim \varepsilon^{-(1-\frac{1}{N})} \; ; \quad d_S = N - \frac{1}{N} \; . \tag{4.26}$$

The total number of resolved border surfaces, or of the domains with regular motion,

$$N_b \sim \varepsilon^{-1} \tag{4.27}$$

does not depend on N, and in this sense is universal.

ACKNOWLEDGMENTS

I would like to express my sincere gratitude to D. L. Shepelyanksy, my main collaborator in the studies of critical phenomena, to F. Vivaldi who attracted our attention to this interesting problem, and to G. Casati, D. Escande, R. MacKay, I. Percival and Ya. G. Sinai for stimulating discussions.

REFERENCES

1. V. M. Alekseev and M. V. Yakobson, Phys. Reports **75** (1981) 287.
2. G. Chaitin, Information, Randomness and Incompleteness, World Scientific, 1990.
3. B. V. Chirikov, F. M. Izrailev and D. L. Shepelyansky, Physica D **33** (1988) 77.
4. B. V. Chirikov, Time-Dependent Quantum Systems, Proc. Les Houches Summer School on Chaos and Quantum Physics, Elsevier, 1990.
5. A. Lichtenberg, M. Lieberman, Regular and Stochastic Motion, Springer, 1983.
6. G. M. Zaslavsky, Chaos in Dynamic Systems, Harwood, 1985.
7. B. V. Chirikov, Phys. Reports **52** (1979) 263.
8. B. V. Chirikov and V. V. Vecheslavov, Astron. Astroph. **221** (1989) 146.
9. G. Casati et al., Phys. Rev. A **36** (1987) 3501.
10. A. A. Chernikov, R. Z. Sagdeev and G. M. Zaslavsky, Physica D **33** (1988) 65.
11. B. V. Chirikov, Foundations of Physics **16** (1986) 39.
12. M. Eisenman et al., Lecture Notes in Physics **38** (1975) 112; J. von Hemmen, ibid., **93** (1979) 232.
13. B. G. Konopelchenko, Nonlinear Integrable Equations, Lecture Notes in Physics **270** (1987).
14. B. V. Chirikov and V. V. Vecheslavov, KAM Integrability, in: Analysis etc., Academic Press, 1990, p. 219.
15. V. I. Arnold and A. Avez, Ergodic Problems in Classical Mechanics, Benjamin, 1968.
16. B. V. Chirikov, Proc. Roy. Soc. Lond. A **413** (1987) 145.
17. A. Rechester et al., Phys. Rev. A **23** (1981) 2664.
18. B. V. Chirikov, D. L. Shepelyansky, Radiofizika **29** (1986) 1041.
19. F. Vivaldi, private communication.
20. I. Kornfeld, S. Fomin and Ya. Sinai, Ergodic Theory, Springer, 1982.
21. R. MacKay, Physica D **7** (1983) 283.
22. B. V. Chirikov and D. L. Shepelyansky, Physica D **13** (1984) 395.
23. R. Artuso, G. Casati and D. L. Shepelyansky, Breakdown of Universality in Renormalization Dynamics for Critical Invariant Torus, Phys. Rev. Lett. (to appear).
24. B. V. Chirikov, D. L. Shepelyansky, Proc 9th Int. Conf. on Nonlinear Oscillations (Kiev, 1981). Kiev, Naukova Dumka, 1983, Vol. 2, p. 421. English translation available as preprint PPL-TRANS-133, Plasma Physics Lab., Princeton Univ., 1983.
25. S. Channon and J. Lebowitz, Ann. N. Y. Acad., Sci. **357** (1980) 108.
26. G. Paladin and A. Vulpiani, Phys. Reports **156** (1987) 147.
27. C. Karney, Physica D **8** (1983) 360.
28. P. Grassberger and I. Procaccia, Physica D **13** (1984) 34.
29. J. Bene, P. Szèpfalusy and A. Fülöp, A generic dynamical phase transition in chaotic Hamiltonian systems, Phys. Rev. Lett. (to appear).
30. B. V. Chirikov and D. L. Shepelyansky, Chaos Border and Statistical Anomalies, in: Renormalization Group, D. V. Shirkov, D. I. Kazakov and A. A. Vladimirov Eds., World Scientific, Singapore, 1988, p. 221.
31. B. V. Chirikov, Intrinsic Stochasticity, Proc. Int. Conf. on Plasma Physics, Lausanne, 1984, Vol. II, p. 761.
32. G. Schmidt and J. Bialek, Physica D **5** (1982) 397.
33. J. Greene, J. Math. Phys. **9** (1968) 760; **20** (1979) 1183.
34. M. Feigenbaum, J. Stat. Phys. **19** (1978) 25; **21** (1979) 669.
35. S. Ostlund et al., Physica D **8** (1983) 303.
36. E. M. Lifshits et al., Zh. Eksp. Teor. Fiz. **59** (1970) 322; ibid (Pisma) **38** (1983) 79; J. Barrow, Phys. Reports **85** (1982) 1.
37. B. V. Chirikov, The Nature and Properties of the Dynamic Chaos, Proc. 2d Int. Seminar, "Group Theory Methods in Physics" (Zvenigorod, 1982), Harwood, 1985, Vol. 1, p. 553.
38. I. Dana et al., Phys. Rev. Lett. **62** (1989) 233.

39. R. MacKay et al., Physica D **13** (1984) 55.
40. J. Hanson et al., J. Stat. Phys. **39** (1985) 327.
41. B. V. Chirikov, Lecture Notes in Physics **179** (1983) 29.
42. J. Meiss and E. Ott, Phys. Rev. Lett. **55** (1985) 2741; Physica D **20** (1986) 387.
43. P. Lévy, Théorie de l'addition des variables eléatoires, Gauthier-Villiers, Paris, 1937; T. Geisel et al., Phys. Rev. Lett. **54** (1985) 616; R. Pasmanter, Fluid Dynamic Research **3** (1988) 320; R. Voss, Physica D **38** (1989) 362; G. M. Zaslavsky et al., Zh. Exper. Teor. Fiz. **96** (1989) 1563.
44. H. Mori et al., Prog. Theor. Phys. Suppl., **99**, 1 (1989).
45. J. Greene et al., Physica D **21** (1986) 267.
46. C. Karney et al., ibid **4** (1982) 425.
47. Y. Ichikawa et al., ibid **29** (1987) 247.
48. P. de Gennes, Scaling Concepts in Polymer Physics, Cornell Univ. Press, 1979.
49. D. Umberger and D. Farmer, Phys. Rev. Lett. **55** (1985) 661; C. Grebogi et al., Phys. Lett. A **110** (1985) 1.

STATISTICAL PROPERTIES OF THE TRANSITION TO SPATIOTEMPORAL CHAOS

S.Ciliberto

Istituto Nazionale Ottica
Largo E.Fermi 6-50125 Firenze-Italy

1 Introduction

It is nowaday well established, both theoretically and experimentally, that the chaotic time evolution observed in many natural phenomena may be produced by the non-linear interaction of a small number of degrees of freedom[1] In spatially extended systems low dimensional chaos is often associated with relevant spatial effects, such as mode competition, travelling waves, localized oscillations[2], but the unpredictable time evolution does not influence the spatial order, that is to say the correlation length is comparable with the size of the system. However an extended system may present a chaotic evolution both in space and time that, instead, implies the presence of many degrees of freedom[3]. The study of the transition from low dimensional chaos to spatiotemporal chaos is a subject of current interest that is not jet completely understood. For example it is important to investigate whether there are general features that are independent of the specific system under study and whether a thermodynamic description may be appropriate[3,4].

The simplest mathematical models, in which the features of the transition to spatiotemporal chaos may be analysed, are systems of coupled maps[5-7], one dimensional partial differential equations[8-12] and cellular automata [13]. These models have a physical relevance because many of the features, that they present, are similar to those observed in experiments on boundary layer flow[14], thermal convection[4,15,16], liquid crystal and surface waves[17].

The transition to spatiotemporal chaos may be characterized either by means of standard tools such as power spectra and correlation functions or by studying the statistical properties of local and global variables. The statistical analysis is clearly very useful in order to compare the properties of spatiotemporal chaos with those of a system near thermal equilibrium[3,4].

In this paper we describe an experiment in which the space time evolution of Rayleigh-Benard convection has been studied to investigate general features of the transition to spatiotemporal chaos and to test methods useful to characterize this transition[4,15]. Rayleigh-Benard convection has been choosen because is one of the simplest pattern forming fluid instabilities in which these problems can be studied. Furthermore the properties of spatiotemporal chaos in fluid dynamics are very useful in order to give more insight into the problem of the transition to turbulence.

The paper is organized as follows. In section 2) we briefly remind the general properties of Rayleigh-Benard convection, and the main features of our experimental apparatus. In section 3) the different space time dynamics observed as function of the control parameter are discussed. In section 4) the results concerning the

Microscopic Aspects of Nonlinearity in Condensed Matter
Edited by A.R. Bishop *et al.*, Plenum Press, New York, 1991

transition to spatitemporal intermittency [14] are reported and compared with those observed in the above mentioned mathematical models. In section 5) we describe the thermodynamic properties that appear above the transition point[4]. Finally conclusions are presented in section 6).

2 Rayleigh-Benard Convection and experimental techniques

To illustrate the general features of Rayleigh-Benard convection, let us consider a fluid layer confined between two horizontal solid plates and heated from below. When the temperature difference between the two plates ΔT is smaller than the threshold value ΔT_c, there is no fluid motion and the heat is transported across the layer only by conduction. In contrast when ΔT exceeds ΔT_c a steady convective flow arises, producing a pattern of parallel rolls with a well defined wavenumber q that is of the order of 3.11/d, where d is the depth of the layer. The roll axis are parallel to the shortest side of the cell containing the fluid.

Increasing ΔT other instabilities, that destabilize the main set of rolls, may appear and finally the fluid motion becomes time dependent. We are interested in studying these transitions and the evolution toward chaotic and turbulent states of the time dependent regimes. Other informations about Rayleigh-Benard convection may be found in standard text books and review papers[18].

In the experiment that we describe in this paper the horizontal fluid layer has an annular geometry. Indeed with this geometry and a suitable choise of the horizontal sizes of the cell, it is possible to construct a pattern that is almost a one dimensional chain of radial rolls (roll axis along radial directions, see also Fig.1) with periodic boundary conditions. These features of the spatial pattern are very useful in order to compare the results of our experiment, with those obtained in the mathematical models mentioned in section 1). Specifically the inner and outer diameters of our cell are 6 cm and 8 cm respectively, whereas the depth of the layer is 1 cm. The working fluid is silicon oil with a Prandtl number of 30 and the computed value of ΔT_c is $0.06°C$. The details of the experimental set up have been described elsewhere[19]

As an example of the spatial organisation of the fluid motion inside our cell, at different values of $\eta = \Delta T/\Delta T_c$, we report in Figs.1a),1b), the shadowgraphs of the convective patterns seen from above. Dark regions correspond to the hot currents rising up and white regions to the cold ones, going down. The picture in Fig.1a) is the image of a typical stationary structure at $\eta = 13$, that is rather close to the convective threshold. The picture Fig.1b) is instead a snapshot a chaotic space time evolution, at $\eta = 230$.

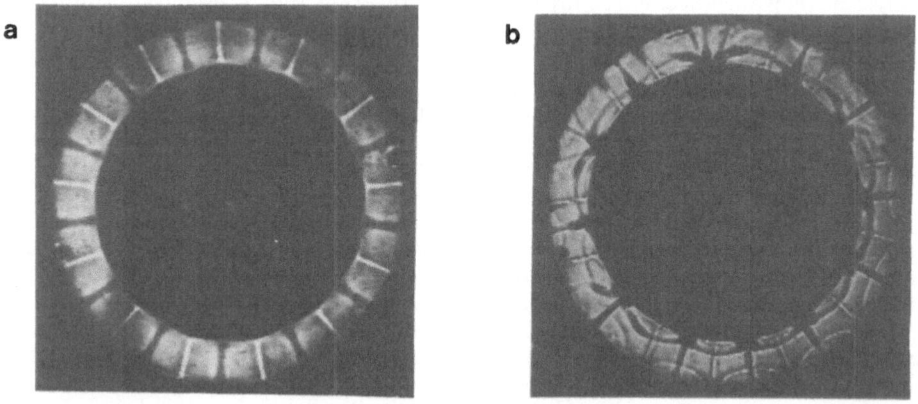

Figure 1. Shadowgraphs of typical spatial patterns. White and dark regions correspond to cold and hot currents respectively. a)Stationary spatial pattern at $\eta = 13$. b) Snapshot of the spatial pattern at $\eta = 230$ in a time dependent regime (space time intermittency).

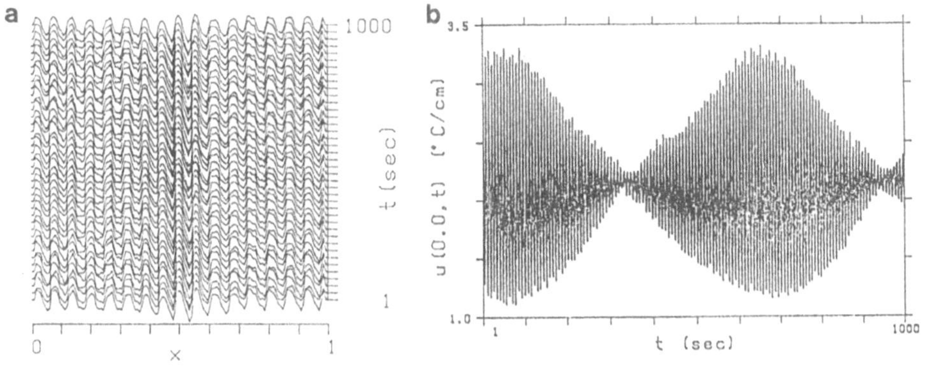

Figure 2.a) Space time evolution of u(x,t) at $\eta = 164$;b) Corresponding time evolution of the point x=0. The vertical scale has been amplified in b) because the time dependent modulation slightly perturbes the spatail pattern shown in a),where the maximum amplitude is roughly $4°C/cm$

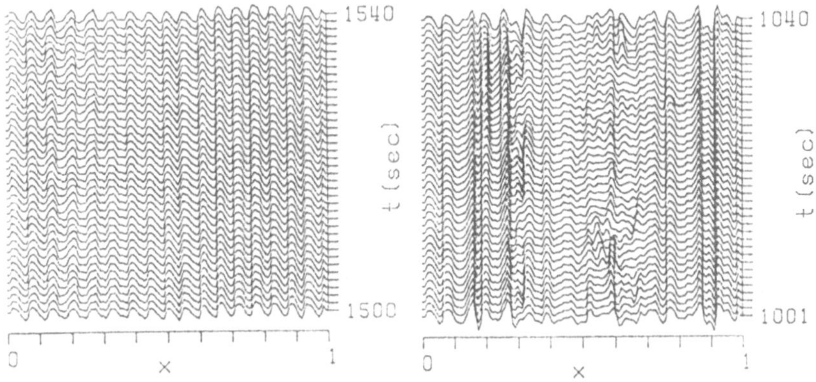

Figure 3. Space time evolutions of u(x,t) at $\eta = 216$ at two different time intervals of 40 sec each.

The shadowgraph technique is very useful to have qualitative informations about the spatial structures, however no quantitative information may be extracted by this method[20]. Thus to quantitatively characterize the space time evolution of the system we measure, on the circle of mean diameter the two horizontal components of the thermal gradient, averaged along the vertical direction. The measurements have been done by means of an other optical method[21]. In what is following we will focus the attention only on the component of the gradient perpendicular to the roll axis. This component is called u(x,t), with $0 < x < 1$, where 1 is the whole length of the circle of mean diameter. The function u(x,t) is sampled at N points in space ($128 < N < 256$). In time dependent regimes u(x,t) is recorded for at least 5000 times at interval of 1 sec, that is roughly 1/10 of the main oscillation period of the system.

3 Spatial patterns

Analysing the fluid behaviour as a function of $\eta = \Delta T/\Delta T_c$, we observe, that for η around 1, the spatial structure has about 22 rolls. This number increases with η and reaches 38 at η around 200. A detailed analysis of the wavenumber selection process has been reported elsewhere[22].

The spatial structure remains stationary for $\eta < 164$ where a subcritical bifurcation to the time dependent regime takes place. For $\eta > 164$ the time evolution is chaotic but, reducing η, the system presents either periodic or quasiperiodic oscillations, and at $\eta = 149$ it is again stationary.In the range $149 < \eta < 200$ the time dependence consists of rather localized fluctuations that slightly modulate the convective structure, which mantains its periodicity.

The space time evolution of u(x,t) and the corresponding time evolution of the point x=0 at $\eta = 164$ are shown in Fig.2a and Fig.2b. In looking at Fig.2b we clearly see that the time evolution is biperiodic. However this time dependent modulation is hardly seen in Fig.2a, because it slightly perturbes the spatial pattern that mantain its original periodic structure. Increasing η the time evolution becomes chaotic but the spatial order is still mantained, and the correlation length of u(x,t) is of the order of 1. The fractal dimension and the orthogonal decomposition [23] indicate that the number of degrees of fredom involved in the dynamics is around 3, thus confirming that this dynamic is produced by a small number of degrees of freedom.

For $\eta > \eta_c = 200$ the correlation length decreases as a function of η, indictaing that the spatial order begins to be destroyed because of the appearence of bursts, detaching from the boundary layer. The snapshot, shown in Fig.1b, correspond to such a regime. It presents, several domains where the spatial periodicity is completely lost (we will refer to them as turbulent) and other regions (that we call laminar) where the spatial coherence is still mantained. The space time evolution of u(x,t) at $\eta = 216$ is shown in Fig.3a),3b) at two different times. We notice that for $1000\,\text{sec} < t < 1040\,\text{sec}$ there are strong oscillations that locally destroy the spatial order whereas for $1500\,\text{sec} < t < 1540\,\text{sec}$ the pattern is again very regular. This mixture of laminar and turbulent regions is also called spatiotemporal intermittency[6,7,8].

The time averaged spatial Fourier spectra at $\eta = 164$, $\eta = 216$, $\eta = 347$ are reported in Figs.4a),4b), and 4c) respectively. The spectrum of Fig.4a) corresponds to the above mentioned bipΩrriodic regime and being the spatial structure still very ordered the spectrum presents well defined peaks. In contrast Fig.4b), corresponding to a value of η that is very close to the threshold for spatiotemporal intermittency presents a broadened third harmonic. This indicate that the most important length scales for this transition are the shortest ones. Finally in Fig.4c) the spectrum, corresponding to a value of η far above the transition point, is totally broadened because the spatial order has been destroyed. Notice the exponential decay at high and the flat region for the small ones. These features are rather similar to those observed in the turbulent regimes of the Kuramoto-Shivanshinsky equation[10].

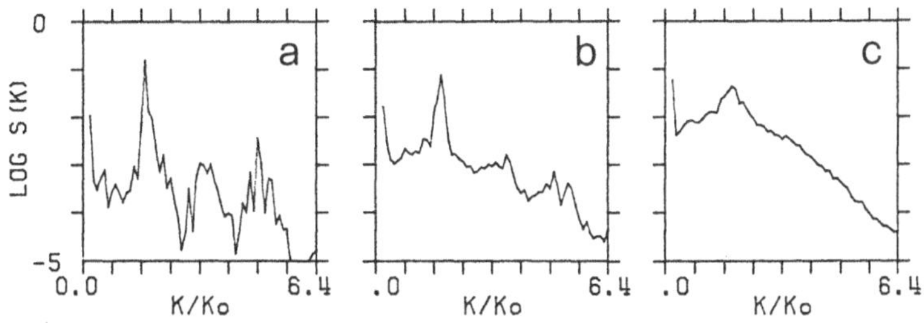

Figure 4. Spatial power spectra at different values of η:a) $\eta = 164$;b) $\eta = 216$;c)$\eta = 348$.

4 The transition to spatiotemporal intermittency

As we discussed in the previous section, the space time evolution of u(x,t) shows that in the turbulent domains the time evolution is characterised by the appearence of large oscillatory bursts. Instead in laminar regions the oscillations remain very weak. Thus the two regions can be identified by measuring the local peak to peak amplitude u_{pp}, for a time interval δ comparable with the mean period of the oscillation, that is:

$$u_{pp}(x,t) = max(u(x,\tau)) - min(u(x,\tau))$$

with $t < \tau < (t+\delta)$. Choosing a cutoff α, setting to 1 all the points in which $u_{pp} > \alpha$ and to 0 all the other points, the space time dynamics is reduced to a binary code in which 1 stands for "turbulent" and 0 for laminar. As an example of such a code we show the spacetime evolution of u(x,t) at $\eta = 216$, in Fig.5a), and $\eta = 248$ in Fig.5b),the black and white regions correspond to turbulent and laminar domains respectively. We remark that the qualitative features of these pictures are rather independent of the precise value of the cutoff. We can easily verify that the code catches the main properties of the dynamics by comparing Fig.5a) with Figs. 4a) and 4b). Indeed we clearly see that at the most oscillating and disordered regions of Figs.4) correspond to black points in Fig.5a) whereas ordered and not oscillating regions are represented by white points.

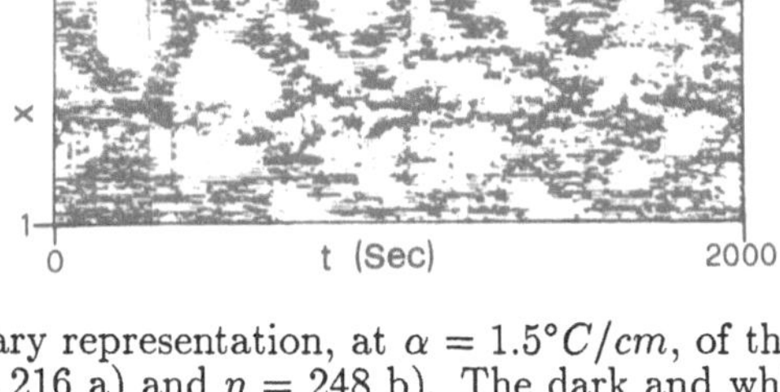

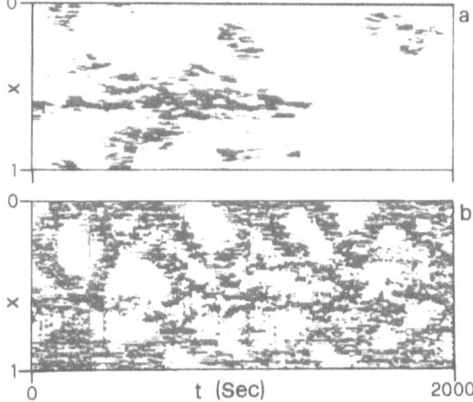

Figure 5. Binary representation, at $\alpha = 1.5°C/cm$, of the space-time evolution of u(x,t) at $\eta = 216$ a) and $\eta = 248$ b). The dark and white area correspond to turbulent and laminar domains respectively.

At $\eta = 216$,Fig.5a), a wide laminar region surrounds completely the turbulent patches that remain localized in space, after their appearence. Furthermore, the nucleation of a turbulent domain has no relationship with the relaxation of another one. In contrast, at $\eta = 248$ Fig.5b),the turbulent regions migrate and slowly invade the laminar ones. This last regim that sets in for $\eta > 245$ is very similar to those obtained in theoretical models [6–8]. The change from the regime of Fig.5a) to that of Fig.5b) is reminescent of a percolation [7], that, indeed, has been proposed as one of the possible mechanisms for the transition to spatiotemporal intermittency.

Following a method also used in numerical models [6–8], we quantitatively characterize [15,19] such a behaviour by computing, over a time interval of 10^4 sec, the distibution P(x) of the the laminar domains of length x. For $\eta < 248$ P(x) decays with a power law.The exponent does not depend within our accuracy, either on α or on η . Its average value is $\mu = 1.9 \pm 0.1$. On the other hand, for $\eta > 248$,the decay of $P(x)$ for $x > 0.1$ is exponential with a characteristic length 1/m.

We find that the dependence of m on η is the following:

$$m(\alpha, \eta) = m_o(\eta) exp(-\alpha/\alpha_o) \tag{1}$$

with $\alpha_o = (0.87 \pm 0.06)°C/cm$ independent of η. The dependence of m_0^2 versus η is reported in Fig.6. The linear best fit for $\eta > 246$ of the points of Fig.6 gives the following result:

$$m_o(\eta) = m_1(\frac{\eta}{\eta_s} - 1)^{\frac{1}{2}} \tag{2}$$

with $\eta_s = 247 \pm 1$ and $m_1 = 117 \pm 2$. This equation shows the existence of a well defined threshold η_s for the appearence of an exponential decay in P(x). Besides we see that the characteristic length $1/m_o$ diverges at $\eta = \eta_s$. In the range $200 < \eta < 400$, P(x) is very well approximated by the following equation:

$$P(x) = (Ax^{-\mu} + B) exp[-m(\alpha, \eta)x] \tag{3}$$

where $m(\alpha, \eta)$ is given by 1) and μ has the previous determined value. A,B are instead free parameters that can be very easily determined. It is possible to fit our experimental P(x),in the range $0.4°C/cm < \alpha < 3°C/cm$, with $A = 10\ B \simeq 4 \cdot 10^3$ for $\eta > \eta_s$ and $B = 0$ for $\eta < \eta_s$.

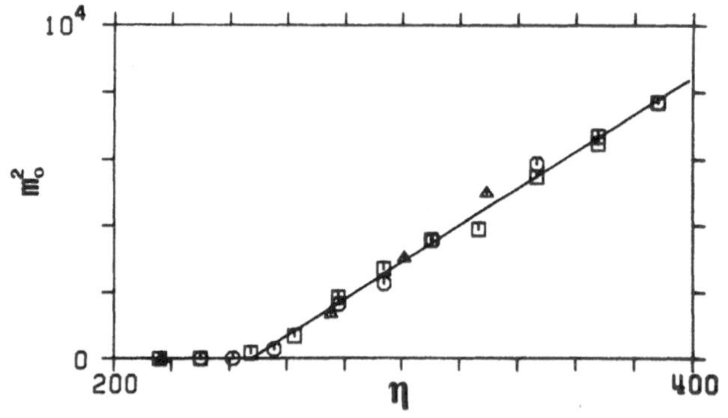

Figure 6. Dependence of $m_o{}^2$ on η ,the different symbols pertain to different sets of measurements done either increasing or decreasing η. The solid line is obtained from eqn.2.

The features of P(x) displayed by equations 2),3)are typical of phase transitions. Therefore, being the transition point η_s very close to the point where the behaviour like that of fig.5b) sets in, we conclude that the transition to this behaviour may be a phase transition. The main features of P(x) and m for $\eta > \eta_s$ qualitatively agree with those obtained in coupled maps [6,7], partial differential equations [7,8], and in a phenomenological cellular automaton model [13] in spatiotemporal intermittent regimes. The presence of a power law decay of P(x) for $\eta_c < \eta < \eta_s$ may be due to finite size effects. Indeed the cellular automaton model of this transition [13], presents the same features when the number of cell is reduced[24].

5 Thermodynamics of spatiotemporal chaos

In the previous section we have seen that the transition to space-time chaos has the properties of a phase transition.

In this section we show that this transition can also be characterized by studying the statistical properties of the Fourier mode amplitudes and of some global quantities, that can be computed from the spatial Fourier spectra $S(k,t)$ [4,25]. Indeed in section 3) we have seen that the time averaged spatial Fourier spectra changes as a function of the control parameter and they become broadened for $\eta > \eta_s$. As a consequence the Fourier spectra are good candidates to study the transition to the space time chaos. A similar approach has been recently proposed also by Hohenberg and Shraiman [3], who suggested to use the dissipation fluctuation theorem to define a temperature of Fourier modes. This kind of description of spatiotemporal chaos has the advantage of dealing with averaged quantities such as the thermodynamic ones of a system near thermal equilibrium[4].

In order to construct an analogy of the transition to space time chaos with the description of a system near thermal equilibrium we need to arise very simple questions. How the energy fluctuations scale as a function of the integration volume? Have the Fourier modes and local amplitudes Gaussian distributions? Thus we are interested in knowing the statistical properties of the fluctuations $W(x,t) = u(x,t) - < u(x,t) >$ ($<>$ denote time average), of their spatial Fourier transform $\tilde{W}(k,t)$, of the energy E(t) and of a suitably defined entropy $S(t)$. The energy is defined in the following way:

$$E(t, N_v) = \sum_{i=0}^{N_v} u^2(x_i, t)$$

with $2 < N_v < N$ where N is the total number of spatial points. The total energy is $E(t) = E(t,N)$. The dependence of the energy on N_v shows how the root mean square(r.m.s.) value, $\Delta E(N_v)$, of the energy fluctuations scales as a function of N_v, that is the volume of integration.

The spectral entropy [25] is instead defined in the following way:

$$\sigma(t) = \frac{S(t)}{S_o} = \frac{-1}{S_o} \sum_{k=1}^{N/2} \Phi_k(t) \log(\Phi_k(t))$$

where $\Phi_k(t) = |\tilde{u}(k,t)|^2 / E(t)$, $S_o = \log(N/2)$ is the equipartition value of $S(t)$ and $N/2$ the total number of Fourier modes. The parameter σ is 1 at the equipartition and 0 when only 1 mode is excited. Thus $\sigma(t)$ is very useful to see when the system is ordered or disordered. It is important to stress that $E(t)$ and $\sigma(t)$ are not exactly an energy and an entropy but they behave like these two thermodynamic quantities.

Also the distributions $P(W)$, of the local fluctuations W, and $P(\tilde{W})$ of the Fourier mode amplitudes $\tilde{W}_r, \tilde{W}_i$ (i and r denote real and immaginary part), change near the transition point for spatiotemporal chaos η_s. The deviation of a generic distribution $P(v)$ from a gaussian may be studied by computing:

$$M_j(v) = \frac{\Psi_j}{\Psi_2^{j/2}},$$

where $\Psi_j(v - v_a)$ is the moment of order j of the variable $(v - v_a)$ and v_a is the mean value of v. For a Gaussian M_3 (the skewness) and M_4 (the flatness) are equal to 0 and 3 respectively. The results for $P(\tilde{W}_r)$ are shown in Figs.7a), 7b) as a function of η for several values of k (The imaginary part $\tilde{W}_i$ behaves in the same way). The statistical accuracy in the calculation is of the order of ± 0.3, and it has been computed using standard methods. The skewness, Fig.7a), is always close to zero because the distributions of our data are always rather symmetric with respect to the mean. In contrast the flatness, that accounts for the tails of the distributions, changes considerably as a function of η, see Fig.7b). $M_4(\tilde{W})$ tends to 3 for almost all the modes for $\eta > \eta_s$ confirming the transition to a Gaussian distribution. We point out that the same transition does not occur in $M_4(W)$ for all the spatial points, indicating that the local dynamics has not, in general, a Gaussian distribution. As an example $P(W)$ and $P(\tilde{W}_r)$, at $\eta = 248$, are shown in Fig.8a) and 8b). The fact, that the Fourier mode amplitudes have Gaussian distributions whereas the local dynamics does not, has been also reported in ref.26 and widely discussed in refs.3,27. The reason of this effect is that the small k Fourier modes are coarse grained variables of the system because they imply an average over many correlation lengths[3,27].

We now analyse how the energy fluctuactions scale as a function of N_v. For $\eta < 200$ the relative fluctuations $\delta E = \Delta E(N_v)/E(N_v)$ do not follow a well defined law as a function of N_v. In contrast, for $\eta > \eta_C$, we find that δE decreases as a function of N_v as a power law N_v^μ, that extends in the range $2 < N_v < N$. The exponent $\mu(\eta)$ tends asymptotically to $-1/2$. The value of μ indicates that above η_s the spatial points are statistically independent and $E(t)$ behaves, as function of the number of points, as an addittive thermodynamic quantity.

All these findings go toward a thermodynamical description of the transition to spatiotemporal chaos, in which the Fourier modes may be considered as an ensemble of non interacting degrees of freedom. An important question is how a "generalized temperature" of the system may be defined[27]. The main difficulty arises from the fact that $\Psi_2(\tilde{W})$, is not constant as a function of k but presents a high-frequency cutoff. The spectrum of fluctuations shows that only the modes with $K < 3 \cdot K_o$ reach the equipartition. This phenomenon , that occours in the chaotic behaviour of the Kuramoto-Shivanshinsky equation, makes the definition of the temperature a very difficult and still unsolved problem, because it is not clear what can be done with the modes whose fluctuations decrease exponentially as a function of k. An approach in this direction has been done by Zalesky[12].

Instead, P. Hohenberg and B. Shraiman[3], suggested to use the dissipation fluctuation theorem to define the temperature of the system. This implies the

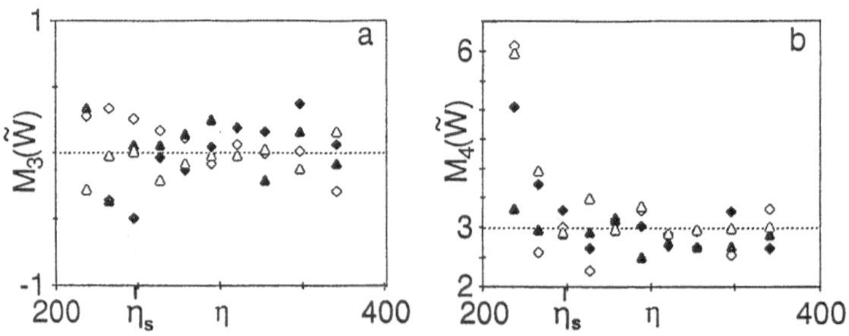

Figure 7. $M_3(\tilde{W}_r)$, a) and $M_4(\tilde{W}_r)$, b), versus η for different values of K: ■, $K/K_o = 1$; ◇, $K/K_o = 1.5$; ▲, $K/K_o = 2$; △, $K/K_o = 3$.

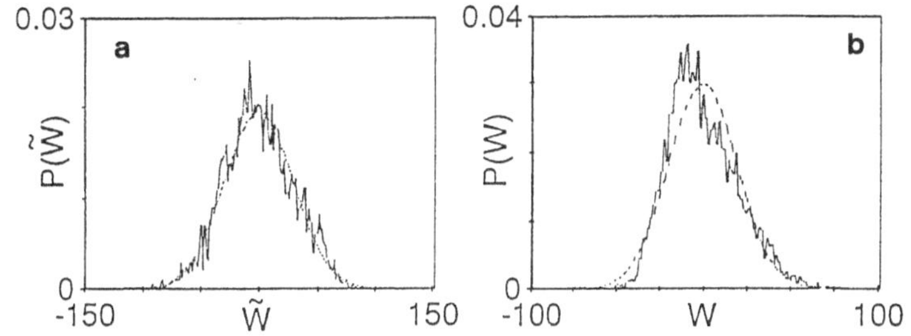

Figure 8. Distributions of $\tilde{W}_r(k,t)$ a) and $W(x,t)$ b) at x=0.5 and $k/k_o = 2$ at $\eta = 300$. Dashed lines correspond to Gaussian fits.

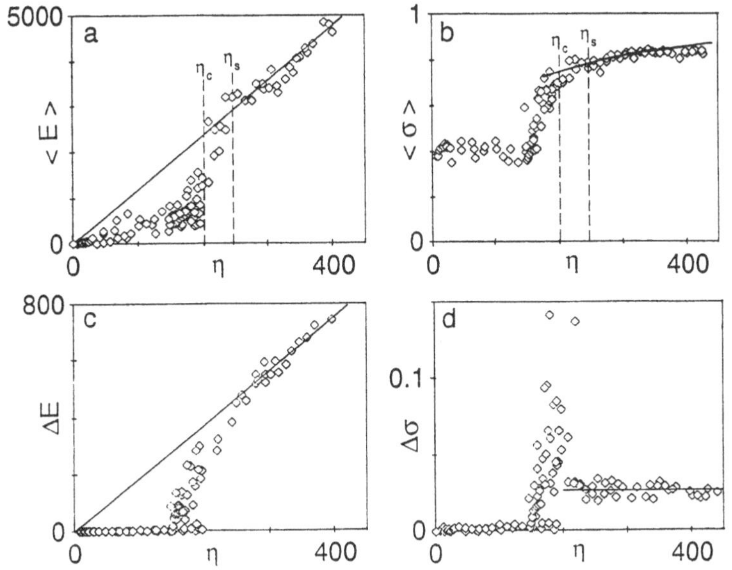

Figure 9. Dependence on η of the mean values of E, a), σ, b),and of the r.m.s. values of their fluctuations. ΔE, c), $\Delta\sigma$, d).

knowledge of the linear response function of the system, that may be measured by perturbing it with a very small signal. This approach is certainly very interesting, and several tests have been done in our experiment, but no relevant result has been obtained. Indeed, it is not simple to extract a very small signal (the response to the perturbation) from the natural fluctuations of the system. Even in the case in which this is possible, the errors, in the calculation of the response function, may be very large.

So we propose here an approach that is rather similar to the one of ref.3, but it uses the natural fluctuations of the system. We know[28] that, for a thermodynamic system at constant pressure and volume, the r.m.s fluctuations of energy and entropy are proportional to $K_B^{\frac{1}{2}}C_v^{\frac{1}{2}}T$ and $K_B^{\frac{1}{2}}C_p^{\frac{1}{2}}$ respectively, where C_v and C_p are the specific heats at constant volume and constant pressure, T is the temperature of the system and K_B is the Boltzmann constant.

To check these points we report in Fig.9, the mean values of E, σ and the r.m.s. values ΔE, $\Delta\sigma$ of their fluctuations, as a function of η. We see that $< E >$, Fig. 9a), and $< \sigma >$, Fig. 9b), are monothonically increasing as a function of η. The behavior of σ, above $\eta_c = 200$, indicates that the power spectrum shape does not change as a function of η. From Fig. 9d) we immediately realise that $\Delta\sigma$ increases by a considerable amount near η_c. In Fig. 9d), we also notice that $\Delta\sigma$ is almost constant above η_s, as a consequence we can make the hypothesis that C_p, of our "thermodynamic system" is constant above η_s. As we cannot distinguish, in our system, between a constant volume and a constant pressure process, we assume $C = C_p \simeq C_v$. Such an hipothesis has to be verified a posteriori. In Fig. 9c), we see that, for $\eta > \eta_s$, ΔE grows linearly as a function of η (solid line in Fig. 9c). As a consequence the ratio $(\Delta E/\Delta\sigma)$ may be considered proportional to the "generalized temperature" ($\tilde{T} = r\eta$) of the system for $\eta > \eta_s$ where the Fourier mode amplitudes have a Gaussian distribution. From the data we obtain $r = 73 \pm 1$.

In order to demonstrate that our definitions are self-consistent we construct, for $\eta > \eta_s = 248$, a free energy F:

$$F = -C\ r\ \eta\ \ln(r\eta) + (\sigma_o + C)\ r\ \eta$$

where $\sigma_o = 0.817 \pm 0.005$, $C = 0.165 \pm 0.005$. From this free energy we may compute $< E >$, $< \sigma >$ and C as a function of η, via appropriate thermodynamic relationships[28]. The solid lines, shown in Fig.10a) and 10b), are the result of the calculations, that are in agreement with the experimental points. This verifies all the hypothesis done to define the "generalized temperature" of the system.

The equivalent of the Boltzmann constant may be also computed using $\Delta\sigma$ and C. The result is the following: $K_B = (\Delta\sigma)^2/C = (4.4 \pm 0.2) \cdot 10^{-3}$.

6 Conclusion

The transition from low dimensional chaos to weak turbulence has been investigated in Rayleigh-Benard convection in an annular geometry.

The onset of spatiotemporal intermittency, in our cell, displays features of a phase transition that is reminescent of a percolation. This result has been obtained by reducing the space-time dynamics to a binary code, that catches the relevant features of the phenomenon[15,19]. A cellular automaton model, whose transition probabilities have been obtained from the experiment, confirms the exsistence of a phase transition with the exponents of the percolation[13,24]

The transition to space time chaos is accompanied by the appearence of gaussian distribution for the Fourier mode amplitudes and by a relevant changes of the distibutions of global quantities such the energy and the entropy. Besides the energy fluctuations decrease when the integration volume increases. A "generalized temperature" has been defined by using the energy and entropy fluctuations. The existence of a free energy shows the self consistency of our definitions[4]. An open problem is now to understand the meaning of a transition between a regime that displays thermodynamic properties (space time chaos) and an other that does not have these features.

An other important problem to solve in spatially extended systems is the estimation of the number of degrees of freedom. It is clear that the calculation of the fractal dimension is not feasible in such a complex regimes [29]. However one could try to obtain a rough estimation by using the orthogonal decomposition [23] that has also the advantage of giving some informations about the most important spatial structures involved in the dynamics. This method has been applied to the data obtained from the experiment here described and it gave very promising results.

This work has been partially supported by GNSM and by CEE contract number sci-0035-c(cd)

References

1. For a general review of low dimensional chaos see for example: J. P. Eckmann, D. Ruelle, Rev. Mod. Phys. 1987; P. Berge, Y. Pomeau, Ch. Vidal, L'Ordre dans le Chaos (Hermann, Paris 1984).
2. A. Libchaber, C. Laroche, S. Fauve, J. Physique Lett. 43, 221, (1982); M. Giglio, S. Musazzi, U. Perini, Phys. Rev. Lett. 53, 2402 (1984); M. Dubois, M. Rubio, P. Berge', Phys. Rev. Lett. 51, 1446 (1983); S. Ciliberto, M. A. Rubio, Phys. Rev. Lett. 58, 25 (1987); S. Ciliberto, J. P. Gollub, J. Fluid Mech. 158, 381 (1984); S. Ciliberto, Europhysics Letters 4, 685 (1987).
3. P.C.Hohenberg,B.Shraiman, Physica 37D, 109 (1989).
4. S. Ciliberto, M. Caponeri, Phys. Rev. Lett. 64, 2775 (1990).
5. G. L. Oppo, R. Kapral Phys. Rev. A 3, 4219 (1986).
6. K. Kaneko, Prog. Theor. Phys. 74, 1033 (1985); J. Crutchfield K. Kaneko in "Direction in Chaos", B. L. Hao (World Scientific Singapore 1987); R. Lima, Bunimovich preprint.
7. H. Chate', P. Manneville, Phys. Rev. Lett. 54, 112 (1987); Europhysics Letters 6,591(1988);Physica D 32, 409 (1988)
8. H. Chate', B. Nicolaenko, to be published in the proceedings of the conference: "New trends in nonlinear dynamics and pattern forming phenomena", Cargese 1988;
9. B.Nicolaenko, in " The Physics of Chaos and Systems Far From Equilibrium", M.Duong-Van and B.Nicolaenko, eds. (Nuclear Physics B, proceedings supplement 1988).
10. Y. Pomeau, A. Pumir and P. Pelce', J. Stat. Phys. 37, 39 (1984)
11. A. Pumir, J. de Physique 46,511 (1985).
12. S.Zalensky,Physica 34 D, 427 (1989).
13. F. Bagnoli, S. Ciliberto, A. Francescato, R. Livi, S. Ruffo, in "Chaos and complexity" , M. Buiatti, S. Ciliberto, R. Livi, S. Ruffo eds., (World Scientific Singapore 1988); F. Bagnoli, S. Ciliberto, A. Francescato, R. Livi, S. Ruffo , in the proceedings of the school on Cellular Automata, Les Houches (1989).
14. M. Van Dyke, An Album of Fluid Motion (Parabolic Press, Stanford, 1982); D. J. Tritton, Physical Fluid Dynamics (Van Nostrand Reinold, New York, 1979), Chaps.19-22
15. S.Ciliberto,P.Bigazzi,Phys.Rev.Lett. 60, 286 (1988).
16. F. Daviaud, M. Dubois, P. Berg, Europhysics Lett. 9, 441 (1989); P. Berg, in " The Physics of Chaos and System Far From Equilibrium", M.Duong-van and B. Nicolaenko, eds. (Nuclear Physics B, proceedings supplement 1988).
17. P. Kolodner, A. Passner, C. M. Surko, R. W. Walden, Phys. Rev. Lett. 56, 2621 (1986); A. Pocheau Jour. de Phys. 49, 1127 (1988)I; I. Rehberg, S. Rasenat , J. Finberg, L. de la Torre Juarez Phys. Rev. Lett. 61, 2449 (1988); N. B. Trufillaro, R. Ramshankar, J. P. Gollub Phys. Rev. Lett. 62, 422 (1989); V. Croquette , H. Williams Physica 37 D, 300 (1989).
18. S. Chandrasekar, Hydrodynamic and Hydromagnetic Stability, Clarendon Press, Oxford 1961; F. H. Busse, Rep. Prog. Phys. 41 (1978) 1929; Ch. Normand, Y. Pomeau, M. Velarde Rev. Mod. Phys. 49, 581,(1977).

19. S. Ciliberto in " Dynamics and Stochastic Processes " R. Vilela Mendes ed.(Springer 1989); S. Ciliberto in " Quantitative Measures of Complex Dynamical System", N. Abraham, A. Albano eds.(Plenum 1990).
20. W. Merzkirch, Flow Visualisation, Academic Press, New York 1974.
21. S. Ciliberto, F. Francini,F. Simonelli, Opt. Commun. 54, 38 (1985).
22. S. Ciliberto, M. Caponeri, F. Bagnoli, Nuovo Cimento D39, 6 (1990).
23. S. Ciliberto, B. Nicolaenko submitted for publication; A. Newell, D.Rand, D.Russell, Physica 33 D, 281 (1988); , N. Aubry, P. Holmes J. L. Lumley, E. Stone, J. Fluid Mech. 192, 115 (1988).
24. F. Bagnoli, Thesis of the University of Florence.
25. R. Livi, M. Pettini, S. Ruffo, M. Sparpaglione, A. Vulpiani, Phys. Rev. A 31, 1039 (1985); M. A. Rubio, P. Bigazzi, L. Albavetti, S. Ciliberto J. Fluid Mech. December (1989).
26. J. P. Gollub,R. Ramshankar, preprint 1989 to be published in: 'New Perspective in Turbulence', S. Orzag, L. Sirovich eds. (Springer 1990).
27. : R. H. Kraichnan, S. Chen, Physica 37D, 160 (1989).
28. : L. D. Landau, E. M. Lifshitz, Statistical Physics, (Pergamon Press 1980).
29. A. Politi, in " Quantitative Measures of Complex Dynamical System", N. Abraham, A. Albano eds.(Plenum 1990).

TIME EVOLUTION OF RANDOM CELLULAR PATTERNS

Kyozi Kawasaki*, Tatsuzo Nagai** and Tohru Okuzono*

* *Department of Physics, Kyushu University 33*
Fukuoka 812, Japan

** *Physics Department, Kyushu Kyoritsu University*
Kitakyushu 807, Japan

A short review is presented of some recent works on a class of models called the vertex models, which is designed for efficient large scale computer simulations of time evolution of cellular patterns. Some new results on a refined version of the model and on elastic and plastic properties of incompressible two-dimensional cellular pattern are also reported.

1.INTRODUCTION

Cellular structure is ubiquitous in nature,[1] ranging from detergent foams in kitchen to the large scale structure in the universe.[2] Its study is also important in technology. Cellular solids[3] and fluid foams[4] are finding ever wider applications in industry. However, despite the long history of cellular structure studies[1,5] our understanding is still rather limited. The reason for this is that this problem belongs to a class of problems most difficult in statistical physics: the problems of randomness with frustrations (constraints) or the problems of compromise between order and chaos.[1] Therefore our problem has something in common with the problems of glasses and turbulence. The frustration here comes from the fact that cells must completely fill up the space: no overlapping or empty regions between cells are permitted. This requirement is expressed in terms of Euler's equations[1]. Its consequence for infinite two-dimensional *topologically* stable cellular systems is that the average edge number per cell is six.

Here we will be mainly concerned with time evolution of cellular structures. Besides its own interest, understanding of time evolution is important for knowing origins of statistical properties of the structure like cell size and face and edge number distributions for specific stable cellular materials[3]. We will also touch on some preliminary results on mechanical properties.

In view of the intrinsic difficulties of the problem, much of the recent studies rely on computer modelling. We have been developing one class of such modelling, that is, the vertex models of cellular structures. The advantages of this class of models are that (i) enormous reduction of the number of degrees of freedom permits simulation of larger systems over longer times than are possible in most other models and (ii) flexibility of construction of the models leaves room for various refinements rendering our modelling closer to reality.

2. VERTEX MODELS

In our previous works we have been developing the so-called vertex models. [6,7,8] The basic idea is to reduce a dynamical system with an infinite number of degrees of freedom describing cell patterns to simpler dynamical systems with finite numbers of degrees of freedom describing a set of vertices which are located at r_i, $i = 1, 2, \cdots$, and are connected by straight edges. There are various versions of vertex models depending upon the degrees of accuracy with which one desires to describe the original cell pattern. In two-dimension[6] four such versions have been proposed so far whose equations of motion are;

Model O
$$\frac{1}{3L}\sum_j^{(i)} |r_{ij}| n_{<ij>} n_{<ij>} \cdot (v_i + \frac{1}{2} v_j) = f_i \quad (2.1)$$

Model O′
$$\frac{1}{3L}[\sum_j^{(i)} |r_{ij}| n_{<ij>} n_{<ij>}] \cdot v_i = f_i \quad (2.2)$$

Model II
$$\frac{1}{6L}(\sum_j^{(i)} |r_{ij}|) v_i = f_i \quad (2.3)$$

Model I
$$\frac{1}{2L}\bar{r}(t) v_i = f_i \quad (2.4)$$

Here $v_i \equiv dr_i(t)/dt$, L a positive constant, $f_i \equiv -\sum_j^{(i)} \hat{r}_{ij}$, $\hat{r}_{ij}$, being the unit vector along the bond directed from the vertices j to i, $n_{<ij>}$ its unit normal, summations are over the three bonds emerging from the vertex i, and $\bar{r}(t)$ is the bond length averaged over the entire system. These equations which involve one central vertex and three others connected by three edges have to be supplemented with rules for the two types of topology changes wherever a pair of vertices come very close. One is the recombination of vertices (T1 process) and another is the annihilation of a triangle (T2 process).[1] The above equations of motion can be derived by balancing the frictional and static forces acting on every vertex.

Three-dimensional vertex models have been also constructed.[8] Here a vertex equation of motion involves one central vertex and four vertices connected to it by four edges which are taken to be straight. The central vertex and each of six pairs formed from outside vertices form six triangles which also enter the vertex equation. The explicit form is given in reference 8. There we mentioned a new difficulty not present in two-dimension. That is, faces in three dimensional cells are not planar in general leading to an ambiguity in triangulation. However, we found by computer simulation that deviations from planarity are only a few percent and the ambiguity mentioned is not a serious problem.[9]

3. A REFINEMENT OF THE VERTEX MODEL

One of the deficiencies of the vertex models proposed so far is their inability to

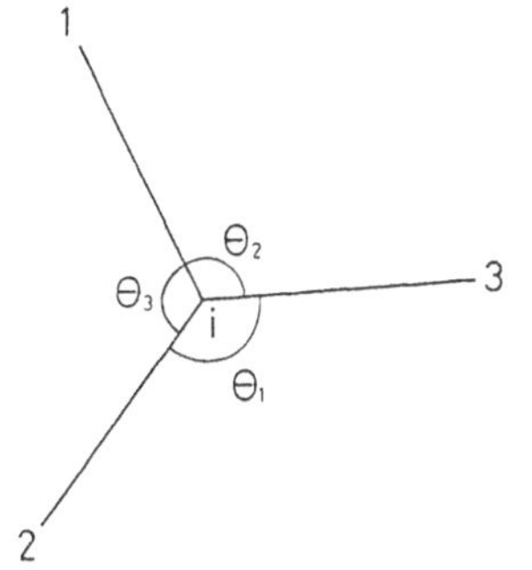

Fig. 1 Definition of the angles θ_j.

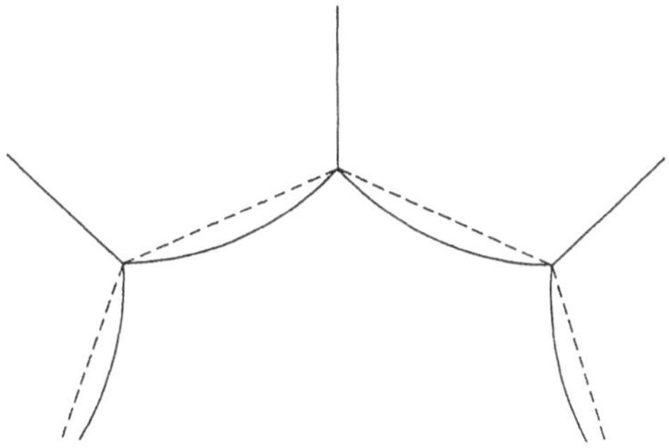

Fig. 2 Part of a regular polygon in the special symmetric configuration.

distinguish among different cellular systems like polycrystalline grains and soap froth. Here we describe a way to remedy this in two-dimension, which at the same time refines the earlier vertex models. The source of inadequacy of the vertex models arises naturally from neglecting of finite curvature of edges which gives rise to corrections to the dissipation rate associated with cell wall movements and to the cell wall free energy. We propose to incorporate both corrections by introducing the correction function $C(\theta_j/\pi)$ to the dimensionless static force as

$$\boldsymbol{f}_i \rightarrow \boldsymbol{f}_i^m \equiv -\sum_j^{(i)} C(\frac{\theta_j}{\pi})\hat{\boldsymbol{r}}_{ij}, \tag{3.1}$$

where θ_j with $j = 1,2,3$ is the angle between the two edges other than the edge j emerging from the vertex i. See Fig.1. The unknown function $C(x)$ can be determined by considering the special cases of isolated expanding or shrinking regular polygons which are bonded to straight edges extending to infinity in a symmetrical fashion as shown in Fig. 2. When the regular polygon is n-sided, we find that the outward component of the force acting on any one of the vertices is,

$$f^m = C(1-\frac{2}{n}) - 2C(\frac{1}{2}+\frac{1}{n})\sin(\frac{\pi}{n}). \tag{3.2}$$

Then the model O equation of motion (2.1) with the r.h.s. modified according to (3.1) yields for the outward component of the vertex velocity v the following:

$$v = LB_M(n)/a, \tag{3.3}$$

where a is the edge length of the polygon and

$$B_M(n) \equiv [C(1-\frac{2}{n}) - 2C(\frac{1}{2}+\frac{1}{n})\sin(\frac{\pi}{n})]/\cos^2(\frac{\pi}{n}). \tag{3.4}$$

The function C for a specific cell system is obtained by solving the functional equation that results from equating $B_M(n)$ with the corresponding quantity $B(n)$ for the specific cell system[6]. This equation may be written in a simpler form in terms of the following quantities:

$$\begin{aligned} x &\equiv n^{-1} - 6^{-1}, & y(x) &\equiv C(1-\tfrac{2}{n}) \\ b(x) &\equiv B(n)\cos^2(\tfrac{\pi}{n}), & d(x) &\equiv 2\sin(\tfrac{\pi}{n}) \end{aligned} \tag{3.5}$$

Thus we obtain the following equation

$$y(x) - d(x)y(-\frac{x}{2}) = b(x), \tag{3.6}$$

with the boundary condition

$$y(0) = 1, \tag{3.7}$$

where x is treated as a continuous variable.

We calculate the correction function $y(x)$ for soap froths and grains. The quantity $B(n)$ for soap froths was given by Eq. (4.19) of Ref. 6:

$$B(n) = -\frac{4}{\sqrt{3}}\sin(\pi x), \tag{3.8}$$

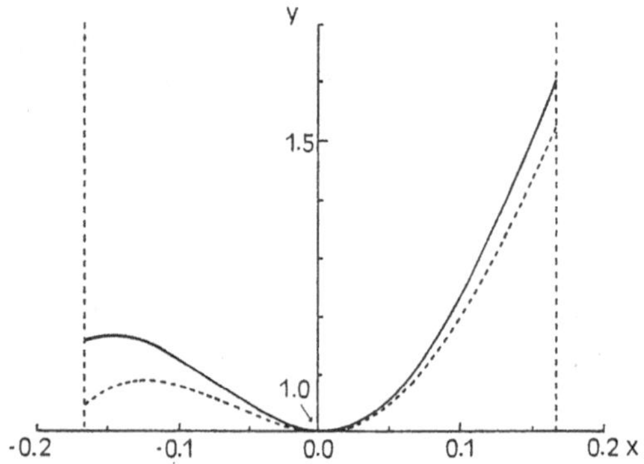

Fig. 3 Correction functions for soap froths (solid line) and for grains (dotted line).

where $x \equiv 1/n - 1/6$. Using Eq. (3.8) we expand Eq. (3.6) into power series with respect to x and obtain the solution $y(x)$ in the form of power series. The result $y(x)$ is shown by the solid line in Fig.3. This shows that the correction function is larger than unity except for $n = 6$. It means that the effect of the correction function is to increase the absolute value of each line tension. This effect is remarkable especially for $n = 3$.

The quantity $B(n)$ for grains was numerically calculated and shown in Fig. 5 of Ref. 6. It can be well fitted by

$$B(n) = b_1 x + b_2 x^2 + b_3 x^3, \tag{3.9}$$

where $b_1 = -7.2552$, $b_2 = -4.4262$ and $b_3 = 8.4353$. We have obtained $y(x)$ for grains by using the same method as that for soap froths. We give the result $y(x)$ by the dotted line in Fig. 3. This is similar to the function $y(x)$ for soap froths but slightly smaller than that.

We carry out computer simulations for the modified model O. Here we use the correction functions obtained above, though they have been derived for the special symmetric configuration and θ_j was replaced by the average over the angles of the cell, which may not be entirely appropriate. We can expect that this is a good approximation because we previously found an evidence that such a special configuration well describes the average behavior of the cell growth in the scaling regime.[7] The initial state is the Voronoi cell network with 3000 cells and the periodic boundary condition is imposed. We show the results for the distribution functions of the number of sides n and linear size R in the scaling regime, $f(n)$ and $g^*(R/\bar{R})$ with $\bar{R}(t)$ the average linear size of cells, in Figs. 4 and 5, respectively. In these figures the solid lines denote the results for soap froths and the dotted lines denote the results for grains. These have been obtained by averaging the distribution functions over 100 independent runs and over 10 time points per run in the scaling regime. We can see that the difference in each distribution between two systems is small and that the widths of both distributions for the modified vertex model are broader than those for the original vertex model.

In Fig. 4 the circles, the double circles and the squares show the experimental results observed by Simpson et al.[10], by Stavans et al.[11] for two-dimensional soap froths and by Simpson et al.[10] for two-dimensional grains, respectively. The agreement between these results and the simulation results for the modified vertex model is excellent.

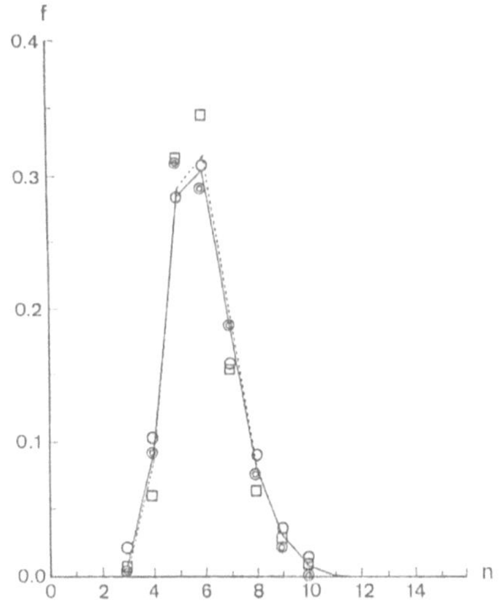

Fig. 4 Distribution functions of the number of cell sides for the modified vertex model (solid, dotted lines) and the experiments (circles, double circles, squares).

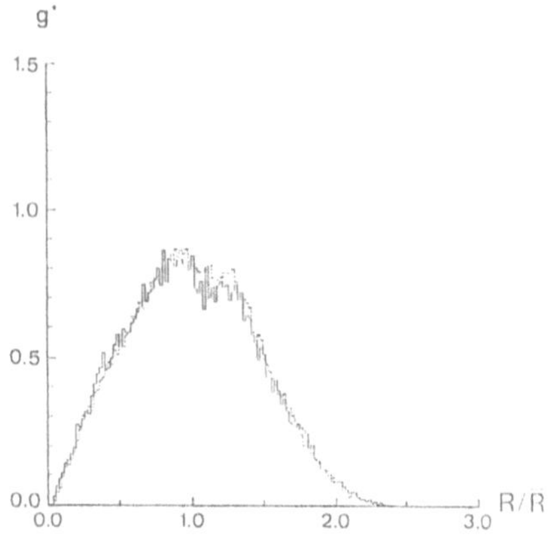

Fig. 5 Distribution functions of linear size of cells for soap froths (solid line) and for grains (dotted line).

4. ELASTIC AND PLASTIC PROPERTIES OF INCOMPRESSIBLE TWO-DIMENSIONAL CELLULAR PATTERN

The first computer simulation study of elastic and plastic properties of random cellular system was performed by Weaire and the coworkers.[12] Here we wish to present some preliminary results on elastic and plastic behavior of our two-dimensional vertex model systems. Since our cellular system is continuously evolving in time, static properties like elasticity cannot be defined in a strict sense. However, when the time evolution is very slow, one can still talk about elastic properties to describe short time response of the system to a deformation. In order to fix the problem more precisely we take a cell pattern, say of a soap froth evolving well into the scaling regime, and after some instant of time completely stop flow of air across cell boundaries, which fixes every cell area. There are still some pattern evolution through continuous vertex motion and T1 processes which will stop on reaching some local equilibrium state where we monitor the cell wall energy density. Next we apply an elongational distortion uniformly throughout the system and again let the system relax through continuous cell wall motion and T-1 processes. On reaching equilibration which still contains some deformation, we again monitor the cell wall energy density. The process can be repeated and we finally find the local equilibrium cell wall energy as a function of the elongational distortion.

The dynamical equation that describes relaxation process with fixed cell areas is obtained using the Lagrange multipliers λ_α for each cell α having the area A_α. That is, the static forces $\boldsymbol{f}_i$ on the r.h.s. of (2.1)-(2.4) are replaced by the following;

$$\boldsymbol{f}_i \to \boldsymbol{f}_i^e \equiv \boldsymbol{f}_i - \sum_\alpha \lambda_\alpha \partial A_\alpha / \partial \boldsymbol{r}_i. \tag{4.1}$$

The partial derivative on the r.h.s. is easily found to be given by

$$\frac{\partial A_\alpha}{\partial \boldsymbol{r}_i} = -\frac{1}{2}\hat{\boldsymbol{z}} \times \boldsymbol{r}_{jk}, \tag{4.2}$$

where $\boldsymbol{r}_{jk} \equiv \boldsymbol{r}_j - \boldsymbol{r}_k$ and (i, j, k) are three vertices of the cell α arranged counterclockwise as shown in Fig. 6 and $\hat{\boldsymbol{z}}$ is the unit vector directed upward perpendicularly to the paper.

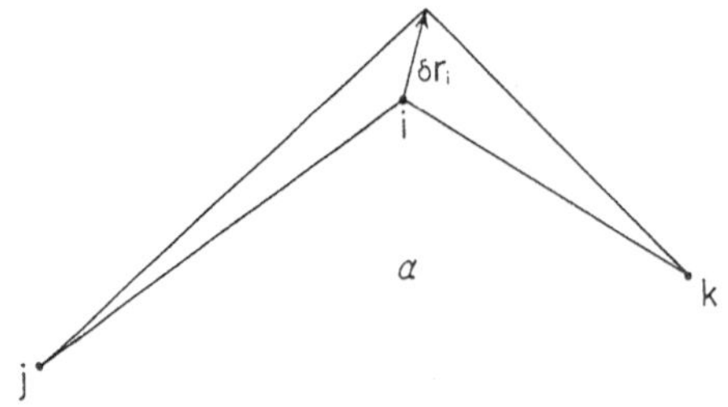

Fig. 6 Arrangement of three vertices (i,j,k) and cell α for Eq. (4.2).

Now, our main concern here is the quasi-static elastic properties and not the relaxation process itself. Therefore, we are permitted to employ any of the vertex equations (2.1)-(2.4) where $\boldsymbol{f}_i$ is replaced by $\boldsymbol{f}_i^e$ everywhere. The λ's are then determined by the requirements

$$\frac{d}{dt}A_\alpha(t) = -\frac{1}{2}\sum_i^{(\alpha)} \hat{\boldsymbol{z}} \times \boldsymbol{r}_{jk} \cdot \boldsymbol{v}_i = 0, \tag{4.3}$$

where the sum is over all the vertices associated with the cell α. This then leads after some algebra to the following equation that determines the λ's by iteration:

$$\lambda_\alpha = -\{2\sum_i^{(\alpha)}\sum_l^{(i)} \hat{z}\cdot(\boldsymbol{r}_{jk} \times \hat{\boldsymbol{r}}_{il}) + \sum_\beta^{(\alpha)}(\boldsymbol{r}_{jk}\cdot\boldsymbol{r}_{km} + \boldsymbol{r}_{li}\cdot\boldsymbol{r}_{in})\lambda_\beta\} \Big/ \sum_i^{(\alpha)} r_{jk}^2. \tag{4.4}$$

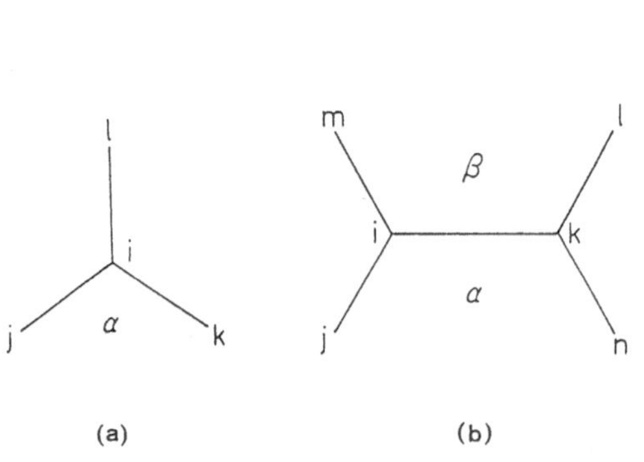

Fig. 7 Arrangement of vertices and cells for Eq. (4.4).

Here the denominator and the first term in the numerator refer to Fig. 7-(a) and the second term in the numerator refers to Fig. 7-(b). The meaning of various summations can be easily read off from the figures. For instance the summation with respect to β is over all the cells adjacent to the cell α. Here it is to be noted that λ_α is indeterminate up to an arbitrary additive number independent of α since the sum of A_α over α is the total area of the system which is always a fixed number.

A local equilibrium will be reached if all f_i^e vanishes. Here note that this does not require dihedral angles to be $2\pi/3$ in contrast to the original dynamical equations (2.1)-(2.4) since f_i^e contains an additional fictitious force arising from the constraints of fixed cell areas. The Lagrange multipliers the λ's are like some external fields acting on each domain, and these fictitious fields exert extra fictitious forces on every vertex.

We should note however that for irregular systems like random soap froth elasticity cannot be strictly defined even for restricted dynamics employed here since some topology changes are inevitable for large systems.[12]

We now describe the computer simulation. As an initial state, or an unstrained state ($s = 0$), we have chosen a mature cellular pattern in the scaling regime obtained by the simulation of the model O (Fig. 8-(a)). The system contains 493 cells and obeys the periodic boundary condition.

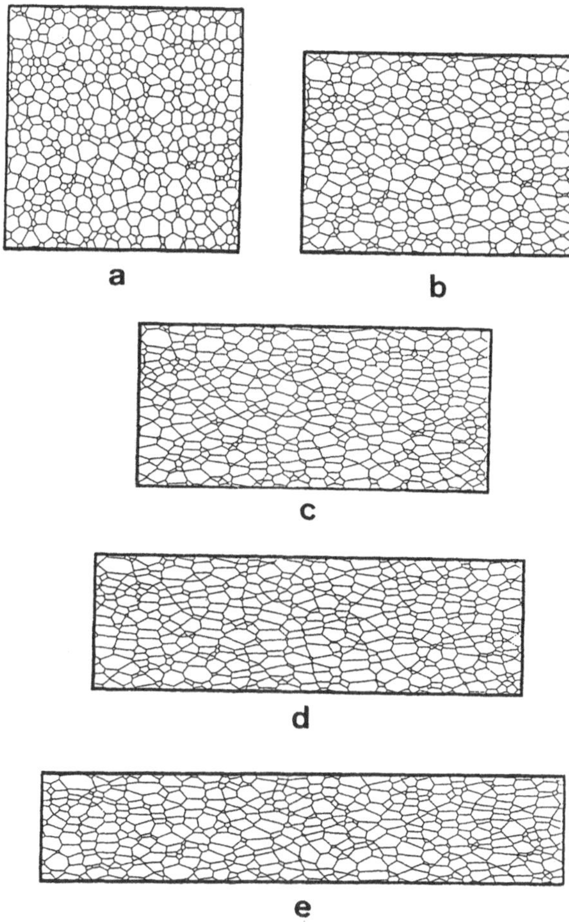

Fig. 8

Cellular structures for some strained states: (a) $s = 0$ (initial state), (b) $s = 20, \varepsilon s = 0.20$, $\varepsilon_{eff} = 0.14$, (c) $s = 40$, $\varepsilon s = 0.40$, $\varepsilon_{eff} = 0.25$, (d) $s = 60$, $\varepsilon s = 0.60$, $\varepsilon_{eff} = 0.28$, (e) $s = 80$, $\varepsilon s = 0.80$, $\varepsilon_{eff} = 0.28$.

We impose an elongational deformation on this system, keeping areas constant. This deformation is given by the affine transformation,

$$x(s+1) = e^{\varepsilon} x(s), \quad y(s+1) = e^{-\varepsilon} y(s), \tag{4.5}$$

where $x(s)$, $y(s)$ are coordinates of the system in the state s and ε is the Hencky strain.

Next, we let the system relax and monitor its energy density. In this relaxation process the equations of motion for vertices are given by Eqs. (2.4),(4.1), (4.2), and (4.4). The energy density of the system, at time t, is defined as

$$E(t) = \sigma \sum_{<ij>} |r_{ij}(t)|/A, \tag{4.6}$$

where σ is the surface tension and A is the total area of the system. If $E(t)$ satisfies the following criterion

$$|E(t'+\delta t) - E(t')|/|E(t')| < 10^{-5} \quad \text{for} \quad t - 10 \cdot \delta t \leq t' \leq t, \tag{4.7}$$

with δt the step size of time, we regard it as constant and the system as being in the corresponding strained state. In the following we choose the units in which we have $\sigma = 1$, $A/N_c = 1$ (N_c is the total number of cells which is constant for the modified dynamics). The cell structures in some strained states are shown in Fig. 8.

In the relaxation process patterns change in two ways, that is, continuous evolution with the same topology and topological change. The former is accompanied by the dissipation of energy. The latter is, in the present case, a change by $T1$ process.

The homogeneous strain of the system produced by the affine transformation (4.5) differs from that of the cellular network in general because the strains of the cellular network are released during the relaxation process. We thus introduce the quantity which expresses the strain of the cellular network by

$$\varepsilon_{eff}(s) = \frac{1}{2}[\ln(\lambda_x(s)/\lambda_x(0)) - \ln(\lambda_y(s)/\lambda_y(0))], \tag{4.8}$$

where $\lambda_x(s)$ and $\lambda_y(s)$ are the average dimensions of cells in the x and y directions, respectively, in the state s. In Fig. 9 we show the energy density for each strain,

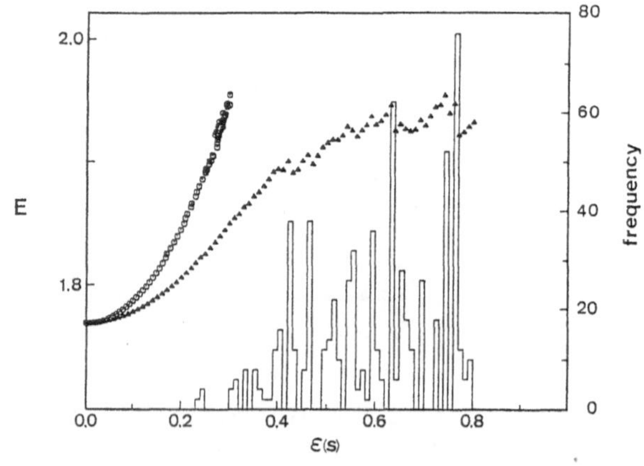

Fig. 9 Energy density vs. strain and frequency of T1 processes. Triangles and circles indicate energy density for εs and $\varepsilon_{eff}(s)$, respectively. Frequencies of T1 processes are also shown by histogram.

Fig. 10 Normal stress difference vs. strain. Solid line and dashed line indicate normal stress for εs and $\varepsilon_{eff}(s)$, respectively.

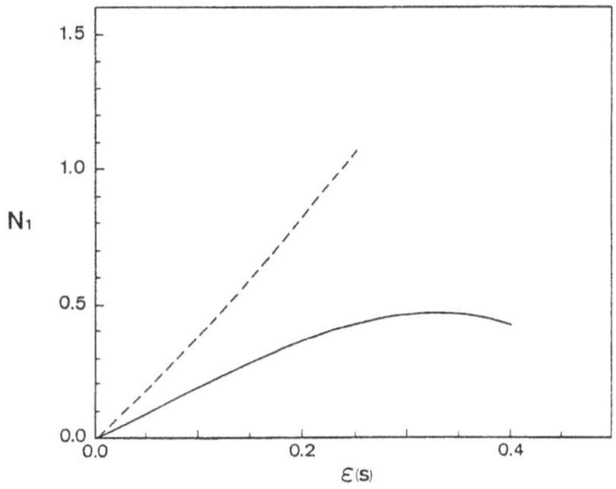

namely, εs (triangles) and $\varepsilon_{eff}(s)$ (circles), where $\varepsilon = 0.01$, $1 \le s \le 80$. The histogram in the same figure shows the frequency of $T1$ processes in each interval $(\varepsilon(s-1), \varepsilon s)$. We can see that $T1$ processes induce large strain recovery.

We present in Fig. 10 the normal stress differences as functions of two kinds of strain for $0 \le s \le 40$. The normal stress difference is defined as

$$\begin{cases} N_1 = \tau_{xx} - \tau_{yy} = \dfrac{dE}{d\varepsilon(s)} \\ \varepsilon(s) = \varepsilon s \quad \text{or} \quad \varepsilon_{eff}(s). \end{cases} \tag{4.9}$$

In this figure, the solid and dashed lines are for εs and $\varepsilon_{eff}(s)$, respectively. We can see that N_1 depends on $\varepsilon_{eff}(s)$ linearly for all s, while it depends on εs linearly in the region of s where no $T1$ process takes place. This suggests that the nonlinearity of the stress-strain curve or the plasticity of the system originates from the change of topology, as pointed out by Ashby and Verrall.[13]

5. DISCUSSION

Other computer modellings that deal with cellular patterns either try to solve the domain wall equations of motion directly[12,14] or map this problem onto kinetic Potts models with large numbers of spin components.[15] The last mentioned models are of some relevance for those interested in nonlinearity because a discrete Potts spin system can be mapped into a nonlinear field theory.[16] Namely, the partition function of the original n-state Potts spin system is expressed as a functional integral of the Boltzmann factor for the field-theoretical Hamiltonian of the n-component real field, whose explicit form is given in ref.16. This Hamiltonian has the symmetry distinctly different from that of the n-component field with continuous symmetry. The single spin flip kinetic Potts spin model then presumably corresponds to the TDGL-type equation without conservation for the n-component field with this field-theoretical Hamiltonian although we have not yet demonstrated this explicitly. However, solution of this equation is not easily obtainable even in the limit of infinite n in contrast to the case with complete continuous symmetry.[17] This is more or less what is expected since the Potts systems have richer structure like domain wall states than the continuously symmetric spin systems. Still one may hope that a study of the TDGL-type equation for a Potts system might yield a useful approximation scheme for statistical dynamical properties of Potts systems such as those tried for other systems.[18]

ACKNOWLEDGEMENT

One of us (KK) would like to acknowledge the Yamada Science Foundation for travel support that enabled him to present the work reported here at the workshop "Microscopic Aspect of Nonlinearity in Condensed Matter".

APPENDIX

Here we discuss about the apparent dilemma and its solution for periodic cellular patterns. The material described here originated from the question explained below concerning the energy of a uniformly deformed periodic hexagonal cellular structure which has been frequently used as a model of soap froth.[19] A small portion of the system containing a single cell is shown in Fig. 11. Here forces are balanced at all the vertices, that is, all the dihedral angles are fixed at $2\pi/3$. Then there are two kinds of edges of lengths b and c as shown in Fig. 11. The area of a hexagon is proportional to $b(b/2+c)$ which may or may not be fixed in this discussion. The edge length per hexagon is $E=2b+c$ which is proportional to the system energy.

We consider an infinitesimal additional uniform deformation where b and c change by δb and δc respectively, resulting in the energy change per cell by $\delta E = 2\delta b+\delta c$. Such a deformation should also be obtained by infinitesimal displacements of vertices at least for finite systems. However, in view of the force balance at every vertex, no net first order change of E in vertex displacements takes place, which is an apparent paradox. The reason for this dilemma is solved by considering the outside boundary. For simplicity we suppose that our system consists of a single cell and half edges that emerge outward from the six cell vertices. In Fig. 11 the system boundary is shown by crosses on edges. Now, the parts of the vertex displacements due to a uniform translation of the system are irrelevant and thus we can assume that the center of gravity of the system remains undisplaced. Then it is easy to show that the x and y components of the displacements of the two vertices marked as A and B in Fig. 11 due to changes δb and δc are given, respectively, by $\delta x_A = 0$, $\delta y_A = \frac{1}{2}(\delta b+\delta c)$, $\delta x_B = \frac{\sqrt{3}}{2}\delta b$, and $\delta y_B = \frac{1}{2}\delta c$. If the system boundaries marked by crosses are fixed in space, this implies for instance that the portion of the vertical edge ending at the vertex A that belongs to the system decreases by $\delta y_A = \frac{1}{2}(\delta b+\delta c)$. Likewise, the portion of the edge between the vertex B and the cross at the upper right corner is shortened by $\frac{\sqrt{3}}{2}\delta x_B + \frac{1}{2}\delta y_B = \frac{3}{4}\delta b + \frac{1}{4}\delta c$. By symmetry we see that the total loss of the edge length of the system under consideration amounts to $4\delta b + 2\delta c = 2\delta E$. This amount is precisely the total change of the edge length of the hexagon. The same argument can be readily extended to larger cellular systems such as that shown in Fig. 12. Therefore we conclude that any change of edge length inside the system is compensated by the "flow" of cell walls across the fixed system boundary, which is consistent with no net change of the total edge length due to infinitesimal vertex displacements. Alternatively, we may imagine that the system boundaries (× in Fig. 12) undergo the same displacements as the nearest vertices inside the system which means no "flow". This results in changes of the total cell wall energy of the outside reservoir, which is equal to the work done by the system on the outside reservoir due to infinitesimal deformation. This is nothing but negative of the elastic energy change of the system.

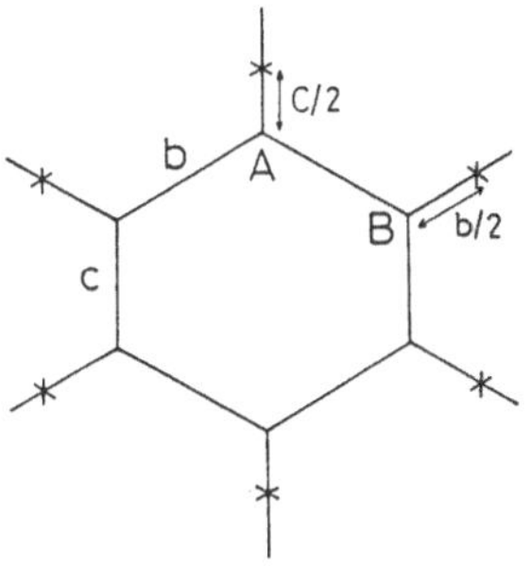

Fig. 11 Small portion of periodic cellular pattern.

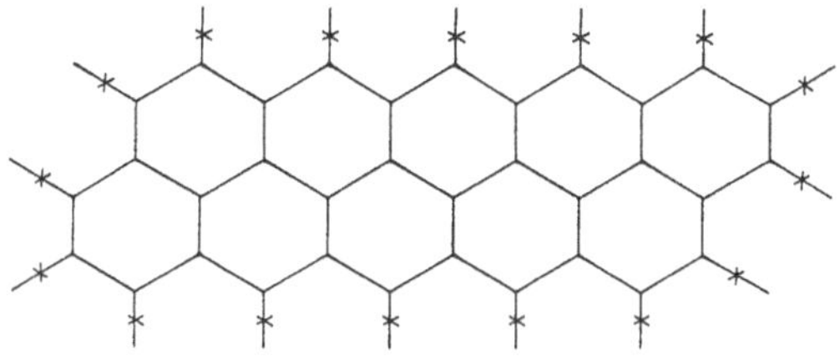

Fig. 12 Periodic cellular pattern.

REFERENCES

1. D. Weaire and N. Rivier, Contemp. Phys. **25** (1984) 59
2. M. J. Geller, *in* "Dark Matter in the Universe", H. Sato and H. Kodama (eds.) Springer, Heidelberg, 1990
3. L. J. Gibson and M. F. Ashby: Cellular Solids, Pergamon Oxford, 1988
4. A. M. Kraynik, Ann. Rev. Fluid Mech. **20** (1988) 325
5. H. V. Atkinson, Acta Metall. **36** (1988) 469
6. K. Kawasaki, T. Nagai and K. Nakashima, Phil. Mag. B60 (1989) 399
7. K. Nakashima, T. Nagai and K. Kawasaki, J. Stat. Phys. **57** (1989) 759
8. K. Kawasaki, T. Nagai and K. Nakashima, *in* "Cooperative Dynamics in Complex Physical Systems", ed. H. Takayama, Springer 1989
   K. Kawasaki, T. Nagai and S. Ohta, *in* "Proceedings of Symposium on Simulation and Theory of Evolving Microstructures and Textures", eds. M. P. Anderson and A. Rollett, 1990, in press
9. T. Nagai, S. Ohta, K. Kawasaki and T. Okuzono, Phase transitions (1990)
10. C. J. Simpson, C. J. Beingessner and W. C. Winegard, Trans. Metall. Soc. AIME **239** (1967) 587
11. J. Stavans and J. A. Glazier, Phys. Rev. Lett. **62** (1989) 1318
12. D. Weaire and J. P. Kermode,Phil. Mag. **B50** (1984) 379
    D. Weaire and T. L. Fu, J. Rheol **32 (3)** (1988), 271
    D, Weaire, Paper presented at Food Science Conference (1988)
13. M. F. Ashby and R. A. Verrall, Acta Metall. **21** (1973) 149
14. H. J. Frost, C. V. Thompson, C. L. Howe and J. Whang, Scripta Met. **22** (1987) 65
15. M. P. Anderson, G. S. Grest and D. J. Srolovitz, Phil. Mag. **B59** (1989) 293 and earlier reference quoted therin
16. R. P. Zia and D. J. Wallace, J. Phys. A: Math. Gen. **8** (1975) 1495
17. G. F. Mazenko and M. Zanetti, Phys. Rev. **32** (1985) 4565
18. T. Ohta, D. Jasnow and K. Kawasaki, Phys. Rev. Lett. **49** (1982) 1223
    K. Kawasaki, Phys. Rev. **A31** (1985) 3880
19. L. W. Schwartz and H. M. Princen, J. Colloid, Interface Sci. **118** (1987) 201

DEFECTS AND ORDER TO DISORDER TRANSITION IN NON-EQUILIBRIUM STRUCTURES

R. Ribotta, A. Joets

Laboratoire de Physique des Solides
Bât. 510, Université de Paris-Sud
91405 Orsay Cedex

ABSTRACT

The role of the defects in the transitions from order to disorder of the ordered states is investigated in a non-equilibrium system. The experimental system is a convective fluid driven to turbulence. Both the stationary and time-dependent homogeneous ordered states may become unstable against localized perturbations which create defects. Then, the defects may contribute either to the disorganization of the states, or they may mediate rapid transitions to fully ordered states of lower symmetry. This role can be understood from the topology and from the instability of the core of the defects and is reminiscent of the displacive transitions in solids.

INTRODUCTION

Nonlinear out-of-equilibrium systems subjected to an increasing external stress may undergo a series of bifurcations to ordered stable states before reaching a full disordered state (space-time chaos). The bifurcations are the counterpart of the phase transitions in systems at thermodynamical equilibrium. These ordered states may be either stationary or time-dependent, or both. Usually one would consider only the homogeneous states, i.e. those which are characterized by an order parameter with constant value all over the space (or time). Next, the bifurcations are always assumed to occur identically at every space scale, as for *diffusive* transitions in solid. However, it is very often found experimentally, as in the case of structural transitions in solids, that these states may become unstable under localized perturbations [1]. One of the consequences, as will be shown hereafter, is the creation of defects in the ordered state. Our purpose is to show that the core of the defects represents the unstable state and consequently it may become more stable under the applied stress and develop in space a localized state with a lower symmetry. As the stress is kept increasing, a new homogeneous stable state can "germinate" and invade the whole space.

Convective flow structures developed in a layer of fluid are examples of such ordered states and we shall illustrate how the defects can mediate the transitions from order to disorder (here turbulence) both in the case of purely stationary and of time-dependent structures. The similarity between those two cases is shown to arise from the symmetry of the considered states. The dynamics of the evolution between different states mediated by the defects under the applied stress, is found to exhibit some similarities with the *displacive* phase transitions in solids.

1 THE EXPERIMENT

The physical model is a layer of an anisotropic fluid[2] (a nematic liquid crystal) subjected to a transverse AC electric field of frequency f. Above a well-defined voltage threshold V_{th} the layer becomes unstable against a convective state characterized by a spatial periodicity along the molecular direction (say **x**). The anisotropy raises the degeneracy in the orientation of the wavevector of the structures that is characteristic of the usual fluids. Then, one has perfectly ordered structures over arbitrarily large distances in the plane of the layer, i.e., the basic state is a "monocrystal". The spatial period $\lambda = 2\pi/k$ is of order 2d, where d is the layer thickness (10 to 50 μm). The frequency is a secondary parameter through which one can select either the stationary state of perfectly ordered convective rolls at low f ($\simeq$ 40 Hz), or a time-dependent state of traveling rolls at a higher f ($\simeq$ 100 Hz).This last state is a novel example of a nonlinear wave [3,4]. The structures are recorded through a polarising microscope and analysed using computer controlled digital processing. In order to study only one-dimensional waves we reduce the extension of the traveling rolls along their axis to a length of order of their diameter. At a given time, a wave is recorded as a periodic array of white dots (the crests of maximum intensity) separated by black spaces (the wave wells) along **x**. The position of this structure is recorded at equal time intervals and the successive recordings are plotted on top of each other. One obtains a 2-D image which is the space-time {x,t} representation of the wave. A progressive wave will then appear as a set of parallel oblique lines tilted by an angle θ onto **y**, and the phase velocity is $v = \tan \theta$. Very close to the threshold the traveling-wave keeps stable and uniform but as the voltage V is slightly increased (by less than 2%), the wave becomes unstable against localized instabilities from which defects usually arise.

2 THE STABILITY DIAGRAM

A stability diagram is obtained by measuring the threshold for a bifurcation, at any frequency[3-6]. At low frequencies the structures are stationary and two typical behaviors are found which depend on the rate of application of the voltage.

a) if typically, $\Delta V/\Delta t < 20$ mV/min., a series of well-defined bifurcations occur as V increases until the full chaotic state (the Dynamic Scattering Mode [3]) is developed inside the whole sample (Fig. 1a). The bifurcated states are of decreasing symmetry and the flows that are associated to each new pattern are studied by tracing small glass spheres immersed in the fluid [4,7]. First, the Normal Rolls are unstable against the Oblique Rolls. Then the tilted rolls are unstable against a periodical pinching, the Skew Varicose. It is found that the pinching produces, locally, a rotation of the vorticity by almost $\pi/2$, and induces stagnation points in the flow [9].

Then, we understand that the sequence of stable bifurcations corresponds to the evolution from a pure "monomode" of rotation around the roll axis in the Normal Rolls, to a Bimodal in the rectangular closed cells. These closed cells correspond to a double mode of two orthogonal rotations around the **x** and **y** axes [4,6,8]. They constitute the last ordered state before the chaos starts developing.

b) if, at the same frequency, the voltage is instead increased at a higher rate, then very soon beyond the first bifurcation to the Normal Rolls, a complex state is developed (fig.1b). For instance if $\Delta V/\Delta t \simeq 10$ mV/$_{sec}$. right above the first threshold, defects are spontaneously created in the whole sample with a low density at start. Simultaneously time-dependence occurs and the defects as well as the structure move in an apparent random mode of motion, inside decorrelated domains thus making this state appearing as "chaotic" (sometimes also named "weak turbulence"). Increasing further the voltage, increases the defects density as well as their average velocity. There, an apparent oscillatory motion of domains of rolls takes place in the plane around a vertical axis [4] and the amplitude of the oscillation increases. Also, the defects undergo a change in shape and may either multiply or anihilate with others (Fig. 2). Finally after some time defined by the stress rate, domains of Bimodal cells form from the modified defects and grow in space, while the states of Oblique Rolls and of

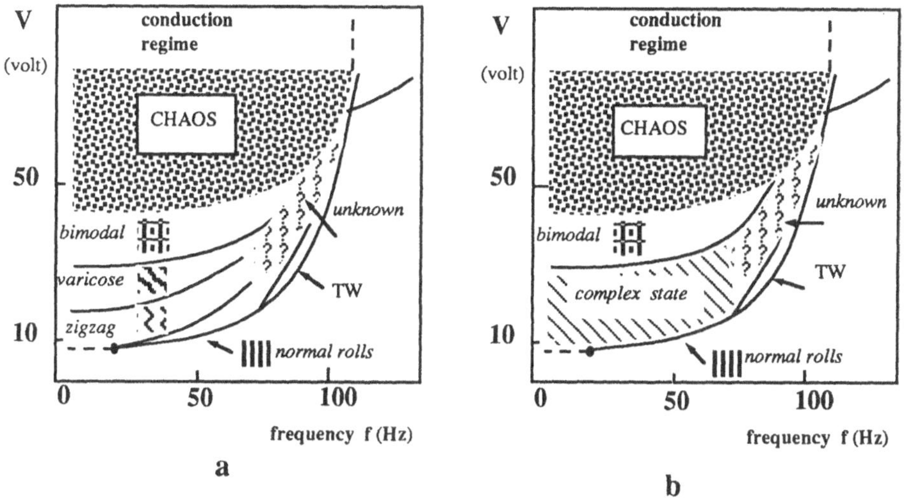

Fig. 1. Bifurcation lines in the voltage–frequency space. a) As V increases slowly the space-time chaos is reached through a series of bifurcations. b) The transition from the normal Rolls to the chaos is mediated by a complex state involving the defects as the voltage is rapidly increased.

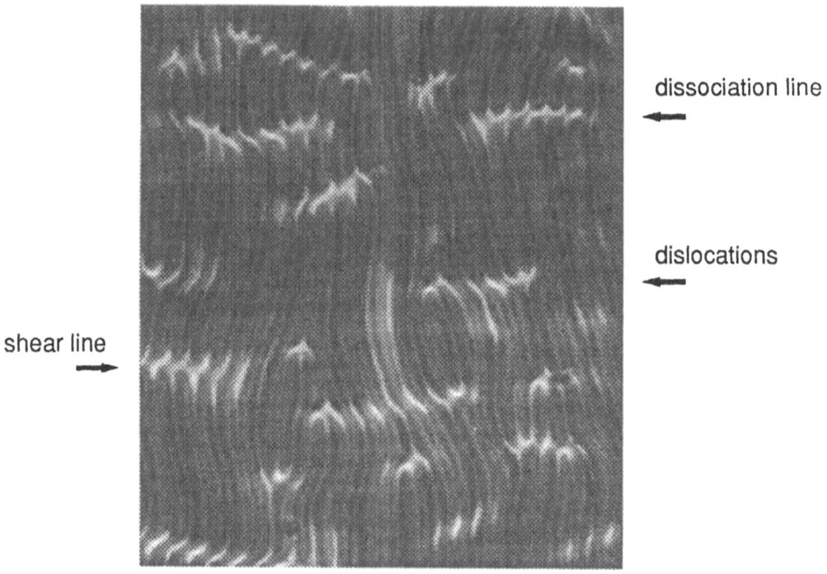

Fig. 2. The complex state developed under rapid increase of the voltage, showing dissociated dislocations.

Varicose are so unstable that they cannot be observed.

This second route to the Bimodal indicates that the defects are involved in a transient process implying localized deformations of the structure. This process is the fastest way by which the Bimodal homogeneous state can be obtained directly from the Normal Rolls, and it can be understood in a large part, by the study of the stability of a defect, under the applied stress.

3 THE PHENOMENOLOGICAL MODELS

The Model for the Stationary Convective Rolls

An homogeneous structure of rolls (here the Normal Rolls) is described by a local quantity, for instance the vertical velocity component :

$$u_z \sim A(x,y,t) \exp i(k_0 x) + c.c. ,$$

where $A = A_0(x,y,t) \exp i\,\varphi$. The structure is homogeneous and stable when $A_0 = C^{te}$. The amplitude $A_0(x,y,t)$ which is the envelope of u(x,t) is solution to a time-dependent nonlinear equation of the Landau-Ginzburg type (the so-called envelope equation [12]). Such an equation can in principle, be derived [13] from the microscopic basic equations of the DGP model [2]. This envelope equation describes the large scale evolution of A in time and in space and takes the form:

$$\partial A/\partial t = \mu A + \Delta A - \beta\,|A|^2 A ,$$

where μ is scaled to the control parameter and ΔA is the diffusion term related to the space gradients in both **x** and **y** directions. In a linear stability analysis of the stationary states, the growth rate s of the most unstable mode is real and the envelope equation coefficients are all real.

A periodic structure of rolls is characterized by a wavevector $\mathbf{k}_0$ along **x**, which is the spatial derivative of the phase $\Psi(x) = k_0 x + \varphi$, so $k_0 = \text{grad}\ \Psi$. The variable A_0 can be considered as the *order parameter* which characterizes the state, as in phase transitions. Its value is zero in the basic unstable state, and takes a finite value in the bifurcated state. If the order parameter is now allowed to vary slowly in space, for instance when modulations of the structure are developed, the total phase variation can be measured along any oriented close contour C encircling the deformed part $\oint_C \text{grad}\ \varphi.ds = \oint_C \mathbf{k}.ds$. In a quasi homogeneous structure, the perfect ordering can be recovered by any small perturbation and the integral is zero.

The lower symmetry homogeneous structures can be constructed by a mere superposition of the higher symmetry states:

-the Normal Rolls are described by: $u_{NR} \sim \mathcal{R}e\{A_0 \exp.\ i(k_x x)\}$,

-the Oblique Rolls by $u_{OR} \sim \mathcal{R}e\{A_1 \exp.\ i(k_x x \pm k_y y)\}$,

-the Skew Varicose is the superposition of two states of symmetrical Oblique Rolls with different amplitudes:

$$u_{SV} \sim \mathcal{R}e\{A_1 \exp.\ i(k_x x + k_y y) + \eta\, A_2 \exp.\ i(k_x x - k_y y)\},\ \text{where}\ \eta < 1.$$

-the Bimodal is obtained in the same way, but with $\eta = 1$.

The Model for the Traveling Rolls

The traveling-wave convective state found [3,4] mainly at high frequencies (see Fig. 1), is a simple example of a nonlinear wave [14]. Waves are ordered structures of space-time propagating either to **x** or **−x** and a local quantity describing them is :

$$u(x,t) = \mathcal{R}e\{A(x,t) \exp.\ i(kx - \omega t) + B(x,t) \exp.\ i(kx + \omega t)\}$$

They are solution to two coupled Complex Landau-Ginzburg amplitude equations [15]:

$$\frac{\partial A}{\partial t} + c\,\frac{\partial A}{\partial x} = \mu A + (1+i\alpha)\,\frac{\partial^2 A}{\partial x^2} - (1+i\beta)|A|^2 A - (1+i\gamma)|B|^2 A$$

$$\frac{\partial B}{\partial t} - c\,\frac{\partial B}{\partial x} = \mu B + (1+i\alpha)\,\frac{\partial^2 B}{\partial x^2} - (1+i\beta)|B|^2 B - (1+i\gamma)|A|^2 B$$

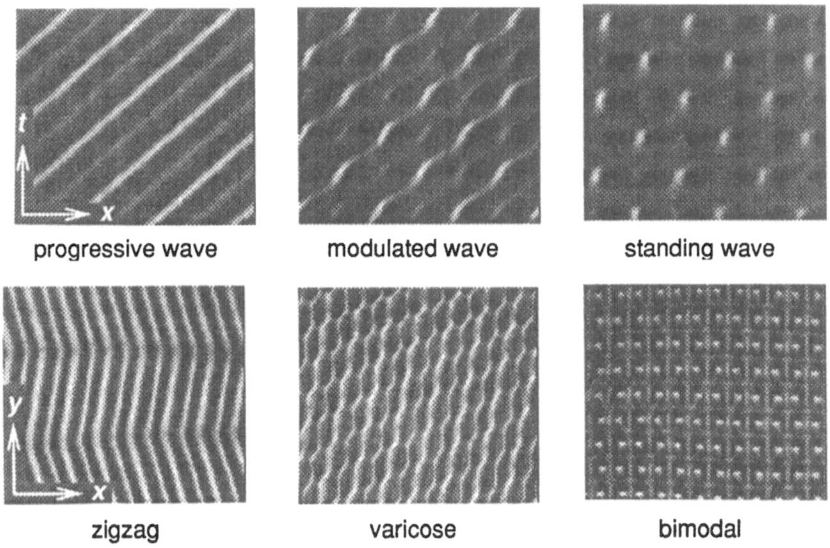

Fig. 3. Structures of the stationary states in the x,y space, and of the 1-D nonlinear wave represented in the x,t space.

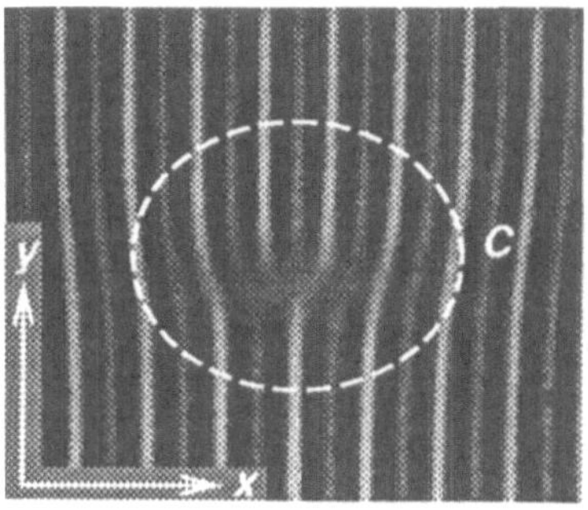

Fig. 4. A dislocation in Normal Rolls. In the upper half plane an extra period is added. The phase jumps by 2π around the core.

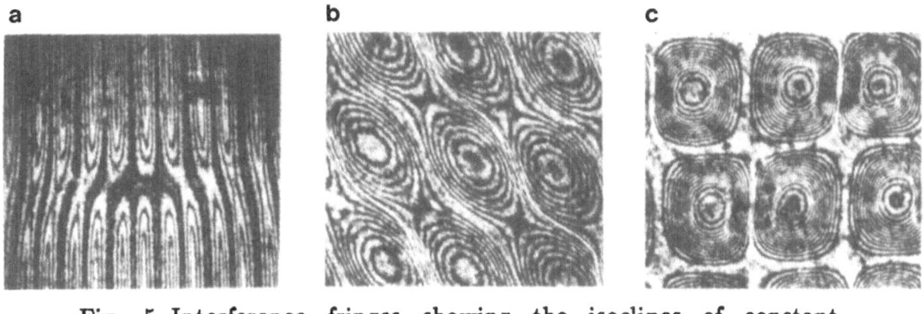

Fig. 5. Interference fringes showing the isoclines of constant velocity in (a) a dislocation, (b) a varicose and (c) a bimodal structure. In the pinched parts the velocity is zero.

This equation contains the basic terms for the instability growth (μA) and the amplitude saturation ($|A|^2A$). The second term is for the diffusion (∂^2A/dx^2) and for the dispersion of the wave ($i\alpha\ \partial^2A/dx^2$). The complex terms of the form $i\beta|A|^2A$ express the coupling between the frequency and the amplitude of the wave, and the crossed terms with the prefactor $(1+i\gamma)$ are for the coupling between the left and right waves. On the left side c is the group velocity.

The solutions are :

i) a uniform right-going wave $|A| \neq 0$ with $|B| = 0$, or left-going: $|B| \neq 0$, $|A| = 0$

ii) a Standing Wave (SW) where A and B are coupled $|A| = |B|$.

The waves are unstable against standing waves when $|\gamma| < 1$ and they may undergo phase instabilities (of the Benjamin-Feir type [12,16]) when $\gamma > 1$, if the condition $(1 + \alpha\beta) < 0$ is fulfilled.

However, we shall be concerned in the following, with the more complex case of non-homogeneous perturbations (i.e. localized ones), which usually trigger instabilities in which both the phase and the amplitude are involved. It is known that nonlinear waves may develop localized states such as solitons [16], kinks [17], or more complex states responsible for the breaking of the waves such as those following a shock [14]. We shall show hereafter that if the system contains some conservation principle as it is the case for kinematic waves, then defects are created following shocks caused by localised instabilities.

Symmetries in the stationary and in the time-dependent states

The nature of the possible defects is determined by the symmetries broken by the bifurcated state and there exists a similarity between the structures and their defects either in the stationary or in the time-dependent states. The phase of the Oblique Rolls is $(k_x x \pm k_y y)$, while that of the progressive wave is $(k_x x \pm \omega t)$. One sees that ωt and $k_y y$ play a similar role reflecting the symmetries $\theta \rightleftharpoons -\theta$ where $\theta = \arctan(k_y/k_x)$, and $v \rightleftharpoons -v$ where v is the phase velocity $v = \omega/k_x$. Also, it is from these states that one can obtain by superposition the lower symmetry ones, namely the Varicose and Bimodal on one hand and the Modulated and Standing Waves on the other hand. For instance the Standing Wave is the superposition of two opposite waves of equal amplitude. The experimental structures shown Fig. 3 reflect those symmetries.

4 DEFECTS IN 1-D AND 2-D STRUCTURES OF ROLLS

A defect is a singularity of the order parameter, which corresponds to a non-zero value for the total phase variation measured along any closed contour C encircling it : $\delta\varphi = \oint_C \mathbf{k}.d\mathbf{s}$. The condition for such a singularity to persist under any small perturbation is that $\delta\varphi = m.2\pi$, where m is an integer. Then the defect is topologically stable, i.e. it may exist. If instead there exist localized distortions of the phase such that the integral is zero, they may disappear under small perturbations, but they also can lead to stable defect. On the other hand it can be shown by the use of the homotopy groups in the order parameter space[18] that in a 2-D real space the only stable defects is the point-defect. A line-defect is stable only in the case of a symmetry-parity breaking, here in the Oblique Rolls. The point-defects are here edge-dislocations while the line-defects are domain-walls.

In the case of ordered structures in 1-D space (here the waves), the above study immediately shows that the point-defect is topologically unstable whereas the line-defect separating two domains of counterpropagative waves is stable (because of the breaking of parity symmetry).

Creation of Dislocations in Normal Rolls

In the almost stationary structure of rolls, it was experimentally found[1] that one of the mechanisms of creation of dislocations pairs would occur after the creation of a strongly localized deformation of the local field due to a modulational instability. The modulation is produced by the sudden application of a relatively small increase of the voltage $\Delta V/V \simeq 2\ \%$. Then dislocations are created in a sequence of three steps in time. At first, after some fractions of a second a modulation of both the phase (the local period) and the amplitude occurs in the **x** direction and strongly localized while constant along **y**. Next, during a longer time up to 100 sec., the local amplitude vanishes around some point centered in the previous localised state, while around it the phase of a few (2 to 5) rolls is locally pinched. Finally, when the pinching reaches its maximum the rolls are cut into two parts which start separating, thus forming a pair of opposite dislocations.

Topology of Dislocations

A point-defect in a Normal Rolls structure is an edge- dislocation in which an extra period (two rolls) is added in one of the two half spaces defined by a line parallel to **x** passing by the defect (Fig. 4).

The structure of the velocity field inside the core of the defect is obtained by using interferometry in monochromatic parallel light [5,9]. The nematic is a uniaxial positive crystal and is thus birefringent. The local birefringence is directly related to the relative tilt angle of the molecular axis Ψ which is a function of the velocity gradients. Then birefringence measured by interferometry can give a direct access to the velocity gradients. Figure 5 shows the distribution of the isoclines (lines of equal tilt angle, i.e. of equal velocity gradient) in the rolls. The density of lines is higher close to the up and down flows and it vanishes out to produce a singularity in the *core* of the defect. There, the tilt angle $\Psi = 0$ as on the up and down lines.

The streamlines inside the core are visualized by tracing individual glass spheres immersed in the fluid [4,7,9]. Close to the core at a distance less than λ the motion which was purely circular in the roll takes now an axial component which makes the trajectory helical. Then at about $\lambda/2$ it suddenly changes plane, and occurs for a short time in a vertical plane almost normal to the previous one, thus indicating the presence of stagnation points S which separate in space, the orthogonal streams taking place over a small volume with dimensions of order a space-period λ. The same flow structure is also found by the same techniques inside the pinching of the varicose instability and also, periodically localized orthogonal rotations are the characteristics of the Bimodal state [6,8,9].

It can be concluded that the core of the dislocation contains the symmetry elements of the (unstable) state which would be more stable for higher values of the external stress. For instance here, the core of the dislocation would represent a Bimodal state, and in its vicinity the continuity with the Normal Rolls is made by a damped Varicose structure. The question of its stability under stress comes naturally as the problem of a local bifurcation to this lower symmetry state.

5 INSTABILITY OF DEFECTS IN THE NORMAL ROLLS

Instability of Dislocations

The core of a defect is equivalent to a local unstable state and it is stabilized by the topological constraint. It can be made unstable by increasing the external stress. Experimentally, the voltage is smoothly applied and one observes an extension L_x of the defect core, which increases with the stress amplitude. The result shown in Fig. 6 indicates that beyond some threshold the length L_x increases sharply. At the same time the width L_y of the deformation field measured along **y** decreases until zero. This region of width L_y would be the analog of a Bloch wall of limited extent connecting two domains of opposite vorticity. This process, whereby the dislocation splits into two halves (the "*partials*") which move apart along x, is named a *dissociation* in analogy with the core dissociation in solids[19]. The part of the structure between the two *partials* is made by a complete pinching of the rolls, and the rolls that now face each other are of opposite vorticity (this *dissociation line* is now a real singular line rather similar to an Ising wall). At this point the voltage is close to the threshold for the Bimodal, and indeed the topology of the *dissociation line* is precisely that of the singular lines parallel to **x** in the Bimodal. Increasing further the voltage, increases the density of such lines until they stack together at equal distances along **y**, thus forming domains of Bimodal structure, after a typical transient time. Fig. 7 shows the rate of area transformed into a Bimodal after a voltage step has been given a value equal to the threshold for the Bimodal.

This rapid process involving the defects and geometrical transformations (translations and rotations) over distances larger than λ is similar to the so-called displacive transitions or Martensitic Transformations [20] in solids operated under a fast temperature gradient, although generally, displacive transitions imply first order transitions (subcritical bifurcations).

Instability of linear defects in Normal Rolls

Line-defects are topologically unstable. *Linear defects* are those which are extended in one direction but which have the topology of point-defects. This is the case for the *dissociation lines* which, as we have just seen, are dislocations in which the core is extended under stress, but where the phase jump is nevertheless 2π, as in a point-defect.

Another case is when a localised modulation of the amplitude and of the phase in form of a pinching, extends linearly over a large distance (Fig. 2). As a result, the geometry is close to that of a dissociated dislocation but there is no defect. Such a localized perturbation shall hereafter be named a "*shear line*". It can be easily unstable under an increasing external stress, thus giving birth to an even number (at least a pair) of opposite dislocations, while the remaining portion of the deformed rolls relaxes back to the basic state of straight rolls. In the same way a *dissociation line* may produce an odd number of dislocations (at least three). In both cases the total topological "charge" is conserved. One has a dynamical process of creation and multiplication of defects, by *dissociation* of initial defects (extension of the unstable localized state) and by relaxation of *dissociation lines* (restabilization).

Regular Patterns of Defects

In some part of the stability diagram close to the traveling rolls the modulational instabilities of the basic structure have a rather low dynamics and they produce modulations periodic in space [1]. There, rolls equally spaced by an integer number of periods $n\lambda$ (typically $n = 2,3,4$) have a larger amplitude and a smaller diameter. We name them "*walls*". Then, as the voltage is increased smoothly, periodic distortions of shear occur in the space between the large amplitude "*walls*", along **y** as in the nucleation process described in *par.*4. However, the splitting does not occur here and the defects creation is aborted. One obtains a perfectly ordered state of periodic modulations of phase and amplitude (*walls*) along **x**, and of periodic *shear lines* along **y** (Fig. 8). As the voltage is further increased, the *walls* disappear, the *shear lines* connect to form continuous lines and one recovers at a sudden, the bimodal state since the *shear lines* are precisely the *dissociated lines* (or equivalently, the basic singular lines of bimodal). So, here too, a purely homogeneous state of lower symmetry (the bimodal) can be developed from defects or localised states that are themselves created from localised instabilities.

Discussion : The core of stable defects has the symmetry elements of the unstable state. This state can therefore be "germinated" from the defects. However, one major question arises here, since such a transformation would imply (in a classical picture of phase transitions), that the bifurcations be subcritical to allow the coexistence of domains with a different order parameter. However, our actual findings indicate that the bifurcations are here rather supercritical (i.e., second-order-like transitions).

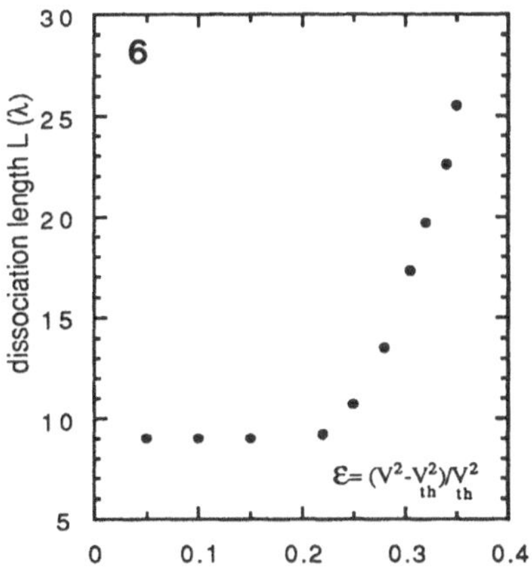

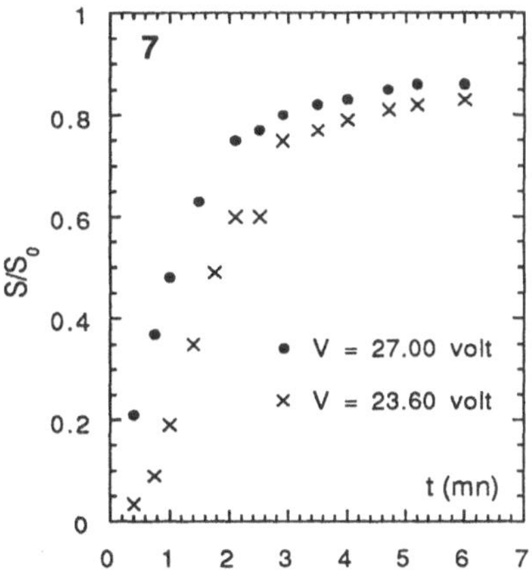

Fig. 6. Core dissociation L_x under an increasing voltage.
Fig. 7. Growing in time of Bimodal domains following the sudden application of a large voltages V.

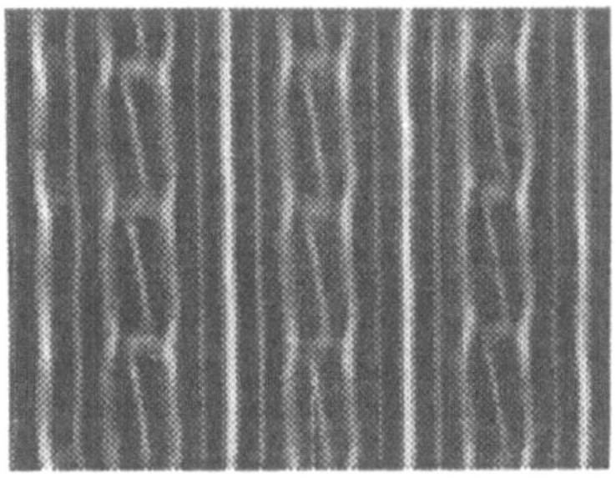

Fig. 8. Regular array of trapped dislocations.

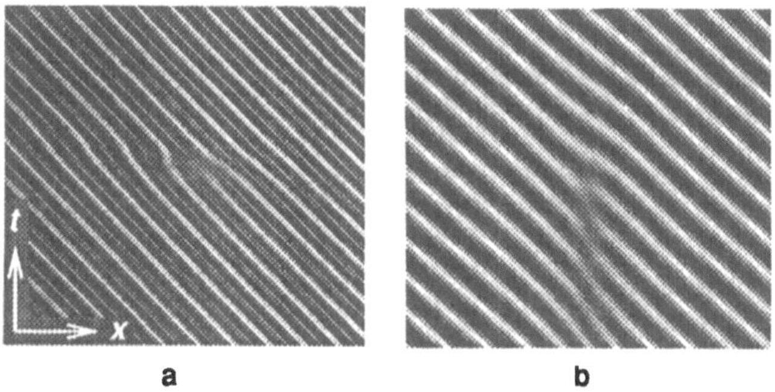

Fig. 9. A space-time dislocation in a nonlinear 1-D wave (a) and its numerical simulation with initial conditions: α=1, β=0.4, γ=1.4, δ=c=0, |A|=Cst, B=0, φ_a= π. th(x/4.4).

6 DEFECTS IN NONLINEAR WAVES

Creation of Dislocations in Waves

In the {x,t} space the point-defect is here too, a dislocation [21], i.e. the sudden loss (or gain) at a given time of one space period. It was shown that space-time dislocations occur after a localised perturbation of the phase [21]. The phase being here $\Phi = (kx-\omega t)$, a local modulation $\Phi(x)$ expresses a localised change in velocity ($v = \omega/k$), i.e. a *shock*. Indeed, it is found that in a shock, the local wavevector undergoes a sudden variation which pulls it beyond some stability limit, for instance of the Eckhaus-Benjamin-Feir type. The restabilization occurs by expelling (in a compressive shock), or creating (in a dilative one) one spatial period. Then, dislocations are created, with a charge related to the type of shock (compressive or dilative).

At the core of a dislocation the amplitude modulus $|A|$ vanishes and because of the coupling between left and right going waves (the two states are equally stable) any decrease in A for instance, favours a correlative increase in B. Numerically, one verifies that dislocations are also solutions to the equations by imposing as initial condition a localised phase jump of 2π [22]. The result is shown on Fig. 9.

The point-defects are topologically unstable whereas the line-defects are stable in the 1-D physical space, and therefore the stability of line-defects can be investigated under stress [15,23]. Line-defects correspond to limits separating two domains of counterpropagative waves, i.e. points (in 1-D) where $|A|=|B|$. Those points are represented in the {x,t} space as line-defects, the *domain-walls* (Fig. 10). A domain-wall in x,t space is either a **sink** when the waves A and B meet at it or a **source** when the waves are "emitted" from it.

Sinks and sources are characterized by the coupling between the two states. At their core $|A| = |B|$, which corresponds to the condition for a Standing Wave at one point [21,22]. Because of the competition between the space gradients and the nonlinearities, the amplitude profile around this point is kink-like and over some finite distance there is a mixing of the two states. A mixing represents in fact a Modulated Wave (MW), the modulation amplitude decreasing with increasing the separation distance from the wall. Fig.10 shows the Modulated Wave in the structure of a source. This structure is quite similar to that of the dislocation core shown in Fig. 4. It is also similar to the structure of a domain-wall in the Oblique Rolls structure because of the same symmetries, mainly. The mechanism of creation of a pair sink–source is close to that of the creation of a pair of space-time dislocations, i.e., a compressive plus a dilative shocks. Numerically, this can be achieved by allowing a state B to grow from noise, while A is locally decreasing to zero due to a local phase jump of 2π [22].

Instability of Sinks and Sources

Space-time defects can be made unstable under small-scale perturbations. Here too, it is the core structure that determines the unstable state of the defect. The core has the symmetry elements of a Standing Wave and is surrounded by a Modulated Wave of decreasing amplitude. It is then expected to be unstable against a localised Standing Wave at least transiently. Two typical cases shall be considered hereafter.

a) Multiplication of sinks and sources[22]

Consider the case where the system is in the uniform wave state but close to the bifurcation to the Standing Wave. In the equations this comes to give γ a value slightly higher than 1. In addition, one must allow phase fluctuations to occur for instance, by satisfying the Benjamin-Feir criterion $1 + \alpha\beta < 0$. Because the core represents the most unstable state a domain of Standing Wave will start developing from it. The finite lifetime of the phase instability makes this state a transient and restabilization occurs either back to the initial defect or to a state with an odd number of defects (to conserve the topological charge). In Fig. 13 a sink gives rise to a source plus two sinks. By numerical simulation of the coupled CLG equations one is able, following this scheme, to mimic the experimental multiplication of defects (Fig.11).

b) Core-widening under stress

The action of an appropriate external stress on a state of progressive waves

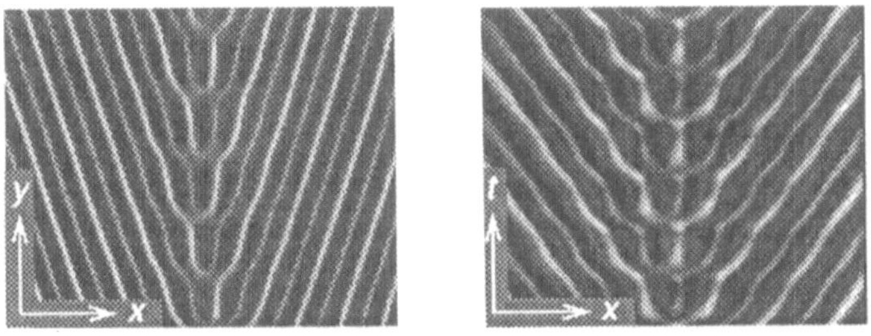

a b

Fig. 10. Domain-walls are line-defects separating "double-states", (a) in the Oblique Rolls, (b) in progressive waves (here a source).

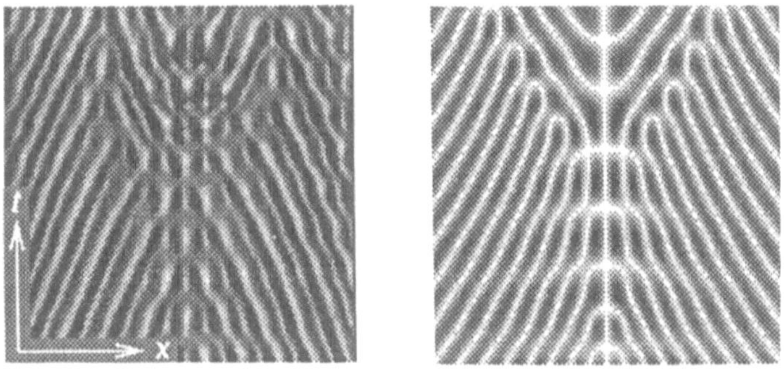

a b

Fig. 11. Instability of a sink giving rise to a sink plus two sources. (a) experiment; (b) simulation with α=1, β=0.4, γ=1.02, δ=c=0, |A|=0.5 sin(2πx/100)+0.5, B=-0,5 cos(2πx/100)+0.5.

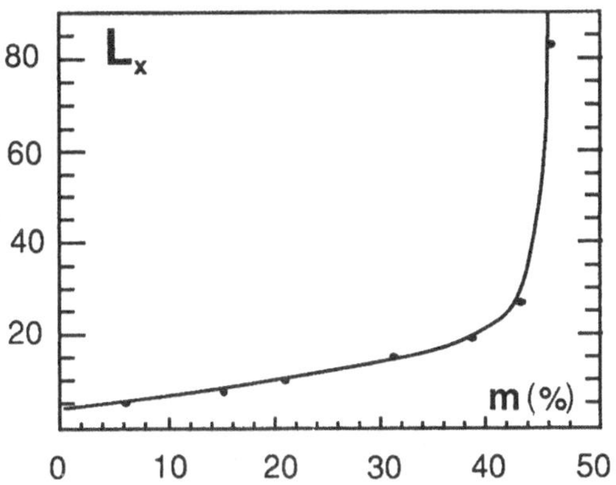

Fig. 12. Core stretching of a source under an increasing stress m that expresses the coupling between the left and right waves.

can make them unstable to standing waves beyond some threshold. For instance applying a modulation of the voltage at frequency 2ω where ω is the frequency of the traveling wave, has for effect a coupling between the two A right and B left-going waves [24]. As the depth of modulation increases, one passes continuously from the progressive wave PW, to a modulated wave MW [25] and a standing wave SW.

Starting from a state of domains of progressive waves A and B separated by a *domain-wall* (say a source), one applies an increasing external forcing. The SW state in the core becomes more and more stable inside the PW state which in turn, becomes more unstable. In space, this gives an extension of the source-width inside the unstable PW state. Experimentally, this variation is found to be nonlinear, diverging as the stable SW state is reached (Fig. 12). Such an instability is quite similar to the dissociation of dislocations described in *par* 4. However, it is not expected from localised instabilities inside states that bifurcate supercritically. In effect, by analogy with solids such a domain-wall widening is rather characteristic of subcritical bifurcations [26] (germination-growth, solidification fronts, transitions induced by grain-boundaries). It has been shown [27] that in the case of subcritical bifurcations also, domain-walls would undergo a widening in a way similar to the germination-growth process in solids.

We notice that both in the case of defects in stationary states, and that of defects in nonlinear waves the spatial extension of the local bifurcation shows a similar type of increase under stress.

CONCLUSION

It has been shown that the transitions either from order to disorder or from disorder to order in a typical non-equilibrium system can be mediated by the defects. The states may be either stationary or time–dependent, and this latter case represents a novel phenomenon in nonlinear waves. A defect results from localized instabilities and its core itself represents a local bifurcated state with a lower symmetry. The instability of a defect may be either transient, giving thus a mechanism of defects multiplication, or steady and the bifurcation to a homogeneous state can be achieved by a continuous "germination" from the core of the defects. However, in analogy with phase transitions in solids, such a behaviour which would be expected when bifurcations are subcritical, raises a serious problem since the bifurcations are here experimentally found to be supercritical. Nevertheless, most of the experimental effects concerning the creation, the structure and the stability of defects, can be reproduced using the generalized Complex Landau-Ginzburg equations. Finally, it is important to notice that a state similar to the complex dynamical state developed after the creation of defects in the basic state is indeed very commonly observed in other non-equilibrium systems when the rate of the external stress is large. Bifurcations to either lower symmetry states or to disorder can be mediated by the defects in a way somehow similar to displacive transitions, whereas the full sequence of bifurcations obtained under the slow application of very small increments would be the analog of a series of diffusive transitions. This rapid complex and apparently disordered process may make unstable, thus unobservable, some homogeneous intermediate states.

We acknowledge fruitful discussions with P. Coullet, J. Friedel, J. M. Ghidaglia, J. Lajzerowicz, L. Lega and A.C. Newell. The numerical simulations on a CRAY2 computer were made possible by the Scientific Committee of the CCVR at Palaiseau. We are also grateful to F. Augier for valuable advices and help during the numerical simulations. This work has been done under DRET contract 87/142.

REFERENCES

1. X.D. Yang, A. Joets, R. Ribotta, p. 194 in *Propagation in Systems far from equilibrium*, J.E. Wesfreid, H.R. Brand, P. Manneville, G. Albinet, N. Boccara Eds., Springer Series in Synergetics, Springer-Verlag, Berlin (1988).

2. E. Dubois-Violette, P.G. de Gennes, O. Parodi, J. Phys. (Paris) 32, 305 (1971); P.G. de Gennes, " The Physics of Liquid Crystals", Clarendon, Oxford (1974).
3. A. Joets and R. Ribotta, Phys. Rev. Lett. 60, 2164 (1988); also in "Propagation in Systems far from Equilibrium" J.E. Wesfreid, H.R. Brand, P. Manneville, G. Albinet and N. Boccara eds., Springer, Berlin (1988);and in the Proceedings of the 12th International Liquid Crystals Conference, Aug. 1988, Freiburg), Liquid Crystals, London (1988).
4. A. Joets, Thesis, Paris VII, 1984 (unpublished).
5. R. Ribotta, in "Nonlinear Phenomena in material Science", p. 489, L. Kubin, G. Martin Eds., Solid State Phenomena, Vol. 324, Trans. Tech. Publications, Switzerland (1988).
6. A. Joets, R. Ribotta, J. Phys. (Paris) 47, 595 (1986).
7. A. Joets, R. Ribotta, Europhysics Lett. 10, 721 (1989).
8. A. Joets, X.D. Yang, R. Ribotta, Physica 23D, 235 (1986).
9. R. Ribotta, A. Joets, J. Phys. (Paris) 47, 739 (1986).
10. R. Ribotta, A. Joets, L. Lei, Phys. Rev. Lett. 56, 1595 (1986).
11. S. Kai, K. Hirakawa, Sol. State Com. 18, 1573 (1976)
12. A.C. Newell and J.A. Whitehead, J. Fluid. Mech., 38, 279 (1969).
L.A. Segel, J. Fluid Mech., 38, 203 (1969).
A.C. Newell, Lect. Appl. Math. 15, 157 (1974).
13. E. Bodenschatz, W. Zimmermann, L. Kramer, J. de Phys. (Paris), 49, 1975 (1988).
14. G.B. Whitham, "Linear and Nonlinear Waves", John Wiley & Sons, New York (1974); J. Lighthill, "Waves in Fluids", Cambridge Univ. Press, Cambridge (1978).
15. P. Coullet, C. Elphick, L. Gil, and J. Lega, Phys. Rev. Lett. 59, 884 (1987) ; P. Coullet, and J. Lega, Europhys. Lett. 7, 511 (1988).
16. T.B. Benjamin and J.E. Feir, J. Fluid Mech. 27, 417 (1967); J.T. Stuart and R.C. DiPrima, R.C., Proc. R. Soc. Lond. A 362, 27 (1978); C.S. Bretherton and E.A. Spiegel, Phys. Lett. 96A, 152 (1983).
17. K. Kawasaki, T. Ohta, Physica 116A, 573 (1982); N. Bekki and K. Nozaki, Phys. Lett. A110, 133 (1985).
18. G. Toulouse, M. Kléman, J. Phys. Lett. (Paris) 37, 149 (1976).
N. D. Mermin, Rev. Mod. Phys. 51, 591 (1979).
19. J. Friedel, Dislocations, Pergamon Press, London (1964).
F.R.N. Nabarro, Theory of Crystal Dislocations, Clarendon Press, Oxford (1969).
20. J.S. Bowles and J.K. Mackenzie, Acta Metal. 2, 129 (1954) ; ibid., 2, 224 (1954); J.W. Christian, "The Theory of Transformations in Metals and Alloys", Pergamon Press, Oxford (1965).
21. A. Joets, R. Ribotta, in "New trends in nonlinear dynamics and pattern forming phenomena: the geometry of nonequilibrium", P. Coullet and P.Huerre (eds.), Nato ASI series B, Plenum Press (1989) (Cargese Workshop Aug. 1988); also in the Proceedings of "Nonlinear coherent structures in physics, mechanics and biological systems" (Paris, June 1988), J. de Phys. C3, 50, 171 (1989).
22. R. Ribotta, A. Joets, in "Partially Integrable Evolution Equations in Physics", R. Conte and N. Boccara eds. Nato ASI Series, Kluwer Acad. Pub., C–310, 279 (1990).
23. J. Lega, Thesis, Université de Nice (1989), unpublished.
24. D. Walgraef, Europhys. Lett., 7, 485 (1988).
25. E. Knobloch, Phys. Rev. A 34, 1538 (1986).
26. J. Lajzerowicz and J. J. Niez in "Solitons in Condensed Matter Physics", A. R. Bishop, T. Schneider eds., Springer Series in Solid State Science, 8, Berlin (1978).
27. P. Coullet, L. Gil, D. Repaux, Phys. Rev. Lett., 62, 2957 (1989).

FROM CHAOS TO QUASICRYSTALS

G.M.Zaslavsky

Space Research Institute
of the USSR Academy of Sciences,
Profsoyuznaya 84/32, Moscow 117810, USSR

INTRODUCTION

It required great effort to find the 5–fold symmetry in the Nature[1]. But may be much more efforts was needed to change anything in our minds. Penrose tiling with 5–fold symmetry has been published years before[2]. And even very wide popularization of it by M.Gardner in Physics Today[3] was not enough to convince a lot of crystallographists that the pentagonal tiling exists. We should not forget the unsuccessful efforts of I.Kepler to solve this problem.

The paradox of History was that the 5–fold symmetry tiling in 2–dimensional case (2D) was known in Moorish art from Alhambra Palace (Granada)[4] (Fig.1). But maybe not less paradoxical is that the pentagonal tiling and arbitrary q–fold symmetrical 2D tiling can be obtained from the dynamical equation of motion of a charged particle in some simple form of electric and magnetic field. The tiling is produced by the only orbit and this orbit reveals to be chaotic. So a chaotic orbit of a particle can display on the phase plane a tiling with arbitrary q–fold symmetry.

Below a very brief comments to the dynamical origin of quasicrystal symmetry (q–symmetry) are given. They are based on the original publication[5,6] and on the reviews[7,8].

KICKED – OSCILLATOR MODEL

The model is given by the Hamiltonian

$$H = {}^1/_2\, p^2 + {}^1/_2\, \omega_0{}^2 x^2 - \epsilon \frac{\omega_0{}^2}{k} T_0 \cos kx \sum_{n=-\infty}^{\infty} \delta(t-nT_0) \tag{1}$$

where ω_0 – oscillator frequency, ϵ – perturbation parameter, T_0 – period of kicks

Microscopic Aspects of Nonlinearity in Condensed Matter
Edited by A.R. Bishop *et al.*, Plenum Press, New York, 1991

Fig. 1. Pentagonal ornament
from Alhambra

and put the mass m = 1. The model describes also the charged particles motion in a magnetic field with hyrofrequency ω_0 and in an electrostatic wave packet which propagates perpendicular to the magnetic field. If $\omega_0 = 0$, $\epsilon\omega_0^2$ finite then (1) is equivalent to the Chirikov–Taylor kicked rotator model.

The equation of motion follows from (1)

$$\ddot{x} + \omega_0^2 x = -\epsilon\omega_0^2 T_0 \sin kx \sum_{n=-\infty}^{\infty} \delta(t - nT_0) \tag{2}$$

Linear term in (1) and (2) can be excluded by a canonical transformation. Therefore even for $\epsilon \ll 1$ the right hand side in (2) doesn't represent a small perturbation.

WEB – MAP

Due to kicks Eq.(2) can be written in a form of the map

$$\hat{M}_\alpha: \quad \begin{aligned} u_{n+1} &= (u_n + K\sin v_n)\cos\alpha + v_n \sin\alpha \\ v_{n+1} &= -(u_n + K\sin v_n)\sin\alpha + v_n\cos\alpha \end{aligned} \tag{3}$$

where the notations are used

$$u = kp/\omega_0,\ v = -kx,\ K = \epsilon\omega_0 T_0,\ \alpha = \omega_0 T_0 \tag{4}$$

and subscript n means that the correspondence value is taken just before the n–th kick.

Map $\hat{M}_\alpha$ may be written in the form of product of rotational operator $\hat{R}_\alpha$ and nonlinear operator $\hat{R}(v)$:

$$(\bar{u},\bar{v}) \equiv \hat{M}_\alpha(u,v) = \hat{R}_\alpha \hat{R}(v)(u,v) = R_\alpha(u,K\sin v) \tag{5}$$

the case $\alpha = \alpha_q$

$$\alpha_q \equiv 2\pi/q \tag{6}$$

and q is integer, is of the main interest. That is resonance of q–th order: q kicks take place during one rotation:

$$\hat{M}_q \equiv \hat{M}_{\alpha_q}: \quad \begin{aligned} u_{n+1} &= \quad (u_n + K\sin v_n)\cos 2\pi/q + v_n \sin 2\pi/q \\ v_{n+1} &= -(u_n + K\sin v_n)\sin 2\pi/q + v_n \cos 2\pi/q \end{aligned} \tag{7}$$

Particular case is $q = 4$:

$$\hat{M}_4: u_{n+1} = v_n; \; v_{n+1} = -u_n + K\sin v_n \tag{8}$$

Parameter K in accordance with (4) is proportional to perturbation.

Map $\hat{M}_q$ in (7) is called web–map. Here are the main properties of $\hat{M}_q$:

(i) There exists an area of chaotic motion in the phase space (u,v) which forms the connected network of finite width called stochastic web. For q = 3,4,6 the web is periodical or of crystal symmetry (Fig. 2). For all others q web is of quasicrystal symmetry (Fig. 3). Inside the meshes of the web there are closed curves or confined thin layers of chaos. It is useful to compare the Fig. 1 with the pentagonal web in the Fig. 3a to check their similarity.

(ii) Random walk process is performed along the channels of a web if any initial condition is taken inside the area of the web. During an infinite time all the web will be displayed. The parts of webs shown in Figs. 2,3 are received as a result of random walking of the only one orbit.

(iii) For a crystal symmetry the web is unremovable. That means the existence of the web for arbitrary small ϵ. For quasicrystal symmetry it is possible to prove the existence of the web for finite (ϵ not too small). The proposition is that the web is unremovable but the size of meshes depends on ϵ: the smaller is ϵ, the larger are sizes of the meshes

a)

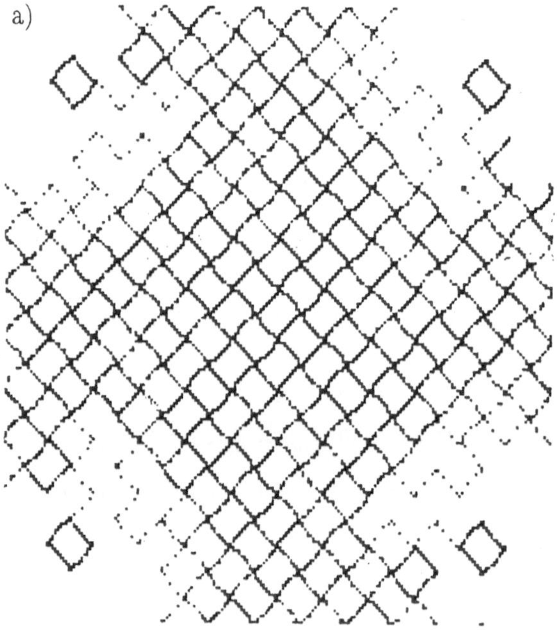

b)

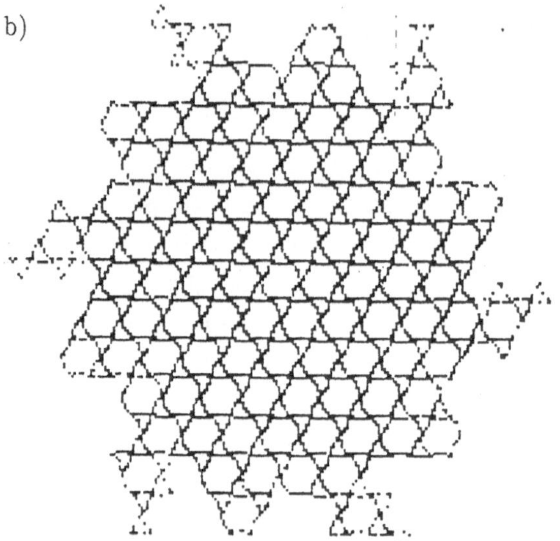

Fig. 2. Two examples of periodical webs: (a) square lattice (q = 4, K = 0.7); (b) cagome lattice (q = 3 or q = 6, K = 0.8).

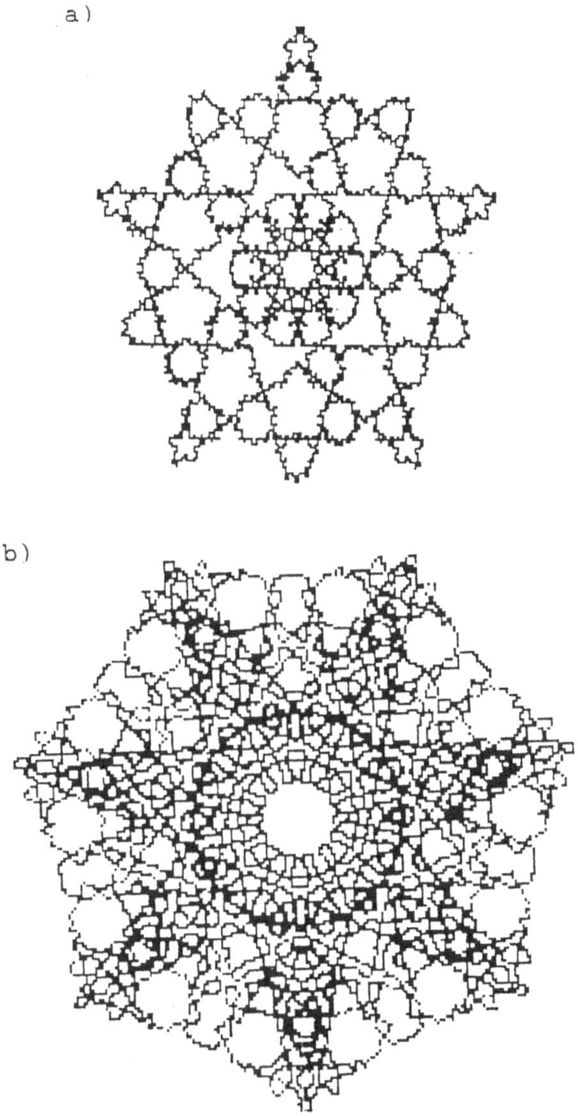

Fig. 3. Two examples of quasicrystal symmetry webs: (a) $q = 5$, $K = 0.7$, Size $= (256\pi)^2$; (b) $q = 7$, $K = 0.5$, Size $= (64\pi)^2$.

(iv) the width δ of a web is exponentially small

$$\delta \propto \text{const} \cdot \exp(-\text{const}/\epsilon) \tag{9}$$

and all constants in (9) are estimated in[9]. The symmetry of webs is controlled by the only parameters q in the web–map $\hat{M}_q$ (7).

(v) A web of symmetry q realizes the correspondent tiling of 2D–plane.

WEB'S SKELETON

From geometrical point of view an infinite web can be considered as a topological object. The real web (Figs. 2,3) may be smoothed in some way without changing its topology. From dynamical point of view a web is invariant manifold in response to the group of motion. Thus the operation of smoothing can be introduced as some dynamical averaging process. For small ϵ (small K) there is natural time–scale of averaging $2\pi/\omega_0$ [5,7]. The way to do this contains two steps. The first step is to exclude the linear term $^1/_2\ \omega_0{}^2x^2$ from the Hamiltonian (1) by a canonical transformation $(u,v) \rightarrow (I,\varphi)$:

$$u = \rho \sin \varphi,\ v = -\rho \cos \varphi,\ I = \rho^2/2$$

with the generation function

$$F = (\varphi - \alpha_q \omega_0 t)I$$

and new Hamiltonian

$$\tilde{H} = \frac{1}{\omega_0{}^2} H + \frac{\partial F}{\partial t} = -K \cos[\rho \cos(\varphi - \alpha_q \tau)] \sum_{n=-\infty}^{\infty} \delta(\tau - n) \tag{10}$$

where dimensionless time is introduced

$$\tau = t/T_0 \tag{11}$$

The second step is a splitting of (10) onto two terms: resonant and nonresonant with a frequency not smaller then ω_0. This second term disappears after averaging and the first term gives

$$H_q = \langle \tilde{H} \rangle = -\frac{1}{q} K \sum_{j=1}^{q} \delta(\vec{R}\,\vec{e}_j) \tag{12}$$

where $\vec{R} = (u,v)$ and $\vec{e}_j$ are unit vectors which form the regular q–rays star.

The function (12) is very typical for quasicrystals if $q \neq 1,2,3,4,6$. If $q = 4$ then

$$H_4 = -\frac{1}{4} K (\cos u + \sin v)$$

i.e. the well known Hamiltonian for electron in a square – lattice potential. For q = 3 and 6

$$H_3 = -\frac{1}{3} K (\cos u + \cos \frac{u+\sqrt{3}v}{2} + \cos \frac{u-\sqrt{3}v}{2}) = 2H_6$$

This expression coincides with stream function for Rayleigh–Bénard convection. The Hamiltonian H_3 realizes also Hexagonal symmetry tiling.

To understand what is happening in the quasicrystal case we consider the distribution of hyperbolic (or saddle) points for $H_q = H_q(u,v)$. The plot of them is shown on Fig. 4 for $q = 5$ and $q = 7$ where ρ_h is the number of saddles in arbitrary units. There are different finite singular orbits (separatrices) which belong to different energy levels. But there is no the only separatrix of infinite size which cover full plane. So there is no web in H_q of quasicrystal symmetry. The web appears due to the second part of Hamiltonian $\tilde{H}$ in (10), i.e. due to the nonstationary perturbation. Though smallness of perturbation it is enough for matching different finite separatrices and to create one connected infinite web. Another way is to consider a layer of small width ΔH in the vicinity of the energy level where there is maximum of hyperbolic points. It is clear from Fig. 4 that such an energy is $H_q \approx 0$ ($q = 7$). The corresponding points, belonging to such a layer, are displayed in Fig. 5. The dark grids in Fig. 5 are skeletons of webs. Their geometrical properties are discussed in[7,8].

Density of states $\rho(E)$ for Hamiltonian $H_q = E$ is another important item for more profound understanding of quasicrystal symmetry. The Fig. 6 reflects the smoothing of van Hove singularities. The larger is q the smoother is $\rho(E)$. From Fig. 6 it is clear that $\rho(E)$ behavior looks like for the liquid if $q = 7$. The origin of this property of smoothing is the distribution of saddles in Fig. 5. But there is no any smoothing in the Fourier spectrum of skeletons in Fig. 5. The spectrum is discrete and for $q = 5$ the spectrum coincides with the original X–rays spectrum from article[1] (see in review[7]). That can be considered as a final manifestation of the linking between the dynamical description of quasicrystal symmetry and its crystallographic consideration.

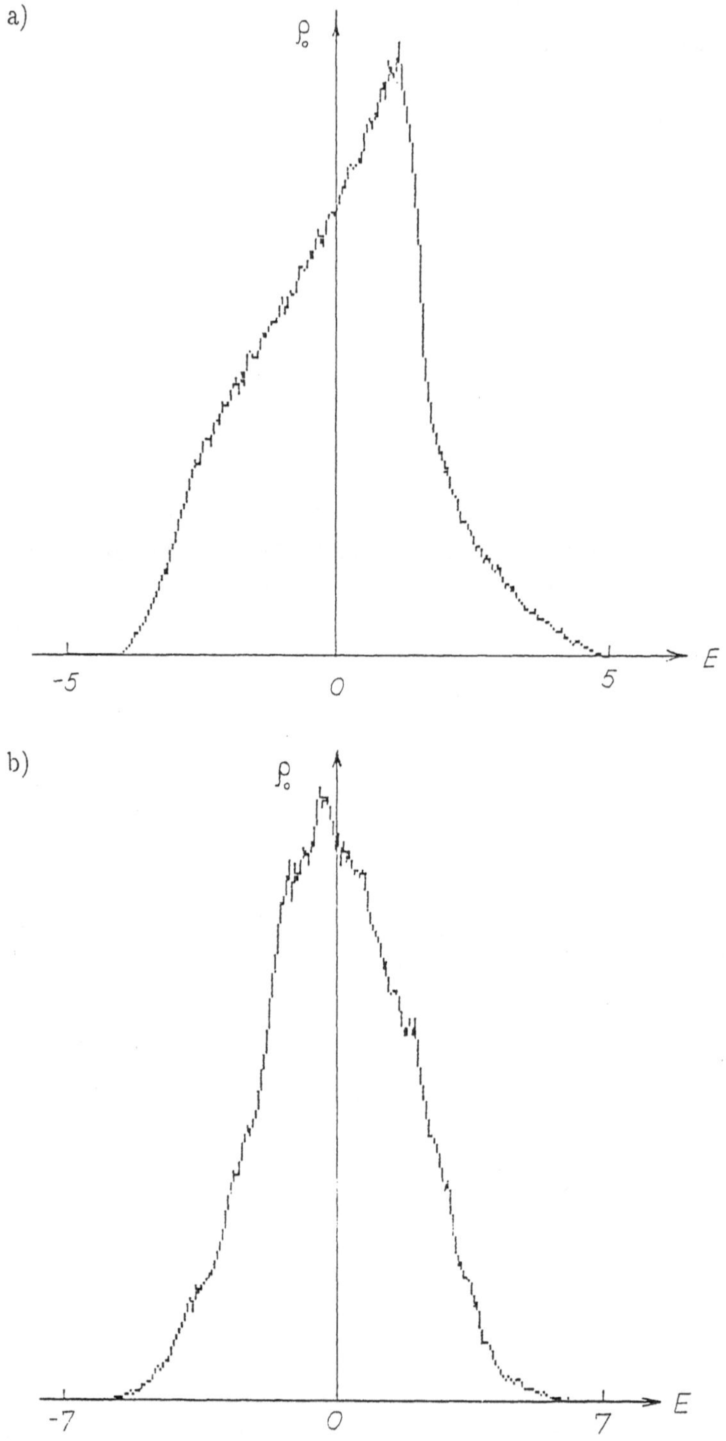

Fig. 4. Distribution of hyperbolic points (saddles) for different energy $E = \check{H}_q$:
(a) $q = 5$, (b) $q = 7$.

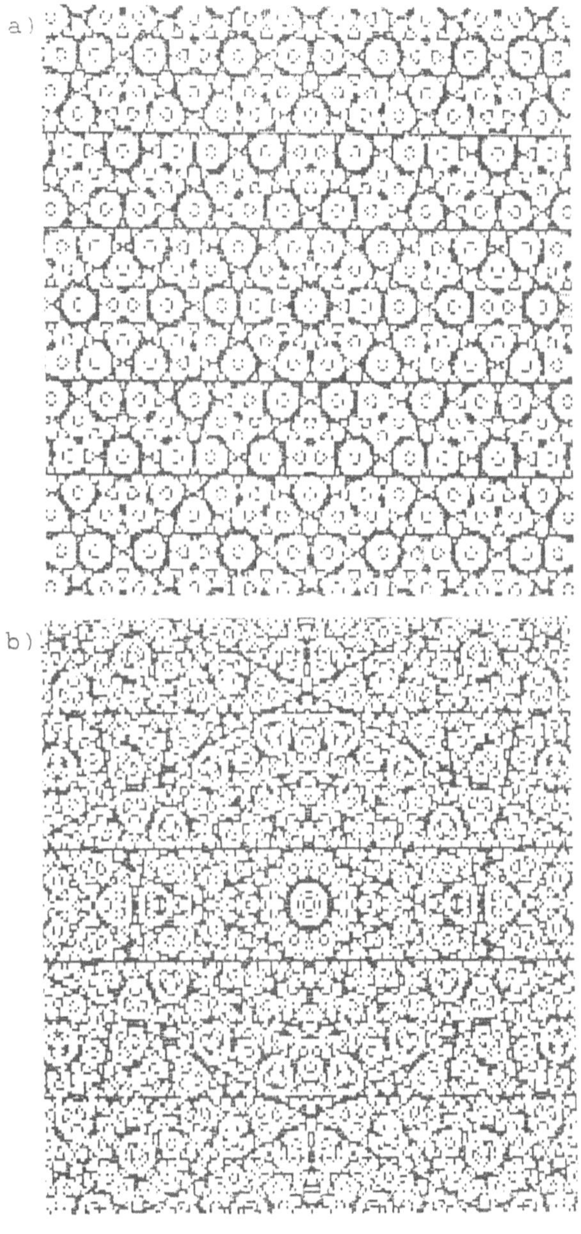

Fig. 5. Two examples of skeletons of quasicrystal webs: (a) q = 5, (b) q = 7.

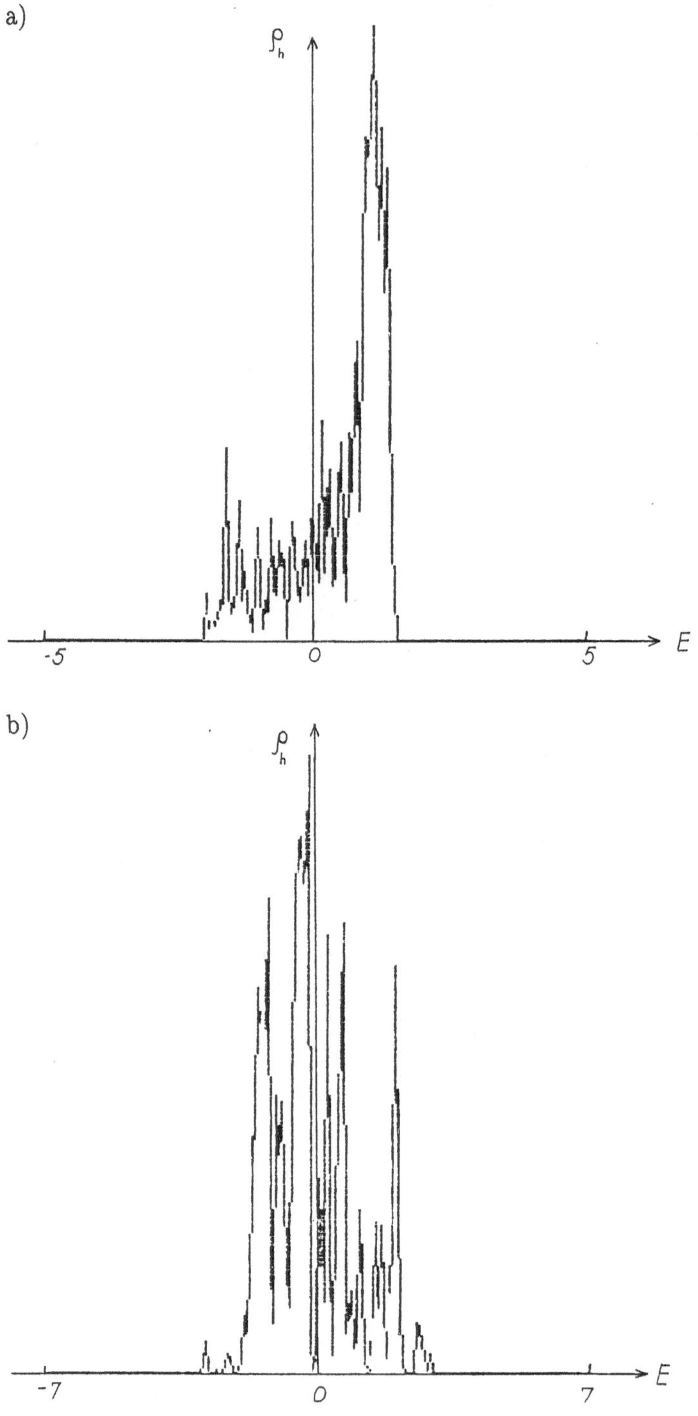

Fig. 6. Density of states versus energy for (a) q = 5, (b) q = 7.

REFERENCES

1. D.Shechtman, I.Blech, D.Gratias, and J.W.Cahn, Phys. Rev. Lett., 53:1951 (1984).
2. R.Penrose, Bull. Inst. Math. Its Appl., 10:266 (1974).
3. M.Gardner, Scientific American, 236:110 (1979)
4. D.Wade, 1976, "Pattern is Islamic Art," Studio Vista, London.
5. G.M.Zaslavsky, M.Yu.Zakharov, R.Z.Sagdeev, D.A.Usikov, and A.A.Chernikov, Sov. Phys. JETP, 64:294, 1986; JETP Lett., 44:451 (1986).
6. A.A.Chernikov, R.Z.Sagdeev, D.A.Usikov, and G.M.Zaslavsky, Phys. Lett, 125:101 (1987).
7. G.M.Zaslavsky, M.Yu.Zakharov, R.Z.Sagdeev, D.A.Usikov, and A.A.Chernikov, Sov. Phys. Uspekhi, 31:887 (1988).
8.. A.A.Chernikov, R.Z.Sagdeev, and G.M.Zaslavsky, Phys. Today, 41:27 (1988).
9. V.V.Afanas'ev, A.A.Chernikov, R.Z.Sagdeev, and G.M.Zaslavsky, Phys. Letts., 144:229 (1990).

A MODEL OF A CATALYTIC REACTION WITH LOCAL INTERACTIONS

F. Bagnoli[1,2], B. Sente[3] and M. Dumont[3,4]

[1] Università di Firenze, Dipartimento di Fisica, largo E. Fermi 2, 50125 Firenze, Italia
[2] INFM, Sezione di Firenze
[3] Université de Mons-Hainaut, Avenue Maistriau 19, B7000, Mons, Belgique
[4] Chercer qualifié, Fonds National de la Recherche Scientifique, Belgique

1. INTRODUCTION

The kinetics of the catalytic oxidation of carbon monoxide on a Pt surface

$$CO^{(g)} + \frac{1}{2}O_2^{(g)} \underset{\text{surface}}{\overset{\text{Pt}}{\rightarrow}} CO_2^{(g)} \tag{1}$$

has recently been investigated by means of Monte Carlo simulations. In this paper we consider the reaction (1) described by the simple irreversible model:

$$CO^{(g)} + * \overset{k_1}{\rightarrow} CO, \tag{2a}$$

$$O_2^{(g)} + *\_* \overset{k_2}{\rightarrow} O\_O, \tag{2b}$$

$$CO\_O \overset{k_3}{\rightarrow} CO_2^{(g)} + *\_*; \tag{2c}$$

where $^{(g)}$ denotes the gaseous species; * stands for an empty site and "_" represents a nearest neighbor (*nn*) bond on the surface (hereafter a square lattice).

The simulations may be performed by using either rate constants k_i[1] or mechanism probabilities p_i. Our approach is based on the Ziff, Gulary and Barshad (ZGB) model[2]. These authors assume an infinite reaction rate constant k_3 and describe the steady state properties of the system in term of the sole bifurcation parameter P_{CO}. This selection probability P_{CO} represents the relative rate of potentially efficacious collisions of **CO** molecules on the surface ($P_{CO} = k_1/(k_1+k_2)$; $P_O = 1 - P_{CO}$).

The bifurcation diagram of the ZGB model shows that the steady state adsorbate undergoes two kinetic phase transitions. Namely a first order (i.e. sudden) transition at $P_{CO} = P_1$ from the maximum reactive state to the poisonning of the surface by **CO** and a second order (i.e. continuous) transition at $P_{CO} = P_2$ where the pure **O** adsorbate changes into a mixed adsorbate. Therefore the reaction proceeds only in the reaction window: $P_2 < P_{CO} < P_1$.

Microscopic Aspects of Nonlinearity in Condensed Matter
Edited by A.R. Bishop *et al.*, Plenum Press, New York, 1991

The main controversial problem about these models concerns the second order phase transition at the finite value $P_{CO} = P_2$. To our knowledge this second order phase transition has never been observed: the reaction rate begins to increase as soon as P_{CO} departs from 0.

At first sight, this second order transition appears as an artefact of the model; namely, the surface is assumed to be made of active sites (of the same kind) which are invariably compatible for the adsorption of both gaseous reactants. In this case the saturation of the surface with **O** necessarily precludes the adsorption of **CO**. Therefore if the Langmuir-Hinshelwood mechanism (2c) is the unique reaction mechanism, it is normal that for low values of P_{CO} the steady state reaction rate vanishes. It is also obvious that if the Eley-Rideal mechanism

$$CO^{(g)} + O \rightarrow CO_2{}^{(g)} + *.$$

is also involved in the model this second order transition desappears (i.e. $P_2 \rightarrow 0$)[5].

The goal of this work is to show that this annihilation of the second order transition may also be obtained within a ZGB-like model involving local interactions between the adatoms. Ehsasi et al.[3] have shown how this bifurcation diagram is modified when desorption and diffusion processes are also involved in the ZGB model. Kaukonen et al.[4] have extended the model by considering a finite reaction probability. The qualitative results obtained within this reduced approach based on probabilities are equivalent to those obtained on the basis of kinetic constants[5].

In section 2 we define our interaction model. In section 3 we decode the probability of occurence of each elementary step in terms of the corresponding kinetic constant.

The results of the simulations are presented in section 4. By introducing repulsive (attractive) local interactions between adatoms of the same (opposite) type we show how the second order transition anneals. Some uniformity tests are used to investigate how local interactions modify the correlation length[6] between adspecies on the surface.

Table 1. Energies and transition probabilities of the various pairs.

pair	energy	possible transitions	probability
_	0	CO_*	$P_{11}=\frac{1}{2}P_{CO}\min(W,1)$
		*_CO	$P_{12}=\frac{1}{2}P_{CO}\min(W,1)$
		O_O	$P_{13}=P_O\min(W,1)$
		_	$1-P_{11}-P_{12}-P_{13}$
CO_* (*_CO)	0	CO_CO	$P_{21}=\frac{1}{2}P_{CO}\min(W,1)$
		*_CO	$P_{22}=\frac{1}{2}D\min(W,1)$
		_	$P_{23}=\frac{1}{2}R_d\min(W,1)$
		CO_*	$1-P_{21}-P_{22}-P_{23}$
O_* (*_O)	0	O_CO	$P_{31}=\frac{1}{2}P_{CO}\min(W,1)$
		O_*	$1-P_{31}$
CO_CO	J_1	CO_*	$P_{41}=\frac{1}{2}R_d\min(W,1)$
		*_CO	$P_{42}=\frac{1}{2}R_d\min(W,1)$
		CO_CO	$1-P_{41}-P_{42}$
O_O	J_2	O_O	1
CO_O (O_CO)	J_3	*_*	P_r
		CO_O	$1-P_r$

2. THE MODEL

In this model each site of the square lattice is either empty (*) or filled with a **O** atom or a **CO** molecule. The oxidation reaction can occur only between adsorbed **CO** and **O**. The adsorption mechanism of **CO** involve a single empty site while the adsorption of O_2 requires a pair of empty *nn* sites. The model is based on a local probabilistic dynamic, using as building blocks the bonds between two adjacents sites (cells) that are updated at a once. The probability of the transition of a pair is determined as a function of the neighborhood. Each transition probability is modulated by a Boltzmann factor *W* depending on the variation of the local configuration energy *H* . The energy of a given configuration is obtained by summing the energy of all the bonds of the central pair with the cells in the neighborhood (seven bonds).

We have six types of pairs, with the energies given in Table 1. The evolution of the pairs is determined by the transition probabilities also reported in Table 1.

In the table R_d stands for the **CO** desorption probability and D for the **CO** diffusion (hopping) probability. As the model is irreversible the former transition probabilities do not completely satisfy the detailed balance principle. In this Monte Carlo simulation when bonds are updated, the new connections are checked. If a **CO_O** pair is found, it is submitted to the next Monte Carlo trial. Note that in a simple Monte Carlo scheme the presence of **CO_O** pairs on the lattice, even if $P_r = 1$, lowers the effective reaction rate. The unit of time corresponds to a cycle of simulation in which each site is visited once, in the average.

3. TRANSITION PROBABILITIES AND KINETIC RATE CONSTANTS

In order to compare our results with experiment we have to link the transition probabilities P_i with the kinetic rate constants k_i involved in the model. The kinetic constant k_3 for the reaction mechanism (2c) is the reciprocal of the reaction time τ_3 , i.e. the mean time spent by the pair **CO_O** on the surface.Using a probabilistic approach to the problem and using a discrete time, we can define a finite probability of transition for the pair **CO_O** obtaining a Markov chain with probability *p* of having the reaction in one time step. The probability that the reaction takes place exactly at time t is $P(t) = p\,(1-p)^t$, and the mean reaction time is $\tau = \sum t\,P(t) = 1/p$.

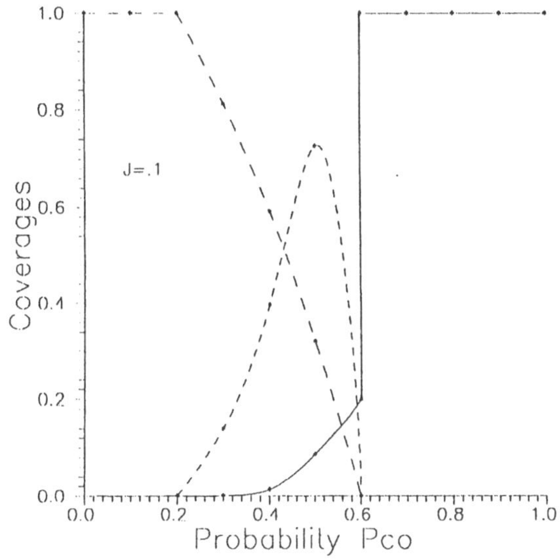

Fig. 1. Phase diagram corresponding to a coupling $J = 0.1$. The solid line represents the **CO** steady coverage, the upmost dashed line the **O** steady coverage, and the middle dashed line the mean CO_2 production.

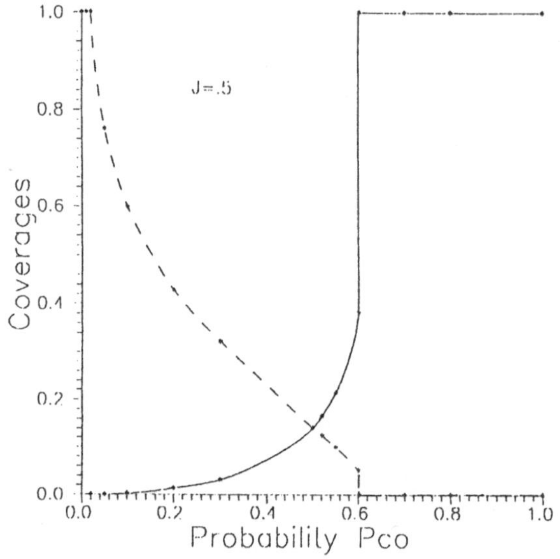

Fig. 2. Phase diagram corresponding to a coupling $J = 0.5$. The solid line represents the **CO** steady coverage, the dashed line the **O** steady coverage.

In this way we have $k = \alpha p$, where α is a constant that relates the Monte Carlo time with the experimental time. With a continuous time, the probability of having the reaction before time t follows a Poisson law:

$$p(t) = 1 - \exp(-k\,t).$$

The simple result $k = \alpha p$ is no longer valid when there are several competing processes. What is still to be done is to measure the effective transition probabilities for the competing reactions (and possibly to elaborate theoretical formulas for them) so to link the Monte Carlo time with the real time, and to confront the behavior of the model with the experimental results.

4. RESULTS

In these preliminar simulations we used a lattice of 50x50 cells, with periodic boundary conditions. The desorption probability R_d and the hopping probability D have been fixed to zero, while the reaction probability P_r was always 1. We also used $|J_1| = |J_2| = |J_3| = J$, with $J_3 = -J_1$. The temperature kT was arbitrarily fixed to 1, and the coupling energies were set to $J = 0, 0.1, 0.5, 1$. The measures were taken when the evolution seemed stabilized, in any case after 500 MCS.

With a coupling $J = 0$ we got the results of Ziff (and not those of Kaukonen), i.e. a second order phase transition for $P_{CO} \approx 0.38$ and a first order phase transition for $P_{CO} \approx 0.52$.

When the coupling increases the second order transition point P_2 goes to zero, and the first order transition point P_1 slightly increases (see fig. 1-2). For a coupling $J \approx 1.$ the second order transition point P_2 corresponds to the origin.

To test such approximation we measured the randomness of the distribution of the species on the lattice. The conclusions are that the correlation length is increasing near the first order transition point ($P_{CO} \rightarrow P_1$).

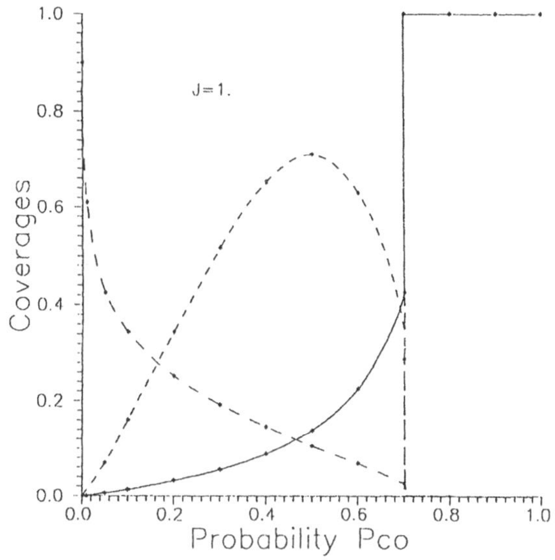

Fig. 3. Phase diagram corresponding to a coupling $J = 1$. Symbols as in fig. 1.

Other statistical tests confirm that the positions of the atoms of given species are strongly correlated just before the first order transition point (for all the couplings) and that for each value of P_{CO} considered the repartition of each species seems to be geometrically uniform.

REFERENCES

1. P.Dufour, M. Dumont, V. Chabart and J. Lion, Comp. Chem. 13: 25 (1989).
2. R. M. Ziff, E. Gulari and V. Barshad, Phys. Rev. Lett. 56: 2553 (1986).
3. M. Ehsasi, M. Matloch, O. Frank and J. H. Block, J. Chem. Phys. 91: 4949 (1989).
4. H. P. Kaukonen and R. M. Nieminen, J. Chem. Phys. 91: 4380 (1989).
5. M. Dumont, P. Dufour, B. Sente and R. Dagonnier, J. Cat. 122: 95 (1990).
6. M. Kolb and Y. Boudeville, J. Chem. Phys. 92: 3935 (1990).

CHAOS IN THE DRIVEN DAMPED PENDULUM STUDIED BY A NONEQUILIBRIUM – TO – EQUILIBRIUM MAPPING

Mads Peter Soerensen

Laboratory of Applied Mathematical Physics
The Technical University of Denmark, Bldg. 303
DK–2800 Lyngby, Denmark

INTRODUCTION

A nonequilibrium–to–equilibrium mapping has been proposed by Eykholt et.al.[1] to analyse the dynamical behaviour of nonlinear damped and driven systems. Particular emphasis is put on space–time complexity and qualitative prediction of routes to chaos. In the above method a stochastic Langevin force is added to the nonlinear system and the distribution of the solution can be determined on the form of a Boltzmann factor. Minimizing the effective Hamiltonian in the Boltzmann factor gives the most probable solution. In reference 1 the distribution of the solution is found from the distribution of the stochastic force by employing a fairly straightforward method. However, this method involves determinants which may be very hard to calculate. In the following it will be shown, how the analysis in reference 1 can be performed in a little less straightforward manner but without very demanding algebra. This later approach will be illustrated by applying it to the driven damped pendulum.

THE DRIVEN DAMPED PENDULUM WITH NOISE

In normalized units the sinusoidal driven damped pendulum with a Langevin force is governed by

$$\begin{aligned} \dot{\psi} &= \phi \\ \dot{\phi} &= -\epsilon\phi - U_{\psi} + \zeta(t) \end{aligned} \tag{1}$$

Here $\psi = \psi(t)$ is the angle from vertical at time t. The overdot indicates derivative with respect to time, and ϵ is the constant damping parameter. In order to shorten the notation the potential $U = U(\psi,t)$ has been introduced where $U(\psi,t) = 1 - \cos\psi - F(t)\psi$ and $U_\psi = \partial U/\partial\psi$. F(t) is an external time dependent driving force. The stochastic force $\zeta(t)$ is Gaussian distributed with the mean value $<\zeta(t)> = 0$ and correlation $<\zeta(t)\ \zeta(t')> = \sigma\ \delta(t-t')$. $< \cdots >$ denotes ensemble average.

Below we shall map the non-equilibrium system (1) onto an equilibrium Hamiltonian system. Following reference 2 this is done by determining the Fokker–Planck equation for the conditional probability function $P_c\ (\phi,\psi,t\,|\,\phi',\psi',t')$ for finding (ϕ,ψ) at time t when (ϕ,ψ) sharply equals (ϕ',ψ') at time $t = t'$. Then expanding P_c for small times and write the distribution $P(\phi,\psi,t)$ in terms of a path integral. The Fokker–Planck equation is determined from the Kramers–Moyal coefficients[2]

$$D_\psi^{(1)} = \frac{1}{1!} \lim_{\Delta t \to 0} \frac{<\Delta\psi>}{\Delta t} = \phi \tag{2a}$$

$$D_\phi^{(1)} = \frac{1}{1!} \lim_{\Delta t \to 0} \frac{<\Delta\phi>}{\Delta t} = -\epsilon\phi - U_\psi \tag{2b}$$

$$D_{\phi\phi}^{(2)} = \frac{1}{2!} \lim_{\Delta t \to 0} \frac{<(\Delta\phi)^2>}{\Delta t} = \frac{\sigma}{2} \quad , \tag{2c}$$

where $\Delta\psi$ and $\Delta\phi$ are found from Eq. (1) by integration from t to $t + \Delta t$ assuming Δt to be so small that (ϕ,ψ) are practically constant, but so big that the stochastic force $\zeta(t)$ exerts numerous impacts in the time interval $[t,t + \Delta t]$. The Kramers–Moyal coefficients $D_{\phi\psi}^{(2)}$ and $D_{\psi\psi}^{(2)}$ vanish. The same is valid for coefficients of order $n \geq 3$ due to the Gaussian distribution of $\zeta(t)$. In reference 2 it is shown that P_c satisfies the Fokker–Planck equation.

$$\frac{\partial P_c}{\partial t} = L_{KM}\ P_c = -\phi \frac{\partial P_c}{\partial \psi} + \epsilon P_c + (\epsilon\phi + U_\psi) \frac{\partial P_c}{\partial \phi} + \frac{\sigma}{2} \frac{\partial^2 P_c}{\partial \phi^2} \tag{3}$$

where the Kramers–Moyal operator L_{KM} is given by

$$L_{KM} = -\frac{\partial}{\partial\psi} D_\psi^{(1)} - \frac{\partial}{\partial\phi} D_\phi^{(1)} + \frac{\partial^2}{\partial\phi^2} D_{\phi\phi}^{(2)} \quad . \tag{4}$$

Also the distribution function $P(\phi,\psi,t)$ satifies Eq. (3).

The small time expansion of P_c reads[2]

$$P_c(\phi,\psi,t+\Delta t|\phi',\psi',t) = (1 + L_{KM}\,\Delta t + O(\Delta t^2))\,\delta(\phi-\phi')\,\delta(\psi-\psi')$$

$$\simeq \exp(L_{KM}\,\Delta t)\,\frac{1}{(2\pi)^2}\int_{-\infty}^{\infty}\int_{-\infty}^{\infty}\exp\{iy_\phi(\phi-\phi') + iy_\psi(\psi-\psi')\}\,dy_\phi\,dy_\psi$$

$$= (2\pi\sigma\,\Delta t)^{-1/2}\,\frac{1}{\Delta t}\,\delta\left(\frac{\psi-\psi'}{\Delta t}-\phi\right)\exp\left\{-\frac{\Delta t}{2\sigma}\left[\frac{\phi-\phi'}{\Delta t}+\epsilon\phi+U_{\psi}\right]^2+\epsilon\Delta t\right\}. \tag{5}$$

By discretizing the time interval [0;t] into N time steps of length $\Delta t = t/N$ and defining $t_n = n\,\Delta t$, $\phi_n = \phi(t_n)$, and $\psi_n = \psi(t_n)$ the distribution $P(\phi,\psi,t)$ can be expressed by the path integral

$$P(\phi,\psi,t) = \int_{-\infty}^{\infty} d\phi_{N-1}\cdot\ldots\cdot\int_{-\infty}^{\infty} d\phi_0\int_{-\infty}^{\infty} d\psi_{N-1}\cdot\ldots\cdot\int_{-\infty}^{\infty} d\psi_0\,\cdot$$

$$P_c(\phi,\psi,t|\phi_{N-1},\psi_{N-1},t_{N-1})\cdot\ldots\cdot P_c(\phi_1,\psi_1,t|\phi_0,\psi_0,0)\,P(\phi_0,\psi_0,0) \tag{6}$$

where $P(\phi_0,\psi_0,0)$ is the initial condition. Insertion of the small time expression in (5) into (6) then gives

$$P(\phi,\psi,t) = \int_{-\infty}^{\infty} d\phi_{N-1}\cdot\ldots\cdot\int_{-\infty}^{\infty} d\phi_0\int_{-\infty}^{\infty} d\psi_{N-1}\cdot\ldots\cdot\int_{-\infty}^{\infty} d\psi_0\,\cdot$$

$$(2\pi\sigma\Delta t)^{-N/2}\,e^{\epsilon t}\prod_{n=1}^{N-1}\delta\left[\frac{\psi_{n-1}-\psi_n}{\Delta t}-\phi_n\right]\cdot$$

$$\exp\left\{-\frac{\Delta t}{2\sigma}\sum_{n=0}^{N-1}\left[\frac{\phi_{n+1}-\phi_n}{\Delta t}+\epsilon\phi_n+U_{\psi_n}\right]^2\right\}P(\phi_0,\psi_0,0)\quad. \tag{7}$$

Integrating out the delta function in (7) and letting $\Delta t\to 0$ we get in the continuum limit for the kernel P_k of the integral in (17)

$$P_k \sim \exp\left\{-\frac{1}{2\sigma}\int_0^t(\ddot\psi+\epsilon\dot\psi+U_{\psi})^2\,d\tau\right\} = \exp\{-\beta\mathcal{H}\} \tag{8}$$

where we have introduced $\beta = 2\epsilon/\sigma$ and the effective Hamiltonian

$$\mathcal{H} = \frac{1}{2\epsilon}\left[\frac{1}{2}\dot\psi^2+\dot\psi(\sin\psi-F)-\epsilon\cos\psi-\epsilon\psi F+\psi\dot F\right]_0^t$$

$$+\frac{1}{4\epsilon}\int_0^t(\ddot\psi^2+\epsilon^2\dot\psi^2-2\dot\psi^2\cos\psi+\sin^2\psi$$

$$-2F\sin\psi+2\epsilon\dot F\psi-2\ddot F\psi)\,d\tau\quad. \tag{9}$$

In Eq. 9 $U(\psi,t) = 1 - \cos\psi - F\psi$ has been used. The most probable solution $\psi(t)$ of Eq. (1) is that function which minimizes the effective Hamiltonian (i.e. maximize P_k).

Let us now see how the Hamiltonian in (9) can be interpreted following the ideas of reference 1. With $F = \rho_1 \sin(\Omega t)$ the Melnikov–function technique has been applied to Eq. (1) in reference 3 in order to determine the threshold for chaos in the parameter plane (ρ_1,Ω). Fig. 2 in reference 3 shows that above a monotonically increasing function $\rho_1^M(\Omega)$ of Ω, Eq. (1) may exhibit chaos and below no chaotic dynamics appear. In order to fix scales we note that $\rho_1^M(\Omega = 0) \approx 0.2$ and $\rho_1^M(\Omega = 2) \approx 3.0$. Considering only the terms under the integral in Eq. (9) we see that the fifth, sixth, and seventh terms simply lock ψ to the driver. The first, second, and fourth terms tend to damp out any temporal complexity, favouring regular and periodic oscillations. The third term, however, favours enhanced complexity in time, i.e. chaotic dynamics. By increasing ρ_1 at a given drive frequency Ω the third term will lead to larger temporal complexity resulting in chaotic dynamics as predicted by the Melnikov–function technique. That the threshold $\rho_1^M(\Omega)$ is monotonically increasing as function of Ω can also be understood from Eq. (9). Increasing Ω will make the damping terms 1 and 2 in Eq. (9) more important and thereby discourage temporal complexity.

SUMMARY

In reference 1 a mapping from a system of nonlinear ordinary differential equations to an effective Hamiltonian system has been developed for analyzing nonlinear dynamics. Here an alternative approach to the same mapping has been presented which facilitate the analytic computations, illustrated by applying the method to the damped driven pendulum.

REFERENCES

1. R. Eykholt, A.R. Bishop, P.S. Lomdahl, and E. Domany, A Nonequilibrium to – Equilibrium Mapping and Its Application to the Perturbed Sine–Gordon Equation, Physica 23D:102 (1986).

2. H. Risken, "The Fokker–Planck Equation", Springer Verlag, Berlin (1984).

3. M. Bartuccelli, P.L. Christiansen, N.F. Pedersen, and M.P. Soerensen, Prediction of Chaos in a Josephson Junction by the Melnikov–Function Technique, Phys. Rev. B33:4686 (1986).

BIPOLARONIC CHARGE DENSITY WAVES

Serge AUBRY

Laboratoire Léon Brillouin[1]
CEN Saclay, 91191-Gif-sur-Yvette Cedex, France

It is well-known that the ground-state of models with electron-phonon coupling, may be either a charge density wave (CDW) or a superconductor (SC). For describing CDWs, standard theories require low dimensional models and consider that the Born-Oppenheimer approximation is valid. Then, in the incommensurate case and with the help of some extra approximations, the existence of strictly gapless phonons called phasons is well admitted. However, some years ago, we predicted on the basis of numerical calculations that CDWs with phason modes with a non-zero gap could exist[1,2]. More recently, we noted that because of a tunnelling energy gain allowed by the quantum lattice fluctuations, phase defect energies of CDWs could become negative below the transition by breaking of analyticity when the phason gap is zero or very small[2,3,4]. Thus, we suggested a conclusion which is very controversial, that CDWs with a strictly zero phason gap should be always unstable against quantum lattice fluctuations. Although the other part of our talk dealing with the concept of anti-integrability is also relevant for understanding bipolaronic CDWs, the reader is refereed to existing publications[5,6] and in preparation[7]. In this short proceeding, we prefer to discuss the most controversial part of this problem and to present new arguments supporting this conjecture. We first briefly describe the theorem which predicts for large enough electron-phonon coupling, the existence of bipolaronic chaotic states for the Holstein model at any dimension, in the adiabatic limit. The ground-state of this model is thus an ordered insulating bipolaronic structure (CDW) with a strictly non zero phason gap. Early numerical observations in one dimension are thus confirmed and extended[1,2]. Next, we investigate the correction due to the quantum lattice fluctuations and we <u>prove</u> that if the adiabatic ground-state has a strictly zero phason gap, the description of this solution within the Born-Oppenheimer approximation is not consistent for the low frequency phason modes. This exact result sharply contradicts the

[1] Laboratoire Commun CEA-CNRS

Microscopic Aspects of Nonlinearity in Condensed Matter
Edited by A.R. Bishop *et al.*, Plenum Press, New York, 1991

usual beliefs in this field. Although we are not able to describe properly the ground-state in that case, this result seems to corroborate our earlier conjecture that CDWs with strictly zero gap cannot exist. We suggest that the real ground-state could be then a superconducting state.

The Holstein model, is likely the simplest model which exhibits the physical features, we want to understand. But emphasize that although for sake of simplicity, our study is presently focused on this specific model, it will be extendable to more complex models in further works. Let us recall our usual notations: the Holstein Hamiltonian is the sum of three terms

(1) $$H = H_k + H_{ep} + H_p$$

H_k is the Hamiltonian of a single band of non-interacting electrons (within a tight binding approximation).

(2-a) $$H_k = -T \sum_{\langle i,j\rangle, \sigma} c^+_{i,\sigma} c_{j,\sigma}$$

where T is the exchange constant between neighboring sites <i,j> on a d-dimensional square lattice and σ is the electron spin $\pm\frac{1}{2}$ noted $\uparrow$ or $\downarrow$. $c^+_{i,\sigma}$ and $c_{i,\sigma}$ are the creation and annihilation fermion operators of an electron at site i with spin σ respectively. The band width is thus 4 T d. This electron band is neither full or empty and its filling is a fixed arbitrary number (rational or irrational). H_p is the Hamiltonian of quantum phonons corresponding to a dispersionless optical branch

(2-b) $$H_p = \sum_i \hbar\omega_0 \left(a^+_i a_i + \frac{1}{2} \right)$$

where a^+_i and a_i are the creation and annihilation boson operators of phonons at site i respectively. The on-site electron-phonon interaction with constant g is represented by the Hamiltonian

(2-c) $$H_{ep} = g \sum_i n_i \left(a^+_i + a_i \right)$$

where n_i is the electronic density operator at site i

(2-d) $$n_i = c^+_{i,\uparrow} c_{i,\uparrow} + c^+_{i,\downarrow} c_{i,\downarrow}$$

It is convenient to define two dimensionless independent parameters which are:

(3-a) $$\gamma = \frac{\hbar\omega_0}{T}$$

which measures the "quantum character" of the phonons and

(3-b) $$k = \frac{2g}{\sqrt{\hbar\omega_0 T}}$$

which is the reduced electron-phonon coupling constant in the adiabatic (classical phonon) limit. Setting, the position operator

(4-a) $$u_i = \frac{\sqrt{\gamma}}{2} (a_i^+ + a_i)$$

and its conjugate operator (the commutator is $[u_i, p_i] = i$)

(4-b) $$p_i = \frac{i}{\sqrt{\gamma}} (a_i^+ - a_i)$$

the initial Hamiltonian (1) becomes in unit of energy 2T:

(5) $$\hat{H} = \frac{H}{2T} = -\frac{1}{2} \sum_{<i,j>\sigma} c^+_{i,\sigma} c_{j,\sigma} + \frac{k}{2} \sum_i n_i u_i + \frac{1}{2} \sum_i (u_i^2 + \frac{\gamma^2}{4} p_i^2)$$

The adiabatic (or mean field) approximation consists in neglecting the lattice fluctuations or equivalently the kinetic energy $\frac{1}{2} \frac{\gamma^2}{4} p_i^2$ of the atoms in the Hamiltonian (5). Then, operator u_i which commutes with the Hamiltonian $\hat{H}$ can be considered as a scalar variable so that , the electronic degrees of freedom can be eliminated and replaced by an effective potential acting on the lattice. The eigenstates $\{\Psi^i_\mu\}$ of the electrons fulfill the eigen equation

(6-a) $$- \sum_{j:<i,j>} \Psi^j_\mu + k\, u_i \Psi^i_\mu = E_\mu(\{u_i\}) \Psi^i_\mu$$

where the eigenenergies $E_\mu(\{u_i\})$ depends on the variables $\{u_i\}$. (The sum in (6-a) $\sum_{j:<i,j>}$ is done for the nearest neighboring sites j of i). We set

(6-b) $$c^+_{\mu,\sigma} = \sum_i \Psi^i_\mu \, c^+_{i,\sigma}$$

In the ground-state, the eigenstates μ with energy smaller $E_\mu(\{u_i\})$ than the Fermi energy E_F are occupied by two electrons with opposite spins. When the electrons are traced out, Hamiltonian (5) becomes a function $\Phi(\{u_i\}))$ of the scalar variables $\{u_i\}$

(6-c) $$\Phi(\{u_i\})) = \frac{1}{2} \sum_i u_i^2 + \sum_{E_\mu<E_F} E_\mu(\{u_i\})$$

When this approximation is valid, the ground-state of the Holstein model is obtained as the atomic configuration $\{\mathbf{u}_i\}$ which minimizes the effective energy (6-c). This variational form has been studied using the new concept of "anti-integrability".

When the kinetic energy of the electron is neglected (T=0) in Hamiltonian (1), the electron density operator n_i commutes with the Hamiltonian. An eigen-state of this Hamiltonian can be associated with any electronic distribution $n_{i,\uparrow}=c^+_{i,\uparrow}c_{i,\uparrow}= 0$ or 1 and $n_{i,\downarrow}=c^+_{i,\downarrow}c_{i,\downarrow}= 0$ or 1 which corresponds to the given band filling. The ground-state of this Hamiltonian is degenerate (because the phonon branch is dispersionless) and correspond to an arbitrary distribution of electron pairs on the lattice $n_i = 0$ or 2 which is naturally coded as a pseudo-spin configuration $n_i = 2\sigma_i$ with $\sigma_i = 0$ or 1. An electron pair associated with its lattice distortion is called a bipolaron.

For the variational form (6-c), the limit $k \to \infty$, is an "anti-integrable" limit characterized by the fact that the existence of chaotic metastable states is trivial. Then, we can prove rigorously[7,8] that any of these bipolaronic states which exists at $k=\infty$, is a continuous function of the parameter k for k large enough. Thus, giving an arbitrary pseudospin configuration $\{\sigma_i\}$, there exists an eigenstates of the adiabatic Holstein Hamiltonian, such that the electronic density at site i become close to $2\sigma_i$ for $k \to \infty$. In addition, it is proven that these states are insulating and have a non-vanishing phonon gap.

In other words, this result means that one can choose arbitrarily the distribution of the electrons as if they were classical particles and that providing the electronic quantum term H_k in the Hamiltonian be small enough, this eigen state relaxes around the initial distribution and survives to the electronic quantum fluctuations! For the adiabatic Holstein model in d dimensions, we found the exact bound $k>4.76\sqrt{d}$ although numerical simulations in one dimension suggests that the real bound should be smaller roughly by a factor of 2.The proof of this theorem is based on the same ideas as in ref.5 although it becomes technically much more complex. Its publication in a general form [9] is still in preparation although a 1-d version of this theorem can be found in ref.8.

The energy of these bipolaronic states depends on the pseudo-spin configuration. In 1-d models, numerical simulations show that the ground-state does belong to this class of insulating bipolaronic states. Then, it is an incommensurate CDW with the wave-vector $2k_F$ expected from the standard theory of Peierls instability. However, this CDW state does not exhibit any Frohlich conductivity and is insulating. It is in fact to an ordered incommensurate array of bipolarons with physical properties which are quite different from those of the standard CDWs described in the literature. We call this state "Bipolaronic Charge Density Wave". In model with several dimensions, the ground-state of the adiabatic Holstein model, is also a bipolaronic configuration but finding it, is still an open problem.

When k decreases, the numerical analysis of ref.9 suggests that these chaotic bipolaronic states merge together through complex cascades of bifurcations in a similar way as for the Frenkel-Kontorowa model[5b]. In one dimension, the adiabatic Holstein model with incommensurate band filling, exhibits a Transition by Breaking of Analyticity[1,2,9] at which the ground-state becomes the only metastable state of the CDW (and is phase undefectible). Then, a gapless phason mode is restored. We expect a similar behavior for the adiabatic Holstein models at any dimensions but up to now, no numerical simulations were done.

In summary, it is now proven rigorously that there exists bipolaronic CDWs with non-zero phason gap in the adiabatic Holstein model. As we will see in the next, this property will become essential for insuring the stability of the CDW against the quantum lattice fluctuations.

For studying the effect of quantum lattice fluctuation on the adiabatic ground-states, it is convenient to introduce the unitary transformation exp(i **S**) which diagonalizes the fermionic part of the Hamiltonian (5) applied to the whole Hamiltonian. **S** is a self-adjoint operator which has the form

$$\text{(7-a)} \qquad \mathbf{S} = \sum_{n,m,\sigma} s_{n,m}(\{u_i\})\; c^{+}_{n,\sigma}\, c_{m,\sigma}$$

and which fulfills by definition

$$\text{(7-b)} \qquad \exp(i\,\mathbf{S})\left(-\sum_{\langle i,j\rangle\,\sigma} c^{+}_{i,\sigma}\,c_{j,\sigma} + k\sum_{i} n_i\,u_i\right) p_i\;\exp(-\,i\,\mathbf{S}) = \sum_{\mu,\sigma} E_{\mu}(\{u_i\})\,c^{+}_{\mu,\sigma}\,c_{\mu,\sigma}$$

The global action of this unitary transformation has been exactly calculated in the general case by M.Wagner[10]. Four our specific model, it comes out after some tedious calculations.

$$\text{(7-c)} \qquad \exp(i\,\mathbf{S})\;\hat{H}\;\exp(-\,i\,\mathbf{S}) = \frac{1}{2}\sum_{\mu,\sigma} E_{\mu}(\{u_i\})\,c^{+}_{\mu,\sigma}\,c_{\mu,\sigma} + \sum_{i}\frac{1}{2}\left(u_i^{2} + \frac{\gamma^2}{4}\,\tilde{p}_i^{\,2}\right)$$

where

$$\text{(7-d)} \qquad \tilde{p}_i = \exp(i\,\mathbf{S})\;p_i\;\exp(-\,i\,\mathbf{S}) = p_i + i\,k\sum_{\mu\neq\mu',\sigma}\frac{\psi^{\mu'}_i\,\psi^{\mu*}_i}{E_{\mu'}-E_{\mu}}\;c^{+}_{\mu,\sigma}c_{\mu',\sigma}$$

The Born-Oppenheimer approximation which improves the adiabatic approximation, consists in assuming that $\tilde{p}_i = p_i$. Then, by tracing out the fermion operators, the Born-Oppenheimer Hamiltonian becomes

$$\text{(8-a)} \qquad H_{BO} = \Phi(\{u_i\}) + \frac{1}{2}\sum_{i}\frac{\gamma^2}{4}\,p_i^2$$

In general, this Hamiltonian is not solvable. The renormalized frequencies of the phonons are usually obtained by expanding $\Phi(\{u_i\})$ at second order around the adiabatic ground-state $\{\mathbf{u}_i\}$ with respect to $\{\varepsilon_i\} = \{u_i - \mathbf{u}_i\}$

$$\text{(8-b)} \qquad \Phi(\{u_i\}) \cong \Phi(\{\mathbf{u}_i\}) + \frac{1}{2}\sum_{n,m}\frac{\partial^2\Phi(\{\mathbf{u}_i\})}{\partial u_n\partial u_m}\,\varepsilon_n\,\varepsilon_m$$

It comes out readily by differentiation of (6-c)

$$\text{(8-c)} \qquad C_{n,m} = \frac{\partial^2\Phi(\{\mathbf{u}_i\})}{\partial u_n\partial u_m} = \delta_{n,m} - k^2\sum_{\mu\ \mathrm{occ},\mu'\ \mathrm{unocc}}\frac{\psi^{\mu\,*}_{n}\,\psi^{\mu'}_{n}\,\psi^{\mu'*}_{m}\,\psi^{\mu}_{m}}{E_{\mu'}-E_{\mu}} + \mathrm{C.C}$$

C.C means the complex conjugate of the previous term. μ occ means the occupied electronic states given by eq.6-a calculated for the adiabatic ground-state $\{\mathbf{u}_i\}$ with $E_\mu < E_F$. Similarly, μ' unocc means the unoccupied electronic states with $E_F < E_{\mu'}$. Thus, $E_{\mu'} - E_\mu$ is always positive.

Since $\{\mathbf{u}_i\}$ is a minima of $\Phi(\{\mathbf{u}_i\})$, this quadratic form with matrix $\bar{\bar{C}}$ is positive. In addition, it is easily proven that the quadratic form defined by matrix $\bar{\bar{1}} - \bar{\bar{C}}$, is also positive. Thus, $\bar{\bar{C}}$ has the normalized eigen modes $|\nu\rangle = \{\alpha_i^\nu\}$ with eigen values $0 \leq \lambda_\nu^2 \leq 1$

(9-a) $$\bar{\bar{C}} \, |\nu\rangle = \lambda_\nu^2 \, |\nu\rangle$$

Expanding $|\varepsilon\rangle = \{\varepsilon_i\}$ and $|p\rangle = \{p_i\}$ on this basis, we define the new phonon operators

(9-b) $$\upsilon_\nu = \langle \nu | \varepsilon \rangle = \sum_i \varepsilon_i \, \alpha_i^{\nu *}$$

and

(9-c) $$p_\nu = \langle \nu | p \rangle = \sum_i p_i \, \alpha_i^\nu$$

with commutator

(9-d) $$[\upsilon_\nu, p_{\nu'}] = i \, \delta_{\nu,\nu'}$$

which diagonalizes the approximate Born-Oppenheimer Hamiltonian

(10-a) $$H_{BO} \equiv \Phi(\{\mathbf{u}_i\}) + \sum_\nu \frac{1}{2} (\lambda_\nu^2 \, \upsilon_\nu^* \, \upsilon_\nu + \frac{\gamma^2}{4} p_\nu^* p_\nu)$$

Each Hamiltonian

(10-b) $$H_\nu = \frac{1}{2} (\lambda_\nu^2 \, \upsilon_\nu^* \, \upsilon_\nu + \frac{\gamma^2}{4} p_\nu^* p_\nu)$$

corresponds to an harmonic oscillator with ground-state energy $\frac{\gamma \lambda_\nu}{4}$ and

(10-c) $$\langle p_\nu^* p_\nu \rangle = \frac{\lambda_\nu}{\gamma}$$

Thus, within the Born-Oppenheimer approximation, we find the well-known result that the bare phonon frequencies ω_0 of the mode are renormalized and soften. The renormalized phonons have the frequency spectrum $\omega_\nu = \lambda_\nu \, \omega_0$. They were calculated numerically in ref.1-b for the ground-state of the 1-d Holstein model

For proving the consistency of the Born-Oppenheimer approximation, and thus for proving that each phonon ν obtained by this calculation is obtained within a reasonably good approximation, one has to check that the energy contribution of the anti-adiabatic terms $\tilde{p}_i - p_i$ which were neglected in (7-d) are really negligible compared to the phonon energy. Treating these terms as a perturbation, one finds for mode ν, that the perturbation to the momentum p_ν which is :

(11-a) $$\tilde{p}_\nu - p_\nu = i \, k \sum_{\mu \neq \mu', n, \sigma} \frac{\alpha_n^\nu \psi_n^{\mu'} \psi_n^{\mu *}}{E_{\mu'} - E_\mu} c_{\mu,\sigma}^+ c_{\mu',\sigma}$$

has the square fluctuation

$$(11\text{-b}) \qquad f_\nu = <(\tilde{p}_\nu - p_\nu)^2> = 2\,k^2 \sum_{\mu\ \mathrm{occ},\ \mu'\ \mathrm{unocc}} \frac{\left(\sum_n \alpha_n^\nu \psi_n^{\mu'} \psi_n^{\mu*}\right)^2}{(E_{\mu'}-E_\mu)^2}$$

Since $E_{\mu'}-E_\mu$ has an upper bound (the maximum renormalized band width) which fulfills $0 < E_{\mu'}-E_\mu \leq k^2 + 4\,d$ where d is the dimension of the square lattice, we find the exact bound

$$(11\text{-c}) \qquad f_\nu \geq \frac{2\,k^2}{k^2 + 4\,d} \sum_{\mu\mathrm{occ},\mu'\mathrm{unocc}} \frac{\left(\sum_n \alpha_n^\nu \psi_n^{\mu'} \psi_n^{\mu*}\right)^2}{E_{\mu'}-E_\mu}$$

$$= \frac{1}{k^2+4d} <\nu|\,\bar{\bar{1}} - \bar{\bar{C}}\,|\nu> = \frac{1-\lambda_\nu^2}{k^2+4d}$$

Then the consistency of Born-Oppenheimer calculation of this phonon with frequency ν requires that the mean square fluctuation $f_\nu = <(\tilde{p}_\nu - p_\nu)^2>$ be much smaller than $<p_\nu^* p_\nu> = \frac{\lambda_\nu}{\gamma}$ which yields the essential condition

$$(12) \qquad \boxed{\frac{\gamma}{k^2+4d} << \frac{\lambda_\nu}{1-\lambda_\nu^2}}$$

If this condition is fulfilled, the phonon mode ν is very weakly coupled to the fermions and thus remains well defined. If not, due to the "anti-adiabatic correction", this phonon is strongly coupled to the electrons and the Born-Oppenheimer approximation should break down.

There are now two possible situations

1- The renormalized phonons of the adiabatic solution have a finite gap which means that $0 \neq \lambda = \mathrm{Inf}_\nu\, \lambda_\nu$. Then, when γ becomes small enough: $\gamma << \frac{\lambda(k^2+4d)}{1-\lambda^2}$, the condition for the validity of the Born-Oppenheimer approximation becomes always fulfilled in the adiabatic limit. This situation occurs for the bipolaronic chaotic states which were predicted to exist for k large enough. Thus, we conclude that bipolaronic chaotic states physically exists not only at the adiabatic limit but in some domain close to it and are well described within the Born-Oppenheimer approximation.

2- The phonon gap is strictly zero. This situation occurs for 1 dimensional CDWs at small k and always in standard theories of CDW which predicts the existence of a phason mode. It also occurs for the configurations which contains solitons which are unpinned by the lattice. Then, since $\lambda = 0$, condition (12) is not fulfilled for all phonon modes. On contrary, for the low frequency phase fluctuations (or the sliding mode) condition (12) is sharply violated since it is the reversed inequality which is fulfilled.

We have here a rigorous proof that the Born-Oppenheimer approximation cannot be valid in structures where there is a strictly zero phonon gap (phason, soliton...).

We are not yet able to predict what will be the new ground-state in that case. Either, the CDW remains stable except that the calculations of the low frequency phonons have to be corrected (overdamping?). Or the role of the anti-adiabatic terms becomes very drastic and the CDW is destroyed.

We believe more this second conjecture because it corroborates our early conjecture based on the idea that for a CDW with a strictly zero phason gap, when the spatial extension of a configurational defects is large enough, its global energy becomes negative. This is due to the fact that its tunnelling energy gain (which remains comparable to $\frac{\hbar\omega_0}{2}$ in the Born-Oppenheimer approximation) necessarily exceeds its potential energy which goes to zero as the defect size diverges. On this basis, we already claimed that any CDWs with strictly zero gap phason should be unstable against quantum lattice fluctuations.

These different approaches all suggest that the CDW instability essentially concerns long wave length phase fluctuation. The role of the anti-adiabatic terms for this instability appears to be crucial which suggest that the ground-state then could become a superconductor. This instability of the phase fluctuations could be related to Cooper pair formation.

To be physically more precise, let us note that although there cannot exist a strictly zero gap, a significant softening of the phonons in CDWs remains compatible with our prediction. To fix the ideas let us take $\gamma = 10^{-2}$ and k=1, which are typically in the physical range for real CDW system, condition (12) shows that a maximum phonon softening by a factor as large as 100, remains acceptable without violating the condition of validity of the Born-Oppenheimer approximation. This remark explains why phasons branches with "almost" zero gap could be observed in some CDW systems such as $K_{0.3}MoO_3$(Blue Bronze)[11] (Note that up to now, this is the unique example of CDW where a quasi-phason branch has been observed). By analogy with well-known standard structural phase transition of crystals, the bipolaronic structure of blue bronze should be considered of the displacive type where the thermal effects generate the soft mode and the quasi-phason mode together with a central peak close to the CDW transition. However, the absence of any phonon softening and of phason branch, has also been recently observed[12] in other CDW systems $(TaSe_4)_2I$ which otherwise exhibits the same characteristic features as blue Bronze (Non-linear conductivity, current noise...). Note that this experimental fact is in sharp contradiction with the standard theory of CDW. Instead of this feature, a critical central peak has been observed which is characteristic of a bipolaronic structure of the order-disorder type.

Let us end this paper by a short discussion of the opposite anti-adiabatic limit. In the Born-Oppenheimer approximation, the electrons are supposed to be in adiabatic equilibrium with the atoms, which allows one to eliminate the electronic degree of freedom and to get an effective Hamiltonian for the atoms. Conversely, the anti-adiabatic approximation assumes that the atoms are in equilibrium with the electrons. There is an associated unitary transformation which "put the atoms in adiabatic equilibrium with respect the electrons" and which play a role symmetric of the Wagner transformation (7-a) which "put the electrons in adiabatic equilibrium with respect to the atoms".The Lang-Firsov operator $\exp(iS_{LF})$ defined with

(13-a) $$S_{LF} = -i \frac{k}{2} \sum_m n_m p_m$$

yields:

(13-b) $$\hat{H} = \exp(iS_{LF})\hat{\mathcal{H}}\exp(-iS_{LF}) =$$

$$-\frac{1}{2} \sum_{<i,j>,\sigma} \hat{t}_{i,j} c^+_{i,\sigma} c_{j,\sigma} - \frac{k^2}{8} \sum_i (\sum_\sigma c^+_{i,\sigma} c_{i,\sigma})^2 + \gamma \sum_i (a^+_i a_i + \frac{1}{2})$$

with

(13-c) $$\hat{t}_{i,j} = \exp\left(\frac{k}{2\sqrt{\gamma}} \left((a^+_i - a_i) - (a^+_j - a_j) \right) \right)$$

The anti-adiabatic approximation at the lowest order neglects the fluctuations of $\hat{t}_{i,j}$. Setting $<\hat{t}_{i,j}> = t$ decouples the electronic Hamiltonian from the phonons. The obtained model is a negative U Hubbard model studied by Nozières at al[13] who concluded that the ground-state of this model is always a superconductor. It evolves continuously from a BCS type superconductor (small U) to a bipolaronic type superconductor (large U). The anti-adiabatic approximation is valid when $\hat{t}_{i,j}$ is close to unity that is when

(14) $$k^2 << \gamma$$

It is clear that when this condition is fulfilled, the condition for adiabadicity (12) is never fulfilled. The electronic Hamiltonian used for the standard BCS theory, is in fact derivated from an electron-phonon Hamiltonian[14]. The electron-phonon coupling is treated as a perturbation and eliminated at the lowest order through a unitary transformation which in some sense hybridizes the Lang-Firsov and the Wagner unitary transformation. Then, a purely effective electronic Hamiltonian with attractive electron-electron is derivated and treated according to the BCS methods. The domain of validity of this Hamiltonian is restricted to small electron-phonon coupling k and but γ large enough (since this unitary transformation must only involve one phonon process).

In summary, we expect that the phase diagram in k,γ of the Holstein model (1) be separate into two main regions (see a scheme in ref.4-a).

1- The adiabatic (or classical") region where the system is well-described within the Born-Oppenheimer approximation (that is with an effective lattice Hamiltonian). The ground-states are insulating bipolaronic structures or CDWs with non-zero phason gap. This region is expected to share into many domains (possibly infinitely many) where the bipolaronic structures are different.

2- The anti-adiabatic (or quantum) region where the system is well-described by an effective purely electronic Hamiltonian decoupled from the phonons. According to the literature, the ground-state is then believed to a superconductor.

3-There is no region where the ground-state is a CDW with a strictly gapless phason mode.

Acknowledgements: The author is indebted of useful discussions with J.L. Raimbault, P. Quemerais, D. Feinberg and D. Campbell whose questions greatly helped for developing these ideas. This work has been supported in part by the EEC project: SCI*/0229-C(AM)

REFERENCES

[1] LE DAERON P.Y. and AUBRY S. (1983) J. Phys. **C16** pp.4827-4838 and b-(1983) J.Physique (Paris) **C3** pp.1573-1577 (1983)
[2] AUBRY S. and QUEMERAIS P. (1989) pp. 295-405 in *Low Dimensional Electronic Properties of Molybdenum Bronzes and Oxides* Editor Claire SCHLENKER Kluwer Academic Publishers Group
[3] AUBRY S. and QUEMERAIS P.(1989) pp.342-363 in *Singular Behavior & Nonlinear Dynamics* Ed. St. Pnevmatikos, T. Bountis & Sp. Pnevmatikos, World Scientific
[4] AUBRY S., ABRAMOVICI G., FEINBERG D., QUEMERAIS P. and RAIMBAULT J.L (1989) in *Non-Linear Coherent Structures in Physics, Mechanics and Biological Systems* Lectures Notes in Physics (Springer) **353** pp.103-116
AUBRY S., QUEMERAIS P. and RAIMBAULT J.L. (1989) in*Third European Conference on Low Dimensional Conductors and Superconductors* Fisica **21** Supp.3, pp. 98-101 and pp.106-108 (1989) Ed. S. Barisić
[5] AUBRY S. and ABRAMOVICI G. (1990) Physica **D43** pp.199-219 b-further applications of the concept of anti-integrability can be found in AUBRY S. Proceeding of Workshop on *Twist Mappings and their Applications* March 1990, Minneapolis USA
[6] AUBRY S., GOSSO J.P., ABRAMOVICI G., RAIMBAULT J.L, QUEMERAIS. P. (1990) Physica **D** in press
[7] AUBRY S., ABRAMOVICI G., RAIMBAULT J.L. and QUEMERAIS P. *Bipolaronic Chaotic States in the Adiabatic Holstein Model* In preparation
[8] ABRAMOVICI G. (1990) PHD Dissertation University Paris VI (France)
[9] RAIMBAULT J.L. (1990) PHD Dissertation University of Nantes (France)
[10] WAGNER M. (1981) Phys.Stat.sol (b) **107** 617
[11]POUGET J.P. (1989) pp.87-157 in *Low Dimensional Electronic Properties of Molybdenum Bronzes and Oxides* Editor Claire SCHLENKER Kluwer Academic Publishers Group
[12] CURRAT R., MONCEAU P. private communication (experiments performed in Laboratoire Léon Brillouin July 1990)
[13] NOZIERES P. and SCHMITT-RINK S. (1985) J. of Low Temp. Phys. **59** 195
[14] KITTEL C. (1976) *Quantum Theory of Solids* Editors John WILEY & SONS

POLARONS IN QUASI-ONE-DIMENSIONAL MATERIALS

Dionys Baeriswyl

Institut de Physique Théorique
Université de Fribourg, Pérolles
CH-1700 Fribourg, Switzerland

INTRODUCTION

An electron added to a polarizable medium gains energy by distorting its surroundings; at the same time its motion is slowed down as the electron has to drag along its polarization cloud. This combined entity of an electron (or hole) and its polarization cloud is commonly refered to as a polaron. The energy gain due to the interaction with the medium is the polaron binding energy and the increased inertia is associated with an increased polaron mass.

The two most familiar polaron models are the Fröhlich and Holstein models which differ in the nature of the electron-phonon coupling. Fröhlich considered the long-range forces induced by the polarization field of longitudinal optical phonons of an ionic crystal [1], Holstein the short-range coupling to an intramolecular mode of a molecular crystal [2]. In both cases the conduction band of the insulator is singled out to host the additional electron and the polarization energy of the medium is calculated using the (idealised) lattice dynamics of the neutral system. Therefore the effects of the filled valence band states are only implicitly taken into account in terms of screened force constants.

More recently the polaron concept has been used in the context of one-dimensional conductors, which become semiconductors at low temperatures due to the Peierls instability. Here the distinction between valence and conduction band is less rigid. In fact, in the metallic high-temperature phase the addition of electrons simply changes the Fermi surface. Correspondingly, in the semiconducting low-temperature phase the energy gap is expected to occur at a shifted Fermi wavevector in order to allow the added electrons to be accommodated within the valence band. However, this is not the favored state for a single added electron, in particular if the neutral metallic state corresponds to a half-filled band. This has been shown by exact solutions of the full problem, including explicitly the valence band states [3,4].

One might expect that the many-electron problem to be faced in the context of the Peierls instability has nothing to do with the single-electron problem of Fröhlich or Holstein. This is not

Microscopic Aspects of Nonlinearity in Condensed Matter
Edited by A.R. Bishop *et al.*, Plenum Press, New York, 1991

true. In fact it will be shown below that the state of a Peierls semiconductor plus electron can be interpreted as a conventional polaron, provided that the screening effects of the valence band on the elastic forces are fully taken into account. However, in this case the effects of the valence band states on the lattice dynamics are particularly pronounced ("giant Kohn anomaly").

We shall limit ourselves on the adiabatic limit where the ions are assumed to be infinitely heavy and therefore described by classical displacement coordinates. There is only one important parameter in this case, the dimensionless coupling constant, and there is no distinction between localization and self-trapping. Treating the ions quantum-mechanically immediately restores the translational invariance. Thus even if there is self-trapping the polaronic eigenstates of the Hamiltonian will be generalized Bloch states [5] and form "polaron bands". The effect of self-trapping simply corresponds to band narrowing or mass enhancement, but not to localization [6]. In the quantum description the finite phonon frequency introduces a second relevant parameter, and it is very important not only to distinguish between weak and strong coupling, but also between adiabatic and non-adiabatic regimes [7].

The paper is organized as follows. In section 2 the polaron states of the SSH model are shown to be equivalent to conventional polarons. Section 3 discusses the fate of polarons in conjugated polymers where the electron-phonon coupling is weak and, correspondingly, the binding energy is small. In section 4 a short summary is given on polarons in metal-halogen chain compounds where the coupling can be rather strong. In the limit of very strong coupling even the ground state of the neutral system is of polaronic nature. In fact, it is then more appropriate to speak of a lattice of bipolarons [8] instead of a Peierls semiconductor.

SSH MODEL

In the adiabatic limit the SSH Hamiltonian [9] is given by

$$H_{SSH} = \frac{1}{2} K \sum_{n} (u_n - u_{n+1})^2 - \sum_{n\sigma} t_{nn+1} \left(c^+_{n\sigma} c_{n+1\sigma} + c^+_{n+1\sigma} c_{n\sigma} \right) , \tag{1}$$

where $c^+_{n\sigma}$ and $c_{n\sigma}$ are, respectively, creation and annihilation operators for electrons at site n and with spin σ, and u_n are displacement coordinates for the ions. The resonance integrals $t_{n,n+1}$ are expanded to linear order in the displacements

$$t_{n,n+1} = t_0 + \alpha \left(u_n - u_{n+1} \right). \tag{2}$$

For an electron density of one electron per site the ground state is dimerized,

$$u_n = (-1)^n \; u , \tag{3}$$

and a gap $2\Delta_0$ opens at the Fermi energy. The gap parameter Δ_0 is given by

$$\Delta_0 = 4\,\alpha\,u = W\,e^{-1/2\lambda} \tag{4}$$

where W is of the order of the bandwith and the dimensionless coupling constant

$$\lambda = \frac{2\alpha^2}{\pi t_0 K} \tag{5}$$

has been assumed to be small, $\lambda << 1$. Therefore the ground state is a semiconductor with a full valence band and an empty conduction band.

The question what happens if electrons are added or removed has played an essential role in the theory of conducting polymers. The corresponding studies have been usually based on models treating both the electronic and the lattice degrees of freedom explicitly [3,4,10]. Here we want to show that the particular case of a single electron can be described rather well as a conventional polaron provided that the screening of the elastic forces by the valence-band electrons is properly taken into account. In the present weak-coupling case it is convenient to take the continuum limit for the staggered order parameter

$$\Delta_{n,n+1} \equiv 2\alpha(-1)^n\,(u_n-u_{n+1}) \rightarrow \Delta(x)\,. \tag{6}$$

The elastic energy associated with distortions with respect to the homogeneous ground state, where $\Delta(x) = \Delta_0$, is found to be [11]

$$\Phi\{\Delta(x)\} = \frac{1}{2}\int dx \int dx'\, [\Delta(x)-\Delta_0]\, D(x-x')\, [\Delta(x')-\Delta_0] \tag{7}$$

where the Fourier transform of D(x) is given by

$$D(q) = \frac{2}{\pi\hbar v_F}\left(1 + \frac{1}{12}\xi^2 q^2 - \frac{1}{120}\xi^4 q^4 + \ldots\right)\,, \tag{8}$$

$\xi = \hbar v_F/\Delta_0$ beeing the coherence length, $v_F = 2t_0 a/\hbar$ the Fermi velocity and a the lattice constant. The renormalized elastic forces differ in two important respects from the elastic term in Eq. (1): they are much softer than the original forces (by a factor $2\lambda << 1$) and of longer range ($\xi >> a$).

We add now an electron to the system. If the lattice is fixed in its homogeneous configuration, $\Delta = \Delta_0$, the lowest available energy level for the electron is at the bottom of the conduction band, $\varepsilon = \Delta_0$. If we allow now the lattice to relax the electronic level will be lowered. For a given configuration $\{\Delta(x)\}$ the electronic structure is given by the solutions of the Bogoliubov-de Gennes equations [10,11]

$$\varepsilon\, u(x) = -i\hbar v_F \frac{d}{dx} u(x) - \Delta(x)\, v(x) \tag{9a}$$

$$\varepsilon\, v(x) = i\hbar v_F \frac{d}{dx} v(x) - \Delta(x)\, u(x) \tag{9b}$$

where u(x) and v(x) are the wavefunctions for right- and left-moving electrons, respectively. For small displacements, $|\Delta(x)-\Delta_0| \ll \Delta_0$, the polarization of the valence band states is already fully taken into account in terms of the screened dynamical matrix of Eq. (7). Therefore we have only to calculate the lowest positive eigenvalue of Eq. (9). Introducing the function f = u + iv we obtain from Eq. (9) the second-order differential equation

$$\left[-(\hbar v_F)^2 \frac{d^2}{dx^2} - \varepsilon^2 + \Delta^2 + \hbar v_F \frac{d\Delta}{dx}\right] f = 0 . \tag{10}$$

Therefore we have to find the minimum of

$$\frac{\varepsilon^2-\Delta_0^2}{\Delta_0^2} = \int dx\, f(x) \left[-\xi^2 \frac{d^2}{dx^2} + \frac{\xi}{\Delta_0}\frac{d\Delta}{dx} + \frac{\Delta^2-\Delta_0^2}{\Delta_0^2}\right] f(x) \Big/ \int dx\, f^2(x) \tag{11}$$

for a given configuration $\{\Delta(x)\}$ in order to determine the energy $\varepsilon\{\Delta(x)\}$ of the added electron. The polaron energy is then simply given by the minimum of

$$E = \Phi\{\Delta(x)\} + \varepsilon\{\Delta(x)\} . \tag{12}$$

The three equations (7), (11) and (12) are completely analogous to the corresponding relations for the convenional polaron problem. The polaron energy consists of an elastic term Φ and the electronic energy ε which depends on the displacement pattern of the lattice through Eq. (11).

In order to illustrate that the arguments presented yield more than a satisfactory formal analogy we use a simple variational Ansatz to discuss the (electron) polaron of the SSH model. We write

$$\frac{\Delta(x) - \Delta_0}{\Delta_0} = -\delta \,\mathrm{sech}\,(x/l) \tag{13a}$$

$$f(x) = A \,\mathrm{sech}\,(x/l)\,[1 + \gamma \tanh (x/l)] \tag{13b}$$

and treat the indentation amplitude δ, the extent l and the asymmetry γ as variational parameters. (A is simply a normalization constant and drops out). In order to simplify the analysis we neglect the dispersion in Eq. (8) (this is a good approximation as long as l is larger than ξ). It is then straightforward to carry out the integrations in Eqs. (7) and (11), and we find

$$\Phi/\Delta_0 = (2/\pi)\, \delta^2 (l/\xi) \tag{14}$$

for the elastic energy and

$$(\varepsilon^2-\Delta_0^2)\,/\,\Delta_0^2 = \left[c_1\,(\xi/l)^2 - c_2\delta - c_3\delta(\xi/l) + c_4\delta^2\right] / \left[1 + \gamma^2/3\right] \tag{15}$$

for the expectation value (11) with coefficients

$$c_1 = \frac{1}{3}(1 + \frac{7}{5}\gamma^2),\ c_2 = \frac{\pi}{2}(1 + \frac{1}{4}\gamma^2),\ c_3 = \frac{\pi}{8}\gamma,\ c_4 = \frac{2}{3}(1 + \frac{1}{5}\gamma^2)\ . \tag{16}$$

The result of the minimization of the total energy $\Phi + \varepsilon$ with respect to δ, l and γ is given in Table 1 and compared with the exact result of the full calculation where the polarization of the valence band is taken into account self-consistently [3,4,11]. In view of the present simplified analysis the agreement is satisfactory. The remaining discrepancies are presumably mostly due to the harmonic approximation in (7) and less to the choice of the variational Ansatz.

Table 1. Extent l, indentation δ, gap level ε and polaron energy E_p both for the exact solution of the full problem and the present variational solution of the reduced problem.

	l/ξ	δ	ε/Δ_0	E_p/Δ_0
Exact	$\sqrt{2} = 1.414$	$\sqrt{2} - 1 = 0.414$	$\frac{1}{\sqrt{2}} = 0.707$	$2\sqrt{2}/\pi = 0.9003$
This work	1.874	0.309	0.816	0.9297

It is tempting to consider this state as a large polaron since its extent is much larger than the lattice constant (in polyacetylene ξ is of the order of 6a) and the binding energy is much smaller than the bandwidth W. However a glance at Eqs. (7) and (11) shows that the appropriate scales are not the lattice constant a and the bandwidth W, but rather the coherence length ξ and the gap parameter Δ_0. On these scales the present solution corresponds to a small polaron.

The SSH Hamiltonian or related effective single-particle models have been widely used for interpreting experimental data in conjugated polymers [9]. For such models it is easy to show [12] that a second added electron preferably occupies the same localized orbital as the first and the resulting relaxed state, the "bipolaron", has a lower energy than two separate polarons. The polaron and bipolaron concepts play an essential role for describing spectroscopic and magnetic properties, especially in polymers with a non-degenerate ground state [9,13]. For the sake of completeness we mention that in the case of trans-polyacetylene and generally for systems with a doubly degenerate ground state kink solitons are energetically prefered with respect to polarons, if topologically possible; correspondingly, at least on a single chain, a bipolaron is expected to decay into a kink-antikink pair.

Unfortunately the SSH model, although very appealing, is too simple to yield an adequate description of conjugated polymers. The most important drawback is the omission of explicit electron-electron interactions [14]. It is true that including them preserves the symmetry of the ground state - bond alternation is even enhanced by Coulomb effects - but the excited states are strongly modified by correlation effects.

In the pevious section we have treated the polaron problem as an effective single-particle problem, where the electrons in the valence band are simply taken into account by renormalized (possibly anharmonic) elastic forces. If Coulomb interactions are explicitly included such a reduction is no more possible, and we have to face the full many-body problem of N+1 electrons on N sites. Indeed, the added electron is coupled to the other electrons not only indirectly through the lattice degrees of freedom but also directly through the electron-electron interactions. To be specific we model these interactions by an on-site term (U) and a nearest-neighbor term (V), i.e.,

$$H_{int} = U \sum_m n_{m\uparrow} n_{m\downarrow} + V \sum_m n_m n_{m+1} \tag{17}$$

where $n_{m\sigma} = c^+_{m\sigma} c_{m\sigma}$ and $n_m = n_{m\uparrow} + n_{m\downarrow}$. Unfortunately the full problem defined by the Hamiltonian $H = H_{SSH} + H_{int}$ requires advanced techniques, and many important questions are still open. A major difficulty is also that there is no agreement about the appropriate values of the various parameters [14].

Here we restrict ourselves on a qualitative discussion of Coulomb effects. The case of kink solitons has been extensively studied in the past [14]. It is clear that they remain topologically stable since the ground state remains dimerized. Recently it was found that, within the Hartree-Fock approximation, a charged kink remains also energetically stable, i.e. $E_K < \Delta_0$, for reasonable values of the parameters in Eqs. (1) and (17) [15]; however, in order to reduce the positive Coulomb energy, the kink expands with respect to its size ξ in the SSH model.

For discussing the stability of a polaron it is sufficient to consider the shallow limit (δ very small, l very large in Eqs. (14) and (15)). The relevant energies are the elastic energy proportional to $\delta^2 l$, the gain in electronic energy proportional to δ and the Coulomb energy proportional to l^{-1}. Thus we write

$$\frac{E-\Delta_0}{\Delta_0} = \left[\frac{1}{2} A(l\delta)^2 - B\, l\delta + C\right] / l \tag{18}$$

with positive constants A, B, C. For fixed l the minimum is at $l\delta = B/A$, yielding an energy

$$\frac{E-\Delta_0}{\Delta_0} = \left[-B^2 / (2A) + C\right] / l. \tag{19}$$

If this expression is positive the minimum is obtained for $l \to \infty$; the (shallow) polaron is unstable. If the expression (19) is negative the energy decreases with decreasing extent l; the stable state will be a small polaron. We notice that for the present one-dimensional case this "dichotomy of self-trapping" [16,17] is generated by the Coulomb term. An estimate of this term on the basis of the exact polaron solution of the SSH model shows that for reasonable values of parameters the expression (19) is negative, indicating that the addition of Coulomb interactions does not destabilize the polaron in conjugated polymers.

Up to this point we have considered only a single, well-ordered chain. The real materials usually are both (partially) crystalline and not very pure. Therefore it is important to include the effects of both interchain coupling and impurities, in particular as the relevant energy, the polaron binding energy, is very small. Various authors have pointed out that in a perfectly ordered array of chains the polaron is easily destabilized by transverse delocalization [17-20]. This effect may be counterbalanced by disorder destroying the coherence between chains [15]. Unfortunately, while disorder can effectively stabilize the polaron, it also tends to pin it such that coherent motion of polarons becomes unlikely.

The combined effects of disorder and Coulomb interactions can be important for understanding the magnetic susceptibility as a function of doping. Roughly speaking, the (localized) occupied states close to the Fermi level will be "polarons" with spin, whereas the deeper states will be spinless "bipolarons". The effect of a magnetic field is, on the one hand, to align the spins of the polarons, on the other hand, to break some of the bipolarons into polarons (with aligned spins). The first effect gives rise to a Curie term, the second to a Pauli-like contribution [12].

MX CHAIN COMPOUNDS

A particular class of materials, chain compounds consisting of transition metal ions M (M = Pt,

Pd or Ni) and bridging halogens X (X = Cl, Br or I), well known since decades to chemists [21], has been recently rediscovered by physicists [22]. The chains are well separated from each other by organic ligands. These materials are semiconductors despite the fact that formally there is one valence electron per unit cell. One of the interesting characteristics is that changing the metal or halogen ions does not affect the crystal structure but strongly modifies the electronic properties. If the MX chain compounds are viewed as Peierls semiconductors then these substitutions can be modeled in terms of a changing strength of the electron-phonon coupling. Thus the PtCl compound appears as a strong-coupling Peierls insulator or rather mixed valence compound with an optical gap of the order of 2.9 eV, whereas PtI corresponds to a weak-coupling Peierls semiconductor with a gap of 1.7 eV, quite similar to that of polyacetylene. This interpretation is confirmed by structural investigations, showing that in the PtCl compound the Cl sublattice is strongly dimerized with a displacement in chain direction of the order of 0.3 A, whereas the dimerization in the PtI compound is much weaker [21].We have argued above that in conjugated polymers the electron-phonon coupling arises from a modulation of the hopping integral $t_{n,n+1}$. In the MX chain it is the Coulomb interaction between the halogen ion and the valence electron on the metal ion which yields the dominant coupling mechanism. It is conceivable that this interaction has to be modeled as a Fröhlich-type long-range coupling between the LO phonon mode and the electronic charge, but the models investigated so far contain only short-range electron-phonon (and sometimes also electron-electron) interactions. To be specific we consider explicitly only a single electronic level at the M sites together with the displacements of the ions in chain direction, denoted by u_n for the M atoms and v_n for the x atoms. In the adiabatic limit the Hamiltonian is written as [23]

$$H = \frac{1}{2}K\sum_{n}\left[(u_n-v_n)^2 + (u_n-v_{n+1})^2\right] + \sum_{n\sigma}\varepsilon_n c^+_{n\sigma} c_{n\sigma} - \sum_{n\sigma} t_{n,n+1}(c^+_{n\sigma} c_{n+1\sigma} + c^+_{n+1\sigma} c_{n\sigma}) \tag{20}$$

where we have neglected (explicit) Coulomb interactions. This corresponds to a single-band tight-binding model with two lattice degrees of freedom per unit cell. The coupling betweeen electrons and lattice displacements is modeled as

$$t_{n,n+1} = t_0 + \alpha(u_n-u_{n+1}) \tag{21a}$$

$$\varepsilon_n = \beta\,(v_n-v_{n+1}) \tag{21b}$$

For $\beta = 0$, $\alpha>0$ the model is equivalent to the SSH Hamiltonian and the ground state would correspond to a dimerized M sublattice, i.e. to bond alternation (a "bond-order wave", BOW). On the other hand, for $\alpha=0$, $\beta > 0$ the coupling is between the X sublattice and the electronic charge on the M sites; thus the ground state will be a charge-density wave (CDW) accompanied by a dimerization of the X sublattice. In the general case there is a competition between the two instabilities. Depending on the relative strengths of the two couplings the ground state will be a BOW, a CDW or a phase where both instabilities coexist [23]. Simple considerations show that the coupling β dominates in these materials [23], and that the CDW phase is the stable ground state, in agreement with experiment.

There is good evidence, mostly from optical spectroscopy [24], that local states play an important role also in the MX chain compounds. For describing these states on the basis of the Hamiltonian (20) one has to distinguish several regimes: weak and strong couplings, real and complex order parameters. We restrict ourselves to the physical regime where the coupling β dominates. For small β and $\alpha = 0$ the charge-density wave has a small amplitude and, correspondingly, the electronic gap is small. It follows that the coherence length is large and a continuum description can be used. Although the symmetry of the order parameter is different from that of polyacetylene it is easy to map the two continuum limits onto each other by a simple gauge transformation [25]. Therefore one obtains the same kink solitons, polarons and bipolarons in the two cases. If α is also finite one has to face the problem of a complex order parameter, a problem which has been addressed earlier on general ground [26,27]. Unfortunately no exact solution for this case has been found so far, although the structure of a topological soliton can be interpolated between the two soluble limits of a purely imaginary and a purely real order parameter.

Of particular interest is the case of large β where the electronic density essentially alternates between 0 and 2 (mixed valence, e.g., Pt^{II} and Pt^{IV}). The ground state can then be viewed as a lattice of well localized bipolarons. Correspondingly the local states produced by adding or removing charge or by creating electron-hole pairs are strongly localized, namely on the scale of the lattice constant [23]. Doping then corresponds to a redistribution of charges, leading to something like a bipolaron glass [8].

In the strong-coupling limit the interchain coupling will play a minor role and not affect the polaron stability. Therefore there is more hope to study intrinsic polarons in the MX chain compounds than in conjugated polymers. Unfortunately the mass of these polarons is expected to be enormously large and their mobility negligibly small. The slow decay of photo-induced defects in PtCl compounds [28] is likely to be related to this large polaronic mass enhancement.

SUMMARY

The aim of this contribution was to give a perspective view on the polaron problem in a Peierls semiconductor, using the adiabatic limit. The case of the SSH model was treated in some detail in order to show that there is no principal difference between a polaron in a polyacetylene chain and, say, in a molecular crystal. A simple variational calculation showed that this mapping works. The remaining discrepancies between the single-electron approach and the full treatment including the valence band states were attributed to the harmonic approximation of the screened elastic forces. In the absence of electron-electron interactions the minimum electron-phonon coupling for polaron formation in one dimension is infinitesimal. It was proposed that the addition of Coulomb interactions shifts the critical coupling to a finite value below which there is no polaron and above which the stable state is a small polaron. Besides the conjugated polymers, where the lattice displacements are coupled to the electronic bond order, the MX chain compounds were briefly presented as examples where the coupling involves mostly the electronic charge density. In contrast to the polymers these systems also offer the opportunity to study the strong-coupling regime where even the (neutral) ground state is better viewed as an assembly of bipolarons than as a Peierls semiconductor.

REFERENCES

1. H. Fröhlich, Adv. Phys. 3:325 (1954).
2. T. Holstein, Ann. Phys. 8:325,343 (1959).
3. S.A. Brazovskii and N.N. Kirova, Pisma Zh. Eksp. Teor. Fiz. 33:6 (1981) [Sov. Phys. JETP Lett. 33:4 (1981)].
4. D.K. Campbell and A.R. Bishop, Phys. Rev. B 24:4859 (1981).
5. H. Haken, Quantum Field Theory of Solids, Elsevier (1983).
6. M.P.A. Fisher and W. Zwerger, Phys. Rev. B 34:5912 (1986).
7. D. Feinberg, S. Ciuchi and F. de Pasquale, Int. J. Mod. Phys. B (1990).
8. S. Aubry, G. Abramovici, D. Feinberg, P. Quémerais and J.-L. Raimbault, Lecture Notes in Physics 353:103 (1990).
9. For a review see A. J. Heeger, S. Kivelson, J.R. Schrieffer and W.P. Su, Rev. Mod. Phys. 60:781 (1988).
10. H. Takayama, Y.R. Lin-Liu and K. Maki, Phys. Rev. B 21:2388 (1980).
11. D. Baeriswyl in Theoretical Aspects of Band Structures and Electronic Properties of Pseudo-One-Dimensional Solids, ed. R.H. Kamimura (Reidel, 1985), p.1.
12. D. Baeriswyl, Springer Series in Solid-State Sciences 91: 54 (1989).
13. H. Kiess, ed., The Physics of Conducting Polymers, Springer, in press.
14. For a recent discussion see the review by D. Baeriswyl, D.K. Campbell and S. Mazumdar in Ref. 13.
15. D. Baeriswyl, Synth. Met. (1991), to be published.
16. Y. Toyozawa, Prog. Theor. Phys. 26:29 (1961).
17. D. Emin, Phys. Rev. B 33:3973 (1986).
18. Yu. N. Gartstein and A.A. Zakhidov, Sol. St. Commun. 62:213 (1987); Erratum 65:ii (1987).
19. D. Baeriswyl and K. Maki, Synth. Met. 28:D507 (1989).
20. P. Vogl and D.K. Campbell, Phys. Rev. Lett. 62:2012 (1989).
21. For a review of the early literature on the subject see R.J.H. Clark, Adv. Infrared and Raman Spectr. 11:95 (1984).
22. See the various contributions in the Proceedings of the ICSM, Synth. Met. (1991), to be published.
23. D. Baeriswyl and A.R. Bishop, J. Phys. C 21:339 (1988).
24. L. Degiorgi, P. Wachter, M. Haruki and S. Kurita, Phys. Rev . B 42:4341 (1990), and references therein.
25. Y. Onodera, J. Phys. Soc. Japan 56:250 (1987).
26. S. Kivelson, Phys. Rev. B 28:2653 (1983).
27. W.P. Su, Sol. St. Commun. 48:479 (1983).
28. S. Kurita, M. Haruki and K. Miyagawa, J.Phys. Soc. Japan, to be published.

SOLITONS IN CHARGE DENSITY WAVE CRYSTALS

S. Brazovskii and S. Matveenko

L.D.Landau Institute for Theoretical Physics
Kosygina, 2, GSP-1, Moscow, 117940, USSR

INTRODUCTION

Electronic properties of quasi onedimensional conductors with charge density waves (CDW) are peculiar for two reasons. First is related to translational degeneracy of CDW ground state which leads to the phenomena of Fröhlich conductivity[1,2]. It shows itself in a giant dielectric susceptibility, in nonlinear and nonstationary effects (see proceedings and the reviews[3−7]). The second reason is related to a strong interaction of CDW deformations with normal electrons, which leads to their fast selftrapping in the course of conversion to various kinds of solitons[8,9]. Extensive experimental studies are devoted to the Frölich conductivity (see reviews by Fleming, Gill, Grüner, Jerome, Nad', Ong, Schlenker et all[3−6]). At the same time physics of solitons is mainly studied at another substance - polyacetylene[10,11], where effects of CDW sliding are absent for the case of two-fold commensurability.

In recent years experimental studies of CDW turned to low temperatures and/or to materials with well developed crystal structure of CDW and corresponding dielectrization of electronic spectra. There are so called blue bronzes and tetrachalcogenides of transition metals[3−6,12]. In this respect and also due to the above mentioned discrepancy in theoretical studies we were motivated to consider a microscopical picture for conversion of normal current carriers to solitons and for their subsequent development into CDW phase slips. Of special importance is to trace how the injected at the junction electrons finally evolve into deformation of CDW in the volume. For isolated one-dimensional chain of CDW this process is clear in principal. Every electron or hole with energy E near the edges of $\pm\Delta$ of the gap 2Δ spontaneously evolves in the nearly amplitude soliton[8,9,13−15] while the original electron is trapped at the local level near the gap center $E = 0$. In this process which goes through a time $\omega_{ph}^{-1} \sim 10^{-11}s$ (where ω_{ph} is the phonon frequency at the CDW wave vector $Q = 2k_F$) the energy $\simeq 0.3\Delta$ is released. During some large collisional time the pairs of amplitude solitons develop into 2π phase slips while trapped electrons disappear in an extended continuum below the gap. As a result the rest part $\simeq 0.6\Delta$ of the original energy Δ is released. But already for a quasi one-dimensional system of weakly bound chains with low 3d ordering temperature $T_c \ll \Delta$ the picture becomes much more complicated, since at $T < T_c$ the very existence of topological solitons is questionable[8,9,14,16−18].

In this publication we will consider effects of long range structural deformations and Coulomb fields due to the CDW crystal media adaptation to various solitons, which are born in the process of electronic conversion. We will study the static deformations in 3-d ordered CDW media with point solitons, interactions of solitons between each other and with impurities. We will consider the effects of screening by normal carriers and of selfscreening in a gas of solitons. These results permit to outline the microscopical picture of subsequent steps for the conversion of normal current to the Fröhlich current of CDW sliding, which is described at the Conclusion. A short presentation for some of these ideas was given[9]. More extensive studies and the theory for the soliton's aggregation or dislocations will be presented elswhere[19].

SOLITONS IN QUASI ONE - DIMENSIONAL MODEL OF CDW CRYSTAL

Quasi one-dimensional system of CDW is characterized by the wave number $Q \simeq 2k_F$, the gap 2Δ in electronic spectrum, the length $\xi_0 = v/\Delta$, where k_F and v are the momentum and the velocity at the Fermi level (The Plank constant is supposed to be unity $\hbar = 1$). The CDW deformation has the form

$$\eta_n(x) = Re\Delta_n(x)\exp(iQx)\ , \quad \Delta_n(x) = |\Delta_n(x)|\exp(i\varphi_n(x))$$

where $|\Delta_n(x)|$ and $\varphi_n(x)$ are amplitudes and phases on chains n and the dependence on the coordinate x along the chain corresponds to the disturbed states. In the ground state $|\Delta_n| \equiv \Delta,\ \varphi \equiv 0$. Here and in the following we suggest for simplicity that at the equilibrium all φ_n are equal, i.e. the CDW wave vector is $\vec{Q} = (Q, 0, 0)$. As we will see in Chapter 3 general results are model independent.

In a system of weakly bound chains , $T_c \ll \Delta$, selftrapping of electrons on the energy scale Δ and the length scale ξ_0 develops independently at any chain. Then topological solitons are created which are incompatible to the definition of the long range order. Adaptational deformations must appear to equalize the values of the order parameter on all chains. These deformations are characterized by a low energy $T_c \ll \Delta$ and a large length $l \sim v/T_c \gg \xi_0$, i.e. which permits a description in terms of phases φ_n at given amplitudes $|\Delta_n| \equiv \Delta$.

The presence of $\pm 2\pi$ - soliton[15,16] at a chain $n = j$ which corresponds to selftrapping of two electrons (+) or two holes (-) is taken into account by boundary conditions

$$n = j\ :\ \varphi_j(+\infty) - \varphi_j(-\infty) = \nu\pi\ ;\quad n \neq j\ :\ \varphi_n(+\infty) = \varphi_n(-\infty) \tag{1}$$

at $\nu = \pm 2$. The presence of an amplitude soliton corresponding to selftrapping of one electron (+) or one hole (-) at a point x_j is taken into account by a local condition[14]

$$\varphi_j(x_j + 0) - \varphi_j(x_j - 0) = \mp\pi, \qquad \varphi_j(\pm\infty) = \varphi_j(\pm\infty) = 0$$

The case is equivalent[9,14] to the condition (1) at $\nu = \pm 1$. These conditions show that electric charges are equal to νe.

The energy functional for the system of chains has the form

$$\mathcal{H} = \int dx \Big\{ \sum_n \Big[\frac{v}{4\pi}\Big(\frac{\partial\varphi_n}{\partial x}\Big)^2 - \sum_m J_{nm}\cos(\varphi_n - \varphi_m) + \frac{1}{\pi}\frac{\partial\varphi_n}{\partial x}\Phi_n \Big] - \int \frac{(\nabla\Phi)^2}{8\pi e^2} d\vec{r} \Big\} \tag{2}$$

where $\Phi = \Phi(x, \vec{r})$ is the electric potential relative to one electron, $\Phi_n = \Phi_n(x, \vec{r}_n)$, $\vec{r} = \vec{r}_n$ is a coordinate of an n-th chain. Smooth dependence of Φ_n on n is suggested. Presence of a soliton on a chain j is taken into account by the condition (1) Independent of the presence or distribution of solitons defined by sets of conditions (1) the following conservation laws for quantities averaged over the sample's crossection takes place

$$\frac{\partial^2}{\partial x^2}\bar{\Phi} - \kappa^2\bar{\Phi} = const$$

$$\frac{v}{2}\sum_n \frac{\partial\varphi_n}{\partial x} + \bar{\Phi} = const \tag{3}$$

$$\kappa^2 = \frac{8e^2}{vs} = r_d^{-2}, \quad \bar{\Phi} = \int \frac{\Phi d\vec{r}^2}{s}$$

where s is the area per one chain, r_d is the screening radius in a metallic, without CDW, phase.

Consider a model which permits to study the whole region of lengths $|x| \gg \xi_0$ at arbitrary n. We suggest that the matrix J_{mn} in (2) connects a large number of chains. Then we may consider phases ϕ_n of neighboring chains at $n \neq j$ to be close and we go to the continual description by substitutions

$$\sum_n \rightarrow \frac{1}{s}\int d\vec{r}^2, \quad \varphi_n(x) \rightarrow \varphi(\mathbf{r}), \quad \mathbf{r} = (x, \vec{r})$$

$$J = \sum_m J_{mn} \sim T_c^2/v\ ; \quad 2\pi J/vs = \alpha$$

We arrive at the following form for the energy functional

$$\mathcal{H} = \int \frac{dx\, d\vec{r}^2}{s}\left\{\frac{v}{4\pi}\left[\left(\frac{\partial\varphi}{\partial x}\right)^2 + \alpha(\frac{\partial\varphi}{\partial\vec{r}})^2\right]\right.$$

$$\left. + \frac{\Phi}{\pi}\frac{\partial\varphi}{\partial x} - \frac{s}{8\pi e^2}(\nabla\Phi)^2 + H_s\right\} \tag{4}$$

where

$$H_s = \sum_j s\delta(\vec{r} - \vec{r}_j)\left[\frac{v}{4\pi}\left(\frac{\partial\varphi_j}{\partial x}\right)^2 - J\cos(\varphi_j - \varphi) + \frac{e}{\pi}\Phi\frac{\partial\varphi_j}{\partial x}\right] \tag{5}$$

The sum in (5) goes over those chains, which contain solitons.

Variation of the energy functional (4), (5) over φ, φ_j, Φ gives equilibrium conditions for the system with solitons.

$$\frac{v}{2}\hat{\Delta}\varphi(x,\vec{r}) - \frac{\partial\Phi(x,\vec{r})}{\partial x} - \sum Js\delta(\vec{r} - \vec{r}_j)\sin(\varphi_j(x) - \varphi(x,\vec{r})) = 0 \tag{6}$$

$$\frac{v}{2}\frac{\partial^2\varphi_j}{\partial x^2} + \pi J\sin(\varphi_j(x) - \varphi(x,\vec{r}_j)) - \frac{\partial\Phi(x,\vec{r}_j)}{\partial x} = 0 \tag{7}$$

$$\frac{2}{v\kappa^2}\Delta\Phi(x,\vec{r}) + \frac{\partial\varphi(x,\vec{r})}{\partial x} + \sum_j s\delta(\vec{r} - \vec{r}_j)\frac{\partial\varphi_j(x,\vec{r})}{\partial x} = 0 \tag{8}$$

where

$$\Delta = \frac{\partial^2}{\partial x^2} + \frac{\partial^2}{\partial\vec{r}^2}\ ; \quad \hat{\Delta} = \frac{\partial^2}{\partial x^2} + \alpha\frac{\partial^2}{\partial\vec{r}^2} = \frac{\partial^2}{\partial\hat{\mathbf{r}}^2}\ ; \quad \hat{\mathbf{r}} = (x, \vec{r}/\alpha^{1/2})$$

Let's consider one soliton at the chain $j = 0$, $\vec{r}_j = 0$ Before taking into account the electric field Φ equations (6) - (8) acquire the form

$$\hat{\Delta}\varphi + \frac{\partial^2\varphi_0}{\partial x^2}s\delta(\vec{r}) = 0 \tag{9}$$

$$-\ l^2\frac{\partial^2\varphi_0}{\partial x^2} + \sin(\varphi_0 - \varphi(0)) = 0, \quad l^2 = s/\alpha \tag{10}$$

The solution of equation (9) relates functionally $\varphi(x,0)$ and $\varphi_0(x)$ as

$$\varphi(0,x) = \frac{l^2}{4\pi}\int \frac{\partial\varphi_0^2(y)}{\partial y^2}\frac{dy}{[|x-y|^2+l^2]^{1/2}} \tag{11}$$

System of equations (10) - (11) defines selfconsistently functions $\varphi_0(x)$ and $\varphi(0,x)$ which describe the phases on the central and on the surrounding chains. Then the phase $\varphi(x,\vec{r})$ is found in the whole volume with the help of equation (9). We can see that equations (10) - (11) indeed have a solution for topological soliton with the scale l which satisfies the conditions (1). Unlike a conventional sine-Gordon equation the asymptotics at $|x| \gg l$ is not exponential but rather has a power law. Indeed the major counterpart to the integral in equation (11) comes from the region $|y| \leq l$. By partial integration in (11) and taking into account (1) we find at $|x| \gg l$

$$\varphi_0(x) - 2\pi\theta(x) \simeq \varphi(0,x)\, sgnx \simeq -\frac{l^2}{4x^2} + \frac{3l^2\ln(|x|/l)}{8\alpha x^4} \tag{12}$$

The second term in (12) is a correction which comes from the region $y \sim x$. Taking into account the Coulomb interaction we would find analogously that

$$\varphi_0(x) \simeq \varphi(0,x) \sim 1/x \tag{13}$$

A comprehensive study for field's distributions at large distances will be given in next chapters on the basis of model independent approach.

INTERACTION OF SOLITONS AT LARGE DISTANCES

Now we are mostly interested not in the local structure of solitons which is model dependent but in the distributions of fields φ, Φ at large distances and in soliton's interactions at $|\mathbf{r}| \gg l$. For these kinds of problems the presence of solitons at a given chains may be taken into account by using asymptotical conditions (1) which can by written as

$$\int \frac{\partial\varphi_j}{\partial x}dx = \nu\pi \tag{14}$$

The main counterpart to the integral (14) comes from the vicinity of soliton $\xi_0 \ll |x-x_j| < l$, since the power laws of (12) and (13) are fast enough for convergency of (14). At large distances the condition (14) may be written as

$$\frac{1}{\pi}\frac{\partial\varphi_j}{\partial x} = \nu_j\delta(x-x_j) \tag{15}$$

where ν_j is an index ν from (1) for the soliton at the point $\mathbf{r_j}$.

The system is described by a model independent Hamiltonian (4) where the solitonic part $\mathcal{H}_s$ should be presented as a Hamiltonian for some point sources.

$$\mathcal{H}_s = \int d^3\mathbf{r}\rho_s(\mathbf{r})V(\mathbf{r})$$

$$\rho_s = \sum_j \nu_j s\delta(\mathbf{r}-\mathbf{r}_j)\ ;\ V = \frac{v}{2}\frac{\partial\varphi}{\partial x} + \Phi \tag{16}$$

where ρ_s is a weighted density of solitons, equal to injected charge density, V is a potential of soliton's interaction with long range fields φ and Φ.

The Hamiltonian (16) is derived in the following way. The term Φ analogously to (4) describes the Coulomb interaction of the soliton charge, which is considered now as a point charge in (15). The first term in (16) provides as we will see the rigorous conservation law (3) and permits to take into account correctly the charge distribution in the system. Indeed integrating the equation (3) term by term

$$\sum_{n\neq j}\delta\varphi_n = -\sum_j \delta\varphi_j\,, \quad \delta\varphi = \varphi(+\infty)-\varphi(-\infty)$$

we arrive at the paradox: every term in a L.H.S. part is equal to zero, while the R.H.S. part is finite. The compensation comes evidently from a divergency at large n and x , where the continuum approximation is valid. Any finite number of terms in the sum over n may be neglected, so the definition of the charge density in the continuum limit is not violated.

We notice, that the potential V for interaction of solitons appears to be exactly the same invariant combination of fields φ and Φ as the potential for normal electrons near the CDW edge, where V plays a role of the electrochemical potential.

By variating the functional (16) over φ and Φ we arrive at equations

$$v\hat{\Delta}\varphi + 2\frac{\partial\Phi}{\partial x} - \pi v\frac{\partial}{\partial x}\rho_s(\vec{r}) = 0 \tag{17}$$

$$\frac{2}{v\kappa^2}\Delta\Phi + \frac{\partial\varphi}{\partial x} + \pi\rho_s(\vec{r}) = 0. \tag{18}$$

Integrating equations (17), (18) over perpendicular plane we arrive at equations for averaged over perpendicular cross-section quantities defined as in (3):

$$\frac{2}{v\kappa^2}\frac{\partial^2\bar{\Phi}(x)}{\partial x^2} + \frac{\partial\bar{\varphi}(x)}{\partial x} + \sum_j \delta(x-x_j) = 0 \tag{19a}$$

$$\frac{\partial^2\bar{\Phi}(x)}{\partial x^2} - \kappa^2\bar{\Phi}(x) = const \tag{19b}$$

We can see from (19) that in the volume $\bar{\Phi} = const$, average electric field is zero, while for the junction problems we should omit exponentially growing solutons and neglect exponentially decreasing solutions on a microscopical scale κ^{-1}. Equations (19) are exact for the discrete model (2) which have been assumed at our construction of Hamiltonian (16).

Without Coulomb field the solution of equations (17) - (18) for a soliton at the origin would have the form

$$\varphi_s(\mathbf{r}) = -\frac{s}{4\alpha}\frac{x}{|\hat{\mathbf{r}}|^{3/2}} \tag{20}$$

It can be easily seen that

$$\bar{\varphi}(x) = -\frac{\pi}{2}sgn\ x$$

so that the first condition (19) is satisfied. A distribution of the charge density in the media is given by

$$\rho(\mathbf{r}) = e\delta(\mathbf{r}) + \frac{e}{\pi}\frac{1}{s}\frac{\partial\varphi(x)}{\partial x}$$

$$= e\delta(\mathbf{r}) + \frac{e}{4\pi\alpha}\left[\frac{1}{|\mathbf{r}|^3} - \frac{3x^2}{|\hat{\mathbf{r}}|^5}\right] \tag{21}$$

For the interaction energy of two solitons with indexes ν_1, ν_2 we easily find from (16) - (20)

$$W(\mathbf{r}) = \pm \frac{v}{2}\frac{\partial\varphi}{\partial x} = \nu_1\nu_2\frac{vs}{4\alpha}\left[-\frac{1}{|\hat{\mathbf{r}}|^3} + \frac{3x^2}{|\hat{\mathbf{r}}|^5}\right] \tag{22}$$

We notice that interaction of solitons changes the sign at a cone surface

$$r/|x| = (\alpha/2)^{1/2}, \quad r = |\vec{r}|$$

For solitons of the same charge ($\nu_1\nu_2 > 0$) which appear due to the selftrapping of similar kind of carriers the attraction takes place in the perpendicular sector of the cone: $r > |x|\,(\alpha/2)^{1/2}$.

Accounting for the Coulomb interaction essentially changes the distribution of the fields. For the functions $\Phi(x,\vec{r})$, $\varphi(x,\vec{r})$ we easily find from (17), (18) the following equations

$$\hat{K}\varphi_s(x,\vec{r}) = \pi\, s\,(\kappa^2 - \Delta)\,\frac{\partial}{\partial x}\,\delta(x,\vec{r}) \tag{23}$$

$$\hat{K}\Phi_s(x,\vec{r}) = -\,4\pi e^2\alpha\Delta_\perp\delta(x,\vec{r}) \tag{24}$$

where

$$\hat{K} = \hat{\Delta}\Delta - \kappa^2\frac{\partial^2}{\partial x^2}\,; \quad \Delta_\perp = \frac{\partial^2}{\partial\vec{r}^2}$$

Equations (23), (24), can be solved explicitly in two cases when the operator $\hat{K}$ ca be factorized: a region of smooth dependences and for a special case $\alpha = 1$. For gradients being small in comparison to κ i.e. in the region where the continuum model is definitely applicable we obtain

$$\hat{K} \simeq (\alpha^{1/2}\frac{\partial^2}{\partial\vec{r}^2} - \kappa\frac{\partial}{\partial x})\,(\alpha^{1/2}\frac{\partial^2}{\partial\vec{r}^2} + \kappa\frac{\partial}{\partial x})$$

Then we can easily find that

$$\Phi_s \simeq \frac{e^2}{2|x|}\exp(-\frac{r^2}{d|x|}) \tag{25}$$

$$\varphi_s(x,\vec{r}) \simeq \frac{s}{2dx}\exp(-\frac{r^2}{d|x|}) \tag{26}$$

where $d = 4\alpha^{1/2}/\kappa$. Notice a very unusual inverce exponential dependence on x in (25), (26) which is evident from analogy between longwave limit of $\hat{K}$ in (23), (24) and the operator for a two-dimensional diffusion in "time" x. A peculiar feature of this dependence is that gradients in (25), (26)

$$d\frac{\partial}{\partial x} \sim (d\frac{\partial}{\partial\vec{r}})^2 \sim \frac{r^2}{x^2}$$

do not drop at large distances as one would expect but rather decrease only in the longitudinal sector $\alpha^{1/2}|x| > |r|$ where solutions (25), (26) are applicable. In this region the Coulomb potential dominates in the total interaction of solitons (16) $V \simeq \Phi$

We see that the interaction changes from a power low repulsion $\sim e^2/2|x|$ at $|x| > r^2/d$ to exponentional attraction at $r/\alpha^{1/2} < |x| < r^2/d$ and reaches exponentially small positive values

$$\sim \frac{e}{r}\exp(-r/d)$$

at the validity limit.

In the perpendicular sector $r < x\alpha^{1/2}$ at $r \gg d$ we can neglect all interactions with exception for a case of the weak Coulomb interaction at reletively strong interchain coupling

$$d^2 \gg l^2 = s/\alpha \text{ , i. e. } \alpha^2 \gg e^2/v\epsilon$$

at small distances $l < r < d$

The field's distributions in the whole region of x, $\vec{r}$ are conveniently demonstrated by an exact solution of equations (23), (24) at $\alpha = 1$. In this case the operator $\hat{K}$ is factorized and we find the following exact solutions of equations (23), (24).

$$\Phi_s(x, \vec{r})|_{\alpha=1} = \frac{e^2\kappa}{4} \frac{\exp(-\tilde{r})}{\tilde{r}} [ch\tilde{x} + sh\tilde{x}\frac{\tilde{x}}{\tilde{r}^2}(1 + \tilde{r})] \tag{27}$$

$$\varphi_s(x, \vec{r})|_{\alpha=1} \simeq \frac{-\kappa^2 s}{16} \frac{\exp(-\tilde{r})}{\tilde{r}} [sh\tilde{x} + ch\tilde{x}\frac{\tilde{x}}{\tilde{r}^2}(1 + \tilde{r})] \tag{28}$$

where $\tilde{x} = \kappa x/2$, $\tilde{r} = \kappa|\mathbf{r}|/2$ The interaction energy of solitons with indexes ν_1, ν_2 is found from (16), (27), (28):

$$W(\mathbf{r}) = \nu_1\nu_2\frac{e^2\kappa}{8}\exp(-\tilde{r}) \Big[2\tilde{x}\Big(\frac{1}{\tilde{r}^2} + \frac{1}{\tilde{r}^3}\Big)sh\tilde{x} +$$

$$\Big(\frac{1}{\tilde{r}} - \frac{1}{\tilde{r}^2} - \frac{1}{\tilde{r}^3} + \frac{\tilde{x}^2}{\tilde{r}^3} + \frac{3\tilde{x}^2}{\tilde{r}^4} + \frac{3\tilde{x}^2}{\tilde{r}^5}\Big)ch\tilde{x}\Big] \tag{29}$$

At $|x| \gg r$ the expressions (27) - (29) correspond to dependences (25) - (26) at $\alpha = 1$ as it should be. In a whole region of large distances $\tilde{r} \gg 1$ the main factor in (27) - (29) is

$$\frac{1}{\tilde{r}}\exp(|\tilde{x}| - \tilde{r})$$

We see that interactions fall down exponentially on going from the chain axes $\vec{r} = 0$ but they grow exponentially in the sector $|x| < r$ by going from the plane $x = 0$. The interaction (29) becomes negative at very short distances $|\mathbf{r}| < \kappa^{-1}$ where the free phase interaction (22) is recovered as it should be. Out of this spot we always have $W > 0$ and the repulsion along $\vec{r}$ takes place. We may speak about the short range binding of solitons only at weak Coulomb coupling $e^2/v \ll 1$ when the width $1/\kappa$ is larger then the soliton length $l \sim s^{1/2}$ We should keep in mind that interactions in the perpendicular sector which decrease exponentially at a microscopical $\sim d$ scale may be disturbed by many factors . For example there is attraction due to exchange of localized electrons for solitons with $\nu = 1$. This process is especially important for the irreversible conversion $\pi + \pi \to 2\pi$.

Due to a very large anisotropy of diffusion coefficients for solitons one should first consider the dependence $W_{\|}(x) = W(x, \vec{r} = const)$ at a given $\vec{r}$. We see that the function $W_{\|}$ has a minimum W_{min} with a negative energy (30) at $x = 0$ which is exponentially shallow in $\vec{r}$. There is a large region of attractive forces $\partial W/\partial x < 0$ for the positive energy $W > 0$ at $x < x_{max} \sim r^2/d$, which is separated by the large barrier $W_{max} \sim e^2d/r^2$ from the repulsion region at $x > x_{max}$. The minimum deep and the barrier weight grow with decreasing r. At $r = 0$ extrema are absent and the repulsion of a non-screened Coulomb type twice weakened (the first term in (29)) dominates. At small finite r W_{min} is negative.

EFFECTS OF SCREENING AND COMMENSURABILITY

At large distances the principal role is played by Coulomb screening due to residual carriers and (in commensurate case) by the pinning energy. Both effects do not influence essentially the structure of solitons but should be taken into account for the long range fields Φ and φ.

In real CDW systems we may find two mechanisms for screening. One of them is related to a redistribution of the soliton gas density which interacts due to (26) simultaneously to the fields Φ and φ. Another one is related to the residual carriers. For example a weakly filled band like in $NbSe_3$ or a low activated electronic band may exist which does not participate in formation of CDW i.e. does not interact with $\partial\varphi/\partial x$. Along with the screening we shall consider possible effects of commensurability[2,7] of CDW with undelying lattice. Taking into account both effects we will find new terms $\sim \varphi^2$ and $\sim \Phi^2$ in the Hamiltonian (16). Thus effects of electronic screening and commensurability are taken into account by the substitution in equations (17) - (18) like

$$\Delta \to \Delta - \lambda^2 , \ \hat{\Delta} \to \hat{\Delta} - \epsilon^2 ; \ \lambda^2 = 4\pi e^2 \partial n_e / \partial\mu ; \quad \lambda \ll \kappa ; \quad \epsilon \ll \kappa$$

where n_e and μ are the concentration and the chemical potential of free electrons and ϵ^2 is proportional to the pinning energy. Solution for modified equations (17), (18) in Fourier representation takes the form

$$\Phi_{\mathbf{k}} = \frac{4\pi e^2(\alpha k_\perp^2 + \epsilon^2)}{\hat{K}(\mathbf{k})}$$

$$\varphi_{\mathbf{k}} = \frac{i\pi k_{\parallel}(\kappa^2 + k^2 + \lambda^2)}{\hat{K}(\mathbf{k})} \tag{31}$$

where

$$\hat{K}(\mathbf{k}) = (\mathbf{k}^2 + \lambda^2)(\hat{\mathbf{k}}^2 + \epsilon^2) + \kappa^2\mathbf{k}_\parallel^2 \simeq \kappa^2\mathbf{k}_\parallel^2 + \alpha\mathbf{k}_\perp^4 + \beta^2\mathbf{k}_\perp^2 + \kappa^2\mathbf{q}^2 ; \tag{32}$$

$$\mathbf{k} = (\mathbf{k}_\parallel, \vec{\mathbf{k}}_\perp); \quad \beta^2 = \epsilon^2 + \alpha\lambda^2 ; \quad \mathbf{q} = \epsilon\lambda/\kappa$$

Taking into account only lowest order terms in $k_\parallel$, $k_\perp$, ϵ, λ as shown in formula (32) we find from (31) the following solutions

$$\varphi_s \simeq \frac{s\kappa^2}{4\beta^2}\frac{\partial}{\partial x}\frac{1}{R^*}\exp(-qR^*) ; \quad \Phi_s \simeq \frac{vs\kappa^2}{8\beta^2}(\epsilon^2 + \alpha\Delta_\perp)\frac{1}{R^*}\exp(-qR^*) \tag{33}$$

$$R^* = (x, \ \mathbf{r}/(\alpha^*)^{1/2}) ; \quad \alpha^* = \beta^2/\kappa^2 \tag{34}$$

Formula (33) is valid if the Fourier transform of (31) is convergent at small enough $k_\perp$ when we can neglect the term $\alpha k_\perp^4$ in (32). It takes place at the distances

$$R^* > d/\alpha^* \tag{35}$$

Otherwise we return to dependences (25), (26).

It follows from (33), (35) that the Coulomb part in the interaction potential V dominates over the phase part following the parameter $\kappa^2/\beta^2 \gg 1$. At $qR^* < 1$ we obtain from (33), (34)

$$V(\mathbf{r}) = \frac{\alpha vs}{8}\left(\frac{\kappa}{\beta}\right)^4 \frac{2x^2 - r^2\kappa^2/\beta^2}{(x^2 + r^2\tilde{\kappa}^2/\beta^2)^{5/2}} \tag{36}$$

We see from (33), (36) that distributions of fields φ and Φ are similar to the fields of dipole and quadrupole sources correspondingly for an effective Coulomb potential. There are two significant differences in comparison to similar results (20) -(22) for the model without physical Coulomb interactions. First of all the effective anisotropy is strongly enhanced.

$$\alpha \rightarrow \alpha^* = \beta^2/\kappa^2 \ll \alpha$$

Secondly for $\epsilon \neq 0$ the finite radius of screening q^{-1} appears. Both of these effects extend the sector of solitonic attraction at large distances (35) practically to the whole sector $|x| > r/\alpha^{1/2}$ except for a narrow longitudinal one at $|x| > r/(\alpha^*)^{1/2}$

In the sector $|x| < r/\alpha^{1/2}$. the effects of ϵ and λ are not important in the presence of fast (exponential on the scale κ^{-1}) dependences (30). We can easily see it with the help of a more general exact solution avalable at $\alpha = 1$, $\lambda = \epsilon$. For this special case the kernal $\hat{K}$ factorizes again and we arrive at exact equations of type (27), (28) with the substitution of $\exp(-\tilde{r})$ for $\exp(-\mu\tilde{r})$, where $\mu^2 = 1 + 2\beta^2/\kappa^2$. Eventually it brings a screening factor $\exp(-q|x|)$ to formula (25), (26).

The selfscreening in the gas of solitons can be shown to give the similar results.

INTERACTION OF SOLITONS WITH DEFECTS

Let's consider a long range interaction of solitons with defects which pin the CDW we consider an isolated defect at the point $\mathbf{r}_0$ with a minimun pinning energy at $\varphi(\mathbf{r_0}) = \theta$. The term which describes this energy must be added to the functional (16).

$$\delta W = \int dx\, d\vec{r}\, \frac{c}{2}(\varphi(\mathbf{r}) - \theta)^2 \delta(\mathbf{r} - \mathbf{r}_0) \tag{37}$$

Variation of the functional (4) with the additional term (37) gives us another local source at the right hand side of the equation (17):

$$(-\pi/v)Cs(\varphi(\mathbf{r}_0) - \theta)\delta(\mathbf{r} - \mathbf{r}_0) \tag{38}$$

The energy for a bilinear continuous functional with the point sources (16) and (38) can be calculated in a general form, which does not depend explicitly on the functional form (16). We arrive at

$$W = \frac{1}{2}V(\mathbf{r}_j) - \frac{C}{2}\theta(\varphi(\mathbf{r}_0) - \theta) \tag{39}$$

and after the selfconsistent definition of $\varphi(\mathbf{r}_0)$ we obtain

$$W = \frac{\tilde{C}}{2}(\varphi_s(\mathbf{r}_0 - \mathbf{r}_j) - \theta)^2 \tag{40}$$

Here $\varphi_s(\mathbf{r})$ is a corresponding solution for single soliton problems which have been considered above. The constant $\tilde{C}$ differs from C in (37) by some renormalisation due to selfenergy of the long range deformations introduced by the isolated defect without soliton.

Let's find distribution of the fields φ and Φ in the presence of a defect. Variation of the functional (5), (16) (37) gives us, in analogy to (23), (24) the following equations.

$$\hat{K}\Phi = -4\pi e^2\alpha\Delta_\perp\delta(\mathbf{r} - \mathbf{r}_j) - (8\pi e^2/v)C(\varphi - \theta)\frac{\partial}{\partial x}\delta(\mathbf{r} - \mathbf{r}_0) \tag{41}$$

$$\hat{K}\varphi = \pi s(\kappa^2 - \Delta)\frac{\partial}{\partial x}\delta(\mathbf{r} - \mathbf{r}_0) + (2\pi s/v)C(\varphi - \theta)\Delta\delta(\mathbf{r} - \mathbf{r}_0) \tag{42}$$

Comparing the r.h.s. of equations (41), (42) and the r.h.s. of equations (23), (24) we are able to express the solution for the problem (41), (42) via the solutions φ_s, $\mathbf{\Phi}_s$ for the single soliton problems without defects.

$$\Phi(\mathbf{r}) = \Phi_s(\mathbf{r}-\mathbf{r}_j) - C(\varphi(\mathbf{r}_0)-\theta)\varphi_s(\mathbf{r}-\mathbf{r}_0) \tag{43}$$

$$\varphi(\mathbf{r}) = \varphi_s(\mathbf{r}-\mathbf{r}_j) - \frac{sC}{2\alpha v}(\varphi(\mathbf{r}_0)-\theta)\Phi_s(\mathbf{r}-\mathbf{r}_0) + $$

$$+ \frac{2C}{\kappa^2 v}(\varphi(\mathbf{r}_0)-\theta)\frac{\partial}{\partial x}\varphi_s(\mathbf{r}-\mathbf{r}_0) \tag{44}$$

It is interesting to notice that the main contribution to the solitonic potential V comes from the Coulomb field Φ in spite of the fact that a defect interacts explicitly with the phase only. Meanwhile the field Φ induced by a defect coincides with the field Φ_s induced by the soliton.

To analyze the interaction (39) it is natural as well as for the two soliton problem to underline the dependence

$$W_{\parallel}(x) = W(x, \vec{r} = const)$$

Remember that $\varphi_s(x,\vec{r})$ is an odd function of x and it has the extrema at $x = \pm x_m(\vec{r})$. In general case $\theta \sim \pi$ formula (40) describes the attraction towards $x_m(\vec{r})$ from the region $x_m(\vec{r}) < x < \infty$ (We suppose here that $\theta > 0$). The perpendicular forces in the valley $x = x_m(\vec{r})$ are directed to the defect. At small θ the absolute minima extends from the microscopical vicinity of defect to a closed line $\mathcal{L}$ which is defined by an equation

$$\varphi_s(\mathbf{r}-\mathbf{r}_0) \equiv \theta$$

For a special case $\theta = 0$ the attraction towards the plane x=0 from the region $|x| < x_m$ and the repulsion in perpendicular directions take place. This case can be found only for coherent crystallographic positions of defects relative to CDW which may probably happen for a special case of mobile impurities (see[20] and references therein).

At large distances the interactions of both $\pi-$ and $2\pi-$ solitons with a defect differ only by a coefficient $\nu = 1,\ 2$ in φ_s and the effect of attraction is general. But the bound state at the defect will be different in principal since the π-soliton contains a nuclei where the amplitude goes through zero and consequantly the binding energy will be $\sim \Delta$. Probably at low temperatures every dopant forms a neutral complex with a π-soliton which plays the role of the neutral defect studied above. Bound states of charged impurity with solitons were considered recently[21]. This reference also contains important references on Coulomb interaction in CDW.

CONCLUSIONS

At the present publication we have solved a number of stationary problems which are required preliminary for describing the conversion of a normal current to the Frölich one of CDW. We considered the formation of π and 2π solitons and we constructed the model to describe the structure of solitons in crystallografically ordered media of CDW.

Then we found the model independent approach to describe interactions of solitons with long range Coulomb and deformational fields. On this bases we calculated the interactions of isolated solitons to each other and with impurities. Also we studied the combined effects of weak commensurability and screening or selfscreening by the residual concentration of electrons or solitons. For all cases the regions are found where the solitons are attractive to the common perpendicular plane. In the case of screening we may trace also an attraction between solitons in their common plane. These results show a tendency towards the aggregation of solitons in some complexes which are equivalent to dislocational loops. These problems will be described elswhere.

ACKNOWLEDGEMENT. One of the authors (S.B.) acknowledges the hospitality of the Institute for Scientific Interchange Foundation in Torino, Italy. We are grateful to N.Kirova for the help in editing the manuskript.

REFERENCES

1. J. Fröhlich, Proc. Roy. Sos.,1954, **A223**, 292.
2. P.A. Lee, T.M.Rice and P.V.Anderson, 1975, Solid State Commun., **17**, 1089.
3. Electronic properties of quasi one-dimensional compounds, ed. P.Monceau, Reidel Publ. Co, 1985 .
4. Low dimensional Conductors and Superconductors, ed.D. Jerome & L. G. Caron NATO ASI Series B : Physics-Vol. **155**. Plenum Publ. Co., New York, 1987.
5. Low dimensional electronic properties of molybdenum bronzes and oxides (ed. C.Schlenker).Reidel Publ. Co. 1989.
6. Charge Density Waves in Solids, ed. L. Gor'kov & G. Grüner, Elsvier Sci. Publ., Amsterdam, 1990.
7. G. E. Grüner, & A. Zettl, Phys. Reports, 1985, **119**, 117.
8. S. Brazovskii, and N. Kirova, 1984, Electron Selflocalization and Periodic Superstructures in Quasi One Dimensional Dielectrics., in: Soviet Scientific Reviews, sec.A: Physics, **6**, ed. I.M. Khalatnikov. (Harwood Academic Publishers).
9. S. Brazovskii , Charge Density Waves in Solids, ed. L.Gor'kov, & G.Grüner, Elsvier Sci. Publ., Amsterdam, 1990.
10. Yu Lu, Solitons and Polarons in Conducting Polymers. World Scientific Publ. Co., 1988.
11. A. J. Heeger, S. Kivelson , J. R. Schrieffer, W. P. Su. 1989, Rev. Mod. Phys.
12. C. Schlenker , J. Dumas , C. Escribe-Filippini , M. Boujida. 1990 , Physica Scripta (to be published).
13. S. Brazovskii. 1978, Pis'ma Zh. Exp. Teor. Fis., **28** , 677 (Sov. Phys. JETP Lett., **28**, 606).
14. S. Brazovskii. 1980, Zh. Exp. Teor. Fis., **78**, 677 (Sov. Phys. JETP , **51**, 342.)
15. S. Brazovskii, and S. Matveenko. 1984, Zh. Exp. Teor. Fis., **87**, 1400 (Sov. Phys. JETP , **60**, 804).
16. S. Brazovskii, N. Kirova and V. Yakovenko, 1983, Journal de Physique, Suppl., Colloque C3, **44**, C3-1525.
17. T. Bohr, S. Brazovskii. 1983, J. Phys. C, **16**, 1189.
18. D. Baeriswyl, K. Maki. Phys. Rev., 1983, **B28**, 2068.
19. S. Brazovskii, S. Matveenko. Zh. Exp. Theor. Phys., 1990; J. de Physique. 1990, to be published.
20. J. C. Gill. Europhysics Lett., 1990.
21. S. Barisic, I. Batistic. J. de Physique, 1989, **50**, 2717.

HYDRODYNAMICS OF CLASSICAL AND QUANTUM LIQUID WITH FREE SURFACE

I.M.Khalatnikov

L.D.Landau Institute for Theoretical Physics
USSR Academy of Sciences
117940 GSP-1 Moscow ul.Kosygina 2

ABSTRACT

Nonlinear effects of interaction of the second sound with the surface waves in the superfluid He are studied. It is shown that the standing surface waves, excited by the second sound on a surface of the superfluid *He*, are stationary.

1. HYDRODYNAMICS OF CLASSICAL LIQUID

Let us start our study with the simplest case of the potential isentropic motion of the liquid. Then as is known, it is possible to introduce the potential of the velocity $+\alpha$ so that the velocity v be defined by the relation:

$$\vec{v} = \vec{\nabla}\alpha \tag{1.1}$$

(Here ∇ is understood to be a vector, $\vec{\nabla}$. The momentum corresponding to a unit volume of the liquid equals

$$\vec{j} = \rho\nabla\alpha. \tag{1.2}$$

As has been shown in our paper [1] Formula (1.2) is a particular case of the general expression for the momentum density

$$\vec{j} = -\sum p\nabla q \tag{1.3}$$

where p and q are the canonically conjugated variables (momentum p with the coordinate q) following from the general symmetry properties of the system. In a liquid to the group of translations corresponds the generator ∇, and its density, according to (1.3), determines the momentum. Thus, in accordance with (1.2) the variables - p and α are the momentum and coordinate, respectively. For a general case of the non-isentropic and non-potential motion, according to [1] for the density and momentum we have

$$\vec{j} = (\rho\nabla\alpha + s\nabla\beta + f\nabla\gamma) \tag{1.4}$$

where f and γ are the Klebsch variables and β is a coordinate, conjugated to the momentum $- s$. In (1.4) there is an extra function related with a certain hidden gauge group (for detail see [2]).

Coming back to a simplest case of the potential motion, write out the action for our system. Having in mind a further investigation of the interaction of surface and bulk modes in the liquid, we shall take into account gravitational and capillar forces.

Recall the formula, determining the Hamiltonian:

$$H = \sum p\dot{q} - L \tag{1.5}$$

Then for the action we have

$$A = \int dt \int^{\zeta(x,y,t)} dV(-H + \sum p\dot{q}) =$$

$$= \int dt \int^{\zeta} dV \left(- \frac{\rho \vec{v}^2}{2} - \varepsilon(\rho) - \rho g z - \rho\dot{\alpha}\right) - \int dt\, k \int dS. \tag{1.6}$$

Here k is the surface tension, dS is an element of the surface of the liquid, $\varepsilon(\rho)$ is the energy density of the liquid; it is assumed that the liquid fills the half-space $(z = \zeta,\ z = -\infty)$.

Calculating variation of the action (1.6) with (1.1) taken into account, we get

$$\delta A = \int dt \int^{\zeta(x,y,t)} dV \left\{\left[- \frac{(\nabla\alpha)^2}{2} - \frac{\partial\varepsilon}{\partial\rho} - \dot{\alpha} - gz\right] \partial\rho + \right.$$

$$\left. + (\dot{\rho} + div\, \rho\vec{v})\partial\alpha\right\} + \int dt \int dxdy \left[- \rho vN \sqrt{1+(\nabla\zeta)^2}\, \partial\tilde{\alpha} + \dot{\zeta}\rho\delta\tilde{\alpha} - \right.$$

$$\left. - \partial\zeta \left(\rho\tilde{\alpha} + \frac{\rho}{2}(\nabla\tilde{\alpha})^2 + \varepsilon + g\zeta - k\, div \frac{\nabla\zeta}{\sqrt{1+(\nabla\zeta)^2}}\right)\right]_{z=\zeta} \tag{1.7}$$

Here

$$vN = (v_z - v_x\zeta_x - v_y\zeta_y)/\sqrt{1 + (\nabla\zeta)^2}$$

∇ and Δ are $2d$ operators in the x,y-plane,

$$\tilde{\alpha} = \alpha|_{z=\zeta}.$$

Equating variations of the bulk part of the action to zero, we get the equations of motion

$$- \frac{(\nabla\alpha)^2}{2} - \mu - \dot{\alpha} - gz = 0 \quad \mu = \frac{\partial\varepsilon}{\partial\rho} \tag{1.8}$$

and the continuity equation

$$\dot{\rho} + div\, \rho\vec{v} = 0 \quad \vec{v} = \nabla\alpha. \tag{1.9}$$

Similarly, from the surface part of the action, we find

$$\dot{\zeta} = vN\sqrt{1+(\nabla\zeta)^2} \tag{1.10}$$

$$\rho\dot{\tilde{\alpha}} + \rho\frac{(\nabla\tilde{\alpha})^2}{2} + \varepsilon + \rho g\zeta - k\,div\frac{\nabla\zeta}{\sqrt{1+(\nabla\zeta)^2}} = 0 \tag{1.11}$$

It follows from (1.11) and (1.8) that at $z=\zeta$ on the surface of the liquid we have

$$p + k\,div\frac{\nabla\zeta}{\sqrt{1+(\nabla\zeta)^2}} = 0, \tag{1.12}$$

where the pressure is

$$p = -\varepsilon + \mu\rho. \tag{1.13}$$

Before passing over to the Hamiltonian formalism of the equations, recall that for the bulk part of the liquid a canonical pair is - ρ (momentum) and α (coordinate). For the surface part the problem of choice is more difficult, since $\dot{\zeta}$ first emerged not in the expression for the action A but in expression (1.7) for its variation. We think it naturally entails from (1.7) that an appropriate ζ and momentum ψ; the momentum being determined by its variation

$$\delta\psi = \rho\delta\tilde{\alpha} \tag{1.14}$$

This fact can be interpreted as manifestation of non-holonomity. There is no other explanation of this fact but the correctness of the choice is confirmed by the fact that eqs. (1.10) and (1.11) are obtained by us in the Hamiltonian form.

Actually, the Hamiltonian of the system (equ.(1.6)) equals

$$H = \int^{\zeta} dV\left(\rho\frac{(\nabla\alpha)^2}{2} + \varepsilon + \rho gz\right) + k\int dS. \tag{1.15}$$

Write down variations of H

$$\delta H = \int^{\zeta} dV\left[\delta\rho\left(\frac{(\nabla\alpha)^2}{2} + \mu + gz\right) + \delta\alpha\left(-\,div\,\rho\nabla\alpha\right)\right] +$$

$$+ \int dxdy\left[\delta\zeta\left(\rho\frac{(\nabla\tilde{\alpha})^2}{2} + \varepsilon + \rho g\zeta\right) +\right.$$

$$\left. + \rho v_N\sqrt{1+(\nabla\zeta)^2}\,\delta\tilde{\alpha} - k\,div\frac{\nabla\zeta}{\sqrt{1+(\nabla\zeta)^2}}\,\delta\zeta\right]. \tag{1.16}$$

v_N the projection of the velocity onto the normal to the surface

$$v_N = (v_z - v_x\zeta_x - v_y\zeta_y)/\sqrt{1+(\nabla\zeta)^2}\,.$$

The Hamiltonian equations ensuing from (1.16), for the bulk, part are

$$\dot{\alpha} = \frac{\delta H}{\delta(-\rho)} = -\left(\frac{(\nabla\alpha)^2}{2} + \mu + gz\right) \tag{1.8'}$$

$$(-\dot{\rho}) = -\frac{\delta H}{\delta\alpha} = div\ (\rho\nabla\alpha) \tag{1.9'}$$

and, for the surfacial part,

$$\dot{\zeta} = \frac{\delta H}{\rho\delta\tilde{\alpha}} = v_N\sqrt{1 + (\nabla\zeta)^2} = v_z - v_x\zeta_x - v_y\zeta_y \tag{1.10'}$$

$$\dot{\psi} = \rho\dot{\tilde{\alpha}} = -\frac{\delta H}{\delta\zeta} = -\left(\rho\frac{(\nabla\tilde{\alpha})^2}{2} + \varepsilon + \rho g\zeta - k\ div \frac{\nabla\zeta}{\sqrt{1 + (\nabla\zeta)^2}}\right) \tag{1.11'}$$

Eqs. (1.8')-(1.11') exactly coincide with eqs. (1.8)-(1.11) which virtually confirms the correct choice of pairs the of canonically conjugated variables.

For an arbitrary non-isentropic and non-potential motion, as is known, it is necessary to introduce 3 pairs of canonically conjugated variables (see (1.4))

$$\begin{matrix} p & q \\ \rho & \alpha \\ s & \beta \\ f & \gamma \end{matrix}$$

And by analogy with the above derivation, one can repeat all calculations and obtain an overall system of hydrodynamic equations. For the surface part the canonically conjugated variables are the coordinate ζ and momentrum ψ, determined by their variations

$$\delta\psi = \tilde{\rho}\delta\tilde{\alpha} + \tilde{s}\delta\tilde{\beta} + \tilde{f}\delta\tilde{\gamma} \tag{1.17}$$

(the tilda labels the values of the functions on the surface $z = \zeta$).

To study the nonlinear phenomena, e.g., processes of transformation of bulk waves into surface waves one should write down an expansion of the Hamiltonian up to the cubic terms with respect to the canonical variables.

2. HYDRODYNAMICS OF QUANTUM SUPERFLUID LIQUID

The problem of canonical formalism of the equations of hydrodynamics of superfluid liquid has been studied in detail in [1].

Recall, that the equations of hydrodynamics are obtained from the Hamiltonian

$$H = \int^{\zeta} dV \left(\frac{\rho v_s^2}{2} + \vec{p}\vec{v}_s + \varepsilon(\rho, s, \vec{p}) + \rho gz\right) + k\int dS \tag{2.1}$$

where the superfluid velocity v_s is determined by the potential α

$$v_s = \nabla\alpha$$

the momentum with respect to the motion of normal and superfluid parts is determined by the potentials β, f and γ

$$\vec{p} = s\nabla\beta + f\nabla\gamma \tag{2.2}$$

and, consequently, the total momentum j is

$$\vec{j} = \rho\nabla\alpha + s\nabla\beta + f\nabla\gamma \tag{2.3}$$

the energy density is determined by the identity

$$d\varepsilon = Tds + \mu dp + (\vec{v}_n - \vec{v}_s, d\vec{p}). \tag{2.4}$$

In the Hamiltonian (2.1) we have taken into account gravitational and capillar forces on the surface. According to (2.3) the pairs of canonically conjugated variables are

$$(-\rho,\alpha),\ (-s,\beta),\ (-f,\gamma). \tag{2.5}$$

To study the acoustic and surface waves it is not necessary to take into account the Klebsch variables (f,γ). Write out the hydrodynamic equations in the form of the Hamiltonian without them taking into account:

$$\dot{\alpha} = -\frac{\delta H}{\delta\rho} = -\left(\frac{v_s^2}{2} + \mu + gz\right) \tag{2.6}$$

$$\dot{\rho} = \frac{\delta H}{\delta\alpha} = -\,div(\rho\vec{v}_s + \vec{p}) \tag{2.7}$$

$$\dot{\beta} = -\frac{\delta H}{\delta s} = -T - \vec{v}_n \cdot \nabla\beta \tag{2.8}$$

$$s = \frac{\delta H}{\delta\beta} = -div\ (\vec{v}_n s). \tag{2.9}$$

To describe surface motion we introduce canonical variables ζ and ψ, where the momentum ψ is determined by its variation similarly to (1.14).

$$\delta\psi = \tilde{\rho}\delta\tilde{\alpha} + \tilde{s}\delta\tilde{\beta} \tag{2.10}$$

(note that the tilda points to the fact that all functions are taken at $z=\zeta$).

$$\rho\dot{\tilde{\alpha}}+s\dot{\tilde{\beta}}=-\frac{\delta H}{\delta\zeta} = -\left(\frac{\rho v_s^2}{2}+\vec{p}\vec{v}_s+\varepsilon+\rho gz\right)\Bigg|_{z=\zeta} +k\ div\frac{\nabla\zeta}{\sqrt{1+(\nabla\zeta)^2}} \tag{2.11}$$

$$\dot{\zeta} = \frac{\delta H}{\delta\psi} = v_{nN}\sqrt{1+(\nabla\zeta)^2} = \frac{j_N}{\rho}\sqrt{1+(\nabla\zeta)^2} \tag{2.12}$$

where the subscript N labels the component of v_n normal to the surface.

Note we do not take into account the interaction of liquid with vapour in the boundary conditions (2.12). Such account would have considerably complicated the results without changing them qualitatively.

It follows from eqs. (2.6) and (2.11) that at $z=\zeta$ there is a constraint

$$p + k\, div \frac{\nabla\zeta}{\sqrt{1 + (\nabla\zeta)^2}} = 0 \qquad (2.13)$$

3. EQUATIONS OF HYDRODYNAMICS FOR SUPERFLUID VELOCITY WITH THE SURFACIAL WAVES TAKEN INTO ACCOUNT IN THE QUADRATIC APPROXIMATION

As independent variables, let us take the pressure p and temperature T. Then from the definition of the pressure

$$p = -\varepsilon + Ts + \mu\rho + (\vec{p}, \vec{v}_n - \vec{v}_s) \qquad (3.1)$$

in the approximation, quadratic over velocities, we get

$$\rho\delta\mu = -s\delta T + \delta p - (p, \delta(\vec{v}_n - \vec{v}_s)) =$$

$$- s\delta T + \delta p - \frac{\rho_n}{2} (\vec{v}_n - \vec{v}_s)^2. \qquad (3.1)$$

δ denotes deviation of the values of the functions from the equilibrium values.

Combining eqs. (2.6) and (2.8) and the boundary conditions (2.13), we get ($\sigma=s/\rho$) at $z = \zeta$

$$-(\dot{\tilde{\alpha}} + \sigma\dot{\tilde{\beta}}) = \frac{v_s^2}{2} - \frac{\rho_n}{2\rho}(\vec{v}_n - \vec{v}_s)^2 + \frac{\rho_n}{\rho} \vec{v}_n(\vec{v}_n - \vec{v}_s) - \frac{k}{\rho} \Delta\zeta + g\zeta$$

$$- \frac{1}{2} \frac{\delta\sigma}{\delta T} (\delta T)^2 - \frac{\delta\sigma}{\delta p} \delta p \delta T + \frac{1}{2} \frac{\delta}{\delta p} \frac{1}{\rho} (\delta p)^2. \qquad (3.3)$$

It is clear that the velocity-quadratic terms are equal to the kinetic energy

$$\left(\rho_s \frac{v_s^2}{2\rho} + \rho_n \frac{v_n^2}{2\rho}\right).$$

In the approximation when one can ignore the thermal expansion in superfluid liquid ($\delta\rho/\delta T=0$) there occurs a complete division of th first (δp) and second sounds (δT). For the first sound from (3.3) we find

$$- \dot{\tilde{\alpha}} = \frac{1}{2} (\nabla\tilde{\alpha})^2 - \frac{1}{2c_1^2} \dot{\tilde{\alpha}}^2 - \frac{k}{\rho} \Delta\zeta + g\zeta. \qquad (3.4)$$

Here $c_1^2 = \frac{\delta p}{\delta\rho}$ is the velocity of the first sound.

If this equation is averaged over the time exceeding the time of the acoustic oscillations we find that

$$\bar{\zeta} = 0$$

i.e., the average oscillations of the surface in the running acoustic wave have the order higher than 2. This is a well-known result (L.Landau, and E.Lifschitz, Mechanics of the Continuous Media).

For the second sound from (3.3) we get

$$- (\dot{\tilde{\alpha}} + \sigma_o \dot{\tilde{\beta}}) = \frac{\rho_o}{2\rho_n} \sigma^2 (\nabla\tilde{\beta})^2 - \frac{1}{2} \frac{\delta\sigma}{\delta T} \dot{\tilde{\beta}} - \frac{k}{\rho} \Delta\zeta + g\zeta. \quad (3.5)$$

Average this expression for the time much larger than the period of oscillations in the second sound and take into account that the velocity of the second sound c_2 is defined by the relation

$$c_2^2 = \frac{\rho_s}{\rho_n} \frac{\sigma^2}{\left(\frac{\partial\sigma}{\partial T}\right)} \quad (3.6)$$

Then we again see that

$$\bar{\zeta} = 0 \quad (3.7)$$

in the running wave of the second sound.

This assertion holds also for calculations performed with the accuracy up to the $\gamma = \left(\frac{\partial\rho}{\partial T} \frac{T}{\rho}\right)$ quadratic terms.

The statement that $\zeta=0$ naturally holds only for the running wave. In the standing wave one can from (3.5) find a non-zero value of ζ. But for second sound we can make a more strong statement. For second sould $\dot{\tilde{\alpha}} + \sigma_0 \dot{\tilde{\beta}}$ is identically equal to zero $(j = \rho\nabla\alpha + s\nabla\beta = 0)$. Therefore from (3.5) follows that instead of (3.7) we have

$$\zeta = 0$$

for running wave of the second sound.

This result qualitatively differs from the result of R.Sorbello [3]. (See also [4]).

4. STANDING SURFACE WAVES, EXCITED BY THE SECOND SOUND

Let us briefly dwell upon the analysis of the experimental results of J.Olsen et al. [4], who have observed excitation of standing surface waves by the second sound in superfluid helium. To carry out this analysis we need equations of hydrodynamics with the accuracy up to the second order terms. Then we shall retain only nonlinear terms generated by the second sound; these terms will play the role of the driving force in the linear equations. Introduce

$$\varphi = \alpha + \sigma_o \beta$$

Then the boundary conditions on the surface according to (2.12) and (3.5) are written with the necessary accuracy as

$$- \dot{\tilde{\varphi}} = \frac{1}{2} \frac{\partial\sigma}{\partial T} (u_2^2 (\nabla\tilde{\beta})^2 - \dot{\tilde{\beta}}^2) - \frac{\kappa}{\rho} \Delta\zeta + g\zeta \quad (4.1)$$

$$\dot{\zeta} = \frac{\partial\tilde{\varphi}}{\partial z} \quad (4.2)$$

Then the equations of motion in bulk, we need, are found from (2.6)-(2.9)

$$\dot{\varphi} + \frac{1}{\rho_o} c_1^2 \delta\rho + \frac{1}{2} \frac{\partial\sigma}{\partial T} (u_2^2 (\nabla\beta)^2 - \dot{\beta}^2) + \frac{1}{2} j u_2^2 (\nabla\beta)^2 \frac{\partial\sigma}{\partial T} = 0 \qquad (4.3)$$

$$\dot{\rho} + \rho_o \Delta\varphi - \rho_o \frac{\partial\sigma}{\partial T} div (\dot{\beta}\nabla\beta) = 0 \qquad (4.4)$$

(where $\gamma = \frac{\partial}{\partial ln\rho} ln(\rho_s/\rho_n)$).

For simplicity, consider a linear standing wave, dependent on one coordinate x (in experiment [4] there has been cylindrical symmetry). The standing wave of the second sound is then defined by the expression for the potential

$$\beta = \beta_o cos(kx)\ cos(\omega t),\quad \omega = u_2 k \qquad (4.5)$$

Remember that the potential β is related to the temperature variation as

$$\dot{\beta} + \delta T = 0 \qquad (4.6)$$

From Eqs.(4.1) and (4.9), if the bulk motion is neglected (which can be done since the velocity of surfacial waves is much smaller than the velocity of the second sound), after inserting (4.5) into them, we get

$$\zeta = \zeta_o cos(2kx) \qquad (4.7)$$

$$\tilde{\varphi} = -\varphi_o \frac{sin(2\omega t)}{2\omega} \qquad (4.8)$$

where the amplitudes ζ_o and φ_o equal

$$\zeta_o = \frac{1}{4} \frac{\partial\sigma}{\partial T} \beta_o^2 \omega^2 \frac{1}{\Lambda(2k)},\quad \Lambda(2k) = g + \frac{\kappa}{\rho} (2k)^2 \qquad (4.9)$$

$$\varphi_o = \frac{1}{4} \frac{\partial\sigma}{\partial T} \beta_o^2 \omega^2 \qquad (4.10)$$

This result considerably differs from the results obtained in [3]. We see that in the main approximation the relief of the surface is time-independent. Therefore the picture of the relief on the surface, obtained in [4] by potographing with long-time exposition is not averaged over time but is stationary and time-independent. Formula (4.8) is also very important; it follows from this formula that the potential $\tilde{\varphi}$ on the surface oscillates with time but, naturally, does not depend on the coordinate (since the second sound in this approximation cannot excite the motion of the liquid as a whole for which the gradient of the potential $\tilde{\varphi}$ is responsible). Oscillation of $\tilde{\varphi}$ somewhat resembles the Josephson effect. Uncomplicated calculations, taking into account the bulk equations (4.3) and (4.4), yield a non-stationary addition to the expression (4.7) for ζ, which has the order of u_s^2/u_2^2, where

$$u_s(k) = \sqrt{k^{-1}\Lambda(k)} \qquad (4.11)$$

is the velocity of the surfacial wave.

Let us give the result for this non-stationary addition to

$$\zeta_1 = -\frac{1}{S} \beta_o^2 \frac{\partial\sigma}{\partial T} cos(2\omega)\ cos(2kx) \qquad (4.12)$$

In more detail these problems will be investigated in the article written together with V.L.Pokrovsky, submitted for publication.*

Acknowledgement: I wish to thank Prof. T.M.Rice for his hospitality during my stay at ETH and Prof. J.L.Olsen for very stimulating discussions.

REFERENCES

1. I.M.Khalatnikov, Zh. Eksp. Teor. Fiz. 23: 169 (1952).
2. V.L.Pokrovskii and I.M.Khalatnikov, Pis'ma Zh. Eksp. Fiz. 23: 653 (1976); [JETP Lett. 23: 599 (1976)]; Zh. Eksp. Teor. Fiz. 71: 1974 (1976); [JETP, 44: 1036 (1976)].
3. R.S.Sorbello, Surface Profile of Second. Sound Standing Waves in Superfluid Helium. J. Low Temp. Phys. 23: 411 (1976).
4. J.Olsen, Standing Surface Waves on Helium Induced by Second Sound. Observations and Calculations. Journ. of Low Temp. Phys.,61: 17(1985).

---

* The author is sincerely thankful to S.Korshunov for useful discussion of the derivation of Formula (4.12).

COLLECTIVE TRANSPORT IN CHARGE AND SPIN DENSITY WAVES

Kazumi Maki

Department of Physics
University of Southern California
Los Angeles, CA 90089-0484 USA

Resumé

After a brief introduction on the subject, we review recent development on the theoretical understanding on the collective transport (Fröhlich conduction) in charge and spin density waves in quasi-one dimensional conductions. Especially the theoretical prediction on the depinning electric field in spin density wave (SDW) is compared with the recent observation in SDW of $(TMTSF)_2$ PF_6.

INTRODUCTION

As is well known, the normal electron in metals is unstable in general in the presence of the interaction. The best known example is the superconductivity which arises from an attractive interaction between a pair of electrons in the vicinity of the Fermi surface as formulated in the theory of Bardeen, Cooper and Schrieffer.[1] A more exotic instability in the one-dimensional electron system has been considered by Peierls[2] and Fröhlich[3] even before the emergence of the BCS theory of superconductivity. Exotic since the theory concerns the one-dimensional system, which does not exist in nature. This instability leads to charge density wave (CDW) or spin density wave (SDW) state depending on the nature of the interaction. Further, when the wave vector associated with CDW or SDW is incommensurate with the underlying crystal lattice, Fröhlich[3] predicted the superconductivity in CDW or SDW. Later, it was shown by Lee, Rice and Anderson[4] that CDW (or SDW) does not superconduct, since any crystal defects or impurities can pin down the CDW and SDW. The limitation of the one dimensional model is gradually removed when it is discovered that a number of quasi one-dimensional metallic compounds exhibit the CDW and SDW instability at low temperatures. Indeed, these instabilities take place in real three-dimensional systems when the Fermi surface satisfies the nesting condition.

As to the Fröhlich conduction, it was Bardeen[5] who first recognized its relevance in interpreting the nonohmic conductivity[6] observed in CDWs of the quasi-one dimensional inorganic compound $NbSe_3$. Since then a large class of inorganic and organic compounds which exhibit the Fröhlich conduction have been discovered. Excellent reviews[7 ~ 9] on this subject are available now.

More recently, the nonohmic conductivity is observed[10 ~ 13] in a number of SDWs in Bechgaard salts ($(TMTSF)_2$ X where TMTSF means tetramethyl tetraselenafulvalinium and X is cation). In this review, we shall describe the recent development on this subject from a theoretical point of view.

Microscopic Aspects of Nonlinearity in Condensed Matter
Edited by A.R. Bishop *et al.*, Plenum Press, New York, 1991

MODEL

For definiteness, we shall consider a quasi-one dimensional Fröhlich Hubbard model given by

$$H = \sum_{p\alpha} \xi(p)\, C^{+}_{p\alpha} C_{p\alpha} + ig \sum_{pq} \left(\frac{1}{2}\omega_q\right)^{\frac{1}{2}} C^{+}_{p+q,\alpha}\, C_{p\alpha} \left(b^{+}_{q} + b_{-q}\right) + \sum_{q} \omega_q\, b^{+}_{q} b_q + U \sum_{q} n_{q\uparrow}\, n_{-q\downarrow} \tag{1}$$

where

$$\xi(p) = -2t_a \cos ap_1 - 2t_b \cos bp_2 - 2t_c \cos cp_3 - \mu \tag{2}$$

with $t_a > t_b > t_c$.

Here, $C_{p\alpha}{}^{+}$ and $C_{p\alpha}$ are the electron creation and annihilation operators with momentum p and spin α and $b_q{}^{+}$ and b_q are the phonon creation and annihilation operators. In the limit $g \to 0$, Eq(1) reduces to the anisotropic Hubbard model first considered by Yamaji[14]. When $t_a >> t_b >> t_c$, $\xi(p)$ is simplified as[14]

$$\xi(p) = v(p_1 - p_F) - 2t_b \cos bp_2 - \epsilon_0 \cos(2bp_2) \tag{3}$$

with

$$\epsilon_0 = -\tfrac{1}{2} t_b^2 \cos ap_F\, (t_a \sin^2 ap_F)^{-1} \tag{4}$$

and ϵ_0 describes the degree of the imperfect nesting.

Within mean field approximation, the ground state of the present system is either CDW or SDW. For $g^2 > U$, we will have CDW while for $U > g^2$ SDW.

In the following, we shall neglect the phonon term and concentrate on SDW. For a nesting vector $Q = (2p_F, \pi/b, \pi/c)$ the quasi-particle Green's function in SDW is given by

$$G^{-1}(\omega_n, \vec{p}) = i\omega_n - \eta - \xi\rho_3 - \Delta\rho_1\sigma_3 \tag{5}$$

where $\eta = \epsilon_0 \cos(2bp_2)$ and ω_n is the Matsubara frequency and ρ_i's are the Pauli matrices operating on the spinor space formed by the right-going and the left-going electrons.

The corresponding gap equation is given by [14,15]

$$1 = N_0 U \pi T \sum_{n=0}^{n_c} \int_0^{2\pi} \frac{d\chi}{2\pi} \left((\omega_n - i\eta)^2 + \Delta^2\right)^{-\frac{1}{2}} \tag{10}$$

where $\chi = 2bp_2$.

In particular, the transition temperature is given by

$$1 = \frac{1}{2} N_0 U \, 2\pi T \sum_{n=0}^{n_c} (\omega_n^2 + \epsilon_0^2)^{-\frac{1}{2}} \tag{11}$$

which means that T_c for nonvanishing ϵ_0 is given in that for $\epsilon_0 = 0$ by

$$\epsilon_0 = \Delta(T_c/T_{co}) \tag{12}$$

where $\Delta(T/T_c)$ is the BCS energy gap at temperature T. In particular, T_c vanishes within this model when $\epsilon_0 = \Delta_0$ the energy gap at T = 0K. It is well known that the CDW transition temperatures of $NbSe_3$ and the SDW transition temperature of $(TMTSF)_2PF_6$ decreases with pressure.[16, 17] Such pressure dependences are well described with the present model, if we assume ϵ_0 increases linearly with pressure.[14, 18] Further, the present model describes [19 ~ 21] also the field induced SDW discovered by Kwak et al.[22] in $(TMTSF)_2ClO_4$ and $(TMTSF)_2PF_6$ under pressure. Finally, we add that the present model accounts[18] for the large ratio of $2\Delta_a / k_B T_c \simeq 11 \sim 13$ determined for CDWs in $NbSe_3$ by the electron tunneling technique[23,24] where $2\Delta_a$ is the apparent energy gap. A typical electron density of states in CDW (or SDW) is shown in Fig. 1, where the apparent energy gap is given by $\Delta_a = \Delta(T) + \epsilon_0$.

Common to both CDW and SDW is the appearance of the phason with dispersion[25].

$$\omega^2 = (1+\bar{U})\, v^2 q_1^2 + v_2^2 q_2^2 + v_3^2 q_3^2 \tag{13}$$

which corresponds to the sliding motion of the SDW. Here $\bar{U} = N_0 U$ and v, v_2 and v_3 are anisotropic Fermi velocity given by

$$(v, v_2, v_3) = (2t_a a \sin a p_F, \sqrt{2}\, t_b b, \sqrt{2}\, t_c c) \tag{14}$$

For Bechgaard salts, we have typically $v_2/v \simeq 10^{-1}$, $v_3/v \simeq 10^{-3}$, while for $NbSe_3$ we have [26] $v_2/v \simeq 1/4$ and $v_3/v \simeq 10^{-1}$. In CDW, ω^2 in the l.h.s. of Eq(13) has to be multiplied by m^*/m where m^* and m are the phason mass and the band electron mass. One of the remarkable features of Eq(13) is that this dispersion is independent of T the temperature.

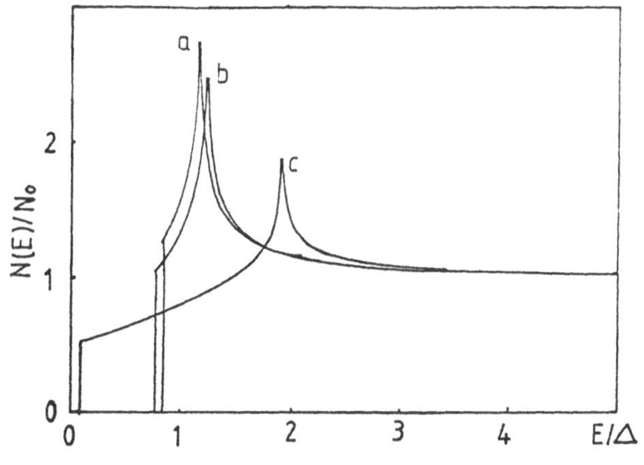

Fig. 1 Electron density of states $N(E)/N_0$ are shown as function of energy E for $\epsilon_0/\Delta = 0.16$ (a), 0.23 (b) and 0.93 (c). (a) and (b) corresponds to that in SDW of Bechgaard salts, while c that in CDWs of $NbSe_3$.

In SDW, we have in addition the spin wave mode with dispersion[27]

$$\omega^2 = (1 - \bar{U})\, v^2 q_1^2 + v_2^2 q_2^2 + v_3^2 q_3^2 \tag{15}$$

These constitute the Goldstone modes in SDW. However, both the translational symmetry and the spin rotation symmetry of the system are in general broken due to crystalline defects, impurities, the magnetic dipole interaction and the spin orbit scattering and the Goldstone modes acquire nonvanishing energy gaps. We shall look into more details of the phason dynamics.

PHASE HAMILTONIAN

The phase Hamiltonian, which describes the phason dynamics of SDW, is given by[28]

$$H(\phi) = \int d^D x \left\{ \frac{1}{4} N_0 f \left[\left(\frac{\partial\phi}{\partial t}\right)^2 + \tilde{v}^2 \left(\frac{\partial\phi}{\partial x}\right)^2 + v_2^2 \left(\frac{\partial\phi}{\partial y}\right)^2 + v_3^2 \left(\frac{\partial\phi}{\partial z}\right)^2 \right] - enfQ^{-1}\, \phi E \right\} + V_{pin}(\phi) \tag{16}$$

where $\tilde{v}^2 = (1+\bar{U})v^2$, $Q = 2p_F$ and $n = 2p_F\, (\pi bc)^{-1}$ is the electron density and f is the condensate density. Further, Eq(16) has to be supplemented by the charge and the current associated with slow spatial temporal distortion of the phase of the order parameter ϕ;

$$n_c = enf\, Q^{-1} \frac{\partial\phi}{\partial x} \tag{17}$$

$$j_c = -enf\, Q^{-1} \frac{\partial\phi}{\partial t} \tag{18}$$

Equations (17) and (18) satisfy the charge conservation

$$\frac{\partial n_c}{\partial t} + \frac{\partial j_c}{\partial x} = 0 \tag{19}$$

which tells that the charge carried by the condensate is conserved at all temperatures. The charge conservation for the condensate is only broken in the presence of time dependent topological defects like phase vortices.[29] When the phase vortices lie in the y - z plane in the direction $\hat{e}_i$ and moving with velocity v_i, the right hand side of Eq (19) is replaced by

$$2\pi enQ^{-1} f \sum_i (\vec{v}_i \times \hat{e}_i)_x\, \delta(\vec{x} - \vec{x}_i(t)) \tag{20}$$

where the sum is over the moving vortices. The condensate density f is in general a complicated function[25] of ω and vq, where ω and q are the frequency and the wave vector associated with the phase ϕ. However, in the adiabatic limit (i.e. $|\omega|$, vq, $\ll \Delta_o$) f takes a few limiting values

$$f = \begin{cases} f_0 & \text{for } \omega > vq_1 \\ f_1 & \text{for } \omega < vq_1 \\ 1 & \text{for } \omega = vq_1 \end{cases} \tag{21}$$

In particular, in the limit of small ϵ_0, f_1 (the static condensate density) has the same temperature dependence as the superfluid density in a BCS superconductor.

$$f_1 = \rho_S(T)/\rho \tag{22}$$

For $\epsilon_0 = 0$, f_0 and f_1 are evaluated numerically and shown in Fig. 2.

Finally pinning potential $V_{pin}(\phi)$ for a SDW is given as[25, 30]

$$V_{imp}(\phi) = -\left(\frac{\pi}{2} N_0 V\right)^2 \Delta(T)\tanh\left(\frac{1}{2}\beta\Delta(T)\right) \sum_i \cos\left(2(\vec{Q}\cdot\vec{x}_i + \phi(\vec{x}_i))\right) \tag{23}$$

if it is due to the impurities where V is the impurity potential and the sum over i extends over the impurity sites. When $\vec{Q}$ is on the other hand commensurate with the crystal lattice we have commensurability potential

$$V_c(\phi) = -4(UW^2)^{-1}\,\Delta^4(T)\cos(4\phi(\vec{x})) \tag{24}$$

where $W = 4t_a$ the total band width and we assumed $N = 4$ corresponding to the 3/4 filled electron band in the Bechgaard salts. We note that the above commensurability potential is one order of magnitude larger than that estimated in Ref. 4.

So far, we considered only the case of the SDW. In a CDW, first $(\partial\phi / \partial t)^2$ in Eq(16) has to be multiplied by $m^*/m \sim 10^3$. Second, the pinning potential due to impurity is simply given by

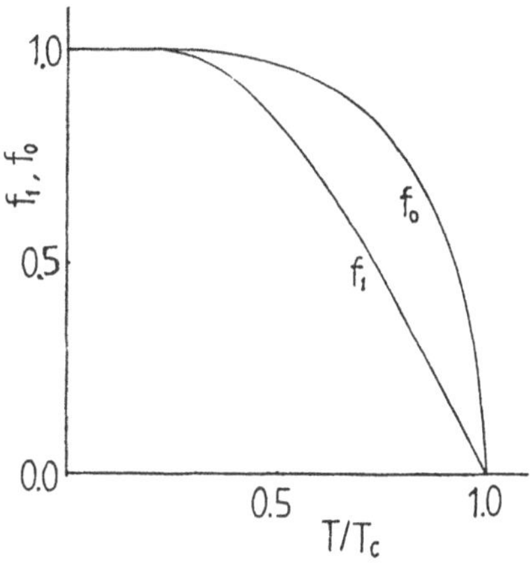

Fig. 2 The dynamical and the static condensate density f_0 and f_1 are shown as function of the reduced temperature for $\epsilon_0/\Delta_0 = 0$.

$$V_{\text{pin}}(\phi) = -2N_0 V\lambda^{-1}\Delta(T)\sum_i \cos\left(\vec{Q}\cdot\vec{x}_i + \phi(\vec{x}_i)\right) \tag{25}$$

while if it is due to the commensurability, we have a similar expression as Eq(24) except U has to be replaced by $2g^2$ when N ≐ 4. However, when N = 3 is more relevant as in TTF-TCNQ[31] we will have

$$V_c(\phi) = -\frac{2}{3}\left(g^2 t_a\right)^{-1}\Delta^3(T)\cos(3\phi(\vec{x})) \tag{26}$$

Again the coefficient of Eq(26) is one order of magnitude larger than that estimated in Ref 4. We note that the relevant commensurability potential for sliding CDW and SDW should be N = 3 or 4. When N = 2 as in transpolyacetylene, the pinning potential is too large and there will be no sliding CDW. Even when N = 3, the commensurability potential dominates the pinning. While when N = 4, $V_c(\phi)$ in Eq(24) will give a threshold electric field of the order of milliVolt/cm which can barely compete with the impurity pinning. When N = 5, the commensurability potential is in most cases completely negligible.

In the following, we shall consider application of the phase Hamiltonian.

DEPINNING ELECTRIC FIELD

In general, the CDW (or SDW) is pinned by impurities and/or commensurability. First, let us consider the CDW (or SDW) is pinned by impurity. Following Fukuyama, Lee and Rice[32], we distinguish two limiting cases, the strong pinning and the weak pinning case.

a) Strong Pinning Limit

In the strong pinning limit, the individual impurities pin the local phase $\phi(\vec{x}_i)$ of the order parameter. The resulting pinning energy is proportional to the impurity concentration n_i and given by

$$E_{\text{pin}} = \begin{cases} 2n_i N_0\, V\lambda^{-1}\Delta(T) \\ n_i\left(\frac{\pi}{2}N_0 V_2\right)^2\Delta(T)\tanh\left(\frac{1}{2}\beta\Delta(T)\right) \end{cases} \tag{27}$$

for CDW and SDW respectively.

Then the threshold electric field is given by [28]

$$E_T^S(T) = \begin{cases} (Q/e)\,(n_i/n)\,2\lambda^{-1}N_0 V\Delta(T)\,f_1^{-1} \\ (Q/e)\,(n_i/n)\,(\pi N_0 V)^2\,\Delta(T)\tanh\left(\frac{1}{2}\beta\Delta(T)\right)f_1^{-1} \end{cases} \tag{28}$$

for CDW and SDW respectively.

Equation (28) tells that $E_T^S(0)$ for a CDW (say, in $NbSe_3$) is larger than that in $(TMTSF)_2PF_6$ roughly by a factor of 10^2 for similar impurity concentrations, since $N_0V \simeq 0.1$ and $\Delta_{CDW}(0) / \Delta_{SDW}(0) \simeq 10$.

As it turns out, Eq(28) cannot describe the observed $E_T(T)$ in CDWs of $NbSe_3$ at low temperatures which decreases exponentially with increasing T for example. In the presence of the thermal fluctuation[33], the cosine potential as given in Eq(25) has to be replaced

$$\cos\left(\vec{Q}\cdot\vec{x}_i + \phi(\vec{x}_i)\right) \rightarrow \exp\left[-\frac{1}{2}\langle\phi^2\rangle\right]\cos\left(\vec{Q}\cdot\vec{x}_i + \phi(x_i)\right)$$

$$= e^{-\frac{T}{T_0}}\cos\left(\vec{Q}\cdot\vec{x}_i + \phi(\vec{x}_i)\right) \tag{29}$$

with $T_0 = Ct_bt_c/\Delta_0$ (30)

and C is a constant of the order of unity. Therefore in general $E_T{}^S(T)$ given in Eq(28) has to be multiplied by e^{-T/T_o}. Then Eq(28) predicts that as T approaches T_c $E_T(T)$ in a CDW diverges like $(T_c - T)^{-1/2}$, while $E_T(T)/E_T(0)$ in a SDW approaches a constant value 1.33 (In SDW, we neglected the thermal fluctuation since T_o is much larger than T_c the transition temperature.) Before making the comparison, we shall consider the other limits.

b) Weak Pinning Limit

When the pinning potential is small or the impurity concentration is small, a single impurity cannot pin the local phase $\phi(x_i)$. In this circumstance, the pinning energy is calculated by optimizing the spatial distortion and the collective impurity potential. Unlike the strong pinning limit, the pinning potential also depends on the sample size.[34, 36] For the 3D samples (i.e. we mean all the sample dimensions are larger than the corresponding Fukuyama, Lee Rice length L, $(v_2/v)L$, and (v_3/v) L where L will be defined shortly) we obtain the threshold field

$$E_T^{W3}(0)/E_T^S(0) = \begin{cases} 2n_i\eta^{-2}\left(\frac{9}{\pi}(N_0V)\frac{bc}{\lambda\xi}\right)^3 \\ \frac{1}{4}\left(\frac{3}{4}\right)^3 n_i\,\eta^{-2}(N_0V_2)^6\left(\frac{3\pi bc}{\xi}\right)^3 \end{cases} \tag{31}$$

for CDW and SDW respectively where $\eta = v_2v_3/v^2$ and $\xi = v/\Delta_o$. On the other hand, the temperature dependence of $E_T^{W3}(T)$ is given by

$$E_T^{W3}(T)/E_T^{W3}(0) = \left(E_T^S(T)/E_T^S(0)\right)^4 \tag{32}$$

for both CDW and SDW.

For CDW assuming $\lambda = (N_0V) = 0.1$, $\xi = 300$ A°, b = 10 A°, C = 15 A°, we will have $E_T^{W3}(0)/E_T{}^S(0) = 6.066 \times 10^{-22}\, n_i$. This implies that the pinning changes from weak pinning to the strong pinning when n_i is 0.1 percent of the electron density. For smaller n_i, the CDW is in the weak pinning limit. In SDW, on the other hand, the system is always in the weak pinning limit when $N_0V \sim 0.1$

The corresponding Fukuyama-Lee-Rice length is now given by

$$L(T) = \begin{cases} \frac{\pi}{6}\,(v/eE_T(T))^{\frac{1}{2}} \\ \frac{\pi}{3}\,(v/eE_T(T))^{\frac{1}{2}} \end{cases} \tag{33}$$

for CDW and SDW respectively. For $E_T(T) = 1$ mV/cm we will obtain $L(T) \approx 10^2$ μm, if we assume $v \approx 3 \times 10^7$ cm/sec. This FLR length is consistent with the one deduced by McCarten et al.[34] but clearly inconsistent with an earlier analysis by Tucker et al.[35] For 2D CDW or SDW (i.e. when one of sample dimensions which we will call t is smaller than $(v_2/v)L$),[34, 36] the threshold field is given by

$$E_T^{W2}(0)/E_T^S(0) = \begin{cases} \frac{6}{\pi}\,(N_0V)\,(bC/\lambda\eta_2\xi t) \\ \frac{3\pi}{4}\,(N_0V)^2\,(bc/\eta_2\xi t) \end{cases} \tag{34}$$

for CDW and SDW respectively where now $\eta_2 = v_2/v$ or v_3/v. It appears whenever the system is 2D, the system should be always in the weak pinning limit, since the ratio Eq(34) is always less than unity as long as t is not less than say 1 μm.

The temperature dependence of $E_T^{W2}(T)$ is given by

$$E_T^{W2}(T)\ /\ E_T^{W2}(0) = \left(E_T^S(T)\ /\ E_T^S(0)\right)^2 \tag{35}$$

Combining Eqs(31), (32), (34) and (35), we obtain

$$\left(E_T^{W2}(T)t\right)^2\ /\ E_T^{W3}(T) = \begin{cases} \frac{4}{27}\,\alpha v \left(\frac{\eta}{\eta_2}\right)^2 e^{-1} \\ \frac{16}{27}\,\alpha v \left(\frac{\eta}{\eta_2}\right)^2 e^{-1} \end{cases} \tag{36}$$

for CDW and SDW respectively. This ratio is independent of kind of impurity, the impurity concentration and the temperature. Therefore, the constancy of the above ratio will demonstrate the validity of the Fukuyama, Lee, Rice approach.

When only $E^S(T)$ and $E_T^{W3}(T)$ are available, we also will have

$$\left(E^S(T)\right)^4\ /\ n_i^2\ E^{W3}(T) = \begin{cases} (4/3e)^3\,\alpha v \\ (16/3e)^3\,\alpha v \end{cases} \tag{37}$$

for CDW and SDW. Again this ratio is independent of kind of impurities and temperature.

Finally, when both transverse dimensions of the sample are smaller, the corresponding FLR length we have 1D CDW or SDW with[36]

$$E_T^{W1}(0) / E_T^S(0) = \begin{cases} \frac{3}{4} n_i^{-\frac{1}{3}} a^{-\frac{2}{3}} \left(\frac{6}{\pi} (N_0 V)\, bc\, (\lambda \xi)^{-1} \right)^{\frac{1}{3}} \\ \frac{3}{4} n_i^{-\frac{1}{3}} a^{-\frac{2}{3}} \left(\frac{3\pi}{4}\, bc\, \xi^{-1} (N_0 V)^2 \right)^{\frac{1}{3}} \end{cases} \tag{38}$$

for CDW and SDW respectively where a is the cross sectional area. Further, the temperature dependence of $E_T^{W1}(T)$ is given by

$$E_T^{W1}(T) / E_T^{W1}(0) = \left(E_T^S(T) / E_T^S(0) \right)^{\frac{4}{3}} \tag{39}$$

In Fig. 3, we show the threshold electric field measured for CDWs in $NbSe_3$ by Fleming[37], together with our theoretical expression for $E_T(T)$. We find that in this particular sample, we can describe $E_T(T)$ in the first CDW with T_C = 144 (CDW I) in terms of the 2D weak pinning model, while $E_T(T)$ of the second CDW with the strong pinning model.[28] More generally, the present theory describes the temperature dependence of the threshold electric field of CDWs compiled by Monceau.[7] Most of data are described either by the 3D weak pinning or the strong pinning theory. Only exceptions to this general rule are the threshold electric field in orthorhombic TaS_3 and blue-bronze $K_{0,3}M_0O_3$.

Coming back to SDW in $(TMTSF)_2PF_6$, three distinct types of the temperature dependence of $E_T(T)$ has been observed[12, 13] as shown in Fig. 4. The behavior O and ● are observed for pristine samples while the behavior ▵, □ and ▴ are observed for the samples irradiated with X ray. As seen from Fig. 4, the behavior a and b are described the threshold field due to the commensurability and the 3D weak pinning limit respectively, whole the behavior ▵, □ and ▴ the strong pinning theory.

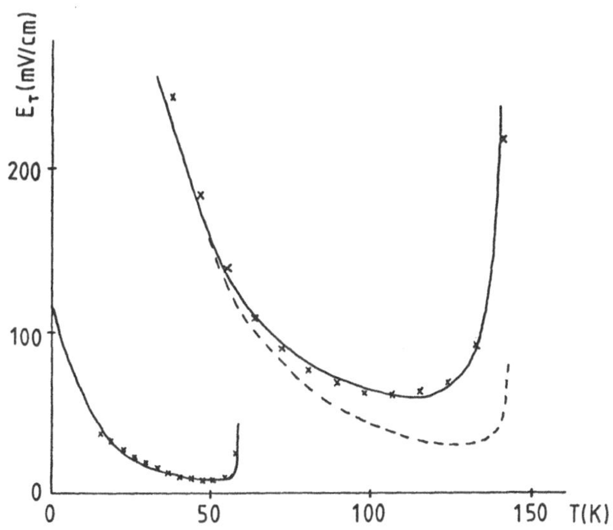

Fig. 3 The threshold electric field of CDWs in $NbSe_3$ taken from Ref. 37 is compared with the theoretical result. E_T of CDWI (with T_c = 144K) is well-described by two dimensional weak pinning model while that of CDW II (with T_c = 59K) is by strong pinning model.

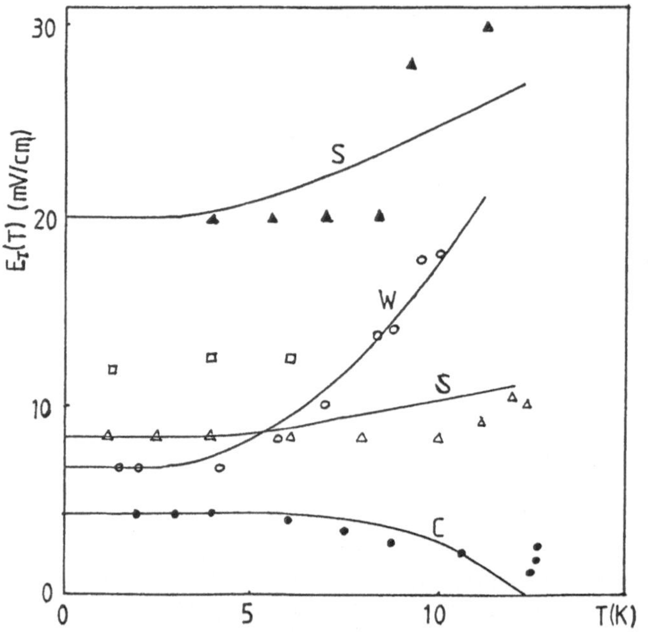

Fig. 4 The threshold electric field of SDWs in $(TMTSF)_2PF_6$ taken from Refs. 12 and 13 are compared with theoretical result data and O and ● are taken from pristine samples, while Δ, □, and ▲ are from X-ray irradiated samples. The curve S, W, and C are theoretical results for the strong pinning limit, three-dimensional weak pinning limit and the pinning due to commensurability.

Therefore, except a few exceptions, the present theory describes quantitatively the observed threshold electric field both in CDW and SDW. This establishes the validity of both the phase Hamiltonian and the Fukuyama-Lee-Rice approach. Since the phase Hamiltonian is based on mean field theory, we stress that mean field theory can describe as well the CDW and SDW in quasi-one dimensional systems. In other words, we believe that all fluctuation contributions to the system can be analyzed within perturbation theory.

CONCLUDING REMARKS

We have discussed the phason dynamics in terms of the phase Hamiltonian with particular attention to the depinning electric field. From comparison with experimental data on the threshold electric field in CDW of $NbSe_3$ and SDW in $(TMTSF)_2PF_6$, we have established the validity of both Fukuyama, Lee, Rice (FLR) approach to the pinning problem and mean field approach which is used to obtain $H(\phi)$ from a microscopic model. It is hoped that this will clarify the controversy surrounding the origin of the depinning field at least for these special cases discussed here.

We believe also the present approach can be used to describe quantitatively both the nonohmic conductivity and the microwave conductivity in both CDW and SDW. But these works are left to be done. In general, we believe that mean field theory is used successfully not only for a superconductor but also for CDW and SDW as well, since these systems are essentially three-dimensional, though their electronic properties are quite anisotropic.

ACKNOWLEDGEMENTS

It is a pleasure to thank Attila Virosztek for collaboration, Xiaozhou Huang for calculating Fig. 1 and Silvia Tomić for useful discussions. I thank also W. Kang et al. for providing some of experimental data before publication. I thank the hospitality of Max-Planck Institut für Festkörperforschung at Stuttgart where most part of this review is written. This work is supported by the National Science Foundation under grant number DMR 86-11829 and DMR 89-15285.

REFERENCES

1. J. Bardeen, L. N. Cooper and J. R. Schrieffer, Phys. Rev. 108 1175 (1957)
2. R. E. Peierls, Quantum Theory of Solids (Oxford University Press, Oxford, 1955) p. 108
3. H. Fröhlich, Proc, Roy, Soc. London A223 296 (1954)
4. P. A. Lee, T. M. Rice and P. W. Anderson, Solid State Commun 14 703 (1974)
5. J. Bardeen, in "Quasi-One Dimensional Conductors I", Lecture Notes in Physics, edited by S. Barisić, A. Bjelis, J. R. Cooper and B. Leontić (Springer-Verlag, Berlin 1979) Vol. 95 3
6. P. Monceau, N. P. Ong, A. M. Portis, A. Meerschaut and J. Rouxel, Phys. Rev. Lett. 37 602 (1976)
7. P. Monceau, in Electronic Properties of Inorganic Quasi-One Dimensional Compounds, Part II edited by P. Monceau (Reidel, Dordrecht 1985)
8. G. Grüner and A. Zettl, Physics Reports 119 117 (1985)
9. G. Grüner, Rev. Mod. Phys 60 1129 (1988)
10. S. Tomić, J. R. Cooper, D. Jérome and K. Bechgaard, Phys. Rev. Lett. 62 462 (1989)
11. T. Sambongi et al., Solid State Commun. 72 817 (1989); K. Nomura et al., ibid 72 1123 (1989)
12. W. Kang, S. Tomić, J. R. Cooper and D. Jérome, Phys. Rev. B41 4862 (1990)
13. W. Kang, S. Tomić, D. Jérome, Phys. Rev. Lett. (submitted)
14. K. Yamaji, J. Phys. Soc. Jpn 51 2787 (1982), 52 1361 (1983)
15. Y. Hasegawa and H. Fukuyama, J. Phys. Soc., Jpn 55 3978 (1986)
16. D. Jérome, A. Mazaud, M. Ribault and K. Bechgaard, J. Phys. Lett. (Paris) 41 195 (1980)
17. A. Briggs, P. Monceau, M. Nuñez-Regueiro, J. Peyrard, M. Ribault and J. Richard, J. Phys. C 13 2117 (1980)
18. X. Z. Huang and K. Maki, Phys. Rev. 40 2575 (1989)
19. K. Yamaji, J. Phys. Soc. Jpn 54 1034 (1985); Synth. Met 13 29 (1986)
20. D. Poilblanc, M. Héritier, G. Montambaux, and P. Lederer, J. Phys. C 19 L321 (1986)
21. A. Virosztek, L. Chen, and K. Maki, Phys. Rev. B 34 3371 (1986)
22. J. F. Kwak, J. E. Shirber, R. L. Greene, and E. M. Engler, Phys. Rev. Lett. 46 1296 (1981)
23. A. Fournel, J. P. Sorbier, M. Konczykowski, and P. Monceau, Phys. Rev. Lett. 57 2199 (1986)
24. T. Ekino and J. Akimitsu, Jpn. J. Appl. Phys 26 Supple. 625 (1987)
25. A. Virosztek and K. Maki, Phys. Rev. B37 2028 (1988)
26. R. E. Thorne and J. McCarten, Phys. Rev. Lett. 65 272 (1990)
27. K. Maki and A. Virosztek, Phys. Rev. B36 511 (1987)
28. K. Maki and A. Virosztek, Phys. Rev. B 39 9640 (1989); ibid 41 557 (1990) and ibid 42 655 (1990)
29. N. P. Ong and K. Maki, Phys. Rev. B 32 6582 (1985)
30. P. F. Tua and J. Ruvalds, Phys. Rev. B 32 4660 (1985)
31. R. C. Lacoe, J. R. Cooper, D. Jérome, F. Creuzet, K. Bechgaard, and I. Johannsen, Phys. Rev. Lett. 58 262 (1987)
32. H. Fukuyama and P. A. Lee, Phys. Rev. B17 535 (1978); P. A. Lee and T. M. Rice, Phys. Rev. B19 3970 (1979)

33. K. Maki, Phys. Rev. B33 2852 (1986)
34. J. McCarten, M. Maher, T. L. Adelman, and R. E. Thorne, Phys. Rev. Lett. 63 2841 (1989)
35. J. R. Tucker, W. G. Lyons, and G. Gammie, Phys. Rev. B38 1148 (1988)
36. J. Bardeen, Phys. Rev. Lett. 64 2297 (1990)
37. R. M. Fleming, Phys. Rev. B22 5606 (1980)
38. K. Maki and A. Virosztek, Phys. Rev. B39 2511 (1989)

PROPAGATING COMPRESSION WAVES IN A DAMPED ac-DRIVEN CHAIN OF KINKS

Boris A. Malomed

P.P. Shirshov Institute for Oceanology of the USSR Academy of Sciences
23 Krasikov Street, Moscow, GSP-7, USSR

ABSTRACT

It is demonstrated that ac drive may support stable propagation of a solitary compression wave ("supesoliton") in a chain of weakly interacting kinks governed by a damped sine-Gordon model, while the chain as a whole remains quiescent. The supersoliton is described as a propagating soliton in an ac-driven damped Toda lattice. It may propagate at resonant velocities determined by the drive's frequency, provided the driving amplitude exceeds a certain threshold value to compensate dissipative losses. Physical realizations of this effect are interpreted in terms of long Josephson junctions (LJJ's) and commensurate charge-density-wave (CDW) systems: Application of ac bias current to a chain of fluxons in LJJ may give rise to dc voltage, and application of ac external electric field to a chain of charged phase solitons in the CDW system may generate dc current.

INTRODUCTION

The ac-driven damped sine-Gordon (SG) equation,

$$u_{tt} - u_{xx} + \sin u = -\alpha u_t - f \sin(\omega t), \qquad (1)$$

is one of fundamental dynamical models of solid state physics. Important applications are long Josephson junctions, where $u(x,t)$ and f are, respectively, the magnetic flux trapped by the junction and amplitude of ac bias current, and commensurate charge-density-wave (CDW) systems, $u(x,t)$ being a phase mismatch of the CDW, f being the amplitude of ac voltage applied to the system. The same model (1) describes a chain of interacting atoms adsorbed by a sinusoidal substrate and subject to the action of the external ac electric field. In all the applications, α stands for a phenomenological loss constant.

A fundamental topological-soliton solution to the unperturbed equation (1) ($\alpha = f = 0$) is the 2π kink:

$$u_k = 4\tan^{-1}\left[\exp(x - z)\right], \qquad (2)$$

z being the kink's coordinate. The kink represents a fluxon (magnetic flux quantum) in LJJ, a charged phase soliton in the CDW system, and a dislocation in the adsorbed atomic chain.

An interesting physical object is a periodic chain of kinks. In the present paper, only the case of a large chain's spacing b (a rarefied chain) is analyzed. In more accurate terms, it will be assumed $e^{-b} \ll 1$. To derive equations governing evolution of disturbances in the rarefied chain, let us recollect that a solitary kink may be regarded as a classical particle with the mass m = 8 subject to the action of the friction and driving forces corresponding, respectively, to the first and second terms on the right-hand side of Eq. (1),[1] and the weak overlapping between the kinks gives rise to exponential repulsion forces.[2] With regard to this, it is straightforward to derive the equation of motion for the coordinate of the n-th kink in the chain:

$$\ddot{z}_n = 4\left(\exp\left[-(z_n - z_{n-1})\right] - \exp\left[-(z_{n+1} - z_n)\right]\right) - \alpha\,\dot{z}_n + (\pi f/4)\cdot\sin(\omega t)\,. \qquad (3)$$

The change of scale

$$t = \tfrac{1}{2}e^{b/2}T,\ \Omega = \tfrac{1}{2}e^{b/2}\omega\,,\ A = \frac{1}{4}e^{b}\alpha\,,\ F = (\pi/16)e^{b}f \qquad (4)$$

transforms Eq. (3) into the ac-driven damped Toda-Lattice (TL) equation for the variable $y_n \equiv z_n - bn$, which has the meaning of displacement of the n-th kink from its equilibrium position:

$$\frac{d^2}{dT^2}y_n = e^{-(y_n - y_{n-1})} - e^{-(y_{n+1} - y_n)} - A\frac{d}{dT}y_n + F\sin(\Omega T)\,. \qquad (5)$$

In what follows, the renormalized friction and drive constants A and F will be treated as small parameters, and the analysis will be developed within the framework of the perturbation theory.

The chain of the interacting kinks may support propagation of a compression wave. In terms of the effective TL equation (5), a solitary compression wave corresponds to a TL soliton, which, in the zeroth approximation (A = F = 0), has the well-known form:

$$y_n(T) = \ln\left(1 + (\xi^{-2} - 1)\,\xi^{2(n - n_o - VT)} \,/\, \left[1 + \xi^{2(n-n_o-VT)}\right]\right), \qquad (6)$$

where ξ is a parameter taking values $-1 < \xi < +1$, n_o is an arbitrary phase constant, and

$$V = \xi^{-1}(1 - \xi^2)/\ln(\xi^{-2}) \qquad (7)$$

is the soliton's velocity. A characteristic size of the TL soliton (7) is $\Delta n \sim 1/\ln(\xi^{-2})$. As it follows from Eq. (7), the shift of the displacement y_n at $n = -\infty$ relative to that at $n = +\infty$ is

$$\Delta y = \ln(\xi^{-2}), \qquad (8)$$

i.e., the soliton may indeed be interpreted as a solitary compression wave.

The weak dissipation causes a slow decay of the TL soliton into nonsoliton (dispersive) waves. In the limit $T \to \infty$ the parameter ξ^2 takes the limit value $\xi^2 = 1$, and the soliton disappears. The dissipation-induced decay of the TL soliton has been recently investigated in detail, by means of numerical and qualitative methods, in Ref. 3. An important result of numerical simulations reported in Ref. 3 is that the decaying soliton with the slowly varying parameter ξ remains close in form to the unperturbed one (6).

It is typical in various physical problems that dissipative damping of a soliton may be compensated by an external dc drive applied to the system, and the condition of the complete compensation uniquely determines parameters of the stabilized soliton.[1] In the same time, an external ac drive cannot support a stable propagation of a kink in a damped system (ac drive may support a quiescent oscillating soliton phase-locked to it, e.g., a nonlinear-Schrödinger soliton[4] or a SG breather [5]). In the present work I will demonstrate that this is, nevertheless, possible in discrete systems, TL being the simplest example.

ac-DRIVEN PROPAGATION OF THE DAMPED TL SOLITON

First of all, let us calculate the total energy E_{diss} dissipated at the n-th site of the lattice during the passage of the soliton (6):

$$E_{diss} = A\int_{-\infty}^{+\infty} dT(dy_n/dT)^2 = A|\xi|^{-1}\left[(1+\xi^2)\ln(\xi^{-2})-2(1-\xi^2)\right]. \qquad (9)$$

To find the quantity (9), the expression for the velocity V in terms of the parameter ξ [see Eq. (7)] has been employed. Next, it is necessary to find the total energy input E_{in} from the ac drive at the n-th site during the passage of the soliton:

$$E_{in} = F\int_{-\infty}^{+\infty} dT \sin(\Omega T)\,(dy_n/dT) = \pi F \operatorname{csch}\left(\pi\Omega|\xi|/(1-\xi^2)\right) \times \cos\left((\Omega/V)(n-n_0)\right). \qquad (10)$$

As one sees from Eq. (10), in a general case the quantity E_{in} oscillates as a function of the site's number n, so that there is no net input of energy from the ac drive. However, there appears the energy input if the velocity takes the resonant values

$$|V| = V_N \equiv \Omega/2\pi N, \quad N = 1, 2, 3, \ldots . \qquad (11)$$

To analyze the slow evolution of the soliton under the joint action of the dissipation and ac drive in the near-resonant situation, it is natural to approximate the soliton by the unperturbed wave form (6) with the velocity exactly equal to a resonant value, the parameter ξ being related to it according to Eq. (7), while the parameter n_0 is a slowly varying function of time. To deduce a corresponding evolution equation for $n_0(T)$, I will employ the energy-balance equation.

The soliton's kinetic energy is

$$E_{kin} = \frac{1}{2} \sum_{n=-\infty}^{+\infty} (dy_n/dT)^2 \, . \tag{12}$$

Inserting Eq. (6) with the slowly varying $n_o(T)$ into Eq. (12), one finds in the first approximation

$$E_{kin} = K(\xi)\,(V^2 + V\frac{d}{dT}n_o), \tag{13}$$

where

$$K(\xi) \equiv (1-\xi^2)^2 \ln^2(\xi^{-2}) \sum_{n=-\infty}^{+\infty} \xi^{4n}(\xi^2+\xi^{2n})^{-2}(1+\xi^{2n})^{-2}.$$

On the other hand, it is evident that the mean rate of change of the soliton's energy is equal to $V(E_{in} - E_{diss})$, provided that the rapid dependence of E_{in} upon n has been removed as the resonant value (11) was substituted for V. Equating this expression to the time derivative of Eq. (13), one arrives at the following evolution equation for n_o:

$$K(\xi_N)\,\frac{d^2}{dT^2}n_o = F\,L(\xi_N)\cos(2\pi N n_o) - E_{diss}(\xi_N)\,, \tag{14}$$

where $L(\xi) \equiv \pi\,\mathrm{csch}(\pi\Omega\,|\xi|\,/(1-\xi^2))$, and ξ_N is implicitly defined by Eq. (7) with V replaced by V_N. Evidently, Eq. (14) is nothing but the equation of motion of a particle with the mass $K(\xi_N)$ in the harmonic potential in the presence of the additional constant force equal to $-E_{diss}(\xi_N)$. The character of motion in this mechanical problem is well known. If F exceeds the threshold value $F_{thr}^{(N)} = E_{diss}(\xi_N)/L(\xi_N)$, i.e.,

$$F_{thr}^{(N)} = (A/\pi)\,|\xi_N|^{-1}\sinh\left(\pi\Omega\,|\xi_N|\,/(1-\xi_N^2)\right) \times \left[(1+\xi_N^2)\ln(\xi_N^{-2}) - 2(1-\xi_N^2)\right], \tag{15}$$

there are two equilibrium values of n_o defined as follows:

$$\sin(2\pi N n_o) = \pm\sqrt{1 - \left[E_{diss}(\xi_N)/F\,L(\xi_N)\right]^2}\,, \tag{16}$$

one stable and one unstable. These equilibria correspond to compensation of the dissipative losses and energy input, $E_{diss} = E_{in}$. As concerns oscillations admitted by the effective equation of motion (14) in a vicinity of the stable equilibrium, they are subject to damping induced by the additional lossy term $\sim \frac{d}{dT}n_o$ that appears when the slow dependence of n_o upon T is taken into account in Eq. (9).

PHYSICAL INTERPRETATION

Let us proceed to physical interpretation of the fact that the ac drive may support stable motion of the TL soliton in the weakly damped system. From the viewpoint of the underlying chain of kinks, the propagating compression wave is a "supersoliton", in terms of Ref. 6. To realize its physical meaning, let us recall that in the LJJ theory the spatially averaged quantity $-\langle u_t \rangle$ gives the mean dc voltage across the junction. Using Eqs. (2), (8), and (11), one concludes that the dc voltage carried by the ac-driven supersoliton propagating at the resonant velocity V_N is

$$U_N = (\Omega/lN)\ \ln(\xi_N^{-2}), \tag{17}$$

where l is the overall length of the system.

As it follows from the definition of the renormalized frequency Ω [see Eq. (4)], it is natural to expect that Ω is large (recall the chain's spacing b is assumed sufficiently large). In this case, Eq. (11) yields large resonant velocities V_N. For $V \gg 1$, Eq. (7) can be inverted, and eventually one finds an approximate expression for ξ_N:

$$\xi_N = (2\pi N/\Omega)\ \ln(\Omega/2\pi N)^2 \ . \tag{18}$$

Inserting Eq. (18) into Eqs. (15) and (17), one can find the threshold value of the ac-drive's amplitude and voltage in the explicit form:

$$F_{thr}^{(N)} = 2A\Omega\ln(\Omega/2\pi N), \tag{19}$$

$$U_N = 2\Omega\,(lN)^{-1}\ln(\Omega/2\pi N). \tag{20}$$

The ac-current - dc-voltage characteristic of the ac-biased damped LJJ with the trapped chain of fluxons, determined by Eqs. (19) and (20), is shown in Fig. 1. The fact that the ac bias current applied to the junction induces dc voltage is sometimes called the reverse Josephson effect (see, e.g., Ref. 7). However, the one-fluxon reverse Josephson effect, studied in Refs. 7 - 9, is of altogether different nature: It is based on shuttle oscillations of a solitary fluxon in a finite-length LJJ, each reflection of the fluxon from a junction's edge reversing the fluxon's polarity.

In application to the chain of charged solitons in the CDW system, the same effect means that ac voltage (external field) applied to the system may give rise to dc electric current carried by the super-soliton. Analyzing kinematics of the motion of the supersoliton, it is easy to see that its effective electric charge is the soliton's charge q times Δy [recall Δy is defined by Eq. (8)]. Thus the supersoliton's effective charge depends on the resonance number N via the dependence $\Delta y(\xi_N)$. At last, the current carried by the supersoliton is given by the above expressions (17) and (20) times the soliton's charge q. This implies that the dc-current - ac-voltage characteristic of the CDW system corresponding to the motion of a supersoliton has the same form as shown in Fig. 1, with the current and voltage axes interchanged.

The same problem can be considered for a quasi-one-dimensional damped antiferromagnet in an external ac magnetic field. The corresponding perturbed SG equation for an orientation angle u(x,t) is [10]

$$u_{tt} - u_{xx} + \sin u = -\alpha u_t - (\pi/2)f\ \sin(\omega t)\cdot\sin(u/2), \tag{21}$$

f being proportional to the field's amplitude. In this model, the 2π kink (2) corresponds to a domain wall. If one considers propagation of a disturbance in a chain of domain walls,

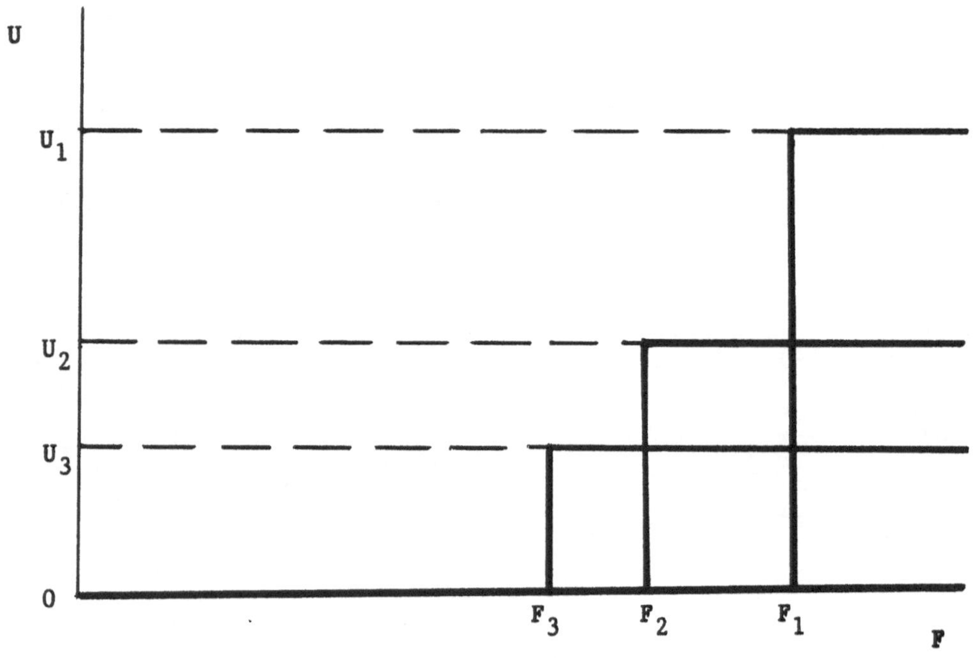

Fig. 1
The dc voltage U, induced by the motion of the driven "super-soliton" (solitary compression wave in the rarefied chain of fluxons), vs. the amplitude F of the bias current (ac drive) for different values of the resonance number N, see Eq. (11).

it will be described by the same equations (3) and (5) with only difference that the multiplier sin(u/2) in the last term of Eq. (21) gives rise to the additional factor $(-1)^n$ in front of the driving terms in Eqs. (3) and (5). Accordingly, the same factor appears in front of the expression (10) for the energy input. Consequently, propagation of the compression soliton is possible at the velocities

$$|V| = \Omega/(2N+1)\pi\ , \ N = 0,1,2,3,\ \ldots, \tag{22}$$

cf. Eq. (11). Thus Eq. (22) gives us another possible spectrum of the resonant velocities. In all other expressions obtained above, e.g., Eqs. (19) and (20), 2N must be replaced by (2N+1).

CONCLUSION

It is important to note that the compression solitons corresponding to different values of the resonance number N in Eqs. (11) or (22) have different velocities, therefore they may collide. Since the unperturbed TL equation is exactly integrable, the soliton-soliton collisions in this model are purely elastic, hence one may expect that in the slightly perturbed TL model (5) they must be quasielastic too. In the same time, strongly inelastic interactions, e.g., fusion of the collidiong solitons into a new one, or their transformation into another pair of solitons may be possible if the initial solitons' velocities are sufficiently small, or if the perturbing terms are sufficiently large. In any case, it would be interesting to consider statistics of a "rarefied gas" consisting of the supersolitons with different velocities.

A natural generalization of the problem considered in the present paper is to analyze a possibility of the stable propagation of a periodic cnoidal wave in the ac-driven damped TL system (5) (the cnoidal wave may be regarded as a periodic array of the compression solitons). Work in this direction is in progress.

The underlying SG models (1) or (21) have been reduced to a corresponding perturbed TL equation in the limiting case of the rarefied kink chain. In the opposite limiting case of a sufficiently dense chain, one may analyze evolution of collective excitations in the chain in the long-wave approximation, assuming a characteristic size (wave length) of the excitation to be much greater than the chain's spacing.[6] As it has been demonstrated in Refs. 6, in this approximation the collective excitations are governed by the driven damped D'Alembert equation [in Refs. 6, a so-called elliptic SG equation has been derived, which is the D'Alembert equation plus a nonlinear term induced by a spatially periodic inhomogeneity added to the model (1)]. It is evident that the ac-driven damped D'Alembert equation cannot support propagation of a solitary pulse. Thus, a steadily moving compression wave in the ac-driven damped kink chain does not exist in the long-wave (dense-chain) approximation. A meaningful problem is to develop a numerical crossover between the limiting cases of the rarefied and dense kink chain.

I am indebted to Luis Bonilla for valuable discussions.

REFERENCES

1. D.W. McLaughlin and A.C. Scott, Perturbation analysis of fluxon dynamics, Phys. Rev. A18:1625 (1978).
2. V.I. Karpman and N.A. Ryabova, On a mechanism of soliton bunching in Josephson junctions, Phys. Lett. A105:72 (1984).
3. G.J. Brinkman and T.P. Valkering, Soliton decay in a Toda lattice caused by dissipation, J. Appl. Math. Phys. (ZAMP) 41:61 (1990).
4. D.J. Kaup and A.C. Newell, Solitons as particles, oscillators, and in slowly varying media: A singular perturbation theory, Proc. Roy. Soc. L. A361:413 (1978).
5. O.H. Olsen and M.R. Samuelsen, Hysteresis in rf-driven large-area Josephson junctions, Phys. Rev. B34:3510 (1986).
6. A.V. Ustinov, Supersoliton excitations in inhomogeneous Josephson junctions, Phys. Lett. A 136:155 (1989); B.A. Malomed, Superfluxons in periodically inhomogeneous long Josephson junctions, Phys. Rev. B41:2616 (1990); B.A. Malomed, V.A. Oboznov, and A.V. Ustinov, "Supersolitons" in periodically inhomogeneous long Josephson junctions, Zh. Eksp. Teor. Fiz. 97:924 (1990).
7. J.J. Chang, Simple theory of the reverse ac Josephson effect, Phys. Rev. B38:5081 (1988).
8. M. Salerno, M.R. Samuelsen, G. Fillatrela, S. Pagano, and D. Parmentier, A simple map describing phase-locking of fluxon oscillations in long Josephson tunnel junctions, Phys. Lett. A137:75 (1989).
9. B.A. Malomed, Oscillations of a fluxon in periodically inhomogeneous long Josephson junction, Phys. Rev. B41: 2037 (1990).
10. A.M. Kosevich, B.A. Ivanov, and A.S. Kovalev, "Nonlinear Waves of Magnetization", Naukova Dumka, Kiev (1983) (in Russian).

SOLITONS IN QUANTUM SPIN CHAINS

Hans-J. Mikeska, Seiji Miyashita* and Gerald Ristow

Institut für Theoretische Physik
University of Hannover, D 3000 Hannover, Germany

1. INTRODUCTION

In this contribution we will discuss the analog of classical solitons in quantum spin chains. Strictly speaking, we will deal with the analog of solitary excitations, being interested in the manifestations of solitons in real physical systems more than in the properties of solitons in a rigorous mathematical sense. From a physical point of view, solitons in magnetic systems are more or less literally equivalent to domain walls and we will use frequently this intuitive picture. The motivation for our interest is that the role of solitons in realistic spin chains has been established as an important one in the last decade both experimentally and theoretically (see several recent reviews, e.g. Izyumov 1988, Mikeska and Steiner 1990). The overwhelming majority of theoretical approaches in this context is based on a treatment of spin chains in the classical limit, although systems of experimental interest have rather low values of the spin magnitude S: S = 1/2 in $CsCoCl_3$ and CHAB, S = 1 in $CsNiF_3$ and S = 5/2 in TMMC. Therefore interest has grown to clarify the role of solitary excitations as elementary excitations also for spin chains with finite and even small values of S. In this article we will deal mostly with spin chains with S = 1/2 and our main example will be the Ising chain in a transverse field; however, at first sight, solitary excitations in classical spin chains appear to differ significantly from those in spin chains with spin S = 1/2, and therefore also the approach to the classical limit, varying S from 1/2 to ∞ is of substantial interest.

In the following we first give a survey of the description of solitary excitations as domain walls (DW) and compare classical to quantum systems (section 2), we then consider solitary excitations in Ising-like spin chains, reviewing recent work by the authors (Mikeska, Miyashita and Ristow 1990) for S = 1/2 (section 3) as well as presenting new results (Mikeska and Miyashita 1990) for arbitrary S, considering in particular the transition to the classical limit (section 4). In section 5 we finally describe the results of Mikeska, Costa and Fogedby (1989) giving the corrections to the classical limit for S >> 1 in a semiclassical approach. Concluding remarks are given in section 6.

---

* Permanent address: Department of Physics, College of Liberal Arts, Kyoto University, Kyoto 606, Japan

Microscopic Aspects of Nonlinearity in Condensed Matter
Edited by A.R. Bishop *et al.*, Plenum Press, New York, 1991

2. SOLITARY EXCITATIONS AS DOMAIN WALLS: CLASSICAL VS. QUANTUM PICTURE OF ELEMENTARY EXCITATIONS

In this section we want to describe typical examples of solitary excitations in classical as well as in S=1/2 chains. The most direct approach is to start from a classical model with two equivalent ground states, e.g. from an energy functional

$$E\{\mathbf{S}_n\} = -J \sum_n \mathbf{S}_n \mathbf{S}_{n+1} - D \sum_n (S_n^x)^2. \tag{1}$$

We introduce the continuum approximation and describe the classical spin vectors $\mathbf{S}_n$ by two angles $\theta(z,t)$ and $\phi(z,t)$,

$$\mathbf{S}_n = S\,\{\sin\theta\cos\phi,\ \sin\theta\sin\phi,\ \cos\theta\}. \tag{2}$$

Minimization of the classical energy gives (z is the coordinate along the chain in units of the lattice constant)

$$\frac{d^2\phi}{dz^2} = \frac{D}{J} \sin 2\phi, \qquad \theta = \frac{\pi}{2}. \tag{3}$$

$\theta = \pi/2$ of course defines only one of a continuum of equivalent planes. (3) is the static sG equation and the soliton solution of interest to us is

$$\phi = 2 \operatorname{arctg} \left\{ \pm \sqrt{\frac{2D}{J}} \, (z - z_0) \right\}. \tag{4}$$

We see that the soliton mediates between the two equivalent ground states of the chain, it is thus completely justified to call it a domain wall; in particular it is localized at an arbitrary lattice site and exponentially approaches the limiting behaviour $\mathbf{S} = \pm S\,\mathbf{e}_x$ at infinity. There exists a characteristic length $\xi = (J/2D)^{1/2}$, governing this spatial localization of the soliton. For many, but not all Hamiltonians, which support solitary excitations of this type, there exists an interesting dynamics of these solitary excitations (see Mikeska and Steiner 1990) - this, however, is not our primary interest here.

To see how this behaviour typical of classical domain walls changes when a quantum system is considered, we choose as an equivalently simple example the S = 1/2 Ising model in a transverse field with Hamiltonian (note the rotation of spin components with respect to the usual notation to have consistency with eq. (1) and for later convenience)

$$H = -2J \sum_n S_n^x S_{n+1}^x - \Gamma \sum_n S_n^z . \tag{5}$$

For $\Gamma = 0$ the ground state has all spins parallel and is twofold degenerate, $S_n^x = +\frac{1}{2}$ or $S_n^x = -\frac{1}{2}$. The lowest excitation in a chain with open ends is a domain wall located at an arbitrary position m, obtained by overturning all spins S_n^x with n > m. This is thus an extremely localized excitation. For $\Gamma \neq 0$ these N degenerate domain wall states with energy J are broadened into a band (Villain 1975) as shown in fig. 1 and the localization is lost. Also included in figure 1 is the spin wave continuum centered at energy 2J; it develops from the states obtained by overturning a single spin and can evidently be considered equivalently as formed of states in the two domain wall sector. To see that the domain wall

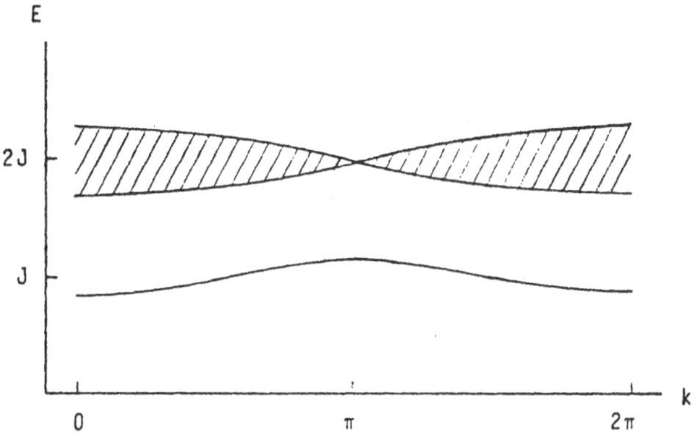

Fig. 1. Energy spectrum of 1 DW and 2 DW states

(soliton) has to be considered as the fundamental elementary excitation of the system we look at the exact result for the free energy of the pure Ising chain with N spins ($\beta = 1/k_BT$)

$$F/N = -k_BT \ln \left(1 + e^{-\beta|J|} \right) - \frac{k_BT}{N} \ln (1 + \gamma). \tag{6}$$

γ depends on the boundary conditions:

$\gamma = \pm \operatorname{tgh}^N \frac{1}{2}\beta|J|$ for periodic (antiperiodic) boundary conditions

$\gamma = 1$ for the open chain

At low temperatures we have $F/N \sim \exp(-\beta|J|)$, i.e. an activated behaviour with the characteristic energy of **one** domain wall (even when periodic boundary conditions force domain walls to come in pairs), in complete agreement with the corresponding phenomenological calculation. We thus conclude that the domain wall (soliton s or antisoliton s') is the fundamental excitation, whereas the spin wave (which we want to call "quantum spin wave" or LSM (Lieb, Schultz and Mattis 1961) spin wave to distinguish it from the semiclassical, Holstein-Primakoff (HP) spin wave, see below) is an ss' state. This picture is in complete agreement with the results for the excitation spectrum as obtained from Bethe's ansatz (Johnson, Krinsky and McCoy 1973) and actually applies likewise even to the isotropic Heisenberg antiferromagnet, where the domain walls degenerate to gapless excitations (Faddeev and Takhtajan 1982).

3. SOLITARY EXCITATIONS IN ISING-LIKE SPIN CHAINS

In order to study in more detail the soliton related properties of perturbed Ising spin chains we will use in this section both a more formal approach, introducing and exploiting soliton creation and annihilation operators, as well as a more physical approach based on calculations in the one domain wall subspace.

For spin chains with S = 1/2 it is possible and useful (Mikeska, Miyashita and Ristow 1990) to replace spin operators by operators D_m^+, D_m, which have the following properties:

i) D_m^+, D_m are fermion operators.

ii) D_m^+, D_m are constructed from a variant of the Jordan-Wigner transformation and are related to spin operators in the following way:

$$D^+_m = \frac{1}{2}\,(c^+_{m+1} + c_{m+1}) + \frac{1}{2}\,(c^+_m - c_m)$$

$$c_m = \exp\left(i\pi \sum_{n=1}^{m-1} a^+_n a_n \right) a_m, \qquad a_m = \frac{1}{2}\,(S^x_m + iS^y_m)\ . \tag{7}$$

iii) When D^+_m (D_m) ($m = 1....N-1$) is applied to an eigenstate of the unperturbed Ising model with free ends (i.e. states specified by the spin projection σ_n at every site $n= 1...N$), it results in turning around all spin projections with $n \le m$ and gives a phase factor $(\sigma_{m+1} \pm \sigma_m)/2$. Thus D^+_m (D_m) creates (annihilates) a domain wall between sites m and m+1. It is therefore justified to call these operators domain wall creation (annihilation) operators. The operators $D_0 \equiv D_N$, D^+_0, which are needed for completeness, on the other hand, mediate between the two degenerate groundstates $|+>$ and $|->$ with all spins up resp. down of the Ising model:

$$(D^+_0 + D_0)\ |\pm\rangle = \pm|\pm\rangle$$
$$(D^+_0 - D_0)\ |\pm\rangle = \pm|\mp\rangle.$$

In terms of these domain wall operators the Hamiltonian of the Ising model in a transverse field, eq.(5) above, reads (we take $J > 0$ for definiteness)

$$H = -\frac{1}{2}J(N-1) + J\sum_{n=1}^{N-1} D^+_n D_n \;-\frac{1}{2}\,\Gamma \sum_{n=1}^{N} (D^+_{n-1} D_n - D^+_{n-1} D^+_n + \text{h.c.}) \tag{8}$$

Formally, of course, nothing is gained beyond the conventional fermion description (Lieb, Schultz and Mattis 1961): H is a free fermion Hamiltonian and can be diagonalized exactly. From a physical point of view, however, the formulation of eq.(8) is more interesting: H is seen to involve domain wall transfer between adjacent sites as well as domain wall pair creation and pair annihilation. When we replace the external transverse field by transverse exchange interactions, also domain wall interactions occur. The exact eigenstates therefore are linear combinations of states with an arbitrary number of domain walls.

We refer to the work of Mikeska, Miyashita and Ristow (1990) for an application of this description to calculate and to discuss the domain wall content of the ground states of these models and to investigate the domain wall aspects of states like $D^+_m|\pm\rangle$. Here we only want to emphasize that the representation of the domain wall creation operator by spin operators, eq.(7) above, makes clear why the direct excitation of domain walls or solitons in scattering experiments is so difficult: D^+_m is a strongly nonlinear combination of spin operators and therefore practically impossible to realize. Using linear combinations of spin operators, a single domain wall can only be created at the end of an open chain where we have e.g.

$$S^y_1 = -i\ (D^+_1 - D_1) \tag{9}$$

We now want to discuss the shape of the quantum domain wall, i.e. the average value of S^x_n. To obtain exlicit expressions to be compared to the

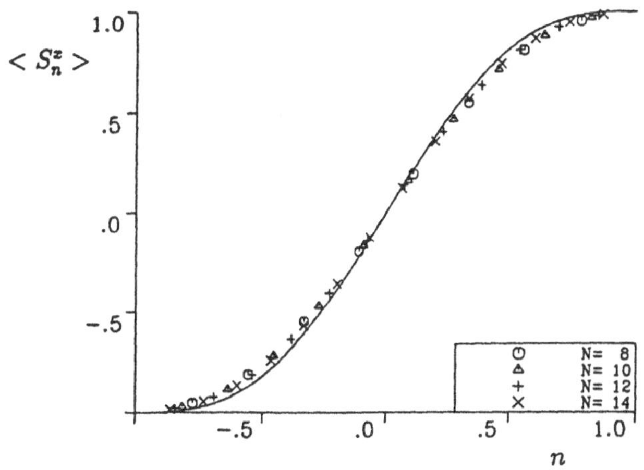

Fig. 2. $<S^x_n>$ for the lowest 1 DW state in the transverse field Ising model ($\Gamma/J = .1$). 1DW approximation (solid line) and finite chain results.

classical result we have to break the symmetry $S^x_n \rightarrow -S^x_n$; in order to sort out the appropriate linear combinations of the degenerate eigenstates we introduce additional external boundary fields fixing S^x_n to ±1 at the left, resp. right end of our open chain. An analytic expression for $<S^x_n>$ can be found when we restrict ourselves to the one domain wall subspace; this is justified for sufficiently small perturbations to the pure Ising case, where the eigenstates very clearly group into bands which can be classified by the number of domain walls. The results of this diagonalization in the one domain wall subspace are the following (Mikeska, Miyashita and Ristow 1990):

i) The domain wall is always centered at the middle of the chain.
ii) The transition from -1 to +1 is much smoother than in the classical case.
iii) The domain wall shape is independent of the strength of the perturbation Γ.
iv) The results as shown in Fig.2 agree with the results of a numerical diagonalization for finite chains - this means that the restriction to the one domain wall subspace is completely sufficient for the values of the coupling considered.

Equivalent results are obtained for different perturbations of the Ising model, in particular transverse interactions for one or both of the remaining spin components.

Summarizing these results we conclude that domain walls or solitons in a quantum spin chain appear to have quite different properties when compared to the corresponding classical excitations: Domain walls are not localized at an arbitrary lattice site and do not possess a characteristic length describing an exponential approach to the asymptotic behaviour. These strong differences, however, should be understood as following directly from the finite quantum mechanical transition amplitude between neighbouring lattice sites related to deviations from the ideal Ising model: Whereas the localization of the basic DW state in the pure Ising chain corresponds to the classical situation of a narrow localized wall, it is only the quantum mechanical eigenstate which via quantum transfer processes shows delocalization.

4. DOMAIN WALLS FOR ARBITRARY S: TRANSITION TO THE CLASSICAL LIMIT

In view of the results of the last two sections it is interesting to investigate the approach to the classical limit by extending the above calculation to arbitrary, in particular large values of S. We do this restricting ourselves again to the one domain wall subspace with boundary conditions $S^x_n = \pm S$ at the left, resp. right boundary, which for arbitrary values of S is defined by the following states

$$\{S^x_n\} = \{S^x_1 = S,\ S^x_2 = S, \ldots\ldots S^x_{m-1} = S,\ S^x_m = S',\ S_{m+1} = -S. \ldots\ldots S^x_N = -S\}. \qquad (10)$$

with n = 0,1...N+1, m = 1,2....N and $-S \leq S' \leq S$. In the Ising limit these states are degenerate with respect to the site variable m as well as with respect to the value of S', i.e. the spin value in the middle of the domain wall: There are 2NS + 1 states with energy $4JS^2$. Diagonalization of the transverse field lifts this degeneracy, leading to a band character of the one DW states (degeneracy with respect to m) as well as to an internal structure or splitting into subbands (degeneracy with respect to S'), see fig.3. The energy of the lowest subband can be calculated explicitly for large values of S (and $N \to \infty$) and is obtained as

$$E_{1,0}(k) = 4JS^2 - \Gamma S - \frac{1}{2^{2S}}\, 4\Gamma S \cos k \qquad (11)$$

To obtain this result we have applied an approximate approach valid for large S, which is based on the fact that unperturbed states at neigbouring sites couple for S' = S - 1 and S' = -S only. This approach will we be described in more detail somewhere else (Mikeska and Miyashita 1990). From this result for the energy of the lowest subband we find the following basic understanding of the transition to the classical limit:

The 1DW states in principle always have band character and thus, strictly speaking, are always delocalized; quantitatively, however, the width of the states with lowest energy goes to zero very strongly with $S \to \infty$,

$$W = 8\Gamma S\, e^{-\alpha S}, \qquad \alpha = 2 \ln 2 \qquad (12)$$

This means that quantum delocalization is less and less important when the classical limit is approached - when we start from a suitable linear combination of states in the lowest subband to prepare a localized DW as an approximate eigenstate, the characteristic time τ for this state to decay to an exact eigenstate diverges as $\tau \sim 1/W \sim \exp \alpha S$. Thus the apparently very different character of the classical domain wall when compared to its quantum counterpart is seen to be related to the noncommutativity of the limits $t \to \infty$ and $S \to \infty$. Considering the subbands higher in energy one finds (Mikeska and Miyashita 1990) that these states can be related to moving solitons in the classical description.

Of course the restriction to the set of degenerate states used above requires further justification, since, contrary to the case of S = 1/2 Ising-like chains, for S > 1/2 there are many states lying between the 1 DW and the 3 DW band. These states can be related to additional spin wave- and breather-like excitations (with energy differences ~ S to the

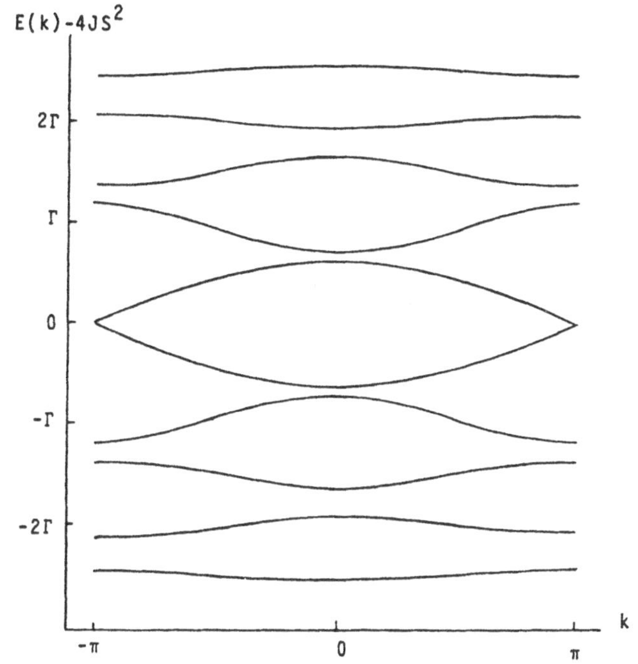

Fig. 3. Splitting of the 1 DW band into 2S subbands for S = 5.

1 DW band). They will be important if one aims at a correct description of the ground state of the chain but they are expected to be of little importance for the description of the qualitative features of the DW-like excitations given above. When located close to the unperturbed DW, they will contribute to a dressing of the shape of the localized DW, which we have not dealt with here.

Finally we want to mention that we also have confirmed some of the results of the preceding analytic discussion by numerical calculations for finite chains with S = 1, 3/2.

5. SEMICLASSICAL APPROACH TO DOMAIN WALLS IN QUANTUM SPIN CHAINS

In this section we want to supplement the results of the two preceding sections by shortly reviewing a calculation which allows to approach the calculation of the domain wall shape starting from the classical limit (Mikeska, daCosta and Fogedby 1989). This calculation has been done for the $CsNiF_3$ Hamiltonian

$$H = -J \sum_n \mathbf{S}_n \mathbf{S}_{n+1} + D \sum_n (S_n^z)^2 - \mu B \sum_n S_n^x . \tag{13}$$

but the qualitative aspects of the results should be independent of this special choice. (The fact that the "domain wall" in $CsNiF_3$ is a 2π soliton and ends up at the same orientation it started with, is more a technical one, related to a factor of 2 in the relevant angular variable). In order to do calculations in the semiclassical limit one starts by replacing the transverse spin components by the angular operator Φ (Villain 1975)

$$S^x_n + iS^y_n = e^{i\Phi_n} \sqrt{\hat{S}^2 - (S^z_n)^2}, \qquad \hat{S}^2 = S(S+1). \tag{14}$$

$$[S^z_m, \Phi_n] = i\delta_{mn}$$

Expanding in 1/S to lowest order, the DW (soliton) is given by the classical solution (for a recent review see Mikeska and Steiner 1990)

$$\Phi \to \phi^{(0)}(z) = 4 \text{ arctg } e^{\pm m(z-z_0)}, \qquad m^2 = \frac{\mu B}{JS}. \tag{15}$$

Higher orders in an expansion in 1/S are obtained starting from the ansatz

$$\Phi_n \to \phi^{(0)}_n + \hat{S}^{-1/2} \varphi_n, \qquad S^z_n \to \hat{S}^{1/2} \chi_n,$$

$$[\varphi_m, \chi_n] = i\delta_{mn}. \tag{16}$$

We call this a local quantization approach, since the quantization axis is taken to be determined by the direction given by the classical soliton. The Hamiltonian and then the soliton energy can be given in an expansion in 1/S, corresponding to the contributions of spin waves, spin wave interactions etc. (explicitly up to second order). When we treat the soliton width parameter m as a variational parameter, a correction of the soliton width m^{-1} is obtained in second order:

$$m^{-1} = \sqrt{\frac{JS}{\mu B}} \left(1 + \frac{9}{16 S^2} \right). \tag{17}$$

The soliton width thus is seen to increase owing to quantum effects. The analogous calculation should apply for the Ising model in a transverse field in the semiclassical limit. The main difference to the $CsNiF_3$ model is that classical TI domain walls are extremely narrow and the continuum approximation therefore is usually not applicable.

One should, however, be aware of the fact that the increase in width reported above is the equivalent of the dressing of the local domain wall from spin waves as mentioned in section 4 above, whereas the quantum delocalization (which depends nonperturbatively on the the small parameter 1/S, see above) is not obtained in this method. Qualitatively, however, we see that the tendency of the various types of results is consistent in that the classically well defined domain wall shape becomes fuzzy and delocalized owing to quantum effects.

6. CONCLUSIONS

We have described that excitations of the domain wall type exist in both classical and quantum magnetic chains, using as our main example the Ising chain with transverse interactions. Whereas classically static domain walls are localized at arbitrary lattice sites, quantum mechanically the possibility of transitions between these degenerate classical solutions becomes important and delocalizes the exact eigenstates. In addition, quantum fluctuations induce a broadening of the basic local shape. The delocalization of an approximate local configuration in the quantum chain takes a strongly increasing longer time $\sim \exp \alpha S$, when the classical limit is approached with increasing values of the spin length S.

From the treatment of the S = 1/2 chain in terms of domain wall creation and annihilation operators we have seen explicitly that the domain wall or soliton is the **basic** excitation in this system, whereas excitations derived from one-spin deviations ("LSM spin waves") should be considered as composite. Domain walls continue to exist as **one** basic excitation for larger values of S (with energy proportional to S^2), supplemented by soliton-antisoliton bound states (with energy proportional to S and turning into breathers, respectively HP spin waves in the semiclassical limit).

Whereas at first sight the exact quantum mechanical eigenstates of a chain which by appropriate boundary conditions is forced to contain an odd number of domain walls, look very different from the classical domain wall, being centered in the middle of the chain and approaching the asymptotic behaviour algebraically instead of exponentially, they should nevertheless be considered as manifestations of a local configuration very similar to the classical one - we do not expect the transfer between different sites which becomes possible in the quantum case to influence strongly the behaviour of correlation functions and thus the experimentally relevant quantities. A quantitative investigation of this important question, however, has to be left to future work.

REFERENCES

Faddeev, L.D., and Takhtajan, L.A., 1981, Phys. Letters, 85A:375

Izyumov, Yu.A., 1988, Soviet Phys. Uspekhi, 31:689

Johnson, J.D., Krinsky, S., and McCoy, B.M., 1973, Phys. Rev., A8:2526

Lieb, E., Schultz, T., and Mattis, D., 1961, Ann. Phys., 16:407

Mikeska, H.J., daCosta, B., and Fogedby, H.C., 1989, Z. Phys., B77:119

Mikeska, H.J., and Steiner, M., 1990, Adv. Phys., to be published

Mikeska, H.J., and Miyashita, S., 1990, in preparation

Mikeska, H.J., Miyashita, S., and Ristow, G., 1990, preprint

Villain, J., 1975, Physica, 79B:1

DYNAMIC MAPS AND ATTRACTOR PHASE-STRUCTURES IN RANDOMLY DILUTE NEURAL NETWORKS

D. Sherrington and K.Y.M. Wong

Department of Physics, University of Oxford
1 Keble Road, Oxford OX1 3NP
United Kingdom

INTRODUCTION

There is currently much interest in attractor neural networks as idealized models of associative memory in the cerebral cortex of the brain [1]. They consist of simple neurons driving one another non-linearly via connecting pathways with mutually interfering characteristics, leading to non-trivial global dynamics. The objective is to train the characteristics of the individual pathways so as to lead to a set of 'basins' in which the global dynamic activity is attracted towards that associated with the patterns which are to be memorized and recallable. For useful memory there should be many attractor basins, each with macroscopic overlap with a single memorized pattern.

In the case of systems with symmetric pathways there exist mathematical mappings between the dynamics of neural networks and the thermodynamics of analagous Hamiltonian systems with disordered and frustrated interactions [1,2], as epitomize spin glasses [3,4]. In the present paper, however, we consider a different situation in which the behaviour of the neural network can be considered in terms of iterative maps such as are favoured in modern dynamical systems analyses.

The networks we consider consist of many binary neurons randomly but dilutely interconnected, without symmetry between pathways. In a mean field sense the global retrieval dynamics of such networks can be studied in terms of iterative maps of the form [5]

$$m(t+1) = f(m(t)). \tag{1}$$

After a brief explanation of the origin of these iterative maps and their use to determine retrieval characteristics, we consider the training of networks so as to optimize various performance criteria.

Microscopic Aspects of Nonlinearity in Condensed Matter
Edited by A.R. Bishop *et al.*, Plenum Press, New York, 1991

THE SYSTEM

Our networks consist of N neurons, each allowed only two states, firing and non-firing, which we characterize by Ising variables $\sigma_i = \pm 1$; $i = 1,\ldots.N$, respectively. They are randomly interconnected by uni-directional pathways and their dynamics are determined by local Boolean operations

$$\sigma_i(t+1) = F_i(\sigma_{i1}(t), \ldots \sigma_{ik}(t)) \tag{2}$$

where $i_1,\ldots i_k$ label the neurons feeding neuron i and the F_i are Boolean functions of their arguments, taking the values ±1. The individual F_i can be either fixed or stochastically chosen at each time-step.

ITERATIVE MAPS FOR OVERLAPS

Our interest is in the global behaviour of networks trained to store sets of random patterns $\{\xi_i^\mu\};\mu=1,\ldots p$, where the ξ_i^μ take the values ±1. The similarity of the actual state of the network compared with the stored patterns can be assessed in terms of the overlaps

$$m^\mu = N^{-1} \sum_i \sigma_i \, \xi_i^\mu. \tag{3}$$

If the distribution of the k is peaked around $c << \ell nN$ the network structure is locally tree-like and correlations are unimportant. We shall also concentrate on $c \to \infty$, $N \to \infty$, $c << \ell nN$. Thus the m^μ at one time-step determine statistically, via the $\{F_i\}$, the m^μ at the next step. If, further, the system starts in a state with only one of the m^μ macroscopic then the dynamics obeys eqn (1) with m equal to that m^μ[5].

Retrieval Properties

From the iterative map (1) the retrieval properties follow readily. Its stable fixed points

$$m^* = f(m^*); \; f'(m^*) < 1 \tag{4}$$

give the overlaps to which systems can iterate. Unstable fixed points

$$m_B = f(m_B); \; f'(m_B) > 1, \tag{5}$$

delimit the basins of attraction. Fixed points are alternately unstable and stable. Thus for a system initialized with $m_B{}^i$ (or − 1 if $m_B{}^i$ does not exist) $\leqslant m \leqslant m_B{}^{i+1}$ (or 1 if $m_B{}^{i+1}$ does not exist) where $m_B{}^i$, $m_B{}^{i+1}$ are consecutive m_B values, the system will iterate to the m^* with $m_B{}^i \leqslant m^* \leqslant m_B{}^{i+1}$. This behaviour is illustrated for two retrieval functions f(m) in Fig. 1.

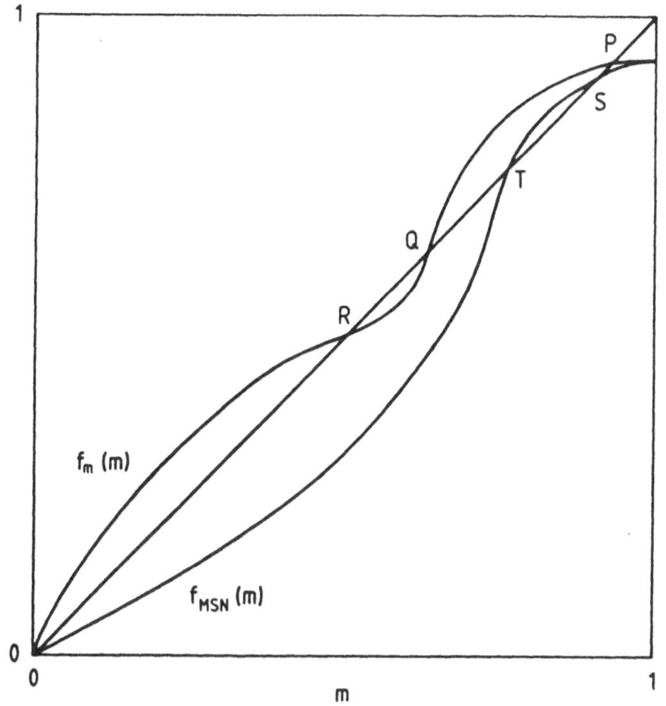

Fig. 1. Two possible retrieval curves f(m). R,P,S are stable fixed points of the corresponding retrieval maps. Q,T are unstable fixed points. For further explanation of the particular curves see the text.

The actual function f(m) depends upon the set of local operators $\{F_i\}$. They in turn depend upon the activity patterns which are to be memorized and in a useful associative memory are chosen such that the resulting global dynamics iterates towards states with finite m^*, ideally as large as possible. With any conceivable rule there is an upper limit on the number of patterns p which can be stored so as to allow each to be retrievable from an initial state within some appropriate basin of attraction; we denote this limit by p_c. In the case of stochastic dynamics p_c depends on the degree of retrieval noise or uncertainty d, reducing with increasing noise. $p_c(d)$ thus provides a phase boundary between retrieval (for $p<p_c(d)$ and non-retrieval (for $p>p_c(d)$.

If there is only one m^* value between 0 and 1 and the immediately lower m_B is equal to zero then the basin of attraction is referred to as wide. Otherwise it is narrow; all the basins illustrated in Fig. 1 are narrow. A transition in which an m^* goes continuously to zero as p increases to a critical $p'_c(d)$, which may or may not be equal to $p_c(d)$, is the analogue of a second order thermodynamic phase transition. If, however, m^* and its lower basin boundary m_B come together as p is varied and disappear as a critical $p_c''(d)$ is crossed then the transition is first order. There exist neural networks with both transition types. Thus, if the curve $f_m(m)$ in Fig. 1 were continuously distorted so as to approach the lower and right-hand axes whilst maintaining the location of its end points, the retrieval overlap association with P would vanish discontinuously (first order) whilst that associated with R would vanish continuously (second order).

OPTIMIZATION AND ADAPTATION

The actual form of f(m) depends on the algorithm used to train the $\{F_i\}$. The degree of stochasticity (or noise) in the retrieval process will, however, be taken to be given and therefore a constraint on the dynamics. There are classes of algorithms which depend upon a small number of control parameters x in their training process. The optimization can be with respect to several possible criteria such as the retrieval overlap, either overall or with respect to an initial overlap, the minimum initial overlap from which retrieval is possible, or the total number of patterns storable in a retrievable form. All these can be studied from the $f_x(m)$.

Let us first consider optimization of the retrieval overlap. This can be obtained from an analysis of the envelope F(m) which touches the set of $f_x(m)$ from above. Over the set of $\{x\}$ the maximum retrieval overlap is obtained from the highest stable fixed point of the retrieval envelope; that is from

$$m_u^* = F(m_u^*) ; \; dF/dm|_{m=m_u^*} < 1 ; \; \max m_u^* . \tag{6}$$

The optimal x for retrieval overlap is then given by the retrieval function $f_x(m)$ for which

$$f_x(m_u^*) = m_u^* . \tag{7}$$

We shall denote the corresponding x (or x's) by x^* (or $\{x^*\}$); ie $f_{x^*}(m)$ touches the envelope F(m) at $m = m_u^*$. Any other x would yield a lower stable fixed point. Fig 2 illustrates this property.

The same envelope suffices to determine the minimum m from which retrieval is possible by a network in the class, in this case by

$$m_{BL} = F(m_{BL}) ; \; dF/dm|_{m=m_{BL}} > 1 ; \; \min m_{BL} . \tag{8}$$

The minimum m from which retrieval is possible to a high overlap is given by

$$m_{BU} = F(m_{BU}) ; \; dF/dm|_{m=m_{BU}} > 1 ; \max m_{BU} . \tag{9}$$

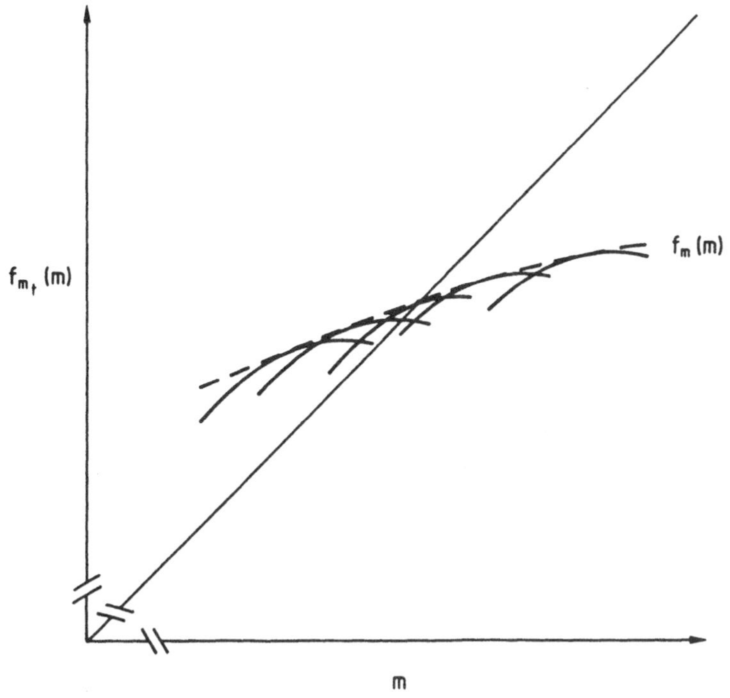

Fig. 2. Schematic plot of family of retrieval curves $f_x(m)$ and their envelope $F(m)$.

The corresponding networks are given analogously to (7) by

$$f_{xB}(m_{BL,U}) = m_{BL,U} \tag{10}$$

(ie $f_{xB}(m)$ touches the envelope F(m) at $M_{BL,U}$) yielding different values of x from those given by (7), thereby illustrating the feature of specialization. If there is more than one unstable fixed point of m=F(m), as is the case if $F(m) = f_m(m)$ in Fig. 1, then there is more than one stable fixed point. In particular, for the example illustrated, if the initial overlap is less than that associated with point Q, the best retrieval overlap one could hope for would be that associated with R. We refer to the two regions as those of strong and weak adaptive retrieval.

If, further, one allows for the possibility that the parameter x can 'self-adapt' during retrieval then the unstable fixed points of m=F(m) delimit the "adaptive retriever" basins and the corresponding stable fixed points give the results of adaptive retrieval.

Optimization of capacity is intimately related to that of retrieval overlap, since at capacity the optimal overlap goes to zero, either continuously or discontinously. Of course, so does the highest of the optimal retriever basin boundaries as given by eqn (9).

Rather than continuing in all generality, let us now turn to a specific class of $\{F_i\}$ rules, namely those corresponding to a synaptic network with sigmoidal rounding of the probabalistic updating rule and given by

$$\sigma_i(t+1) = \mathrm{sign}\,(\Sigma^{i}{}_{k}{}_{j=i_1} J_{ij}\sigma_j + Tz) \qquad (11)$$

where the J_{ij} are the synaptic efficacies and store the memorized information, z is a Gaussian random variable of zero mean and unit variance and T is a measure of the sigmoidal rounding, often referred to as the 'temperature'. Let us choose the $\{J_{ij}\}$ so as to maximize $f(m_t)$ for this system, where m_t is an as-yet arbitrary overlap which will be called the "training overlap" [6]. Considering m_t then as the tuning parameter x, the retrieval function f_{m_t} (m) for the storage of random patterns is given by

$$f_{m_t}(m) = \int d\Lambda\ \rho_{m_t}(\Lambda)\ g_m\ (\Lambda) \qquad (12)$$

$$\text{where } g_m\ (\Lambda) = \mathrm{erf}\ \ [m\Lambda/(2(1-m^2 + T^2))^{1/2}] \qquad (13)$$

$$\rho_x(\Lambda) = \int \frac{dt}{\sqrt{2\pi}}\quad \exp\ (-t^2/2)\ \delta(\Lambda - \lambda_x(t)) \qquad (14)$$

and λ_x (t) is the inverse function of $t_x(\lambda)$ defined by

$$t_x(\lambda) = \lambda - \gamma\ g_x'\ (\lambda) \qquad (15)$$

with γ a constant determined by the condition

$$\int \frac{dt}{\sqrt{2\pi}}\ \exp(-t^2/2)\ (\lambda_x(t) - t)^2 = \alpha^{-1} \qquad (16)$$

where $p=\alpha c$ is the number of stored patterns and the $\lambda_x(t)$ giving the largest value of $g_x(\lambda) - (\lambda_x - t)^2/2\gamma$ is chosen in cases where $\lambda_x(t)$ is multi-valued. The envelope function F(m) is $f_m(m)$ in this case. Clearly it is a function of both T and α.

In Fig. 3 is exhibited the phase diagram [7] for self-adaptive retrieval for the networks of eqn (11). In region I the retrieval envelope f_m (m) has a single stable fixed point and a wide basin of adaptive attraction. On the line DF the stable fixed point of $f_m(m)$ goes continuously to zero. In region II the retrieval envelope has acquired an additional pair of stable and unstable fixed points, as in the upper curve shown in Fig. 1, thus permitting the existence of adaptive networks with both strong and weak retrieval. On lines AE and ED there are discontinuous transitions to the single stable fixed point behaviour of region I. The weaker adaptive retriever disappears continuously on line DB. In region III there occurs adaptation with a high asymptotic overlap but with a narrowed basin of adaptation and a discontinuous transition on DC to a non-retriever region. The dotted line shows for comparison the retrieval-non retrieval boundary for finite-temperature retrieval in the network which is maximally stable at T-0 [7]. A typical map function f_{MSN} (m) for the maximally stable network in the region $\alpha>0.42$ is shown in the lower curve of Fig. 1.

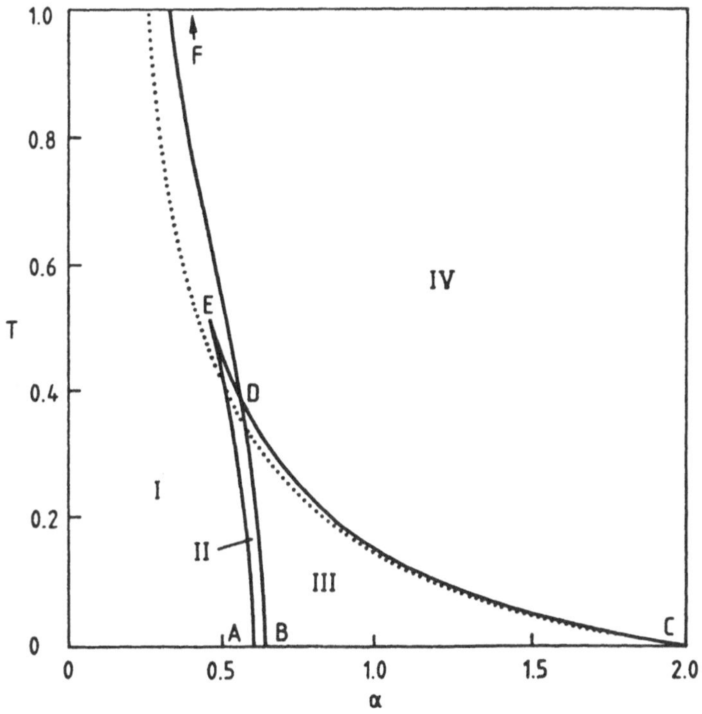

Fig. 3. Retriever phase diagram for adaptation in temperature-storage capacity space of synaptic neural networks with Gaussian retrieval noise. Also shown for comparison is the phase boundary for finite-temperature retrieval of network which is maximally stable at zero-temperature (dotted line). From Ref. 7.

CONCLUSION

In conclusion we note that dilute randomly connected neural networks yield interesting non-linear dynamic maps with several possible attractor phases, separated by transitions of both first and second order. Furthermore, a study of the envelopes of map functions permits optimization of a variety of performance criteria and points to the possibility of attaining optima by a process of self-adaptation. Individual maps describe the dynamics of retrieval of networks with fixed structure, whereas the envelopes of the maps describe the dynamics of adaptation of networks with adiabatically evolving structures. Explicit results for the optimization of adaptive synaptic networks have been presented.

REFERENCES

[1] D.J. Amit, "Modeling Brain Function" (Cambridge Univ. Press, Cambridge 1989)

[2] J.J. Hopfield, Proc. Natl. Acad. Sci. USA 79, 2554 (1982)

[3] D. Sherrington, "Spin Glasses" in "Disordered Solids: Structures and Processes" ed. B. di Bartolo (Plenum, New York 1989) p. 225

[4] D. Sherrington, "Spin Glasses and Neural Networks" in "Neural Computing", eds. C. Mannion and J.G. Taylor (Adam Hilger 1989) p.15

[5] B. Derrida, E. Gardner and A. Zippelius, Europhys. Lett. 4, 167 (1987)

[6] K.Y.M. Wong and D. Sherrington, J. Phys A23, L175 (1990)

[7] K.Y.M. Wong and D. Sherrington, J. Phys A23, 4659 (1990)

[8] D. Amit, M. Evans, H. Horner and K.Y.M. Wong, J. Phys A23, 3361 (1990)

SOLITONS AND TWISTONS
IN CONDUCTING POLYMERS

S. Brazovskii*, N. Kirova*

Institute for Scientific Interchange Foundation
Villa Gualino, Viale Settimio Severo 65, 10133 Torino, Italy

INTRODUCTION

An important property of polyacetylene (PA) is the spontaneous violation of symmetry, which manifests itself in an alternation of bond lengths between the carbon atoms in molecule. Because of twofold degeneracy of the ground state domain walls may appear between the equivalent phases. Such walls in the molecular chain are quasiparticles of the type of topological solitons, or kinks[1,2]. For an isolated chain kink can be located at any point. But in a system of interacting chains a divergence of a pair of kinks or the displacement of one of the kink from the end of the chain corresponds to the nucleation on this chain of another phase, which results in a loss of surface energy - the confinement energy $W = Fx$ (x-the distance between kinks, or the distance from a kink to the end of the chain)[2,3]. Thus the single kink should be localized at the ends of chains, and for charge solitons the confinement to bipolarons is possible[4,5]. However as can be seen from the experimental data[6] the spins are delocalized at least by a distance of 10^2A. Since the interaction between the chains is not very small the plausibility of a soliton mechanism for PA appears questionable.

We shall show that in PA exitations can exist, that transfer the equivalent configurations of the chain independently of kinks[7]. Such exitations are special mechanical deformations of a polymeric backbone - "twistons". Their interaction with kinks results in formation of combined particles (kink-twiston complexes) for which the confinement is absent. The complex can move freely along the chain, but the kink is localized in the region of it's centrum. The degree of localization is determined by quantum or thermal fluctuations.

SCELETAL SOLITON FOR A SYSTEM OF LINEAR MOLECULAR CHAINS

For simplification consider now linear molecular chains (Fig.1). There are: a kink (the topological charge Q=1) on the chain A, undeformed chain B, nonlinear acoustic deformation the sceletal soliton(Q=1) on chain C. The R.H.S.sign of the dimerization for A and C chains relative to B changes to the opposite. If two solitons of any type are simultaneously present on any of the chains, outside of this pear

---

* Permanent address: L.D.Landau Institute for Theoretical Physics A.S.USSR, Kosygina, 2, 117940, Moscow, USSR

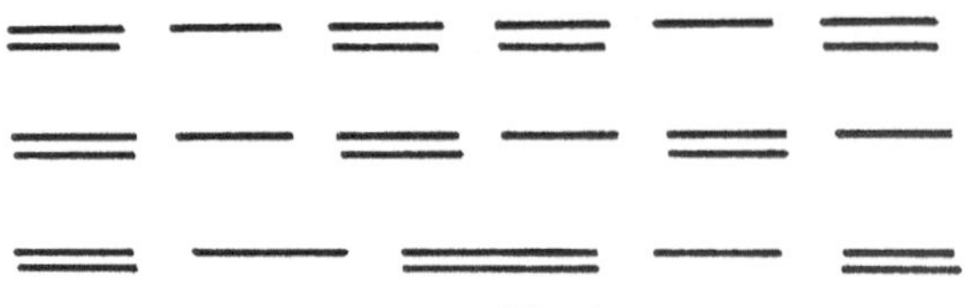

Fig.1

the dimerization remains unchanged. At the same time confinement exist between solitons that make up the pear.

Following three situations are possible:

1. Bound state of two kinks - bipolaron, exiton (Q=0).
2. Bound state of two sceletal solitons (Q=0).
3. Bound state of a kink and sceletal soliton (Q=0).

The model may be formulated in following way. The coordinate of n-th atom in dimerized chain x_n^α is given by the number n and two functions which change gradually with respect to n:

$$x_n^\alpha = na + \frac{a}{\pi}\varphi^\alpha(na) + \eta^\alpha(na)(-1)^n d \tag{1}$$

where a - average interatomic distance; d - dimerization amplitude ($d \ll a$); $\varphi^\alpha(na)$ - acoustic deformation of the backbone; $\eta^\alpha(na)$ - the form of the kink. The characteristic dimensions of a sceletal soliton ξ_1 an kink ξ_0 are large in comparison with the interatomic distance, so we use the continual approximation:

$$\varphi^\alpha(na) = \varphi^\alpha(x) \quad \eta^\alpha(na) = \eta^\alpha(x)$$

In the absence of kink

$$\eta^\alpha(x) = \eta^\alpha = \pm 1 \tag{2}$$

the energy of the chain α can be written as

$$W^\alpha = \int dx\{A(d\varphi^\alpha/dx)^2 + \sum_\beta U_0(\varphi^\alpha - \varphi^\beta) + \sum_\beta U_1(\varphi^\alpha - \varphi^\beta)\} \tag{3}$$

The interaction of the backbones U_0 is invariant with respect to translations $\varphi \to \varphi + \pi$. U_1 appears with dimerization being taken into account invariant with respect to translations $\varphi \to \varphi + 2\pi$ We shall apply the MFA taking into account only the deformations on the chain $\alpha = 0$, $\varphi^0 = \varphi$ and choosing the simplest form of interactions U_0, U_1 compatible with their symetries :

$$\sum_\beta U_0(\varphi^\alpha - \varphi^\beta) = \frac{B}{2}(1 - \cos 2\varphi); \ \sum_\beta U_1(\varphi^\alpha - \varphi^\beta)\} = \pm 2C(1 - \cos\varphi) \tag{4}$$

Here C is related to the confinement force F and to the confinement parameter γ as $F = \gamma\Delta/\xi_0 = 4C$, where $\xi_0 = hv_F/\Delta$ is the kink width, Δ is the dimerization gap. In Eq. (3) signs ($\pm$) stand for the two kinds of dimerization corresponding to the sign in Eq. (2). In the absence of dimerization for sceletal soliton we get

$$\cos\varphi = \pm th\xi, \ \ \xi = x/\xi_1, \ \xi_1 = \sqrt{A/B} \tag{5}$$

In the presence of dimerization the bound state of two sceletal solitons:

$$\cos\varphi_{\pm} = \mp[1 - \frac{2\mu th^2(\sqrt{1+\mu/2}\xi)}{2+\mu-2th^2(\sqrt{1+\mu/2}\xi)}], \; \mu = C/B \tag{6}$$

INTERACTION OF KINK AND THE SCELETAL SOLITON

Let us consider the bound state of the kink and the sceletal soliton. Being light kink must be described as a quantum particle. We are interested in the states which are equilibrium with respect to the classical field φ and stationary with respect to kink's wave function ψ. The variational functional:

$$W = \int_{-\infty}^{\infty} dx\{A(\frac{d\varphi^{\alpha}}{dx})^2 + \frac{B}{2}(1 - \cos 2\varphi) + 2C[1 - \cos\varphi(\int_{-\infty}^{x} \psi^2(y)dy$$

$$- \int_{x}^{\infty} \psi^2(y)dy)] + \frac{h^2}{2m}(\frac{d\psi}{dx})^2 - E\psi^2\} \tag{7}$$

E is the kink eigen energy. By variing functional (7) we arrive at equations

$$\frac{d^2\varphi}{d\xi^2} - \sin\varphi\cos\varphi - \mu\sin\varphi[1 - 2\int_{\xi}^{\infty}\psi^2(\eta)d\eta] = 0 \tag{8}$$

$$\frac{d^2\psi}{d\xi^2} + \psi[\epsilon - \frac{\beta}{2}(\int_{-\infty}^{\xi}(1+\cos\varphi)d\eta + \int_{\xi}^{\infty}(1-\cos\varphi)d\eta)] = 0 \tag{9}$$

where $\epsilon = E/E_0$; $\beta = 4C\xi_1/E_0$; $E_0 = h^2/2m\xi_1^2$ If $\beta < 1$ then in the ground state kink is localized at the sceletal soliton. The amplitude of the quantum fluctuations is

$$l_0 \sim \xi_1\beta^{-1/4} \sim (\frac{\xi_0}{\xi_1})^{3/4}(\frac{m^*}{\gamma m})^{1/4}$$

The character of the thermal fluctuations depends on the ratio ($\nu = 4C_1\xi_1/T$). If $1 < \nu < \beta$ then the kink is found in highly exited state which is still localized at the sceletal soliton $l_T \sim \xi_1/\nu^{1/4} < \xi_1$. If $\nu < 1$ then the kink moves mostly in the region of pure confinement: $l_T \sim T\xi_0/\gamma\Delta > \xi_1$

TWISTONS ON THE PA-CHAIN

A real PA chain differs from a considered linear chain by it's zig-zag structure (Fig.2). Before the dimerization is taken into account there is a two-fold screw axis (glice plane). Permitted defects that transfer *chain* $\rightarrow$ *itself*:

1. translation $\pm 2a$. 2. rotation $\pm 2\pi$. 3. translation $\pm a$ + rotation $\pm\pi$.

On the deformation's plane of the translations φ_{tr} and rotations φ_{rot} there are four equivalent equilibrium positions (Fig.3). Transitions: ($1 \leftrightarrow 2$), ($2 \leftrightarrow 3$), ($3 \leftrightarrow 4$), ($4 \leftrightarrow 1$), ($1 \leftrightarrow 3$), ($2 \leftrightarrow 4$). The sceletal solitons which correspond to the first four transitions we shall call twistons. Dimerization leaves only ($1 \leftrightarrow 3$) or ($2 \leftrightarrow 4$). The sceletal soliton between these positions a trivial translational soliton or the complex of two twistons, which are bound by the same confinement force as

Fig.2. PA-chain.

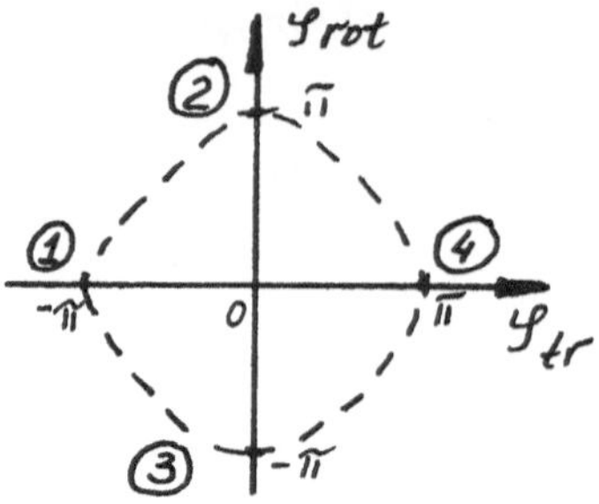

Fig.3. Equilibrium positions on ($\varphi_{rot}, \varphi_{tr}$) plane. Dashed line shows the transitions, corresponding to twistons.

two kinks. The kink on the chain changes the equilibrium points and restores the previous transitions.

If the elastic constants corresponding to the chain rotation and to the change i the angle between the carbons atoms in the chain are of the same order we can assume that on the plain (φ_{tr}, φ_{rot} there is a valley along which the transition takes place from one state into another and we cam back to the above model.

Nevertheless the combined nature of deformation is important:

1. Cis-trans-isomerization proceeds via repeated rotations of the chain segments.

2. The remnant insertions of cis-segments actually carry out the reflection's operation, so that here is only some translational distortion φ_{tr} required to complete the twiston.

Since twistons are topological objects, they can be eliminated only at the chain ends. Complete healing of twistons is prohibited because of interchain crosslinking.

For twistons in PA we obtain the next parameters: energy $W \sim 0.05eV$, length $\xi_1 \sim 200A$, mass $M \sim 10^2 m_e$, amplitudes of quantum fluctuations $l_0 \sim 20A$, amplitude of thermal fluctuations $l_{300K} \sim 50A$. In a result of kink-twiston confinement the backbone defects are marked by such quantum numbers as spinor charge. This effect can provide a new tool to study the chain slipping in polymers.

In conventional low symmetry dense packed phase the twiston-kink complexes can hardly exist until the twiston is ready from the beginning. The questions concerned with the possibility of twiston's formation and the role of solitons as a trigger in crystallografic transformation will be discussed elsewhere [7].

REFERENCES

1. W.P.Su, J.R.Schrieffer, A.J.Heeger. 1979, Phys. Rev. Lett., **42**, 1698.
2. S.Brazovskii. 1980, Sov. Phys. JETP, **51**, 342.
3. S.Brazovskii, T.Bohr. 1980, J.of Phys.C, **16**, 1189.
4. S.Brazovskii, N.Kirova. 1984, Sov. Sci. Rev. A, Phys. Rev., **5**, 99.
5. S.Brazovskii, N.Kirova, V.Yakovenko, 1983, J. de Physique, Colloque C3, Suppl. **44**, C3-1525.
6. Yu Lu. Solitons and polarons in conducting polymers. 1988, World Scien. Publ. Co.
7. S.Brazovskii, N.Kirova. 1988, Sol.St.Commun., **66**, 17.
8. S.Brazovskii, N.Kirova. 1990, Synth. Met., to be publ.

ON THE MOTION OF VORTEX PAIRS IN THE ANISOTROPIC HEISENBERG MODEL

A. R. Völkel, F. G. Mertens,* A. R. Bishop, G. M. Wysin†

Theoretical Division and CNLS, LANL, Los Alamos, NM 87545

INTRODUCTION

We consider a classical 2D Heisenberg model with easy-plane symmetry. Kosterlitz and Thouless[1] showed that such a system has a topological phase transition: at low temperatures there exist bound vortex pairs which start to dissociate above a critical temperature T_{KT}. Just above T_{KT} we can assume that there are only a few free vortices which move ballistically between their interactions. A model of dynamics built on such a "vortex gas" has been constructed assuming a Gaussian velocity distribution.[2] Here we use effective equations of motion for the collective (center-of-mass) vortex variables and compare these analytical results of vortex-vortex and vortex-anitvortex interactions with molecular dynamics simulations of the full spin system.[3] We investigate both ferromagnets (FM) and antiferromagnets (AFM) with an anisotropy parameter λ varying from zero to one.

Theory

Our system is described by the Hamiltonian

$$H = -J \sum_{\langle i,j \rangle} (S_i^x S_j^x + S_i^y S_j^y + \lambda S_i^z S_j^z) , \tag{1}$$

with the sum running over all the nearest neighbor pairs in the plane and a positive exchange parameter J for the FM and a negative one for the AFM. Assuming Landau-Lifshitz dynamics we obtain two different single vortex solutions from the equations of motion:[2] for $\lambda < \lambda_c$ (FM: $\lambda_c \approx 0.72$; AFM: $\lambda_c \approx 0.71$) static vortices are purely in-plane; for $\lambda > \lambda_c$ an additional out-of-plane component develops, allowing a continuous crossover to the isotropic Heisenberg limit ($\lambda = 1$) where the topological excitations are merons and instantons.[4]

Following a general procedure for magnetic systems,[5] and allowing also some damping, an equation of motion of a single vortex in the presence of another vortex in the center-of-mass coordinates and the continuum limit is

$$\mathbf{G} \times \mathbf{v} + \mathbf{D}\mathbf{v} + \frac{\gamma}{m_0} \mathbf{F} = 0 . \tag{2}$$

*University of Bayreuth, 8580 Bayreuth, West Germany

†Kansas State University, Manhattan, KS 66506

Microscopic Aspects of Nonlinearity in Condensed Matter
Edited by A.R. Bishop *et al.*, Plenum Press, New York, 1991

Here **G** and **D** are the gyrovector and the dissipation matrix, respectively, and both contain information about the actual vortex structure. **F** is the static force between the two vortices,[1] γ is the gyromagnetic ratio, and m_0 is the local magnetic moment per unit area.

Simulations

For the simulations we considered a 50×50 square lattice with free boundary conditions. The integration was performed with a fourth-order Runge-Kutta method with time step 0.04 $\hbar/JS$. For $\lambda < \lambda_c$ we initialized the simulations with two static in-plane vortices while for $\lambda > \lambda_c$ the four spins surrounding the vortex cores were given a small z component to guarantee the desired sign of the out-of-plane structure.

Ferromagnet

For $\lambda < \lambda_c$ the vortex is purely in-plane with small z components proportional to the velocity. For this case the gyrovector is vanishing and Eq. (2) predicts a motion of the vortices along straight lines: if the two vortices have equal vorticities they repel each other while a vortex and an antivortex attract each other. However, the simulations were performed on a discrete lattice which acts like a periodic "Peierls-Nabarro" pinning potential.[6] Thus the vortices have minimal energy if they are in the middle of a plaquette of four spins and they have maximal energy if they are at a lattice site. The resulting trajectories show therefore some fluctuations around the straight lines expected from Eq. (2). Moreover, the vortices will stop moving if their mutual distance is so large that the force between them is too weak to push them over the lattice potential. This scenario agrees very well with our simulations (Fig. 1).

For $\lambda > \lambda_c$ the vortices have a large ferromagnetic ordered out-of-plane structure which is extended over several lattice constants and which makes these vortices less sensitive to discreteness effects. The out-of-plane structure also acts like an effective magnetic field on the other vortex described by a nonzero gyrovector. In addition

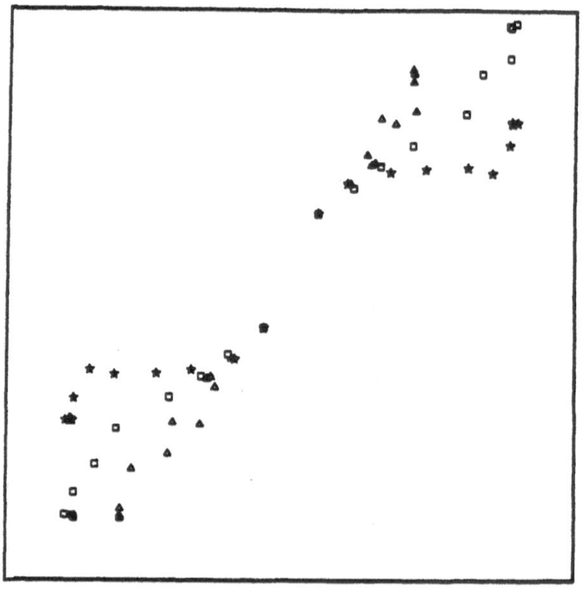

Fig. 1. Trajectories of vortex pairs on a FM with $\lambda < \lambda_c$ (square: $\lambda = 0.0$; triangle: $\lambda = 0.3$; star: $\lambda = 0.6$), $q_1 = q_2 = 1$ and initialized with two static in-plane vortices. Only a part of the lattice is shown: $18.0 \leq x, y \leq 30.0$, and the time between two successive points is four in our units.

to the repulsion or attraction, the motion of the vortices therefore shows additional behavior depending on the product of the vorticity q_j and the sign of the out-of-plane structure p_j of the two vortices $j = 1,2$. For $p_1q_1 = p_2q_2$ the vortices rotate around each other (Fig. 2a) while for $p_1q_1 = -p_2q_2$ they move parallel to each other (Fig. 2b).

Antiferromagnet

In the AFM the static vortices have, for all λ, a local antiferromagnetic order, and deviations from this structure due to the vortex motion are in the same direction for adjacent spins.

For $\lambda < \lambda_c$ this gives almost the same behavior as in the FM for the same λ range, and the simulation yields a picture which is very similar to Fig. 1. For $\lambda > \lambda_c$, however, the behavior is quite different to the FM: here the out-of-plane structure is antiferromagnetically ordered which gives no contribution to the gyrovector. Thus, also for $\lambda > \lambda_c$ the vortices will feel only the static force between them and move on straight lines, but without the strong dependence on the discrete lattice (Fig. 3).

CONCLUSIONS

Considering pairs of unbound vortices in a classical easy-plane Heisenberg model we found that the resulting trajectories depend strongly on the anisotropy parameter λ and the exchange parameter J (FM or AFM). For $\lambda < \lambda_c$ the mainly in-plane structure gives a zero gyrovector in Eq. (2) and therefore the vortices move on straight lines caused by the static force (attraction or equal, repulsion for different vorticities). The discreteness effects are strong in our simulations which were performed at $T = 0$. For $T \gtrsim T_{KT}$, where we expect ballistically moving vortices, the thermal fluctuations should be large enough to cancel this lattice effect. For $\lambda > \lambda_c$

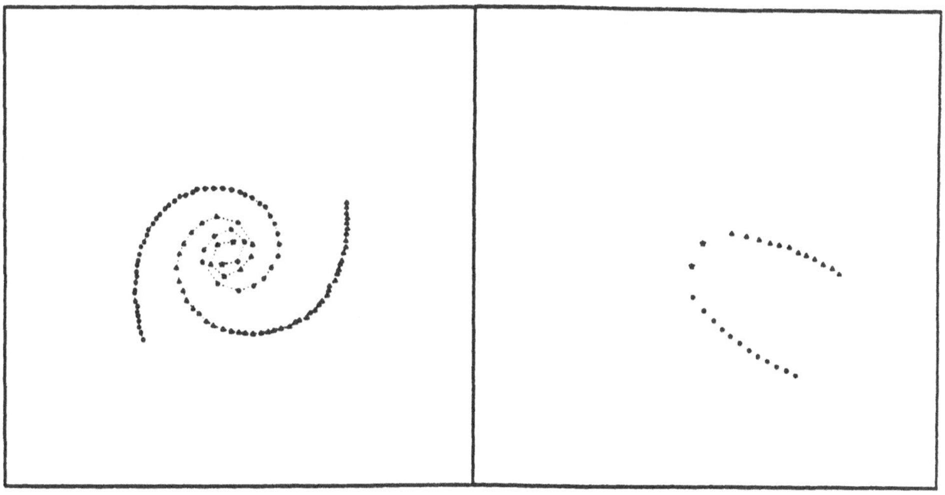

(a) (b)

Fig. 2. Trajectories of vortex-vortex simulations on a FM with $\lambda = 0.9, p_1 = p_2 = 1$ and initialized with two static out-of-plane vortices at positions (23.5, 23.5) and (24.5, 25.5) (star: start positions; circle: vortex 1; triangle: vortex 2); a) $q_1 = q_2 = 1$, the dashed line is a guide to the eye and connects successive points by straight lines; b) $q_1 = -q_2 = 1$.

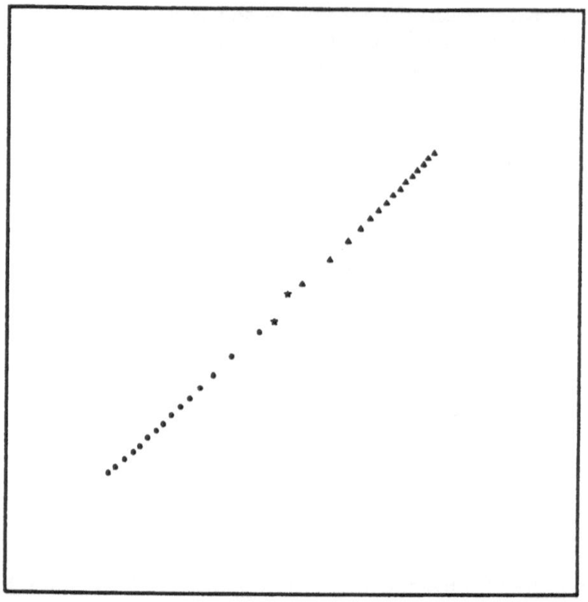

Fig. 3. Trajectories of a vortex-vortex ($q_1 = q_2 = 1$) simulations on an AFM with $\lambda = 0.9$ and initialized with two static in-plane vortices.

we must distinguish between $J > 0$ and $J < 0$. For the FM the out-of-plane structure results in an additional rotation to a translation of the vortices depending on the products $p_1 q_1$ and $p_2 q_2$. This behavior is similar to scenarios in 2D incompressible fluids,[7] superconductors[8] and superfluids.[9] In the AFM for $\lambda > \lambda_c$, however, the out-of-plane structure has no additional effect on the motion and we find only attraction or repulsion due to the static force.

The work was supported by NATO (Collaborative Research Grant 0013/89) and the United States Department of Energy.

REFERENCES

1. J. M. Kosterlitz, D. J. Thouless, J. Phys. **C6** 1181(1973); J. M. Kosterlitz, J. Phys. **C7** 1046 (1974).
2. F. G. Mertens, A. R. Bishop, G. M. Wysin, C. Kawabata, Phys. Rev. **B39** 591 (1989); M. E. Gouvêa, G. M. Wysin, A. R. Bishop, F. G. Mertens, Phys. Rev. **B39** 11840 (1989).
3. A. R. Völkel, F. G. Mertens, A. R. Bishop, G. M. Wysin, Phys. Rev. **B**, in press.
4. A. A. Belavin, A. M. Polyakov, Pis'ma Zh. Eksp. Teor. Fiz. **22** 503 (1975) [JETP Lett. **22** 245 (1975)].
5. A. A. Thiele, Phys. Rev. Lett. **30** 230 (1973); D. L. Huber, Phys. Rev. **B26** 3758 (1982); V. L. Pokrovsky, M. V. Feigl'man, A. M. Tsvelick, in "Spin Waves and Magnetic Excitations," ed. A. S. Borovik-Romanov, S. K. Sinha (Elsevier Science Publishers B. V.), Chapter II (1988).
6. J. Friedel, "Dislocations," Oxford, New York, Pergamon Press (1964).
7. H. M. Wu, E. A. Overman II, N. J. Zabusky, J. Comp. Phys. **53** 42 (1984).
8. I. Halperin, D. R. Nelson, J. Low. Temp. Phys. **36** 599 (1979).
9. W. J. Glaberson, R. J. Donnelly, Prog. Low Temp. Phys., Vol. **IX**, ed. D. F. Brewer (Elsevier Science Publishers B. V., 1986).

OSCILLATIONS DUE TO MANY-BODY EFFECTS IN RESONANT TUNNELING

F. Capasso[a], G. Jona-Lasinio[b] and C. Presilla[b,c]

[a]AT&T Bell Laboratories, 600 Mountain avenue, Murray Hill, N J 07974
[b]Dipartimento di Fisica dell' Universitá "La Sapienza", Roma, I 00185
[c]Dipartimento di Fisica dell' Universitá, Perugia, I 06100

ABSTRACT. *We analyze the dynamical evolution of the resonant tunneling of a cloud of electrons through a double barrier in the presence of the self-consistent potential created by the charge accumulation in the well. The intrinsic nonlinearity of the transmission process is shown to lead to oscillations of the stored charge and of the transmitted and reflected fluxes.*

In recent years there has been renewed interest in the phenomenon of resonant tunneling (RT) through double barriers. The unique capabilities of molecular beam epitaxy make it possible to investigate fundamental questions on RT through simple man-made potentials by controlling the barrier and well parameters down to the atomic scale[1].

In this paper we investigate the dynamics of RT of ballistic electrons in the presence of the potential created by the charge trapped within the barriers[2]. This problem is interesting not only from a technological point of view but also as a test of quantum mechanical non-equilibrium situations in which many particles are involved. The interest of this type of problem was emphasized some years ago by Ricco and Azbel[3] but as it will appear in the sequel the mechanisms involved are significantly more complicated than envisaged by these authors.

We propose the following model[4] to describe the physical situation. A cloud of electrons is created within a contact layer and launched towards embedded semiconductor layers forming a double barrier potential. The experimental set-up we have in mind puts ideally all the electrons in the same high-energy longitudinal (in the direction perpendicular to the double barrier) state while the transversal degrees of freedom are essentially decoupled[1]. As a consequence, at the starting time, the cloud wave function is factorized in a longitudinal and a transversal component. The longitudinal component is a product state of the same single particle state, $\psi(x,0)$. The cloud state is then symmetric with respect to the exchange of the longitudinal degrees of freedom. The correct antisymmetry of the total wave function is ensured by the transversal coordinates.

At low temperature and when the transversal dimensions of the semiconductor device are large with respect to the screening length in the medium, the decoupling between the longitudinal and transversal degrees of freedom is preserved as the time flows. In fact, theorem 5.7 of Ref. 5 guarantees in the mean field approximation (which is reasonable due to the large number of electrons involved) that the longitudinal state remains a product state during its evolution and allows us to write the following self-consistent equation for the single particle state $\psi(x,t)$:

$$i\hbar\frac{\partial}{\partial t}\psi(x,t) = \left[-\frac{\hbar^2}{2m}\frac{\partial^2}{\partial x^2} + V(x) + W(x,t)\right]\psi(x,t) \tag{1}$$

The external potential, V, is assumed, as customary, to be a piecewise constant function (no electric field is applied) representing a double barrier and admits of a resonance in the transmission coefficient which has an energy width Γ_R. The mean field interaction term, W, can be expressed in terms of the electron density $|\psi(x,t)|^2$ via the Poisson equation. This repulsive feedback effect can be reasonably represented by a simple model in which the bottom of the well between the two barriers of V is shifted to a higher energy proportional to the electron charge, $Q(t)$, trapped in it at time t. In this model the intensity of the repulsive potential is assumed proportional also to a parameter, α, which is inversely proportional to the capacitance of the double barrier and thus takes into account the dielectric constant of the medium.

The 1-particle state, $\psi(x,0)$, which is the initial condition in our mean field equation, has been chosen to be a gaussian shaped superposition of plane waves, with energy spread Γ_0, impinging on the double barrier with a mean kinetic energy close to the resonance energy. At the starting time the cloud is localized distant from the double barrier so that no appreciable charge sits in the well, *i.e.* $Q(0) = 0$. The normalized charge in the well, $Q(t)/Q_0$, where Q_0 is a convenient normalization factor depending on the shape of $\psi(x,0)$, has been calculated as a function of time by numerically integrating[6,7] Eq. (1). The results for different choices of the parameter α are shown in Fig. 1 in the case of electron clouds with energy-spread larger, of the same order and smaller than the resonance width.

When the non linear term is neglected, *i.e.* for $\alpha = 0$, the evolution of the trapped charge, as well as the transmitted and reflected fluxes, presents a smooth increase followed by a decrease. The differences between the two extreme cases $\Gamma_0 \gg \Gamma_R$ and $\Gamma_0 \ll \Gamma_R$ for what concerns the decrease, are simply understood in terms the of decay law of a quantum mechanical state which has a lorentzian or a gaussian shape[7,8].

In the case of a real interacting electron cloud, *i.e.* for $\alpha \neq 0$, the evolution of the trapped charge changes drastically and oscillations can appear. A detailed analysis of our global numerical simulations for different values of α and Γ_0 supports the following considerations. Oscillations appear, for an appropriate strength of the interaction term, *i.e.* for α sufficiently high, only when the cloud energy spread is wider or comparable to the resonance width. No oscillations are seen for a nearly monochromatic cloud, *i.e.* $\Gamma_0 = 0.8\ meV$. When α increases, the oscillations, if present, increase in number but decrease in amplitude.

To understand these results, let us first interpret the dependence of the intensity of the trapped charge as a function of the parameters α and Γ_0. We simplify the question by considering a time-average of the charge dynamically present inside the well. The mean charge Q trapped in the well is determined self consistently by requiring that it is proportional to the asymptotically transmitted charge[8], Q_T, which itself depends on Q:

$$Q \propto Q_T = \int_{-\infty}^{+\infty} dk\ |\widetilde{\psi}(k,0)|^2\ T_Q(k) \tag{2}$$

where $\widetilde{\psi}(k,0)$ is the Fourier transform of $\psi(x,0)$ and $T_Q(k)$ is the transmission coefficient of the potential V modified by the charge Q. The self-consistent relation (2) predicts correctly the decrease of the charge trapped in the well as the self interaction, *i.e.* α, is increased as well as the dependence from the cloud energy spread Γ_0[4].

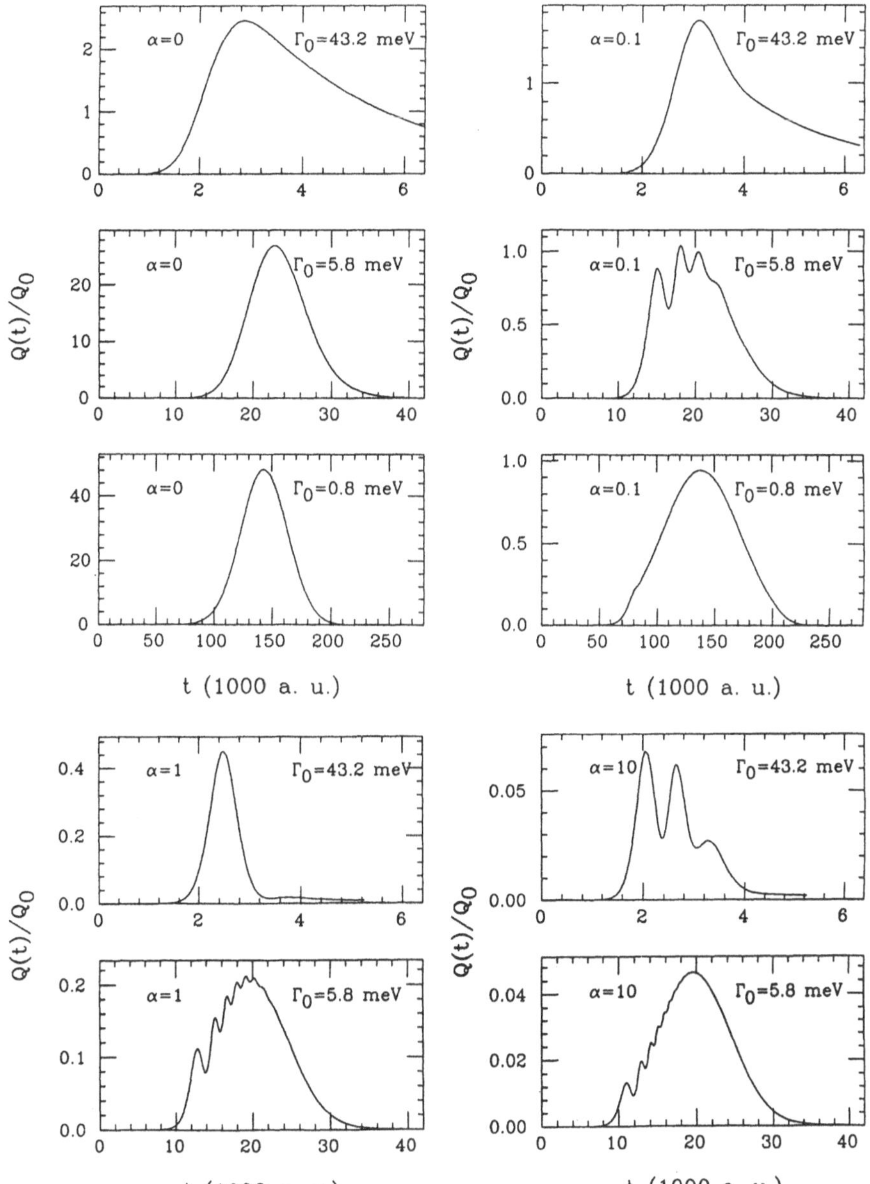

Figure 1. *Time development of the normalized charge trapped in the well for electron clouds with energy spread much wider* ($\Gamma_0 = 43.2\ meV$)*, of the same order of* ($\Gamma_0 = 5.8\ meV$) *and much smaller* ($\Gamma_0 = 0.8\ meV$) *than the resonance width* ($\Gamma_R \simeq 5\ meV$)*. The results are shown for various values of the feedback intensity parameter,* α*. The conversion of the atomic units of time to seconds is* $1\ a.u. \simeq 4.83\ 10^{-17}\ s$.

We then consider the oscillating behavior. Let us assume that this phenomenon is due to the competition of two processes: *(a)* the filling up of the well by the incoming cloud and *(b)* the natural decay of the trapped charge. For the process *(a)* the time scale is of the order of $\hbar/\Gamma_0$. For the process *(b)* a reasonable time scale is $\hbar/\Gamma_Q$, where Γ_Q is the energy spread of the function to be integrated in Eq. (2), *i.e.* the width of the spectral decomposition of the charge present in the well. Oscillations are then expected if a substantial crossover of Γ_Q and Γ_0 is realized for the Q values reached during the time evolution. The analysis of Γ_Q as a function of Q shows[7] that when $\Gamma_0 \ll \Gamma_R$, its value is very close to Γ_0 and nearly independent of Q. As a consequence no oscillations are possible in this case for any value of α. On the other hand, when $\Gamma_0 \geq \Gamma_R$, Γ_Q crosses Γ_0 at a some Q; oscillations are then realized for a sufficiently high value of α. This critical value of α increases with the ratio Γ_0/Γ_R. These predictions agree with the results of the simulations reported above.

Experimentally, the geometry considered here can be implemented by ballistically launching electrons into a double barrier inserted in the thin (< 1000 Å) base of a unipolar transistor[9]. The predicted range of oscillations (≤ 1 ps) should be detectable with electro-optic sampling techniques[10].

REFERENCES

1. For a review on resonant tunneling through double barriers, the reader is referred to *Physics of Quantum Electron Devices*, F. Capasso, ed., Spinger-Verlag, New York, Heidelberg (1990).
2. M. Tsuchiya, T. Matsusue and H. Sakaki, *Tunneling escape rate of electrons from quantum well in double barrier heterostructures*, Phys. Rev. Lett. **59**, 2356 (1987); J. F. Young, B. M. Wood, G. C. Aers, R. L. S. Devine, H. C. Liu, D. Landheer, M. Buchanan, A. S. Springthorpe and P. Mandeville, *Determination of charge accumulation and its characteristic time in double barrier resonant tunneling structures using steady-state photoluminescence*, Phys. Rev. Lett. **60**, 2085 (1988); V. S. Goldman, D. C. Tsui and J. E. Cunningham, *Resonant tunneling in magnetic fields: evidence for space-charge buildup*, Phys. Rev. B **35**, 9387 (1987).
3. B. Ricco and M. Ya. Azbel, *Physics of resonant tunneling. The one dimensional double barrier case*, Phys. Rev. B **29**, 1970 (1984).
4. C. Presilla, G. Jona-Lasinio and F. Capasso, *Nonlinear feedback oscillations in resonant tunneling through double barriers*, Phys. Rev. B (Rapid Comm.) in press.
5. H. Spohn, *Kinetic equations from Hamiltonian dynamics: Markovian limits*, Rev. Mod. Phys. **53**, 569 (1980).
6. W. H. Press, B. P. Flannery, S. A. Teukolsky and W. T. Vetterling, *Numerical Recipes: the Art of Scientific Computing* (Cambridge University Press, Cambridge 1986).
7. C. Presilla, G. Jona-Lasinio and F. Capasso, *Dynamical analysis of resonant tunneling in presence of a self consistent potential due to the space charge*, in *Resonant Tunneling in Semiconductors: Physics and Applications*, L. L. Chang, E. E. Mendez and C. Tejedor Ed.s (Plenum Press, New York 1990).
8. S. Collins, D. Lowe and J. R. Barker, *A dynamic analysis of resonant tunneling*, J. Phys. C **20**, 6233 (1987).
9. M. Heiblum, M. J. Nathan, D. C. Thomas and C. M. Knoedler, *Direct observation of ballistic transport in GaAs*, Phys. Rev. Lett. **55**, 2200 (1985).
10. J. F. Whitaker, G. A. Mourou, T. C. L. G. Sollner and W. D. Goodhue, *Picosecond switching time measurement of a resonant tunneling diode*, Appl. Phys. Lett. **53**, 385 (1989).

COMPUTER SIMULATION OF TUNNELING IN ATOMIC-SIZE DEVICES

H. De Raedt

Institute for Theoretical Physics
University of Groningen
P.O. Box 800
NL-9700 AV Groningen
The Netherlands

INTRODUCTION

Electrons emitted from small atomic-size sources exhibit properties that differ substantially from those emitted by conventional sources. Experimental work on electron field emission from one and three-atom tungsten tips has shown that the low-energy electron beam is strongly focussed. [1-4] Electron sources having these properties would be very valuable for applications to electron holography and electron interferometry. [5-9]

In the past few years, it has become clear that it is possible to manufacture solid-state devices that have properties akin to those of electron guns. [10] Firstly, in order to mimick the vacuum, the electrons in the solid state should be able to travel almost freely, i.e. ballistically. Secondly, the size of the solid state device should be sufficiently small, a requirement which can be met by modern circuit lithography. Although several fundamental problems have to be overcome before such devices can be made to perform as (very fast) transistors, they provide a unique laboratory to explore all kinds of quantum interference effects. [11,12]

All these devices have in common that the electrons are emerging from a small (with respect to the electron wave length) region. More generally, we are faced with a situation where the relevant size of the source is roughly equal to the wave length of the emitted wave. In such a case conventional methods of treating wave mechanics are no longer valid and one has to resort to more elaborate techniques.

In this paper we demonstrate that computer simulation is particularly useful to study this class of meso- and microscopic devices. The approach we follow is to solve numerically the time-dependent Schrödinger equation (TDSE) using a novel numerical method [13] which is based on generalizations of the Trotter-Suzuki product formula. [14,15] The salient features of our method are its unconditional stability, high accuracy and superior efficiency. Typically we solve the TDSE for the initial state being an appropriate Gaussian wave packet. Analysis of the time development of the wave function allows us to uncover the physical mechanism(s) that are responsible for the peculiar properties of the emitted electron waves.

Below we present and discuss results for two model systems, other examples and more details being given elsewhere.[16,17,18] As a first example we consider a model for tunneling of a particle through a one-dimensional time-modulated rectangular barrier, introduced by Büttiker and Landauer [19] (BL) to substantiate one of the expressions for the traversal time. In the second example we model electron field emission from atomic-size tips. We demonstrate that the large diffraction effects, expected to occur due to the smallness of the emission area, are reduced by the presence of the metal-vacuum tunneling potential, leading to possibility of achieving a resolution of the order of the electron wave length (atomic resolution).

TUNNELING THROUGH A TIME MODULATED BARRIER

The question whether a tunneling event is characterized not only by a tunneling probability but also by a tunneling time [20,21] is still controversial, as is most evident from the lively debates in recent literature. In a conservative quantum system, time is not an observable and is therefore not measurable in a strict sense. This fundamental problem can be circumvented by adding to the Hamiltonian a time-dependent term. The model is defined by [19]

$$H = \frac{-\hbar^2}{2m} \frac{\partial^2}{\partial x^2} + V_0(x) + V_1(x) \cos \omega t. \qquad (1)$$

whereby $V_0(x) = V_0\Theta(x)\Theta(d - x)$ represents the static barrier of height V_0 and thickness d, and $V_1(x) = V_1\Theta(x)\Theta(d - x)$ in which V_1 sets the amplitude of the modulation of frequency ω. For practical purposes it is convenient to express all length scales in units of λ_F, the dominant wave length of the initial wave packet, wave vectors in units of $k_F = 2\pi/\lambda_F$, energy in units of $E_F = \hbar^2k_F^2/2m$, ω in units of $E_F/\hbar$ and time in units of $\hbar/E_F$. The model parameters d, V_0, V_1, and $\hbar^2/2m$ are chosen such that the barrier is almost completely reflecting, i.e. opaque.

The BL analysis [19] is based on the following physical picture. The probability that a particle will absorb or emit modulation quanta depends on the modulation frequency ω and the traversal time, i.e. the time [22] during which the particle interacts with the barrier. If the former is much smaller than the inverse of the latter, the particle feels a static potential. In the opposite limit the particle undergoes many cycles of the modulation and can emit or absorb modulation quanta. According to BL there is a crossover from one type of behaviour to the other, occurring at $\omega_{BL}\tau_{BL} \sim 1$. Hence the traversal time τ_{BL} of the transmitted particle can be determined by changing the modulation frequency ω. For model (1), we have $\tau_{BL} = d[m/2(V_0 - E_F)]^{1/2}$, i.e. the time a classical free particle of mass m and kinetic energy $V_0 - E_F$ would need to travel the distance d. In what follows we focus on the frequency-dependent model properties, not on the existence of a traversal time as such.

Figure 1 shows our simulation results [18] for the transmission coefficient

$$T(\omega,\sigma) = \lim_{t\to\infty} \int_d^\infty dx\ |\psi(x,t)|^2, \qquad (2)$$

for different values of V_1/V_0 and the width (σ) of the initial Gaussian packet. As the modulation frequency increases, the transmission coefficient changes by several orders of magnitude, provided the coupling with the external field is sufficiently strong (e.g. $V_1 \geq V_0/100$). There is a crossover from a regime in which the transmission coefficient increases drastically with the modulation frequency ω to a regime where saturation sets in. Our simulations show that the position of the crossover (ω_c) is independent of V_1 and σ (for the values of σ considered here), as confirmed by simulation results for barriers of various thickness and height.

The physical mechanism leading to the crossover is the following. The particle can gain energy by absorbing modulation quanta. If the energy (incident plus absorbed modulation quanta) becomes larger than the barrier height, it does not have to tunnel through the barrier but goes over it. This

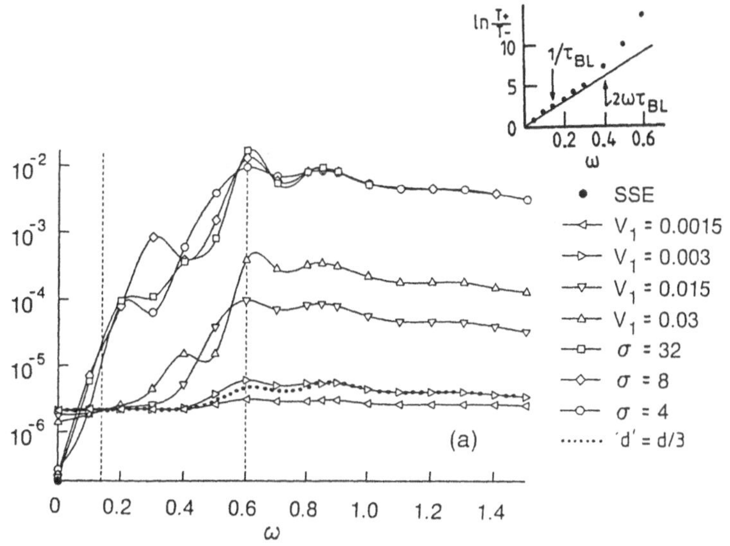

Fig. 1 Transmission coefficient of a time-modulated rectangular barrier as a function of modulation frequency ω. Model parameters are $V_0 = 1.5E_F$, $V_1 = V_0/10$, $d = 5\lambda_F/3$, and $\sigma = 4\lambda_F$ unless specified explicitly by the legend. The reference line closest to $\omega = 0$ corresponds to $\omega_{BL} = 1/\tau_{BL}$. The other reference line gives our estimate of ω_c. SSE denotes the rigorous solution of the stationary Schrödinger equation ($\omega = 0$). Dots represent the results for a barrier modulated over a distance $d' = d/3$ with amplitude $V_1 = 0.003$. Inset: ratio of the side-band intensities, T_+ and T_- versus frequency ω. Dots: Simulation results, solid line: BL prediction.

gives rise to the plateau. At certain energies there are extrema related to the resonances of the barrier. For $|V_1| \leq V_0/100$ the frequency at which the plateau in $T(\omega,\sigma)$ sets in is well described by the condition for the coincidence of the first-order sideband energy $1 + \omega$ with the first-order resonance energy of the static barrier, i.e.

$$\omega_R(V_0,d) = V_0 - 1 + \frac{1}{4d^2} \,. \tag{3}$$

Indeed, there is almost perfect agreement between our numerical estimates for

ω_c and the corresponding value of ω_R. As V_1 increases, both the intensity and the number of observable sidebands increases. The interplay of higher-order sidebands and higher-order resonances becomes very complicated and it is not always possible to identify the onset of the plateau with ω_R.

Defining $\omega_{BL} = 1/\tau_{BL}$ we find that at this particular frequency there is no sign of a crossover, as illustrated in Fig. 1. Accounting for some unknown constant factors we also find no evidence that the BL expression for $1/\tau_{BL}$ is proportional to ω_c. If τ_{BL} would be a measure for how long the transmitted particle interacts with the (weak) modulation, changing the length over which the barrier is modulated should have an effect on the tunneling properties. To test this hypothesis we have repeated some calculations taking $V_1(x) = V_1\Theta(x)\Theta(d' - x)$ and $d' = d/3$. As illustrated in Fig. 1 there are no significant changes in $T(\omega,\sigma)$.

To compare directly with BL we have also computed the side-band intensities T_+ and T_-, as defined by BL, [19] in the parameter regime where the BL analysis applies. From Eq. 7 of Ref. 8 it follows that $\ln(T_+/T_-) = 2\omega\tau_{BL}$. The inset of Fig. 1 clearly shows our simulation data for $\ln(T_+/T_-)$ to be in perfect agreement (without fitting) with the BL prediction, for $\omega < 0.2$ (in their derivation [19] BL have made the assumption that $\omega \ll E_F$ and $E_F = 1$ in our calculations). Our results (Fig. 1) indicate that there is no crossover at a frequency $1/\tau_{BL}$, not in the transmission coefficient nor in T_+/T_-. For $\omega \approx \omega_c$, $\ln(T_+/T_-)$ deviates considerably from the BL result.

To summarize: We conclude that model (1) exhibits a crossover as suggested by BL, but the crossover frequency is not related to $1/\tau_{BL}$. It appears that the two frequency regimes we observe differ from the ones dealt with by BL. In the regime where the modulation amplitude is not small, non-linear effects are important and the physical behavior of model (1) is rather complicated.

ELECTRON EMISSION FROM TETON TIPS

Attempts to describe the peculiar properties of field-emitted electron beams from ultra-sharp, teton and built-up tips require a proper treatment of the full quantum mechanical problem. Earlier calculations have shown that it is possible to explain the small angular spread (of the order of a few degrees, i.e. strong focussing) observed experimentally. These calculations suggest that by changing the geometry of the tip one can tailor the electron beam to considerable extent. In what follows we concentrate on the remarkable properties of tips ending on three atoms. The conditions under which a 3-atom tip can emit three separated waves are not known. In view of the small distance between the atoms (approximately equal to the Fermi wavelenght λ_F) it would be remarkable indeed that the emitted wave packets would not merge. Below we present simulation results for a model containing the essential features of these atomic-size tips, and demonstrate that under certain conditions it may be possible to obtain electron field emission with atomic resolution, i.e. observe well-separated spots. Recent experiments, still to be confirmed by field ion microscopy, indicate that such sources can indeed be build.

To construct a model that is amenable to simulation on present day supercomputers but still contains the essential ingredients it is necessary to work in two spatial dimensions. Instead of three emitting atoms we then have two. The geometry of the model is depicted in Fig. 2. As precise information about the electrostatic potential surrounding the teton tip is lacking, a model potential is constructed which is both simple and has the main ingredients. The curvature of the electrostatic potential surrounding the tip seems to be such that each atom sees a potential which has a minimum along a

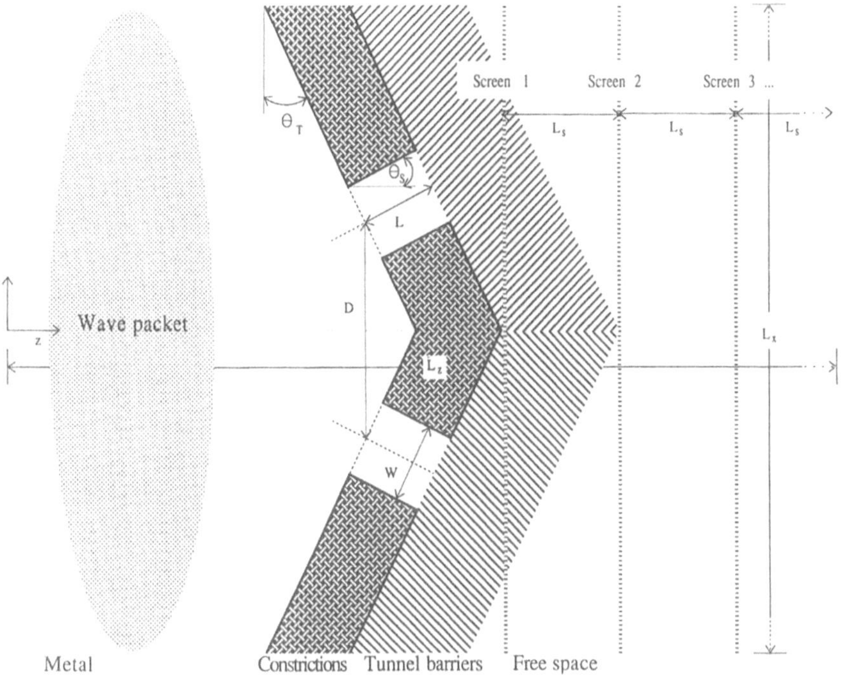

Fig. 2 Geometry of the 2D model of the teton tip used in TDSE simulations.

certain direction, depending on the position of the atom. Moving in a direction perpendicular to this preferential path, the potential increases rapidly. This can be modelled by a constriction which has a size of the atom. For the 2D model it is reasonable to assume that the constrictions are to be placed symmetrically around $y = 0$ (see Fig. 2). Each constriction is titled by a certain angle θ_T. The angle of the slit θ_S is taken to be equal to θ_T. In the simulations the dimensions of the slits are $W = L = \lambda_F/2$ their separation $D = 1.5\lambda_F$, the height of the triangular barrier $V = 1.5E_F$, and the slope corresponds to a field of $0.33V/\text{Å}$, all reasonable values from experimental viewpoint. The metal–vacuum tunnel barrier is assumed to have a plane triangular shape, properly tilted as indicated in Fig. 2.

A convenient choice for the initial wave packet is

$$\psi(x,z,t = 0) \propto \sin\left[\frac{n\pi x}{L_x}\right] \exp(ik_z z) \exp(-(x - x_0)^2/2\sigma_x^2 - (z - z_0)^2/2\sigma_z^2) , \tag{4}$$

i.e. of Gaussian packet moving in the z–direction. The angle of incidence θ_i fixes the value of the "mode" index n and k_z as

$$n\pi/k_F L_z = \sin \theta_i , \tag{5a}$$

and

$$k_z^2 + (n\pi/L_x)^2 = k_F^2 . \tag{5b}$$

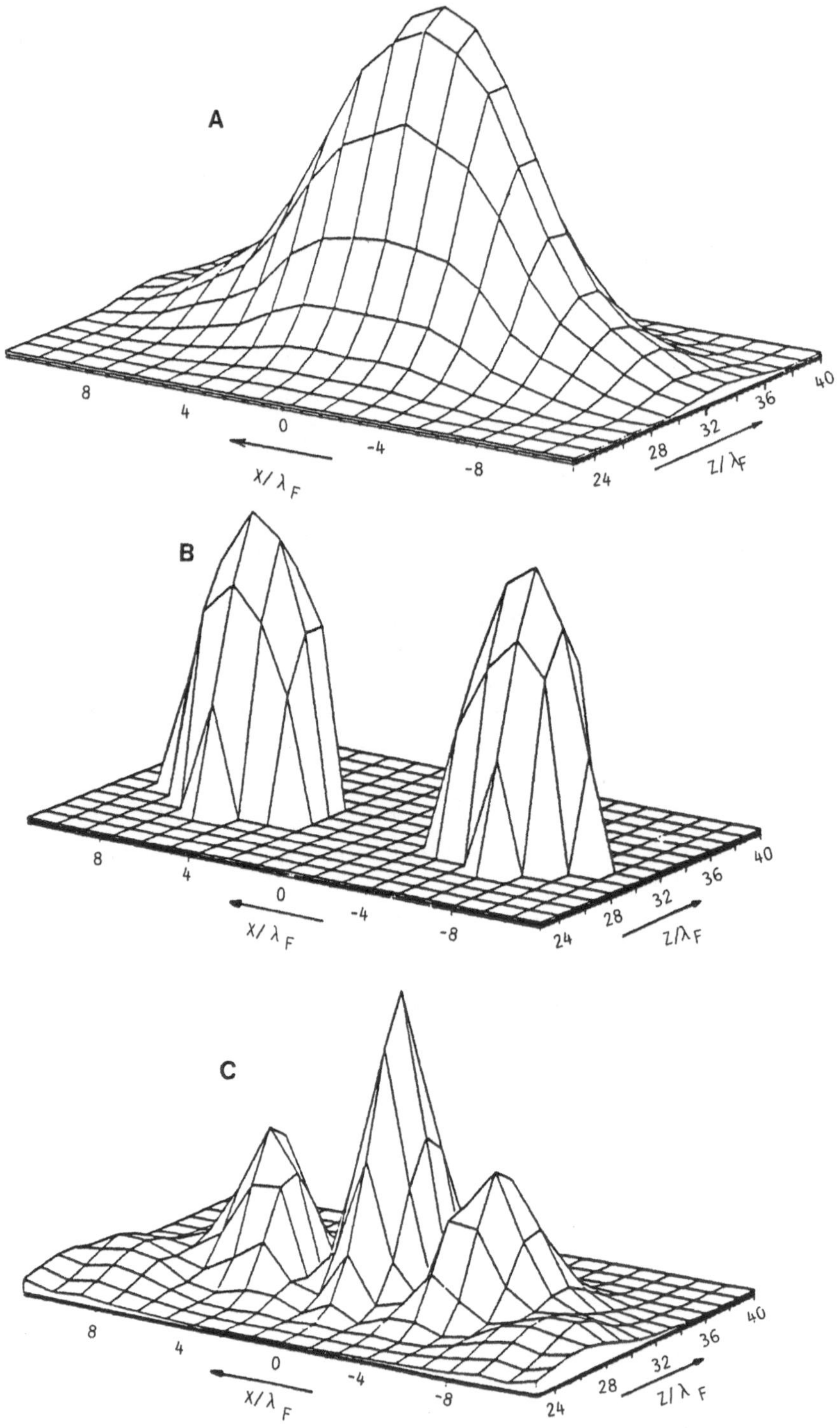

Fig. 3 TDSE simulation of the 2D teton tip model. Real-space probability distribution of the transmitted electron wave for A) $\theta_T = \theta_S = 10°$, B) $\theta_T = \theta_S = 30°$, and C) $\theta_T = \theta_S = 30°$ but without metal-vacuum potential barrier.

In practice $x_0 = 0$, $\sigma_x = 5\lambda_F$, $\sigma_2 = 2\lambda_F$, and z_0 is chosen such that the initial wave packet does not overlap with the constrictions. The size of the simulation box used is $40\lambda_F \times 23\lambda_F$, corresponding to a grid of 501×467 lattice points. At regular time intervals, snap-shots of the probability distribution are taken.

Figure 3.a shows TDSE simulation results for $\theta_T = \theta_S = 10°$. It is clear that only one spot will be observed at a screen far away from the tip. Taking $\theta_T = \theta_S = 30°$, the behavior has changed tremendously as shown in Fig. 3b. Now well-separated spots are observed. At first sight this might be explained quite trivially by purely geometrical reasoning. Tilting the slits more and more has to result in a situation where the incident wave splits in two outgoing pieces. However if this reasoning is correct, turning off the tunnel barrier should not alter the pictures on a qualitative level. As demonstrated in Fig. 3c this is not what happens as without tunnel barrier the transmitted wave looks as one big packet with a lot of structure. On a genuine screen, such a packet would produce one instead of two spots. To understand this behavior in simple terms first note that the two slits are very close ($\approx \lambda_F$) to each other. In the absence of a tunnel barrier, there is a lot of diffraction at the exit plane of a narrow constriction, i.e. the angular spread of a wave emitted by a constriction is large. Without tunnel barrier the two parts of the wave join to form one object. On the other hand if the tunnel barrier is present the focussing is so strong that it is possible to separate the two parts.

Simulations with other initial conditions, triangular potentials of different height and slope have been carried out and all lead to the same conclusion: The tilt angle θ_T necessary to observe two well-separated electron beams is in the range of 20° and 30°.

From experimental point of view the range of the values of the tilt angle needed to separate the spots is reasonable as a teton tip consists of a small microtip on top of a support tip (corresponding to a large angle of tilt, $\theta_T = \theta_S \approx 30°$). Moreover it has to be taken into account that additional focussing of the electron beams occurs in the far-field region, an effect which has not been incorporated in our simulation. Hence the effective tilt angle obtained by measuring the distance between the spots on a screen at a macroscopic distance from the tip, will be much less than the true tilt angle at the apex of the tip.

To summarize: our simulations for the model described above provide a scenario for understanding the experimental observation of several distinct spots of electron intensity in field emission patterns of teton tips.

ACKNOWLEDGEMENTS

The work reported on here is the result of a collaboration with N. García, J.J. Sáenz, Vu Thien Binh, and J. Huyghebaert. I would like to thank R. Leavens for his suggestion to interpret ω_c in terms of ω_R and to perform simulations with $d'' < d$, and for drawing our attention to a number of pertinent papers.

REFERENCES

1. H-W. Fink, IBM J. of Research and Development 30, 460 (1986).
2. H-W. Fink, Physica Scripta 38, 260 (1988).
3. Vu Thien Binh, Surface Science 202, L539 (1988).
4. Vu Thien Binh, J. Microscopy 152, 355 (1988).
5. D. Gabor, Proc. Roy. Soc. London A454, 197 (1949).

6. D. Gabor, Proc. Roy. Soc. London B64, 449, (1951).
7. A. Tonomura, Rev. Mod. Phys. 59, 639 (1987).
8. H. Lichte, Ultramicroscopy 20, 293 (1986).
9. F.G. Missiroli, G. Pozzi, and U. Valdre, J. Phys. E 14, 649 (1981).
10. S. Washburn, Nature 343, 415 (1990).
11. B.J. van Wees et al, Phys. Rev. Lett. 60, 848 (1988).
12. D.A. Wharam et al, J. Phys. C 21, L209 (1988).
13. H. De Raedt, Comp. Phys. Rep. 7, 1 (1987).
14. H. De Raedt, and B. De Raedt, Phys. Rev. A. 28, 3575 (1983).
15. M. Suzuki, J. Math. Phys. (NY) 26, 601 (1986).
16. N. García, J.J. Sáenz, and H. De Raedt, J. Phys., Condens. Matter 1, 9931 (1989).
17. H. De Raedt, N. García, and J.J. Sáenz, Phys. Rev. Lett. 63, 2260 (1989).
18. H. De Raedt, N. García, and J. Huyghebaert, Solid State Comm. (in press).
19. M. Büttiker and R. Landauer, Phys. Rev. Lett. 49, 1739 (1982).
20. C.R. Leavens and G.C. Aers, Phys. Rev. B 39, 1202 (1989).
21. E.H. Hauge and J.A. Støvneng, Rev. Mod. Phys. 61, 917 (1989).
22. In this paragraph we use the term "time" in the same sense as BL.

QUANTUM BALLISTIC TRANSPORT

Y.B. Levinson

Institute of Microelectronics Technology and High Purity Materials, USSR Academy of Sciences, 142432 Chernogolovka, Moscow District, USSR

INTRODUCTION

Quantum ballistic transport occurs in devices microfabricated by lateral patterning of the high-mobility two-dimensional electron gas (2DEG) in a GaAs/ AlGaAs heterojunction. A quantum ballistic device (QBD) is a channel[1,2] (constriction) or a net of channels[3,4,5] with length L shorter than elastic scattering mean free path ℓ (about 10 μm) and width d of the order of Fermi wavelegth λ_F (about 50 nm). The narrow channels are connected with wide contact pads (reservoirs), where the current enters the device and the applied voltage is measured.

The conductivity of a QBD has several distinguishing features: quantization[1,2], waveguide and nonlocal properties[3,4,5]. All these features stem from the fact that the resistance of a QBD is not due to scattering of electrons by random inhomogeneities (impurities and/or phonons) but due to diffraction of the electron wave. The diffraction occurs not only at the junctions of different channels, but also at the entrance to a channel from a contact pad. Therefore the channels cannot be isolated from the reservoirs and the resistance of a QBD is to be calculated taking into account the contact pads explicitly.

In contrast with the previous papers on quantum transport[6,7,8,9] we treat[10] the conducting channels and the contact pads as a single unit and thus we find a suitable approach for effects which occur as electrons go from the reservoir to the conducting channel.

1. FORMULATION OF THE PROBLEM - BASIC EQUATIONS

In the present paper we consider a model of a QBD which is a constriction between two expanding horn-shaped contact pads (Fig.1). We describe the electrons in the QBD in the self-consistent-field approximation.

Microscopic Aspects of Nonlinearity in Condensed Matter
Edited by A.R. Bishop *et al.*, Plenum Press, New York, 1991

When there is no current through the constriction ($J=0$) the electron system is in equilibrium. Each electron is in a potential $U(r)$ which is the sum of the potentials created by the ions and the electrons. Potential U determines electron eigenstates $\psi_n(r)$ and their energies ε_n. States n are occupied in accordance with the Fermi distribution $f_T(\varepsilon_n)$. The equilibrium charge density is

$$\rho(r) = 2e \sum_n f_T(\varepsilon_n)|\psi_n(r)|^2. \tag{1}$$

This density is self-consistent with the electronic contribution to U by the Poisson equation. In the constriction, with $d \lesssim \lambda_F$ the states ψ_n are standing waves across the channel and because the density ρ is spatially nonuniform. Density becomes uniform in the horns far from the constriction where many standing waves contribute to the density (1).

When the device is biased by a voltage V and a current flows through the constriction ($J \neq 0$), a change occurs in the electron density, $\delta\rho(r)$, and in the self-consistent potential, $U(r)$. If we write $\delta U(r) = e\varphi(r)$, then $\varphi(r)$ is just the electrostatic potential which arises from biasing the device. Obviously φ and $\delta\rho$ are related by the Poisson equation

$$\nabla^2\varphi = -(4\pi/\epsilon)\,\delta\rho. \tag{2}$$

Moving far away from the constriction, the cross-sections of the horns increase and, hence, the current density tends to zero. This means that the electron system in a point r far away from the constriction is almost in equilibrium. In an equilibrium system the potential is uniform and hence, $\varphi(r) \to$ const as $r \to \infty$. The asymptotic values far on the left $\varphi(-\infty)$, and on the right, $\varphi(+\infty)$, differ, and the difference is just the applied bias:

$$\varphi(+\infty) - \varphi(-\infty) = V. \tag{3}$$

The electron redistribution due to bias V is found from the equation for the one-electron density operator f (spin neglected, $\hbar=1$):

$$\partial f/\partial t = -i[H,f] - (f-f_0)/\tau = 0. \tag{4}$$

Here

$$H = H_0 + e\varphi\,, \quad H_0 = -\frac{1}{2m}\nabla^2 + U\,, \quad f_0 = f_T(H_0). \tag{5}$$

Equations (2) and (4) for the two unknown quantities f and φ, are the basic equations to describe the QBD.

The relaxtion time τ is responsible for inelastic scattering. We assume of course, $\tau v_F \gg L$. Neverthless, due to the inelastic scattering the remoted parts of the horns (at $r \gg v_F\tau$) are thermal reservoirs. Elastic scatterers, if present, can be included in potential U.

Calculating $\delta f = f - f_0$ for given φ from (4), we find the charge redistribution,

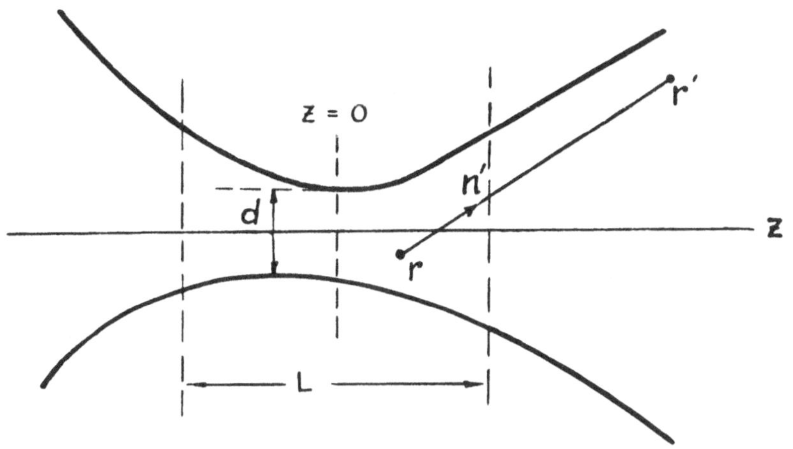

Figure 1

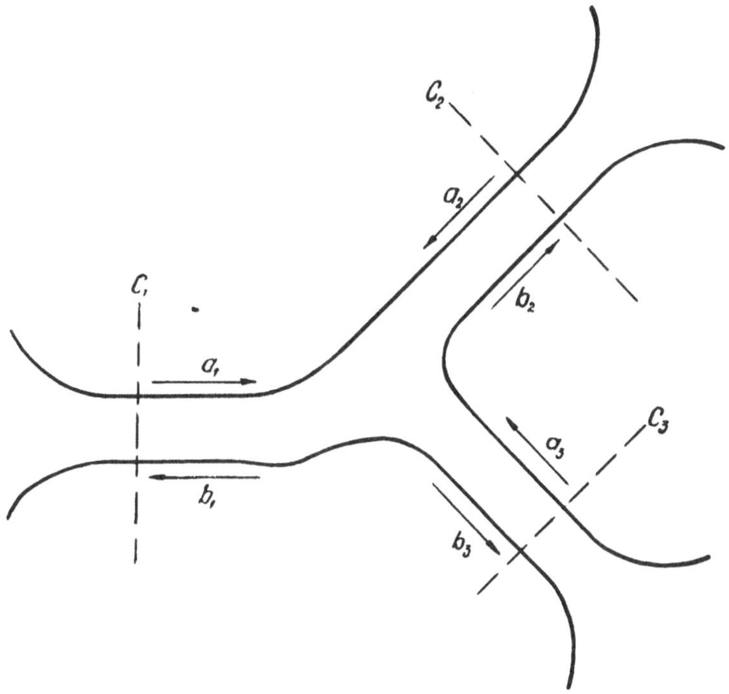

Figure 2

$$\delta\rho(r) = 2e\,(r|\delta f|r) \qquad (6)$$

as a functional of the potential: $\delta\rho\{\varphi\}$. Substituting this functional into (2) we find an equation for seeking φ . This equation is to be solved taking into account boundary condition (3) (and also the 3D dielectric environment of the 2DEG).

Ater φ is known, we come back to δf and calculate the current density

$$j(r) = Sp\{\delta f\cdot \hat{j}(r)\} = \frac{ie}{2m}\{\nabla'(r|\delta f|r') - \nabla'(r'|\delta f|r)\}_{r'=r} \qquad (7)$$

Integrating $j(r)$ over the constriction cross-section, we find total current J induced by applied voltage V .

2. LINEAR APPROXIMATION

The program outlined in sec. 1 can be worked out explicitly in the case of small applied voltage, when $eV \ll \varepsilon_F$. In this case eq. (4) can be linearized in $e\varphi$ and we find

$$\delta f = S[\langle e\varphi\rangle - e\varphi] \equiv \delta f_c + \delta f_e \,. \qquad (8)$$

Here the superoperator S is defined by

$$S v = -i\int_0^\infty ds\, e^{-\eta s}[f_0, \tilde{v}(-s)], \qquad \eta \to +0$$

$$\tilde{v}(t) = e^{iH_0 t}\, v\, e^{-iH_0 t}, \qquad (9)$$

and the average $\langle\varphi\rangle$ by

$$\langle\varphi\rangle = \int_{-\infty}^{0} \frac{dt}{\tau}\, e^{t/\tau}\, \tilde{\varphi}(t)\,. \qquad (10)$$

In the linear approximation all the quantities of interest can be expressed in terms of the retarded Green's function which satisfies the equation

$$[H_0(r) - \varepsilon]\, G_\varepsilon(r, r') = -\delta(r - r') \qquad (11)$$

and the following boundary condition: only outgoing waves are present in the horns at $z \to \pm\infty$.

One obtains for δf_e in terms of the Green's function

$$(r_1|\delta f_e|r_2) =$$
$$= -\int dr\left\{\iint d\varepsilon\, d\varepsilon' \frac{f_T(\varepsilon) - f_T(\varepsilon')}{\varepsilon - \varepsilon' - i\eta}\, g_\varepsilon(r, r_1)\, g_\varepsilon(r, r_2)\right\} e\varphi(r), \qquad (12)$$

where

$$g_\varepsilon(r,r') = -\operatorname{Im} G_\varepsilon(r,r'). \tag{13}$$

It is easy to verify, that

$$f_0 + \delta f_e = f_T(H_0 + e\varphi). \tag{14}$$

Hence, the meaning of δf_e is as follows: if after potential $e\varphi$ were furned on, the electron system remained in equilibrium, the change of the density operator would be δf_e .

It is clear now, that δf_e makes no contribution to current density (7). A current arises only due to term δf_c . This current-carrying contribution to the density operator δf shows a remarkable feature: for $\tau \to \infty$ it depends only on the asymptotic values of potential $\varphi(\pm\infty)$. One finds

$$(r_1|\delta f_c|r_2) = e\varphi(+\infty)(r_1|Q^+|r_2) + e\varphi(-\infty)(r_1|Q^-|r_2) \tag{15}$$

with

$$(r_1|Q^\pm|r_2) = -\int_\pm dr \left\{ \iint d\varepsilon d\varepsilon' \frac{f_T(\varepsilon) - f_T(\varepsilon')}{\varepsilon - \varepsilon'} \cdot \right.$$
$$\left. \cdot \left[1 + i(\varepsilon - \varepsilon')\tau\right]^{-1} g_\varepsilon(r_1, r) g_{\varepsilon'}(r_2, r) \right\}. \tag{16}$$

The $\pm$ on the integral means that the integration is over the half-space $z > 0$ or $z < 0$ (see Fig.1). It follows from the derivation of (15) that only remote values of r of the order of $v_F\tau$ contribute to the integral in (16). We thus write for $r' \to \infty$:

$$G_\varepsilon(r,r') = A_\varepsilon(r,n') \exp(ik_\varepsilon r')/r', \tag{17}$$

where n' is a unit vector in the r' direction and $k_\varepsilon^2 = 2m\varepsilon$. Using this definition of emission amplitude A_ε we obtain

$$(r_1|Q^\pm|r_2) = \frac{1}{2\pi}\int d\varepsilon \left(-\frac{\partial f_T}{\partial \varepsilon}\right) v_\varepsilon \int_\pm do\, A_\varepsilon(r_1,n) A_\varepsilon^*(r_2,n) \tag{18}$$

with $v_\varepsilon = k_\varepsilon/m$. Here the $\pm$ on the integral means that the integration over do is for the directions n in the right and left horn.

3. CHARGE DENSITY AND POISSON EQUATION

In accordance with decomposition (8)

$$\delta\rho(r) = \delta\rho_\infty(r) + \delta\rho_e(r). \tag{19}$$

"Quasiequilibrium" charge $\delta\rho_e$ which is due to δf_e can be calculated from (6) and (12). It is given by

$$\delta\rho_e(r) = -e^2 \int dr' \Pi(r,r')\varphi(r'), \tag{20}$$

where

$$\Pi(r,r') = -2\iint d\varepsilon d\varepsilon' \frac{f_T(\varepsilon)-f_T(\varepsilon')}{\varepsilon-\varepsilon'} g_\varepsilon(r,r') g_{\varepsilon'}(r,r'). \tag{21}$$

Kernel Π is the polarization loop of the screening of the Coulomb interaction, corresponding to RPA. Using the orthogonality relation

$$\int dr g_\varepsilon(r_1,r) g_{\varepsilon'}(r,r_2) = \delta(\varepsilon-\varepsilon') g_\varepsilon(r_1,r_2), \tag{22}$$

one can obtain the following important property of the kernel :

$$\int dr' \Pi(r,r') = \int d\varepsilon \left(-\frac{\partial f_T}{\partial \varepsilon}\right) g_\varepsilon(r,r) = g_T(r) \tag{23}$$

Quantity $g_T(r)$ is the effective density of states at point r, active in transport. The range of kernel Π is λ_F.

Charge $\delta\rho$ which is due to the current-carrying contribution to the density operator δf_c can be calculated from (6) and (15). It depends only on the asympotic values of the potential

$$\delta\rho_\infty(r) = e^2\left[\varphi(+\infty)q^+(r) + \varphi(-\infty)q^-(r)\right]. \tag{24}$$

"Weighting factors" $q^\pm(r)$ are

$$\begin{aligned} q^\pm(r) &= 2(r|Q^\pm|r) = \\ &= \frac{1}{\pi}\int d\varepsilon \left(-\frac{\partial f_T}{\partial \varepsilon}\right) v_\varepsilon \int_\pm do\, |A_\varepsilon(r,n)|^2 . \end{aligned} \tag{25}$$

The integral over angles multiplied by v_ε is the flux of particles of energy ε, which is emitted to the right of left horn by a point source at point r. Hence, factors $q^\pm(r)$ show the extent to which point r is coupled to the right and left contact pads. With the reciprocity principle in mind, we can say that $q^\pm(r)$ is the "probability" that the electron waves emitted from the corresponding thermal reservoirs reach point r.

Introducing charges (20) and (24) into the Poisson equation (2) we obtain the final result

$$\nabla^2\varphi(r) - \frac{4\pi e^2}{\epsilon}\int dr' \Pi(r,r')\varphi(r') = -\frac{4\pi}{\epsilon}\delta\rho_\infty(r). \tag{26}$$

It follows from this equation that $\delta\rho_\infty$ acts as an external charge, while $\delta\rho_e$ is responsible for screening this charge.

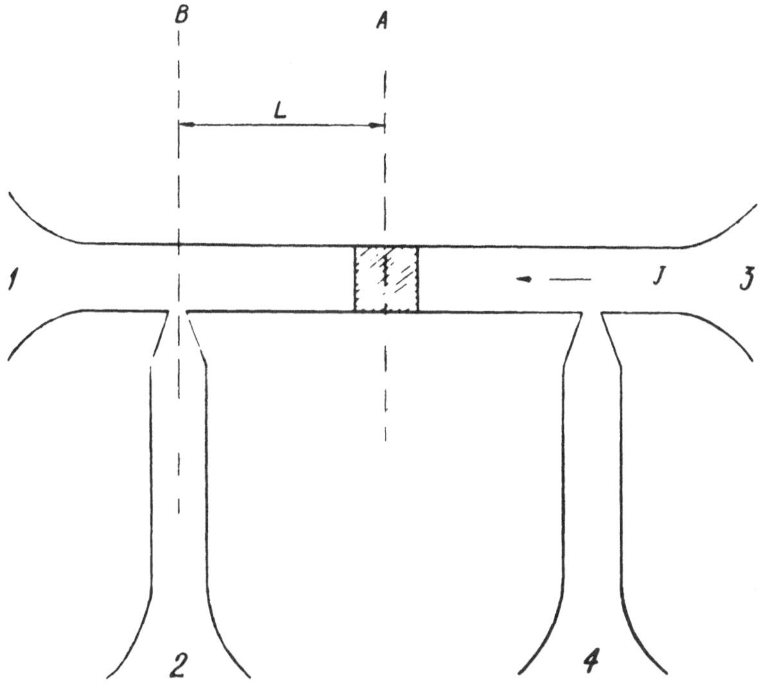

Figure 3.

There is an important distinction between $\delta\rho_\infty$ and $\delta\rho_e$. Only states near the Fermi surface participate in emissivities $q^\pm(r)$, while all the states, both filled and empty ones, contribute to screening kernel Π. This can be seen from (21) and (25).

Equation (26) is invariant with respect to a shift of the potential: $\varphi(r) \to \varphi(r) + \varphi_0$. The invariance can be checked using (23) and the relation

$$q^+(r) + q^-(r) = g_T(r) \tag{27}$$

which follows from (25) and (18) using (22).

Equation (26) can be compared with eq. (45) of the paper of Buttiker[11] for the potential in a one-dimensional channel. In eq. (45) of Ref.[11] the screening is local, while in (26) it is nonlocal over the distances of the order of λ_F. More important is the circumstance that the "exhernal charge" in eq. (45) of Ref.[11] contains, instead of the emissivities $q^\pm$, the wave functions $|\psi_\pm|^2$, for electrons arriving from the left and right reservoirs. For this reason, as was pointed out Ref.[11], the potential in the channel cannot be matched smoothly to the reservoir potential.

The generalization of the results of this section to the case of a multiprobe QBD (more than two contact pads) is straightforward. The right-hand side of the Poisson equation (26) is now

$$\delta\rho_\infty(r) = e^2 \sum_s \varphi_s q_s(r), \tag{28}$$

where φ_s is the potential at the infinity in the horn s and $q_s(r)$ is defined as in (25), the integration being over the directions n in the born s.

4. POTENTIAL DISTRIBUTION IN A QUANTUM BALLISTIC DEVICE

Consider, for example, the QBD shown in Fig. 2. Assume the channels are long ($L >> d$) and single-mode; i.e. there is only transverse quantization level ε_1 below ε_F in each channel. Assume also T=0.

If point r is in a channel far from the channel ends, it is easy to calculate $q_s(r)$ in terms of reflection coeffifients r_s of the channel mode from the channel junction in the center of the device and reflection coefficient R_s of this mode from the matching with the horn. One obtains

$$q_s(r) = A_s |\psi_1(x,y)|^2 \cos(k_1 z + \phi_s) \tag{29}$$

Here $\psi_1(x,y)$ is the electron wave function of the transverse motion for level ε_1 in the corresponding channel (x is in the plane of the 2DEG, y is normal to this plane). The wave-vector of the propagating mode is $k_1^2 = 2m(\varepsilon_F - \varepsilon_1)$, z is along the channel, the amplitudes A_s and phases ϕ_s depend on r_s and R_s

When point r enters the horn, the amplitudes of the oscillations become smaller since in the wide region a large number of propagating modes with different propagation wavevectors along z contribute to q_s, averaging out the oscillations.

When point r is far in a horn, only one of the emissivities $q_s(r)$ is nonzero. For example, if point r is far in horn 1, than

$$q_2(r) = q_3(r) = 0 \quad , \quad q_1(r) = g_T(r) = g , \tag{30}$$

where g is the density of states of unconfined electron gas.

With regard to the kernel $\Pi(r,r')$, we note that even for a single-mode channel it is difficult to evaluate this kernel. This is because all the (nonpropagating) modes, which correspond to the quantization-transverse levels ε_2, ε_3 ... above ε_F contribute to the kernel. It can be verified, however, that far from the ends of the channel the kernel depends on $z-z'$. This is due to interference between states with different energies. The former occurs during the integration in (21), even at T=0. When both points r and r' are far in the horn, the kernel $\Pi(r,r')$ can be calculated as in the case of unconfined electron gas and depends on $r-r'$.

It follows from the properties of $q_s(r)$ and $\Pi(r,r')$ that potential $\varphi(r)$ in a long channel depends on x,y in a rather complicated way and oscillates in z with a period of the order of λ_F. The oscillation amplitude depends on the reflection coefficient from the channel ends. When potential $\varphi(r)$ is averaged over these oscillations it becomes independent on z. It is important to note, that even in the case when the channel is matched to the contact pad adiabatically ($R_s = 0$), the potential oscillates in z due to the reflection from the channel junction ($r_s \neq 0$).

When point r enters the horn, the amplitude of the oscillations becomes smaller and a regular variation of potential $\varphi(r)$ is superimposed on the oscillations. Far in the horn s the potential appoaches asympotic value φ_s. Regular changes of potential $\varphi(r)$ occur also when point r croses the junction of the channels.

The general properties of the potential distribution can be summarized as follows: regular changes of the potential occur in the regions where the electron wave diffracts; in long channels the potential oscillates; in wide regions the potential is constant.

5. CURRENT DENSITY AND CONDUCTANCE

The current density in the QBD can be calculated from (7) and (15). Using (18) we obtain

$$j(r) = (e^2/2m)\left[\varphi(+\infty)\, w^+(r) + \varphi(-\infty)\, w^-(r)\right], \tag{31}$$

where

$$w^\pm(r) = \frac{i}{2\pi}\int d\varepsilon \left(-\partial f_T/\partial\varepsilon\right)\cdot \int_\pm do\left[A_\varepsilon(r,n)\nabla A^*_\varepsilon(r,n) - c.c.\right]. \tag{32}$$

The current density (31) is invariant with respect to a shift of the potential: $\varphi(r) \to \varphi(r) + \varphi_0$. It follows from (15) that under such a shift

$$(r_1|\delta f_c|r_2) \to (r_1|\delta f_c|r_2) + e\varphi_0 (r_1|Q^+ + Q^-|r_2). \tag{33}$$

One can obtain from (16) and (22), that

$$(r_1|Q^+ + Q^-|r_2) = \int d\varepsilon \, (-\partial f_T/\partial\varepsilon) \, g_\varepsilon(r_1, r_2). \tag{34}$$

This matrix is symmetric under $r_1 \leftrightarrow r_2$ and hence the additional term in (33) does not contribute to the current density.

As a result we can see, that the current density in the linear approximation is not influenced by the charge redistribution. This point was brought out by Landauer[12].

In the linear approximation the conductance $G = J/V$ can be calculated in terms of transmission amplitude T , defined from the asymptotic expansion of the Green's function. We have for $r, r' \to \infty$

$$G_\varepsilon(r, r') = T_\varepsilon(n, n') \frac{1}{v_\varepsilon} \frac{1}{r} \exp(ik_\varepsilon r) \frac{1}{r'} \exp(ik_\varepsilon r'). \tag{35}$$

The conductance is given as an average of the transmission amplitude:

$$G = 2\frac{e^2}{h} \int d\varepsilon \left(-\frac{\partial f_T}{\partial\varepsilon}\right) \int_- do \int_+ do' \, |T_\varepsilon(n, n')|^2. \tag{36}$$

Here the integration is carried out over the directions n and n' in the left and right horns, respectively. $|T_\varepsilon(n, n')|^2$ is the probability that an electron plane wave arriving from the left horn along direction $-n$ goes off into the right horn along direction n' . From (36) one can see, that the geometry of the contact pads affects the conductance of a QBD.

The electrical properties of multiprobe QBD are given by the Buttiker's matrix $G_{ss'}$ defined by the relation

$$J_s = \sum_{s'} G_{ss'} \varphi_{s'} , \tag{37}$$

where J_s is the current leaving contact s. Matrix elements $G_{ss'}$, for $s \neq s'$ are given by (36) the integration being now over directions n in horn s and over n' in horn s'. Diagonal matrix elements G_{ss} can be found from the relation

$$\sum_{s'} G_{ss'} = 0 . \tag{38}$$

This relation follows from the invariance of currents J_s under a shift of the potential.

Now we are going to compare our results for the Buttiker's matrix with those obtained by Stone and Szafer[8]. In this paper the QBD is considered as a junction of semiinfinite channels without contact pads. It was assumed that $\varphi(r) = \varphi_s$ when point r is in channel s far from the junction.

To make comparision possible we consider a QBD such that each horn goes over a long waveguide channel, as showen Fig. 2. For such a QBD the diagonal conductances $G_{ss'}$ can be expessed in terms of scattering matrix S for a waveguide junction with truncated horns (between cross sections C_s , Fig. 2) and the reflection coefficients R_s for waves going from the waveguide into the horn. For simplicity assume the waveguides to be single-mode channels. Then starting from (36) we obtain

$$G_{ss'} = 2\frac{e^2}{h} Av \left\{ (1-|R_s|^2)(1-|R_{s'}|^2)\,|(Q^{-1}S)_{ss'}|^2 \right\}. \tag{39}$$

Here scattering matrix S is defined by

$$b_s = \sum_{s'} S_{ss'} a_{s'}. \tag{40}$$

Where a_s and b_s are the amplitudes of the incoming and outgoing waves in channel s, respectively, the phases of the waves being zero at cross section C_s. $|a_s|^2$ and $|b_s|^2$ are the electron fluxes; for such a normalization the matrix S is unitary and symmetric. R_s is the reflection coefficient from cross section C_s. The matrix Q is

$$Q_{ss'} = \delta_{ss'} - S_{ss'} R_{s'}.$$

The matrix elements $S_{ss'}$ and the reflection coefficients R_s depend on energy ε and

$$Av\{(\ldots)\} \equiv \int d\varepsilon \left(-\partial f_T/\partial\varepsilon\right)(\ldots). \tag{41}$$

In the case when the contact pads are adiabatically matched to the channels we have $R_s = 0$. Then

$$G_{ss'} = 2\frac{e^2}{h} Av\, |S_{ss'}|^2. \tag{42}$$

It was this expression (for T=0) that was obtained by Stone and Szafer[8].

Taking into account the contact pads, we can understand why the conductance of a channel without scatterers in it ($|S_{12}|^2 = 1$), even in the case when the contact pads are adiabatically matched with the channel ($R_1 = R_2 = 0$), is not infinitely large, $G_{12} \neq \infty$. The reason is that an adiabatically smooth matching is a nonreflecting transition only for a wave incident from the waveguide; a wave incident from the horn is totally reflected if this wave is not an "adiabatic continuation" of a waveguide mode.

6. RESISTANCE MEASUREMENT BY THE 4-PROBE METHOD

The fact that the potential undergoes oscillations along a conducting channel and varies over the channel cross section force to reexamine[10] the analysis carried out by Engquist and Anderson[13].

Consider a 4-probe QBD in Fig.3. All the channels are assumed to be single-mode, and they are adiabatically matched with the contact pads. For simplicity we assume the QBD to be symmetric about plane A. The hatched square in the middle of the horizontal current channel is a scatterer, whose resistance is to be measured. The vertical channels are potential probes with $J_2 = J_4 = 0$. Hence, $J_1 = -J_3$. By definition, the 4-probe conductance is $G = J_1/\varphi_{42}$. Choosing $-\varphi_4 = +\varphi_1 = V/2$ and $\varphi_3 = -\varphi_1 = V/2$, we easily find from (37) and (38)

$$G = \frac{1}{2}\left\{(G_{12}-G_{14})-(G_{12}+G_{14}+2G_{13})\,\frac{G_{14}+G_{12}+2G_{24}}{G_{14}-G_{12}}\right\} \tag{43}$$

The scatterer of interest is characterized by a reflection coefficient $R=|R|e^{i\rho}$ from plane A. The junction between the current channel and the probe channel is assumed to be symmetric about plane B and is described by the scattering matrix

$$\begin{vmatrix} R' & T' & t \\ T' & R' & t \\ t & t & R'' \end{vmatrix}, \tag{44}$$

(the first and second rows and columns correspond to the left and right sides of the current channel, the third corresponds to the probe channel). To calculate matrix elements $G_{ss'}$, we use (42). The scattering matrix S can be expressed in terms of R and matrix (44). The calculation is quite tedious but straightforward.

Consider a "nonperturbing" probe, i.e. $|t| \to 0$. Then $G_{13} = 1-|R|^2$, while G_{12}, G_{14} and G_{24} are small, G_{24} being a small quantity of the second order. Making use of this relations, we find

$$G = 2\frac{e^2}{h}(1-|R|^2)\,\frac{G_{12}+G_{14}}{G_{12}-G_{14}}. \tag{45}$$

The last factor is a ratio of two small quantities. In the limit $|t| \to 0$ we obtain for T=0:

$$G = 2\frac{e^2}{h}(1-|R|^2)\,\frac{1+|R|\cos\beta}{|R|^2+|R|\cos\beta}. \tag{46}$$

Here $\beta = 2k_1L+\rho$ is the phase accumulated by the electron wave in the current channel during the roundtrip between planes A and B. A similar result (with different β) was obtained by Buttiker[14]. It follows from (46) that the 4-probe resistance of a scatterer in a QBD depends on the position of the voltage probes. The resistance can even be negative.

Engquist and Anderson[13] assert that the oscillating terms in (46) vanish because the reservoir is "wide" and has many modes. Our calculations, however, which explicitly incorporate the presence of wide horns, do not confirm that conclusion.

The oscillating terms can vanish if $T \neq 0$. It can be seen from (42), that all the matrix elements $G_{ss'}$ should be averaged over ε near ε_F in an interval of width T. Only the phase β depends strongly on ε in this interval, because of the relation $L \gg \lambda_F$. If $T \gg v_F/L$ the oscillating terms vanish and one obtains the famous Landauer formula

$$G = 2\frac{e^2}{h}\,\frac{1-|R|^2}{|R|^2}. \tag{47}$$

We conclude that the 4-probe method measures the resistance given by the Landauer formula only at a sufficiently high temperature, only if the probes are sufficiently far from the scatterer, and only if the voltage probes are weakly coupled to the current channel. (Note that for $v_F = 3\times10^7$ cm/s and T=1K we have $v_F/T = 2\mu m$. The nonadiabatic matching of the current channel with the source and drain contact pads can also influence the results of a 4-probe measurement.

REFERENCES

1. B.J. van Wees, H. van Houten, C.W.J. Beenakker, J.G. Williamson, L.P. Kouwenhoven, D. van der Marel, and C.T. Foxon, Quantized Conductance of Point Contacts in a 2DEG, Phys. Rev. Lett. 60: 848(1988)

2. D.A. Wharam, T.J. Thornton, R.Newbury, M.Pepper, H.Ahmed, J.E.F. Frost, D.G. Hasko, D.C. Peacock, D.A. Ritchie, and G.A. Jones, 1D Transport and the Quantization of the Ballistic Resistance, J. Phys. C. 21: L209 (1988)

3. M.L. Roukes, A.Schrer, S.J. Allen, Jr., H.G. Craignead, R.M. Ruthen, E.D. Beebe, and J.P. Harbison, Quenching of the Hall Effect in a 1D Wire, Phys. Rev. Lett. 59: 3011 (1987)

4. G.Timp, H.U. Baranger, P. de Vegvar, J.E. Cunningham, R.E. Howard, R. Behringer, and P.M. Mankiewich, Propagation Around a Bend in a Multichannal Electron Waveguide, Phys. Rev. Lett. 60: 2081 (1988)

5. Y. Takagaki, K. Gamo, S. Namba, S. Takaoka, K. Murase, K. Ishibashi, and Y. Aoyagi, Nonlocal Quantum Transport in Narrow Multibranched Electron Waveguide of GaAs-AlGaAs, Solid State Comm. 68: 1051 (1988)

6. E.N. Economou, and C.V. Soukoulis, State Conductance and Scaling Theory of Localization in 1D, Phys. Red. Lett. 46: 618 (1981)

7.. D.S. Fisher, and P.A. Lee, Relation Between Conductivity and Transmission Matrix, Phys. Rev. B23: 6851 (1981)

8. A.D. Stone, and A. Szafer, What is Measured when you Measure a Resistance ? - The Landauer Formula Revisited, IBM, J. Res. Dev. 23: 384 (1988)

9. H.U. Baranger, A.D. Stone, Electrical Linear-Response Theory in a Arbitrary Magnetic Field: A New Fermi-Surface Formulation. Phys. Rev. B40: 8169 (1989)

10. Y.B. Levinson, Potential Distribution in a Quantum Point Contact, Sov. Phys. JETP 68(6): 1257 (1989)

11. M. Buttiker,
Symmetry of Elecrtical Conduction,
IBM J. Res. Dev. 32: 317 (1988)

12. R. Landauer,
Spatial Variation of Currents and Fields due to Localized Scatterers in Metallic Conduction,
IBM J. Res. Dev. 32: 306 (1988)

13. H.L. Enqguist, and P.W. Anderson,
Definition and Measurement of the Electrical and Thermal Resistance,
Phys. Rev. B 24: 1151 (1981)

14. M. Buttiker,
Chemical Potential Oscillations near a Barrier in the Presence of Transport,
Phys. Rev. B 40: 3409 (1989)

FEW REMARKS ABOUT SMALL CAPACITANCE JOSEPHSON JUNCTIONS

A. Tagliacozzo and F. Ventriglia

Dipartimento di Scienze Fisiche, Università di Napoli
e G.N.S.M. (CNR)
Mostra d'Oltremare Pad.19
I-80125 Napoli, Italy

For most purposes, the current voltage characteristic (I/V) of junctions with superconducting electrodes can be satisfactorily described by means of the so called Resistive Shunted Junction model (RSJ), just in terms of the dynamics of the classical variable φ. This is the difference in the phase of the order parameter between the two superconductors. Dissipation is phenomenologically included as an ohmic term in the equation of motion for φ, by means of a suitably defined resistance R.

Recently Leggett has proposed that quantum effects could be detected in the switch of a superconducting junction out of the zero voltage state (for I very close to the critical current, I_c) [1]. Being the phase difference φ a macroscopically measurable quantity, one cannot safely ignore its interaction with the environment. Therefore, considerable effort has been devoted to the inclusion of dissipation in a quantum mechanical frame[2]. Presently junctions of much smaller area can be constructed ($\sigma \sim \mu m^2$), what implies that the charging energy of the junction capacitance, $E_c = e^2/2C$, can become larger than the Josephson coupling energy E_J [3]. Besides, temperature and external noise can be made so small that charging effects are not overwhelmed by thermal fluctuations.

Recent experiments[4] are making clear that, bejond the usual working conditions (i.e. $E_c < \hbar/RC << k_BT < E_J$) the new range of values for the parameters ($E_c > \hbar/RC > E_J > k_BT$) gives rise to qualitatively new behaviours, which do not seem to fit into the RSJ scheme[5].

We recall in the following the basic assumptions of the description of the Josephson conduction of pairs at very low temperature in terms of the phase φ and of its quantization. We are not going to stress the effects of dissipation; but we try to focus on the qualitative changes which weaken the picture when the capacitance of the junction becomes smaller .

The " phase only " model

The usual thermodynamical formulation of a Josephson Junction of capacitance C starts from the model Lagrangian for the phase φ :

$$\mathcal{L} = \frac{1}{2}C\left(\frac{\hbar}{2e}\right)^2 \dot{\varphi}^2 - U_0(\varphi)$$

where the potential energy is the Josephson coupling $U_0(\varphi) = -E_J(\cos\varphi - 1)$, while the kinetic energy is the charging energy of the junction $\frac{1}{2}CV^2$, rewritten in terms of the phase derivative by use of the second Josephson relation:

$$V = \frac{\hbar}{2e}\dot{\varphi} \tag{1}$$

In a classical formulation it is useful to go over to an hamiltonian formalism. The canonical conjugate variable with respect to the phase is the charge $Q = CV$. The classical partition function will be

$$Z = \int \frac{dQ}{2e} \int \frac{d\varphi}{2\pi} e^{-\beta H_0(Q,\varphi)}$$

with

$$H_0 = \frac{Q^2}{2C} + U_0(\varphi)$$

We only consider the case when the flowing current is lower than the critical current $I_c = 2eE_J/\hbar$, what we call " Josephson conduction". The current bias fixes the phase difference to the value $\overline{\varphi} \neq 0$,according to the first Josephson relation $I_{ext} = I_c \sin\overline{\varphi}$ $(I_{ext} \leq I_c)$.

This constraint can be included in the same way as one does when moving to the gran canonical ensemble for a many particle system:

$$Z(\beta, I_{ext}) = \int \frac{dQ}{2e} \int \frac{d\varphi}{2\pi} \exp -\beta\{H_0(Q,\varphi) - \hbar I_{ext}\varphi/2e\}$$

so that

$$\frac{2e}{\hbar}\frac{\partial \ln Z}{\partial I_{ext}} = \overline{\varphi}$$

This leads to the washboard potential for a current biased Josephson Junction (CBJJ):

$$U(\varphi) = -E_J(\cos\varphi - 1) - \hbar I_{ext}\ \varphi/2e\ . \tag{2}$$

It has the property that its minimum satisfies the first Josephson relation.

Quantum mechanically the partition function can be written in a path integral formulation as:

$$Z \sim \int d\varphi(0) \int_{\varphi(0)=\varphi(\hbar\beta)} \mathcal{D}\varphi(\tau)\ e^{-S_E/\hbar} \tag{3}$$

where S_E is the euclidean action:

$$S_E = \int_0^{\hbar\beta} d\tau \left[\frac{1}{2}C\left(\frac{\hbar}{2e}\frac{d\varphi(\tau)}{d\tau}\right)^2 + U(\varphi(\tau))\right]$$

The paths that make the action an extremum satisfy the equation of motion for an hypotetical particle in the inverted washboard potential:

$$C\left(\frac{\hbar}{2e}\right)^2 \frac{d^2\varphi}{d\tau^2} - \frac{\partial U(\varphi)}{\partial\varphi} = 0,$$

with the boundary condition: $\varphi(0) = \varphi(\hbar\beta)$.

To approximate the functional integral semiclassically, one could include the extremal paths only, together with those paths which deviate only slightly from them (saddle point approximation). It is remarkable that at high temperature $(\beta \rightarrow 0)$ the only periodic paths possible are those in which the particle oscillates around the minima of the inverted potential. However, when the temperature lowers $(\hbar\beta \rightarrow \infty)$, other trajectories can be found, in which the particle comes back to the origin in time $\hbar\beta$ after having visited increasingly wider stretches of the potential.

This qualitative argument gives an idea why the mobility of the particle is non vanishing in the case of a cosine potential at zero temperature and zero current bias, even including ohmic dissipation in the system [6]. Infact, integrating out the coordinates of the environment one is left with an effective action of the form:

$$\begin{aligned} S_{eff} = & \int_0^{\hbar\beta} d\tau \frac{1}{2} C \left(\frac{\hbar}{2e}\frac{d\varphi(\tau)}{d\tau}\right)^2 + U_0(\varphi) + \\ & + \frac{R_Q}{4\pi^2 R}\int_0^{\hbar\beta} d\tau \int_0^{\hbar\beta} d\tau' \left(\frac{\varphi(\tau)-\varphi(\tau')}{\tau-\tau'}\right)^2 \end{aligned} \qquad (4)$$

The mapping of this problem onto that of a one-dimensional coulomb gas of total zero charge shows that there is a localization - delocalization transition when $E_J/E_C \rightarrow 0$ and $R > R_Q$.

The first condition implies that the kinetic energy has a larger fraction in the balance of the total energy, so that similar considerations as above apply. This is the case when the capacitance of the junction (i.e. the "mass" of the particle) is strongly reduced. One can also see that a smaller E_J/E_C ratio implies larger phase fluctuations. Viceversa, because $\omega_J = (2eI_c/\hbar C)^{\frac{1}{2}}$ is the typical frequency of the phase fluctuations, the assumption that $\Delta\varphi >> 1$ implies that is $E_J << E_C$.

However, when is $\Delta\varphi >> 1$, attention should be paid to the fact that the model description should be intrinsecally $2\pi-$periodic.

We are here implicitely assuming that the phase is an extended coordinate and states in which it differs by 2π are distinguishable. Actually the action should be invariant with respect to $2\pi -$ changes. Of course, this is found in a more complete scheme, which arises from a full microscopic picture leading to an effective action which also includes dissipation[7][8]. One obtains, within standard approximations,

$$\begin{aligned} S_{eff} = & \int_0^{\hbar\beta} d\tau \frac{1}{2} C \left(\frac{\hbar}{2e}\frac{d\varphi(\tau)}{d\tau}\right)^2 + U_0(\varphi) \\ & + \frac{1}{2}\int_0^{\hbar\beta} d\tau \int_0^{\hbar\beta} d\tau' \alpha(\tau-\tau') \left[1 - \cos\left(\frac{\varphi(\tau)-\varphi(\tau')}{2}\right)\right] \end{aligned} \qquad (5)$$

where

$$\alpha(\tau) = \frac{R_Q}{R}\left(\frac{2/\hbar\beta}{\sin(\pi\tau/\hbar\beta)}\right)^2$$

in the case of ohmic dissipation. This expression coincides with eq.(4) in the limit of high temperature.

The current bias , as well, has to be included in such a way that the the periodicity is not destroyed.

This can be accomplished by introducing a parameter of the dimension of a charge $Q_{ext} = I_{ext}\tau$. In real time this would represent the total charge transferred by the generator across the junction within the time interval t .

One transforms the washboard potential to:

$$U_{per} = U_0 - \frac{\hbar}{2e} Q_{ext}\dot{\varphi} \tag{6}$$

what adds only a total time derivative to the Lagrangian:

$$\mathcal{L} \to \mathcal{L} + \frac{\hbar}{2e}\frac{d}{dt}\left[Q_{ext}\varphi(t)\right]$$

The presence of the total time derivative has no consequence on the classical equation of motion, of course, but the transformation changes the wavefunction by a phase factor[9]:

$$\phi(\varphi) \to e^{iQ_{ext}\varphi/2e}\phi(\varphi) = \Psi_{Q_{ext}}(\varphi) \tag{7}$$

where ϕ satisfies periodic boundary conditions: $\phi(\varphi + 2\pi) = \phi(\varphi)$.

Because $\Psi_{Q_{ext}}(\varphi)$ is Bloch - like and the potential is periodic, the corresponding energy spectrum is a periodic function of Q_{ext} :

$$E_{Q_{ext}+2e} = E_{Q_{ext}} \quad ,$$

what introduces the semiclassical picture which eventually leads to the prediction of Bloch oscillations of the voltage in the nondissipative case [10][11]. The crucial assumption that makes its direct experimental observation cumbersome is the availability of an ideal current generator, which can transfer charges in a smooth and continuous fashion.

The fact that the states, in which the phase differs by a multiple of 2π are distinguishable, or indistinguishable, certainly follows from the boundary conditions chosen ,i.e. from the experimental arrangement. The issue is related to the fact, whether one can measure the number of flux lines crossing the junction or not. Were the states indistinguishable, then the phase should be defined only in the interval $(0, 2\pi)$.

If this is the case, we face a typical example of quantum mechanics on a multiply connected space. The transition amplitude for the particle has to be written most generally as (Laidlaw. and Morette-de Witt theorem [12]):

$$K(\varphi, t; \varphi', t') = \sum_{\alpha} \chi(\alpha)\; K_{\alpha}(\varphi, t; \varphi', t') \tag{8}$$

where $\chi(\alpha)$ is a one-dimensional unitary representation of the homotopy group which in our case is $\mathcal{Z}$:

$$\chi(\alpha) \to \chi(n) = e^{in\theta}$$

(n integer, θ real) and K_{α} is the kernel restricted to the homotopy class of paths in the configuration space, labeled by α :

$$K_{\alpha}(\varphi, t; \varphi', t') = \int_{\varphi(t)\in\alpha} \mathcal{D}\varphi(t) e^{i\int_{\varphi'}^{\varphi} dt\mathcal{L}/\hbar}$$

This theorem states that paths belonging to different homotopy classes have as complex weights phase factors which have no a priori reason to be the same.

Zwerger et al. [13] investigate whether non trivial effects of quantum coherence could arise in the real time dynamics from the interference between paths with different winding numbers n. They conclude that, for an ohmic system this is not the case at zero temperature. However the question is rather subtle and the matter is still unsettled[14].

Leaving dissipation aside, we stress that there are two other major hypotesis underlying the statement of eq.(3).

First, the charge difference at the junction has to be a continuous variable.

This can be easily seen from the infinitesimal imaginary time propagator $(\epsilon \to 0)$:

$$K(\varphi_i, \tau_{i-1} + \epsilon; \varphi_{i-1}, \tau_{i-1}) \equiv \; < \varphi_i | e^{-\epsilon \hat{H}/\hbar} | \varphi_{i-1} >$$

$$= \int \frac{dQ_i}{4\pi e} e^{iQ_i(\varphi_i - \varphi_{i-1})} \exp\{-\epsilon H(Q_i, \varphi_{i-1})\} + \mathcal{O}(\epsilon^2)$$

$$= \left(\frac{R_Q C}{4\pi^2 i\epsilon}\right)^{1/2} \exp \frac{i}{\hbar}\left[\frac{C}{2\epsilon}\frac{\hbar}{2e}(\varphi_i - \varphi_{i-1})^2 - \epsilon U(\varphi_{i-1})\right] + \mathcal{O}(\epsilon^2)$$

Thus, the expression for the propagator in terms of the Lagrangian $\mathcal{L}(\varphi, \dot{\varphi})$ is obtained via an integration over the continuous variables Q_i. Such an assumption is questionable for $E_J < E_C$ and very low temperature, when the discreteness of the charge transfer across the junction becomes important.

Charging effects

This leads us to our second remark, which is even more fundamental: when the capacitance becomes small and the normal resistance very large, there are electric effects, such as the charging of the capacitance, that are independent of the superconduction and even interfere with it. We refer to electric fields which are not related to the time variations of the superconducting phase.

If this is the case, the electromagnetic field should be added, the dynamics of the order parameter should be included separately and their coupling should be examined [15]. The cost for this is to move from the quantum mechanics of the collective variable φ to a theory of coupled quantum fields, that is the complex field $\psi = n_s^{\frac{1}{2}} e^{i\eta}$ and the e.m. potential $(A_0, \vec{A})$. Here φ corresponds to $\eta_+ - \eta_-$, that is the difference in phases of the order parameter across the junction.

A good starting point is a phenomenological Lagrangian density:

$$\mathcal{L}[\psi, \dot{\psi}; A_\mu, \dot{A}_\mu] = -\frac{1}{16\pi} F^{\mu\nu} F_{\mu\nu}$$

$$+E_J \left\{ \frac{\xi_0^2}{2} \left(\frac{\partial}{\partial ct} + iA_0 \right) \psi^* \left(\frac{\partial}{\partial ct} - iA_0 \right) \psi - \frac{\xi_1^2}{2} \left(\vec{\nabla} - i\vec{A} \right) \psi^* \left(\vec{\nabla} + i\vec{A} \right) \psi \right.$$

$$\left. + |\psi|^2 - \frac{1}{2n_s} |\psi|^4 \right\} , \qquad (9)$$

where $F_{\mu\nu}$ is the electromagnetic field strength tensor.

We take $\xi_0 c^{-1} = \hbar / E_c$ as characterizing the time changes of the order parameter. If we consider a junction of very small cross section ($\sigma^{\frac{1}{2}} <<$ coherence length) and we neglect the magnetic fields, only the spatial variation in the direction orthogonal to the junction is relevant, so that the model can be restricted to $1+1$ dimensions. In this case ξ_1 is of the order of the width of the junction oxide barrier d and the space derivative can be viewed as the continuum limit of the potential $U_0 \approx \frac{1}{2}E_J(\vec{d}\cdot\vec{\nabla}\eta)^2$.

The scalar potential A_0 should also include the chemical potential. Its presence is responsable for local changes of the superfluid density within the " two fluids " picture, due to pair breaking effects. However these resistance producing fluctuations decay in a time of the order of $\tau_r = 4\pi(e/I_c)\cdot(R_Q/R)$, which can be very small if $R_Q/R << 1$ [16]. Within this approximation the chemical potential becomes an inconsequential constant.

Notwithstanding the similarities with a time dependent Ginzburg-Landau (TDGL) theory [17], it is apparent that the equation of motion for the order parameter ψ , which arises from this Lagrangian is not diffusive as it is the case of the TDGL equation, which has been successfully applied in describing most of the features of the resistive transition of the wire to the normal state[18].

Our Lagrangian gives rise to a dynamical model in its own right, which we study at zero temperature. Undamped excitations of frequency ω_J are attributed to the Josephson Junction, unless dissipation is included. It is likely that very thin wires can sustain collective charge density oscillations also,with a sound-like dispersion and weak damping [19].

The Lagrangian of eq.(9) is of course invariant with respect to gauge transformations:

$$\eta \rightarrow \eta + \phi_o^{-1}\chi \quad \text{and} \quad A^\mu \rightarrow A^\mu - i\,\phi_o\, e^{-i\eta}\partial^\mu e^{i\eta} \tag{10}$$

(ϕ_o is the flux quantum unit $= hc/2e$ divided by 2π). Usually the Lorentz gauge is chosen : $\partial_\mu A^\mu = 0$.

Variation of the Lagrangian with respect to A_0 gives the equation for the charge:

$$\Gamma(t,x) = \frac{1}{4\pi}\vec{\nabla}\cdot\vec{E} - 2e\,\frac{E_J}{E_c}\,|\psi|^2\,\frac{\hbar}{E_c}(\dot{\eta} - \frac{2e}{\hbar}A_0) = 0 \tag{11}$$

In the bulk of a superconductor the electric field vanishes so that, classically, the Josephson relation of eq.(1) is recovered. Occasional local fluctuations of the phase could be compensated by changes of the chemical potential. They are uninteresting to us and cut out anyway, due to our assumption that the latter is constant. On the contrary, in the case of a junction between two superconductors an electric field could be present within the oxide layer and the Gauss law of eq.(11), $\Gamma(t,x)=0$,being a physical constraint, should be enforced at each point $x \in (-d/2\,,\,d/2)$, at any time.

We show in the following that this is accomplished by writing the quantum physical state as a superposition of all small gauge transformed states[20].

Let us consider a stationary situation. We call " small gauge transformations " the ones described by gauge functions $\tilde{\chi}(x)$ which vanish at $x = \pm d/2$. Their generator is:

$$\mathcal{U}_{\tilde{\chi}} = \exp\frac{i\sigma}{\hbar c}\left\{\int_{-d/2}^{+d/2} dx \;\left[\Pi_A\frac{d\tilde{\chi}}{dx} + \Pi_{\psi*}(-i\phi_o^{-1}\tilde{\chi}\psi^*) + \Pi_\psi(i\phi_o^{-1}\tilde{\chi}\psi)\right]\right\}$$

where

$$\Pi_A = \frac{\delta\mathcal{L}}{\delta(\partial_o A_1)} = -\frac{F_{01}}{4\pi}\;,\Pi_{\psi^*} = \frac{\delta\mathcal{L}}{\delta(\partial_o\psi^*)} = \frac{1}{2}E_J(\partial_0 - i\phi_o^{-1}A_0)\psi \tag{12}$$

Partial integration of the first term yelds $\Gamma(x)$ times the function $\tilde{\chi}$ and the term at the boundary obviously disappears due to the vanishing of $\tilde{\chi}$ at $x = \pm d/2$. We have

$$\mathcal{U}_{\tilde{\chi}} = \exp i \int dx \ \Gamma(x)\tilde{\chi}(x)$$

Therefore the condition that the operator Γ acting on physical states $|\psi; A>_{phys}$ has to give zero, implies that

$$\mathcal{U}_{\tilde{\chi}} \, |\psi; A>_{phys} = \left|\psi \exp(i\phi_o^{-1}\tilde{\chi}); A + \partial\tilde{\chi} >_{phys} = |\psi; A>_{phys}\right.$$

This can be obtained by superposition of all small-gauge related states:

$$|\psi; A>_{phys} = \int \mathcal{D}\tilde{\chi} \ \ \mathcal{U}_{\tilde{\chi}} \ \ |\psi; A>$$

Thus, the set of all field configurations $\{\psi, A\}$ which are classical vacua, divides into equivalence classes, the members of each class being related to one another by a small gauge transformation.

We stress that we denote by $|\psi; A>$ the quantum state for the two coupled superconductors, without making any attempt to include the details of the inside of the insulating layer. This is only accounted for as a place where the order parameter may vanish, so that boundary conditions have to be provided for the phases of the two metals.

There can be other gauge functions however, which match, modulo 2π, to the gauge chosen in the two superconductors. These are such that $\chi(\pm d/2) = 2\pi m_\pm \phi_o$ ($m_\pm$ are two unequal integers) and are called "large" gauge transformations.

Consequently, in our crude approximations, a "large" gauge transformation is an unitary operator which induces a winding of $2\pi(m_+ - m_-)$ in the phase, what doesn't affect the physics of the superconductors. This is what is called a "phase slip ".

Now, the contribution at the boundary arising from the integration by parts in eq.(12) is non zero, yelding:

$$-i\frac{\sigma}{\hbar c}\int_{-d/2}^{d/2} dx \frac{\partial_1(F_{01}\chi)}{4\pi} = -i\frac{\sigma}{\hbar c}\frac{E(d/2)}{4\pi}\left[\chi(d/2) - \chi(-d/2)\right] \tag{13}$$

Here we have implicitely assumed the geometry of a planar plates capacitor for the junction, what implies that $E(\pm d/2) = 4\pi Q_{ext}/\varepsilon_r\sigma$, where ε_r is the dielectric constant of the insulating layer, so that $Q_{ext}/\varepsilon_r = \overline{Q}$ is the total charge transferred from one capacitor plate to the other.

Thus, a phase slip describes the transfer of the charge $\overline{Q}$ across the junction and only implies a phase factor added to the wave function, if the superconductors are unaffected. In particular, taking $m_+ - m_- = 1$, we have:

$$|\psi(\eta_\pm \rightarrow \eta_\pm + 2\pi m_\pm); \ A>_{phys} \sim e^{2\pi i\overline{Q}/2e} \ |\psi(\eta_\pm); \ A>_{phys} \tag{14}$$

Eq.(14) is the field analogue of eq.(7), that was obtained within one particle quantum mechanics. Of course, transferring pairs between the two superconductors ($\overline{Q} = 2em$) doesn't affect the physical state at all.

If charging effects are present the two physical states do not coincide any longer. This fact hinders the build up process of quantum coherence between states corresponding to different windings, unless the charging is compensated by a time variation of the modulus of the order parameter.

It is remarkable that, at $T = 0$ and in two euclidean dimensions, our model displays classical field configurations which become pure gauges at infinity, belonging to different homotopy classes, labeled by a winding number n.

This happens because, by Wick rotation, the Lagrangian maps into a GL free energy in two space dimensions, for which the vortex solutions exist.

We rewrite the Lagrangian in dimensionless variables [15], measuring the lengths in units of $l_0 = (32\pi n_s C E_J/E_c)^{-\frac{1}{2}}$ and times in units of $t_0 = \hbar/(16\pi n_s e^2 d^2 E_J)^{\frac{1}{2}}$.

One gets:

$$\mathcal{L} - \frac{1}{2}E_J n_s = \frac{2}{g^2}E_J n_s\left\{-\frac{1}{4}F_{\mu\nu}F_{\mu\nu} + \frac{1}{2}(D_\mu\psi)^*(D_\mu\psi) - \frac{g^2}{4}(|\psi|^2 - 1)^2\right\}$$

Here D_μ is the continuation to imaginary time of the usual gauge covariant derivative .

The coupling parameter is $g^2 = (E_c/E_J)/(16\pi C d^2 n_s) = (E_c/E_J)/(4\varepsilon_r n_s \sigma d)$. Because, for very small junctions and very low critical current densities the product $n_s\sigma d$ can become of the order of one, it is likely that g^2 can be made larger than $1/2$. This is the case when GL vortices repel each other and are stable[21]. These correspond to phase slips in imaginary time, which we assume to be located right at the junction where the order parameter vanishes.

An electric field is found inside the oxide layer, while the phase slips by $2\pi m$. Its value at the capacitor plates is

$$E \equiv \frac{4\pi}{\sigma}q_{inst} \cdot m \propto \hbar m/(2el_0t_0),$$

what defines the charge transferred across the junction in one slip of the phase as $q_{inst} \propto (E_J/E_c) \cdot n_s\sigma d \cdot 2e$.

These classical field configurations, making the action S^{inst} stationary and finite, should be considered when evaluating the partition function semiclassically. Inclusion of quantum corrections within the instanton dilute gas approximation lowers the energy of the ground state by an amount proportional to $\exp -S^{inst}/\hbar$.

Up to now we have discussed the case when no current bias is present. A current, flowing across the junction can be accounted for adding an extra term to the Lagrangian density of eq.(9)[22]:

$$\mathcal{L}^{bias} = -\, Q_{ext}(t)\ \frac{1}{2}\epsilon_{\mu\nu}F^{\mu\nu} \tag{15}$$

This term, integrated over space, gives $Q_{ext}(t) = I_{ext}t$ times the electromotive force along the device, so that the bias is included in the same way as in eq.(6).

We have studied the modifications that this term produces in the instanton solution. It is not isotropic in space-time, any longer. Following the same route sketched above, one finds that the quantum fluctuations accounted for by the phase slips, cause the presence of an average electromotive force in the ground state, supplied by the external sources. This implies that there is an intrinsic resistance in the conduction. Infact the current bias, by breaking the time reversal symmetry, introduces irreversibility in the system.

This happens no matter that no conventional dissipation has been included in the system and that these results are in the limit of zero temperature.

The effective resistance that we find is of the order of $\sim \varepsilon_r(E_c/E_J)\cdot R_Q \exp -S^{inst}/\hbar$.

Conclusions

Recent experiments for superconductive junctions with $E_c >> E_J$ and $R > R_Q$ seem to suggest that in ultrasmall Junctions the Josephson current is still well defined but conduction is intrinsecally resistive. The measured resistance is several kilo-Ohms,that is of the order of the quantum resistance R_Q . We have argued that the explanation of this findings goes bejond the " phase only " model. Infact, charging effects require that the electromagnetic field should be quantized together with the order parameter and both treated on the same foot. Given our model Lagrangian, phase slips of 2π associated with transfer of charge at the junction are solutions of the classical equations of motion for the fields. These are responsable, within a semiclassical approximation, of the presence of an electromotive force accompanying the supercurrent in the ground state of the system. It is argued that this can only happen when is $(E_c/E_J)\cdot n_s\sigma d/\varepsilon_r > 1/2$, so that the dilute gas approximation is valid in the evaluation of the partition function at low temperatures.

References

[1] A.J.Leggett in *Percolation,localization and Superconductivity* (ed. by A.M.Goldman and S.A.Wolf), Plenum, New York (1984)

[2] A.O.Caldeira and Leggett, Ann. Phys. (NY) **153**,445 (1984); Erratum Ann.Phys.(NY) **153**,445(1984)
A.J.Leggett, Les Houches, XLVI, (1986),Elsevier Science Publ.; eds. J.Souletie,J.Vannimenus and R.Stora (1987)
H.Grabert, P.Olschowski and U.Weiss, Phys.Rev.**B 36**, 1931 (1987)

[3] L.J.Geerligs, V.F. Anderegg, J.Romijn, J.E.Mooij, Phys.Rev.Lett.**65**,377 (1990)

[4] M.Iansiti,A.T.Johnson,C.J.Lobb and M.Tinkham, Phys. Rev. Lett.**60**, 2414 (1988), Phys. Rev.**B40**, 11370 (1989)

[5] M.Iansiti,M.Tinkham,A.T.Johnson,W.F.Smith, C.J.Lobb , Phys. Rev.**B39**, 6465 (1989)

[6] A. Schmid, Phys.Rev.Lett.**51**,1506 (1983)

[7] V.Ambegaokar,U.Eckern and G.Schön, Phys.Rev.Lett.**48** ,1745 (1982)
U.Eckern,G.Schön and V.Ambegaokar, Phys.Rev. **30**,6419 (1984)

[8] F.Guinea and G.Schön, J.Low. Temp.Phys. **69**,219 (1987)

[9] R.Jackiw,"Topological investigations of quantized gauge theories" in Les Houches, Session XL,(1983), B.S. De Witt and R.Stora eds., Elsevier Science Publ. B.V., 1984

[10] A.Widom, G.Megaloudis, T.D.Clark, H.Prance, R.J.Prance, J.Phys. **A15**,3877 (1982)

[11] e.g. D.V.Averin, K.K.Likharev, J.Low Temp.Phys.**62**,345 (1986); K.K.Likharev and A.B.Zorin, J.Low Temp.Phys.**59**,347 (1985)

[12] L.S.Schulman: "Techniques and applications of path integration", Wiley, New York (1981), chapter 23

[13] W.Zwerger,A.T.Dorsey, M.P.A. Fisher, Phys.Rev.**B34**, 6518 (1986)

[14] D.Loss, K.Mullen, preprint (Urbana 1990)

[15] A.Tagliacozzo, F.Ventriglia submitted to Phys.Lett.**A**

[16] L.Kramer,A.Baratoff Phys.Rev.Lett**38**,518 (1977)

[17] G.Schön, V.Ambegaokar Phys.Rev.**B19**,3515 (1979)

[18] J.S.Langer, V.Ambegaokar Phys. Rev. **164**,498 (1967)
D.E.McCumber and B.I.Halperin, Phys.Rev.**B 1**,1054 (1970)

[19] J.E.Mooij, G. Schön, Phys. Rev.Lett.**55**,114 (1985)

[20] R.Rajaraman "Solitons and Instantons",North Holland Publ. Co. Amsterdam (1982)

[21] L.Jacobs and C.Rebbi, Phys.Rev.**B19**,4486 (1979)

[22] A.Tagliacozzo in Proc. of NATO ASI " Applications of Statistical and Field Theory to Condensed Matter",Evora, May 22 - June 2, 1989, ed.s D.Baeriswyl and A.Bishop,Plenum, New York, 1990

DYNAMICS OF TUNNELING SYSTEMS IN METALS

Ulrich Weiss and Maura Sassetti*

Institut für Theoretische Physik, Universität Stuttgart
D-7000 Stuttgart 80, Germany

I. INTRODUCTION

In recent years, it has been observed in a multitude of systems in physical and chemical sciences that quantum tunneling is strongly affected by dissipative influences of the environment. Dissipation was found to cause novel features such as dissipative phase transitions,[1] exponential suppression[2] and qualitative change of the temperature dependence[3] of tunneling rates. Much of the theoretical efforts have been devoted to the phenomenon of macroscopic quantum tunneling (MQT) at $T = 0$,[2] and at finite temperatures,[3] and to macroscopic quantum coherence (MQC).[4,5] The theoretical predictions for the temperature and damping dependence of MQT have been verified precisely, e.g., in experiments on the decay of the zero-voltage state of a Josephson junction.[6]

In this contribution we focus our attention on the dynamics of light particles like hydrogen atoms or muons in metals. The tunneling of these particles is strongly influenced by the nonadiabatic interaction with conduction electrons. The fermionic effects have been reviewed recently by Kondo.[7] The nonadiabatic effects of the electronic screening cloud lead to an increase of the tunneling rate with decreasing temperature, a behavior first observed for the muon hopping rate in aluminium and copper below $10K$. In this region, the defect tunnels incoherently to a neighboring site. As the temperature is decreased, the defect becomes delocalized with a wave function extending over several interstitial sites. This case corresponds to coherent tunneling of the defect between many sites. A theoretical study of the transition from incoherent to coherent tunneling in a periodic potential in the presence of the electronic screening cloud has been given in Ref. 8.

The simpliest situation for coherent tunneling is the delocalization of a particle in a double well potential. At sufficiently low temperatures, excitations in the two wells can be neglected, and the double well can be truncated to a two-level system formed by two energetically split tunneling eigenstates. In the following we provide a unified view of the two-state dynamics under the influence of conduction electrons. Special attention is devoted to the behavior at very low temperatures where the presence or absence of system-bath correlations in the initial state gives qualitative differences in the evolution of the system at long times. The equilibrium correlation function shows algebraic long

*On leave from Dipartimento di Fisica, Università di Genova, Genova, Italy

time tails at zero temperature, and the system approaches thermal equilibrium always faster for factorizing initial states. Correspondingly, the spectral properties of the system at low frequencies are qualitatively influenced by correlations in the initial state. Recently, the important influence of conduction electrons on defect tunneling has been seen very clearly in neutron-spectroscopy experiments on hydrogen trapped by oxygen in niobium.[9] The proton is trapped in the $Nb(OH)_x$ samples in two nearest-neighbor tetrahedal sites. This represents a two-level system being ideally suited for studying the tunneling dynamics in a metallic environment.

The functional integral method provides a uniform approch to the dynamics of a two-level system under the influence of conduction electrons. In Section II we review the path integral method to this problem and present exact formal expressions for the system dynamics. In Section III we give the exact solution for a special value of the coupling strength. Section IV is mainly devoted to the discussion of the weak damping case, which is especially important for interstitials in metals. In Section V we present new results for thermodynamic properties of the system.

II. DYNAMICS OF TWO-LEVEL SYSTEMS

We consider a defect interacting with conduction electrons and tunneling in a double well with bias energy $\hbar\epsilon$ and with tunneling matrix element Δ_0. In the regime $\hbar\Delta_0, \hbar\epsilon, k_BT \ll \hbar\omega_0$ (but $k_BT/\hbar\Delta_0$ and $k_BT/\hbar\epsilon$ arbitrary), where $\hbar\omega_0$ is the energy of excitation in a single well, the original double well potential problem can be truncated to an effective two-state system.[4,10] The dynamics of the isolated system is then simply described by the pseudospin Hamiltonian

$$H_0 = -\frac{\hbar}{2}(\Delta_0\sigma_x + \epsilon\sigma_z) , \tag{2.1}$$

where the σ's are the Pauli matrices.

At very low temperatures the influence of the conduction electrons is governed by gross features such as the density of low energy excitations off the Fermi surface, so that the fermionic bath is essentially equivalent to a bosonic bath with appropriately chosen spectral density of the coupling strength.[4,7] The simplier spin-boson Hamiltonian reads

$$H = H_0 + \sum_i \left[\frac{p_i^2}{2m_i} + \frac{1}{2}m_i\omega_i^2x_i^2 - \frac{1}{2}c_ix_iq_0\sigma_z\right] , \tag{2.2}$$

where q_0 is a parameter which represents the distance between the two wells.
The effects of the boson bath are in the spectral function[4]

$$J(\omega) = \frac{\pi}{2}\sum_i \frac{c_i^2}{m_i\omega_i}\delta(\omega - \omega_i) , \tag{2.3}$$

and the equivalence with the low energy excitations of a fermionic bath holds for the specific form

$$J(\omega) = \eta\omega\exp(-\omega/\omega_c) = (2\pi\hbar K/q_0^2)\omega e^{-\omega/\omega_c} , \tag{2.4}$$

which is known to cause an ohmic form of dissipation in the classical limit.[2] Here η is the phenomenological ohmic friction coefficient while K is a characteristic dimensionless damping strength introduced by Kondo.[7] The parameter K is identical to the parameter α introduced by Leggett et al.[4] In (2.4) we made the specific choice of an exponential cut-off. We shall assume that ω_c is of the order of ω_0 and that the high frequency modes

($\omega > \omega_c$) are already absorbed into an exponential dressing factor of the tunnel matrix element. In the following, Δ denotes the matrix element renormalized by this factor.

A. Relation to the Kondo systems

Before entering into the discussion of relevant dynamical quantities we now summarize some relations of the spin-boson system described by (2.2)-(2.4) with fermionic systems showing the Kondo effect. These relations follow from well known equivalences between Fermi and Bose operators in one dimension.[11] Here we restrict our attention to the anisotropic Kondo model[12] and to the resonance level model.[12] The simpliest Kondo model describes a single magnetic impurity of spin 1/2 interacting via an exchange scattering potential with a band of conduction electrons. In the so called Fermi liquid regime the very low-lying excitation energies above the Fermi surface are relevant. They give rise to a logarithmic infrared divergence, e.g., in the static magnetic susceptibility, or in the soft X-ray absorption and emission of metals.[7] This is the Fermi surface effect originated by Kondo.[7]

Linearizing the electron's dispersion relation around the Fermi energy ϵ_F and measuring momentum from its reference value k_F the Hamiltonian of the anisotropic Kondo model takes the form[12,4]

$$H_K = \hbar v_F \sum_{k,\sigma} k c^\dagger_{k,\sigma} c_{k,\sigma} + \frac{J_{||}}{4}\sigma_z \sum_\sigma \sigma c^\dagger_\sigma c_\sigma + \frac{J_\perp}{2}\left(\sigma_+ c^\dagger_\downarrow c_\uparrow + \sigma_- c^\dagger_\uparrow c_\downarrow\right) , \qquad (2.5)$$

where $\sigma_\pm = 1/2(\sigma_x \pm i\sigma_y)$. Further, $c^\dagger_\sigma = L^{-1/2}\sum_p c^\dagger_{p\sigma}$ creates localized Wannier states at the origin where the impurity is located. For the relation to the spin-boson problem it is now essential to keep the exchange constants $J_{||}$ and $J_\perp$ as two independent parameters. Then, the equivalence may be expressed by the relations

$$\frac{\Delta}{\omega_c} = \rho J_\perp \cos^2 \delta_K \quad ; \quad K = \left(1 - \frac{2\delta_K}{\pi}\right)^2 , \qquad (2.6)$$

where $\rho = (2\pi\hbar v_F)^{-1}$ is the density of states and where

$$\delta_K = \frac{\pi}{4}\rho J_{||} + O((\rho J_{||})^2) \qquad (2.7)$$

is the scattering phase for the potential $J_{||}/4$. The function $\delta_K(\rho J_{||})$ is non-universal, since, apart from the linear term, it depends on the regularization procedure. The correspondence becomes exact for $\rho J_\perp \ll 1$ and $\rho|J_{||}| \ll 1$, that is for $\Delta/\omega_c \ll 1$ and K near 1.[1] The critical coupling $K = 1$ separates the corresponding ferromagnetic Kondo regime $\rho J_{||} < 0$ from the more interesting antiferromagnetic one, $\rho J_{||} > 0$.

The other model of interest is the resonance level model. The corresponding Hamiltonian describes a *d-level* near the Fermi surface interacting with a band of spinless fermions and has an additional Coulomb interaction U,[12]

$$H_{RL} = \sum_k \epsilon_k c^\dagger_k c_k + \epsilon_d d^\dagger d + V\sum_k\left(c^\dagger_k d + d^\dagger c_k\right) + \frac{U}{2}\sum_{k,k'}\left(c^\dagger_k c_{k'} - c_{k'} c^\dagger_k\right)\left(d^\dagger d - \frac{1}{2}\right) . \qquad (2.8)$$

The equivalence to the spin-boson Hamiltonian (2.2) is given by

$$\frac{\Delta^2}{\omega_c} = \frac{4\rho}{\hbar}V^2 \quad ; \quad K = \frac{1}{2}\left(1 - \frac{2}{\pi}\delta_{RL}\right)^2 ,$$

$$\epsilon = -\epsilon_d/\hbar , \qquad (2.9)$$

where $\delta_{RL} = \pi\rho U + O(U^2)$ is the scattering phase for the potential U. In the absence of the Coulomb interaction, $U = 0$ ($K = 1/2$), which corresponds to the Toulouse limit of the Kondo Hamiltonian, the resonance level model is exactly soluble.[11,13]

B. Exact formal expressions for correlation functions

The dynamical quantities of interest are the functions ($\beta = 1/k_B T$)

$$\begin{aligned} P(t) &= <\sigma_z(t)> \\ C(t) &= \frac{1}{2} <\sigma_z(t)\sigma_z(0) + \sigma_z(0)\sigma_z(t)>_\beta \ . \end{aligned} \tag{2.10}$$

The function $P(t)$ describes the expectation value of σ_z at time $t > 0$ supposing that at all times $t < 0$ the system has been held in (say) the state $\sigma_z = +1$, and the environment is assumed to have came into thermal equilibrium with it. At time zero, the constraint is released, so that for $t > 0$ the dynamics is governed by the Hamiltonian (2.2). The other dynamical quantity $C(t)$ is the symmetrized equilibrium correlation function. While $P(t)$ is the relevant quantity in the macroscopic quantum coherence problem, $C(t)$ is directly related to the neutron scattering function of the system.[14] Within the real-time path integral approach the exact formal expression for $P(t)$ has been given in Ref. 4, and for $C(t)$ in Ref. 13. Let us briefly sketch the derivation. The functions $C(t)$ and $P(t)$ are related to the joint probability $P(\sigma, t; \sigma, 0; \sigma', t_0)$ according to

$$\begin{aligned} C(t) &= \lim_{t_0 \to -\infty} \sum_{\sigma = \pm 1} [P(\sigma, t; \sigma, 0; \sigma, t_0) + P(-\sigma, t; -\sigma, 0; \sigma, t_0)] - 1 \ , \\ P(t) &= 2 \lim_{t_0 \to -\infty} P(+1, t; \sigma = +1, t_0 \le \tau \le 0) - 1 \ , \end{aligned} \tag{2.11}$$

where in the latter relation the system is constrained to the state $\sigma = +1$ of σ_z for times τ where $t_0 \le \tau \le 0$. As the heat bath modes in the functional integral approach are represented by Gaussian integrals, they can be evaluated exactly. Following the lines of Feynman and Vernon[15] for the spin-boson model (2.2), or summing the diagrams that are most divergent at low temperatures in the fermionic model,[7] one finds that the influence of the conduction electrons is expressed by a complex interaction between each pair of tunneling transitions. The pair interaction is given by[7,4]

$$S(\tau) + i\, R(\tau) = K \left[2 \ln \left[\frac{\hbar\beta\omega_c}{\pi} \sinh\left(\frac{\pi\tau}{\hbar\beta} \right) \right] + i\, \pi \mathrm{sgn}(\tau) \right] \ . \tag{2.12}$$

Next, it is convenient to visualize the path integral for the joint probability as a single path integral over the four states of the density matrix. The periods $t_{2j} < \tau < t_{2j+1}$, in which the system is in a diagonal state of the density matrix, are called *sojourns*, and the periods $t_{2j-1} < \tau < t_{2j}$, in which the system is in a nondiagonal state, are called *blips*.[4] We will assign the label $\chi_j = +1$ ($\chi_j = -1$) to the jth sojourn if the system is in the RR (LL) diagonal state of the reduced density matrix, while we assign the label $\xi_j = +1$ ($\xi_j = -1$) to the jth blip if the system is in the nondiagonal state RL (LR). For later convenience we introduce the blip lengths $\tau_j = t_{2j} - t_{2j-1}$ and sojourn lengths $s_j = t_{2j+1} - t_{2j}$.

It is now straightforward to write down the various factors constituting the path integral expression. The amplitude per unit time to switch between a diagonal and nondiagonal state is $\pm i\Delta/2$. The amplitude to stay in a sojourn is unity, while the amplitude to stay in the jth blip with label ξ_j and length τ_j is $\exp(i\epsilon\xi_j\tau_j)$. We will include this factor into the phase of the influence functional. If we define

$$S_{jk} = S(t_j - t_k) \quad ; \quad \Lambda_{jk} = S_{2j,2k-1} + S_{2j-1,2k} - S_{2j,2k} - S_{2j-1,2k-1} \,, \tag{2.13}$$

the modified influence functional for n blips at negative times and m blips at positive times consists of the two factors

$$\begin{aligned} G_{m,n} &= \exp\Big\{-\sum_{j=1}^{m+n} S_{2j,2j-1} - \sum_{j=2}^{m+n}\sum_{k=1}^{j-1} \xi_j\xi_k\Lambda_{jk}\Big\} \,, \\ H_{m,n} &= \exp\Big\{i\sum_{j=1}^{m+n} \xi_j\big[\epsilon\tau_j + \pi K\chi_{j-1}\big]\Big\} \,. \end{aligned} \tag{2.14}$$

It is obvious from the interaction factor $G_{m,n}$, that the blips may be viewed as neutral pairs of *charges* ± 1. The label $\xi_j = +1$ ($\xi_j = -1$) indicates that the charge $+1$ is preceded (succeeded) by the charge -1. For the time ordered integrations over the flip times t_j we introduce the compact notation

$$\int_{t_0}^{t} \mathcal{D}_{m,n}\{t_j\} \equiv \int_0^t dt_{2n+2m} \cdots \int_0^{t_{2n+2}} dt_{2n+1} \int_{t_0}^{0} dt_{2n} \cdots \int_{t_0}^{t_2} dt_1 \;. \tag{2.15}$$

Summing over all possibilities the system can go we obtain

$$P(\sigma, t; \sigma, 0; \sigma', t_0) = \sum_{m=0}^{\infty}\sum_{n=0}^{\infty} \left(-\frac{\Delta^2}{4}\right)^{m+n} \int_{t_0}^{t} \mathcal{D}_{m,n}\{t_j\} \sum_{\{\xi_j\}} G_{m,n} \sum_{\{\chi_j\}''} H_{m,n} \,. \tag{2.16}$$

The sum over arrangements $\{\xi_j\}$ and $\{\chi_j\}$ in (2.16) extends over the possible values ± 1 of the ξ_j and χ_j, where $j = 1, 2, \cdots, m+n$. Further the double prime in $\{\chi_j\}''$ is to mark the constraint $\chi_0 = \sigma'$, $\chi_n = \chi_{m+n} = \sigma$. It is now straightforward to perform the χ_j-sum in (2.16). In the end we find the exact formal expressions

$$\begin{aligned} P(t) &= P_s(t) + P_-(t) \,, \\ C(t) &= P_s(t) + Q(t) \,, \end{aligned} \tag{2.17}$$

where $P_s(t) = 1 + P_+(t)$ and

$$P_\pm(t) = \sum_{m=1}^{\infty} \int_0^t \mathcal{D}_{m,0}\{t_j\} \sum_{\{\xi_j\}} A_m^{(\pm)} G_{m,0} \,, \tag{2.18}$$

$$Q(t) = -\lim_{t_0 \to -\infty} \sum_{m=1}^{\infty}\sum_{n=1}^{\infty} (\tan(\pi K))^2 \int_{t_0}^{t} \mathcal{D}_{m,n}\{t_j\} \sum_{\{\xi_j\}} \xi_1\xi_{n+1} A_{m+n}^{(+)} G_{m,n} \,, \tag{2.19}$$

with

$$\begin{aligned} A_m^{(+)} &= \left(\frac{-\Delta^2\cos\pi K}{2}\right)^m \cos\left(\epsilon\sum_{j=1}^{m}\xi_j\tau_j\right) \,, \\ A_m^{(-)} &= \left(\frac{-\Delta^2\cos\pi K}{2}\right)^m \sin\left(\epsilon\sum_{j=1}^{m}\xi_j\tau_j\right) \,. \end{aligned} \tag{2.20}$$

The expressions for $P(t)$ and $C(t)$ are in the form of a series in Δ^2, i.e., in the form of an expansion in the number of transitions between diagonal states. The factor $G_{m,n}$ is the self-interaction of the blips and the interaction between blips j and k of the type ξ_j and ξ_k. The function $P_s(t)$ ($P_-(t)$) defines the symmetric (antisymmetric) contribution of $P(t)$ under the inversion of the bias $\epsilon \to -\epsilon$. The effects of the system-bath correlations on the dynamics of $C(t)$ are contained in the function $Q(t)$.

If we neglect interactions between the positive and negative time parts in (2.19), which corresponds to the assumption of a factorizing system-bath initial condition at $t = 0$, we find $C(t) = C^{(r)}(t)$, where

$$C^{(r)}(t) = P_s(t) + P_\infty P_-(t) \,, \tag{2.21}$$

and where $P_\infty = P_-(t \to \infty)$ is the equilibrium value of $P(t)$. Because of the omission of initial correlations the function $C^{(r)}(t)$ differs qualitatively from $C(t)$ for low temperatures and long times. (See below).

In recent theoretical studies[14,16,17] the structure factor for neutron scattering, which is directly related to $C(t)$ by Fourier transformation,[14] has been calculated from the function $C^{(r)}(t)$. We shall see below that the low-frequency behavior of the structure factor is modified drastically at low temperatures, if it is calculated from the actual equilibrium correlation function $C(t)$.

III. THE CASE $K = 1/2$

For the special value 1/2 of K the formal expression for $C(t)$ has been summed exactly in Ref. 13. This case is explicitly solvable for arbitrary temperatures and arbitrary bias, since the zeros of the $\cos(\pi K)$-factors in the series (2.18)-(2.20) need to be compensated by divergent factors which arise from the short distance singularity of the attractive interactions between opposite *charges*. Each pair of this type represents a collapsed *dipole* with vanishing moment, and hence does not interact with the other charges. Let us review the results for the unbiased case, $\epsilon = 0$, for which $P_-(t) = 0$, and also $P_\infty = 0$. As to the function $P_s(t)$, the series (2.18) represents a noninteracting gas of collapsed blips with zero moment. Thus, it can be summed to an exponential form, yielding

$$C^{(r)}(t) = P(t) = \exp\Big(-\gamma t\Big) \,, \tag{3.1}$$

where $\gamma = \pi\Delta^2/2\omega_c$ is the effective frequency scale of the problem. Note that (3.1) does not depend on temperature and bias. The correlations resulting from the nonfactorizing initial state of $C(t)$ are in the function $Q(t)$. There follows from the $\tan^2(\pi K)$-factor in (2.19) that each term of the series has one extended blip at negative times and one extended blip at positive times. Inside of these blips there is an ideal gas of collapsed sojourns. Further, the two extended blips are followed in their respective time region by a gas of noninteracting collapsed blips. Again, the resulting expression can be summed exactly, yielding $Q(t) = -F^2(t)$, and hence

$$C(t) = P(t) - F^2(t) \,, \tag{3.2}$$

where

$$F(t) = \frac{\Delta^2}{2\gamma} \int_0^\infty d\tau e^{-S(\tau)} \Big[e^{-\frac{\gamma}{2}|t-\tau|} - e^{-\frac{\gamma}{2}(t+\tau)} \Big] \,. \tag{3.3}$$

The result (3.2) with (3.1) and (3.3) is *exact* for all temperatures and times. It should be mentioned that exactly the same expression is obtained for the equilibrium correlation function of the Toulouse model, i.e., for the Hamiltonian (2.8) with $U = 0$. If we now compare the effects of the initial system-bath correlations we find qualitatively different behavior at long times for low temperatures. While $C^{(r)}(t)$ shows purely exponential behavior, and is independent of temperature, $C(t)$ is temperature dependent. At zero temperature, $C(t)$ decays algebraically,

$$\lim_{t\to\infty} C(t) = -\left(\frac{4}{\pi\gamma}\right)^2 \frac{1}{t^2} . \tag{3.4}$$

At very low (finite) temperatures, the decay law (3.4) holds in the intermediate time region $\gamma^{-1} \ll t \ll \hbar\beta$, while in the asymptotic limit, $t \gg \hbar\beta$, one has exponential decay with a rate given by the Matsubara frequency $\nu_1 = 2\pi/\hbar\beta$.[13] Note that $C(t)$ and $C^{(r)}(t)$ approach zero from opposite sides.

The preparation effects of the initial state also show up in frequency space. While the Fourier transform $\tilde{C}^{(r)}(\omega)$ of $C^{(r)}(t)$ is a pure Lorentzian with the maximal value $2/\gamma$ at $\omega = 0$, the function $\tilde{C}(\omega)$ depends on temperature, and shows at $T = 0$ the low frequency behavior $\tilde{C}(\omega \to 0) = (16/\pi)|\omega|/\gamma^2$. Hence, these functions differ drastically at $T = 0$. As the temperature is increased, the minimum in $\tilde{C}(\omega)$ is levelled off, and at high temperatures $\hbar\beta\gamma \ll 1$, the difference between $\tilde{C}(\omega)$ and $\tilde{C}^{(r)}(\omega)$ becomes negligible. More details are given in Ref. 13.

IV. WEAK DAMPING CASE

The weak damping case $K \ll 1$ is especially interesting, since in the $Nb(OH)_x$ samples the parameter K was found to be very small, namely near $K = 0.05$. Most of the theoretical predictions for the system dynamics have been derived within the so-called noninteracting blip approximation (NIBA),[4,5] in which the interactions Λ_{jk} in the expression (2.14) are neglected. For a *symmetric* system ($\epsilon = 0$), the effect of the interblip interactions is of order K^2, whereas the terms that are kept give nontrivial effects of order K. It turns out that the NIBA is a systematic weak-coupling approximation for a *symmetric* system down to $T = 0$, except for extremely long times. (See below). For *asymmetric* two-state systems the effect of the interblip interactions contributes to the order K at temperatures in the range $\hbar\Delta_e/k_B$ or smaller, so that the NIBA is inadequate in this case. The appropriate generalization, in which the effect of the interblip interactions in linear order in K is systematically taken into account, is given in Ref. 16. Only at sufficiently high temperatures, $k_B T \gg \hbar\Delta_e$, the NIBA is a consistent approximation for arbitrary K. In the following we discuss the symmetric system if not otherwise stated.

A. Dynamics at short and intermediate times

In terms of the self-energy $\Sigma(\lambda)$ the Laplace transform of $P(t)$ has the form

$$\hat{p}(\lambda) = \frac{1}{\lambda + \Sigma(\lambda)} . \tag{4.1}$$

Within the NIBA the self-energy $\Sigma'(\lambda)$, where the prime indicates the NIBA, emerges as

$$\Sigma'(\lambda) = \Delta_e \Big(\frac{\hbar\Delta_e}{2\pi k_B T}\Big)^{1-2K} \frac{\Gamma(K + \hbar\lambda/(2\pi k_B T))}{\Gamma(1 - K + \hbar\lambda/(2\pi k_B T))} . \tag{4.2}$$

Here, Δ_e is the renormalized zero temperature tunneling splitting,

$$\Delta_e = \Delta\Big(\frac{\Delta}{\omega_c}\Big)^{K/(1-K)}\Big[\cos(\pi K)\Gamma(1-2K)\Big]^{1/(2-2K)}\,, \tag{4.3}$$

and $\Gamma(z)$ is the gamma function. The inelastic neutron scattering cross section within the NIBA has been discussed in Ref. 14, and the corresponding behavior of $P(t)$ in Refs. 4 and 5. At temperatures, where $k_BT \ll \hbar\Delta_e/K$, the system shows damped oscillations with frequency Ω and damping rate γ, where for $K \ll 1$,

$$\begin{aligned} \Omega(T) &= \Delta_e\{1 + K\,[Re\,\psi\,(i\Delta_e/2\pi k_BT) - \ln(\Delta_e/2\pi k_BT)]\}\,, \\ \gamma(T) &= (\pi K/2)\Delta_e \coth(\Delta_e/2k_BT)\,, \end{aligned} \tag{4.4}$$

where $\psi(z)$ is the digamma function. Correspondingly, the scattering function has two Lorentzians centered at $\omega = \pm\Omega(T)$ with linewidth $2\gamma(T)$. At higher temperatures the tunneling dynamics can no longer be described in terms of an effective tunneling frequency. However, for high temperatures where $k_BT \gg \hbar\Delta_e/K$, the situation simplifies again. In this region the system tunnels incoherently between the two sites with a rate $\gamma(T) \propto T^{2K-1}$, which is the power law behavior predicted first by Kondo.[7] In this region, the structure factor shows a single quasielastic Lorentzian peak centered at $\omega = 0$ with linewidth $2\gamma(T)$.

B. Long-time behavior of $P(t)$ and $C(t)$

The simple NIBA does not describe the asymptotic behaviors of $P(t)$ and $C(t)$ correctly. In the absence of a bias the functions $P(t)$ and $C(t)$ are the same in this approximation, and at $T = 0$, they show asymptotically a power law behavior,

$$C'^{(r)}(t) = P'(t) \approx -\frac{(1-2K)}{\Gamma(2K)}\frac{1}{(\Delta_e t)^{2-2K}} \qquad \text{for } t \to \infty\,. \tag{4.5}$$

This behavior can be traced back to a branch point in the self-energy $\Sigma'(\lambda)$ at $\lambda = 0$. The asymptotic behavior both of $P(t)$ and $C(t)$ is changed qualitatively by the interblip correlations. Let us first consider $P(t)$ at $T = 0$ and for $K \ll 1$. The quantitative effect of the interblip correlations in the self-energy $\Sigma(\lambda)$ in the limit $\lambda \to 0$ is very difficult, since all frequency scales are coupled together, as is well-known from the Kondo problem.

In the first step, the correlations lead to generalized sojourns, in which the first charge interacts with the last one, and between them there are all possible arrangements of noninteracting blips. These *irreducible* diagrams are either neutral (type $\mathcal{A}$), or have the two exterior charges of the same sign (type $\mathcal{B}$). Observing that the pair interaction at $T = 0$ is given by $S_0(\tau) = 2K\ln(\omega_c\tau)$, the breathing mode integral of the diagrams of type $\mathcal{A}$ and $\mathcal{B}$ respectively are

$$\begin{aligned} a(\lambda) &= \frac{1}{2}\frac{\Delta_e^{2-2K}}{\Gamma(1-2K)}\int_0^\infty d\tau\frac{\cos(\Delta_e\tau)}{\tau^{2K}}e^{-\lambda\tau}\,, \\ b(\lambda) &= \frac{1}{2}\frac{\Delta_e^{2-2K}\omega_c^{4K}}{\Gamma(1-2K)}\int_0^\infty d\tau\cos(\Delta_e\tau)\tau^{2K}e^{-\lambda\tau}\,, \end{aligned} \tag{4.6}$$

where λ is the Laplace variable. The function $\cos(\Delta_e\tau)$ comes from the insertion of the blips. Now, it is important that the $\cos(\Delta_e\tau)$-term regularizes the integrals in (4.6) when $\lambda \to 0$, so that these generalized sojourns are effectively narrow. Putting $a = a(\lambda = 0)$ and $b = b(\lambda = 0)$, we find

$$a = \pi K \Delta_e / 2 \quad ; \quad b = -\pi K \Delta_e / 2 \,, \tag{4.7}$$

where terms of order K^2 have been neglected. Since the b-sojourns are not neutral, they can be arranged into interacting pairs which form a new class of irreducible sojourns of the type $\mathcal{A}$ and type $\mathcal{B}$, called $A(\lambda)$ and $B(\lambda)$, respectively. The $\mathcal{A}$-type pair of b-sojourns is attractive, and we may insert between them an arbitrary number of noninteracting a-sojourns, and any even number of noninteracting b-sojourns. On the other hand, the $\mathcal{B}$-type pair of b-sojourns is repulsive, and we may insert between them again an arbitrary number of a-sojourns, but now any odd number of noninteracting b-sojourns. Summing over all intermediate states we obtain for $K \ll 1$ the expressions

$$\begin{aligned} A(\lambda) &= 8Kb^2 \int_0^\infty d\tau \ln(\Delta_e \tau) \cosh(b\tau) e^{-(\lambda+a)\tau} \,, \\ B(\lambda) &= 8Kb^2 \int_0^\infty d\tau \ln(\Delta_e \tau) \sinh(b\tau) e^{-(\lambda+a)\tau} \,. \end{aligned} \tag{4.8}$$

In the next step, irreducible blip-diagrams are considered. Here again, it is convenient to classify the diagrams according to their exterior charges. If we include all diagrams with attractive exterior charges in the expression $U(\lambda)$, and all diagrams with repulsive exterior charges in the expression $V(\lambda)$, the self-energy assumes the form

$$\Sigma(\lambda) = U(\lambda) + V(\lambda) \,, \tag{4.9}$$

where

$$\begin{aligned} U(\lambda) &= \frac{\Delta_e^{2-2K}}{\Gamma(1-2K)} \int_0^\infty d\tau \frac{G_1(\tau)}{\tau^{2K}} e^{-\lambda\tau} \,, \\ V(\lambda) &= \frac{\Delta_e^{2-2K} \omega_c^{4K}}{\Gamma(1-2K)} \int_0^\infty d\tau G_2(\tau) \tau^{2K} e^{-\lambda\tau} \,. \end{aligned} \tag{4.10}$$

The functions $G_1(\tau)$ and $G_2(\tau)$ are determined by the insertions of noninteracting sojourns of type $\mathcal{A}$ and $\mathcal{B}$ between the two exterior charges. In the diagram, in which the exterior charges are attractive (repulsive), we may insert an arbitrary number of sojourns of type $\mathcal{A}$, and any even (odd) number of sojourns of type $\mathcal{B}$. Summing over all intermediate states we find

$$\begin{aligned} G_1(\tau) &= \exp\left[-(\lambda + a + A(\lambda))\tau\right] \cosh[(b + B(\lambda))\tau] \,, \\ G_2(\tau) &= -\exp\left[-(\lambda + a + A(\lambda))\tau\right] \sinh[(b + B(\lambda))\tau] \,. \end{aligned} \tag{4.11}$$

Inserting (4.11) into (4.10) we obtain in leading order in K for the self-energy

$$\Sigma(\lambda) = \frac{\Delta_e^2}{\Lambda(\lambda)} \,, \tag{4.12}$$

where

$$\Lambda(\lambda) = \lambda + a + A(\lambda) + b + B(\lambda) \,. \tag{4.13}$$

It is now important that the function $\Lambda(\lambda)$ stays finite as $\lambda \to 0$, as follows from (4.7) and (4.8). We find for the self-energy in leading order in K

$$\Sigma(\lambda \to 0) \approx \frac{\Delta_e}{2\pi K^2 |\ln K|}, \tag{4.14}$$

from where we obtain

$$\hat{p}(0) \equiv \hat{p}(\lambda \to 0) = \frac{2\pi}{\Delta_e} K^2 |\ln K| . \tag{4.15}$$

Thus we have shown that for $K \ll 1$ the Laplace transform $\hat{p}(\lambda)$ is regular at the origin. Hence, $P(t)$ shows in fact not algebraic, but exponential decay, as $t \to \infty$. Note that the exponential decay was also found above for the case $K = 1/2$. It is immediately obvious that similar dressing of the U-blip and the V-blip takes place for arbitrary damping strength, so that we expect that $P(t)$ shows asymptotically exponential decay at $T = 0$ for all $K < 1$. Further, it is straightforward to show that for $\epsilon \neq 0$ the function $P_s(t)$ again decays exponentially, and also $P_-(t)$ approaches the equilibrium value P_∞ exponentially fast.

We are now prepared to study the asymptotic behavior of the equilibrium correlation function for $K < 1$. The algebraic long-time tails of $C(t)$ at $T = 0$ arise from the correlations between the positive and negative time parts in the function $G^{(0)}_{m,n}$ defined in (2.14). As $t \to \infty$, the leading contribution under the integral (2.19) is

$$A^{(+)}_{m+n} G_{m,n} \approx G^{(0)}_m G^{(0)}_n \left[-A^{(-)}_m A^{(-)}_n + A^{(+)}_m A^{(+)}_n \left(1 - \sum_{j=1}^{n} \sum_{k=n+1}^{n+m} \xi_j \xi_k \tau_j \tau_k \ddot{S}_0(t) \right) \right] , \tag{4.16}$$

where $S_0(t) = 2K \ln(\omega_c t)$, and where $G^{(0)}_m A^{(\pm)}_m$ and $G^{(0)}_n A^{(\pm)}_n$ are the interaction factors at zero temperature in the positive and negative time parts, respectively. The first term gives the contribution $P_\infty P_-(t)$ in the function $Q(t)$, and the second term describes the algebraic tail of $Q(t)$. Using (2.19) we obtain asymptotically

$$Q(t) \approx P_\infty P_-(t) + \left(\frac{\partial P_\infty}{\partial \epsilon} \right)^2 \ddot{S}_0(t) \tag{4.17}$$

The effects of correlations are then described by the function $\Delta C(t) \equiv C(t) - C^{(r)}(t)$. From (2.17), (2.21) and (4.17) we asymptotically have

$$\Delta C(t) = \left(\frac{\partial P_\infty}{\partial \epsilon} \right)^2 \ddot{S}_0(t) = (2\hbar\chi_0)^2 \ddot{S}_0(t) . \tag{4.18}$$

In the last form we have identified $(\partial P_\infty / \partial \hbar\epsilon)/2$ with the static susceptibility χ_0 at zero temperature. Inserting the above zero temperature expression $S_0(t)$ for the pair interaction we finally get the algebraic long time behavior

$$\Delta C(t) \approx -2K (2\hbar\chi_0)^2 \frac{1}{t^2} . \tag{4.19}$$

This relation holds for arbitrary bias and for all $K < 1$. Note that the bias and the damping parameter K (except of the factor K) enters into (4.19) only implicitly through the static zero temperature susceptibility. For $K > 1$, the Kondo temperature $T_K \equiv 1/(2\pi k_B \chi_0)$ is zero provided we are in the limit $\Delta/\omega_c \to 0$, and the asymptotic expansion discussed here does not exist. In this damping region the NIBA is valid down to $T = 0$. In the absence of a bias, $\chi_0 \to (2\hbar\Delta)^{-1}$ as $K \to 0$, while for $K = 1/2$ one finds[13] $\chi_0 = 2/(\pi\hbar\gamma)$, where γ is the resonance level width $\gamma = \pi\Delta^2/2\omega_c$ discussed above. From the analogy with the Kondo model[12,4] we may also infer the value of χ_0 for $1-K \ll 1$, yielding $\chi_0 = (1-K)^{K/(1-K)}/(4\hbar\Delta_r)$, where $\Delta_r = (\Delta/2\omega_c)^{K/(1-K)}\Delta/2$.

We would like to mention that the zero temperature behavior (4.19) is not purely academic, since it is also valid at finite temperatures in the time region $\hbar\chi_0 \ll t \ll \hbar\beta$. At times $t \gg \hbar\beta$ the equilibrium correlation function approaches the equilibrium value exponentially fast with a rate given by the smallest Matsubara frequency $\nu_1 = 2\pi/\hbar\beta$.

From (4.19) we may also infer the behavior in frequency space near $\omega = 0$. We find

$$\Delta\tilde{C}(\omega \to 0) = 2\pi K(2\hbar\chi_0)^2|\omega| \, . \tag{4.20}$$

This relation is analogous to a relation that has been proven by Shiba[18] for the general Anderson model. While Shiba's derivation is essentially based upon a particle number conservation law, our proof is based upon the exact formal solution.

V. THERMODYNAMIC PROPERTIES

Instead of using the standard imaginary time functional integral approach for the partition function,[19] the thermodynamic properties of the dissipative two-level system can also be traced from the long-time limit of the dynamics. This approach directly renders exact formal expressions for the free energy F and the static susceptibility χ_0 allowing for a systematic study of their low temperature properties. Observing that P_∞ is related to the partition function Z by $\partial \ln Z/\partial\epsilon = \hbar\beta P_\infty/2$ the free energy may be determined by integrating (2.18) for $P_-(t)$ in the limit $t \to \infty$ with respect to the bias. Also, we may find the solution for χ by differentiating (2.18) with respect to the bias, $\chi = (\partial P_\infty/\partial\epsilon)/2\hbar$. Inserting the low temperature expansion of the pair interaction, $S(\tau) = S_0(\tau) + (\pi^2/3)K(\tau/\hbar\beta)^2 + O((\tau/\hbar\beta)^4)$ into the resulting expressions, one obtains for $K < 1$ the exact low temperature behavior

$$\begin{aligned} F &= F_0 - (\pi^2/3)\chi_0(k_BT)^2 + O(T^4) \, , \\ \chi &= \chi_0 + (\pi^2/3)\chi_0''(k_BT)^2 + O(T^4) \, . \end{aligned} \tag{5.1}$$

Here, F_0 and χ_0 are the free energy and static susceptibility at zero temperature, and $\chi_0'' = \partial^2\chi_0/(\partial\hbar\epsilon)^2$. The bias enters only through the nonlinear susceptibility. Observing that the specific heat is given by

$$c(T) = -T\partial^2 F/(\partial T)^2 \, , \tag{5.2}$$

one finds for the Wilson ratio

$$R \equiv \lim_{T\to 0} \frac{c(T)}{k_B^2\chi_0 T} \tag{5.3}$$

the exact relation

$$R = 2K\pi^2/3 \, . \tag{5.4}$$

Thus, the system shows Fermi liquid behavior for $K < 1$. The known results of R for the anisotropic Kondo model ($1-K \ll 1$) and for the Toulouse model ($K = 1/2$) are just special cases of the general formula (5.4). For $K > 1$, the static susceptibility χ_0 diverges. In this damping region, the high temperature expansion for $\chi(T)$ and $F(T)$, which effectively is an expansion in powers of Δ^2, is valid down to $T = 0$.

VI. CONCLUDING REMARKS

The functional integral method which we have presented here has provided a uniform approach to the investigation of the dynamics of a two-level system in a metallic environment. We have discussed the extension of earlier studies to the important case of non-factorizing initial states. In the parameter region of low temperatures $k_BT \ll \hbar\Delta$, long times $t \geq \Delta^{-1}$, and $K < 1$, the series expressions for $P(t)$ and $C(t)$ are exceedingly complex and withstand to a summation by analytic methods. Only for the special value 1/2 of K, the correlation function can be determined exactly for all times and all temperatures. We have discussed that the noninteracting blip approximation is adequate at high temperatures for all K and all times, while at low temperatures, $k_BT < \hbar\Delta$ the NIBA is valid only for the unbiased, weakly damped system in the time region of the order Δ^{-1} or smaller, in which the system shows damped oscillations. We have made a systematic study of the asymptotic time behavior at $T = 0$ and have found that $P(t)$ decays exponentially, while $C(t)$ shows algebraic long-time tails with a power law t^{-2} for the case of a metallic environment.

In recent neutron spectroscopy experiments on trapped hydrogen in niobium covering the entire region of temperatures,[9] the crossover from incoherent to coherent tunneling becomes apparent in a change from quasielastic to inelastic neutron scattering. The quantitative agreement between experiments and theory confirms the nonadiabatic influence of the conduction electrons on hydrogen tunneling.

REFERENCES

1. S. Chakravarty, "Quantum fluctuations in the tunneling between superconductors", Phys. Rev. Lett. **49**:681 (1982); A.J. Bray and M.A. Moore, "Influence of dissipation on quantum coherence", ibid. **49**:1545 (1982).
2. A.O. Caldeira and A.J. Leggett, "Quantum tunneling in a dissipative systems", Ann. Phys. (N.Y.) **149**:374 (1983).
3. H. Grabert, P. Olschowski, and U. Weiss, "Quantum decay rates for dissipative systems at finite temperatures", Phys. Rev. B **36**:1931 (1987), and references therein.
4. A.J. Leggett, S. Chakravarty, A.T. Dorsey, M.P.A. Fisher, A. Garg, and W. Zwerger, "Dynamics of the dissipative two-state system", Rev. Mod. Phys. **59**:1 (1987).
5. U. Weiss, H. Grabert, and S. Linkwitz, "Influence of friction and temperature on coherent quantum tunneling", J. Low. Temp. Phys. **68**:213 (1987).
6. A.N. Cleland, J.M. Martinis, and J. Clarke, "Measurements of the effect of moderate dissipation on macroscopic quantum tunneling", Phys. Rev. B **37**:5950 (1988), and references therein.
7. J. Kondo, "Two-level systems in metals", in "Fermi Surface Effects", Vol. **77**:1 of Springer Series in Solid-State Sciences, edited by J. Kondo and A. Yoshimori (Springer-Verlag, Berlin) (1988).
8. U. Weiss and M. Wollensak, "Dynamics of the dissipative multiwell system", Phys. Rev. B **37**:2729 (1988).
9. H. Wipf, D. Steinbinder, K. Neumaier, P. Gutsmiedl, A. Magerl, and A.J. Dianoux, "Influence of the electronic state on H tunneling in niobium", Europhys. Lett. **4**:1379 (1987); D. Steinbinder, H. Wipf, A. Magerl, A.J. Dianoux, and K. Neumaier, "Non adiabatic low-temperature quantum diffusion of hydrogen in $Nb(OH)_x$" *ibid.* **6**:535 (1988).
10. U. Weiss, H. Grabert, P. Hänggi, and P. Riseborough, "Incoherent tunneling in a double well", Phys. Rev. B **35**:9535 (1987).

11. F. Guinea, V. Hakim, and A. Muramatsu, "Bosonization of a two-level system with dissipation", Phys. Rev. B **32**:4410 (1985).

12. A.M. Tsvelick and P.B. Wiegmann, "Exact results in the theory of magnetic alloys", Adv. Phys. **32**:453 (1983).

13. M. Sassetti and U. Weiss, "Correlation functions for dissipative two-state systems: effects of the initial preparation", Phys. Rev. A **41**:5383 (1990).

14. H. Grabert, S. Linkwitz, S. Dattagupta, and U. Weiss, "Structure factor for neutron scattering from tunneling systems in metals", Europhys. Lett. **2**:631 (1986).

15. R.P. Feynman and F.L. Vernon, "The theory of a general quantum system interacting with a linear dissipative system", Ann. Phys. (N.Y.) **24**:118 (1963).

16. U. Weiss and M. Wollensak, "Dynamics of the biased two-level system in metals", Phys. Rev. Lett. **62**:1663 (1989); R. Görlich, M. Sassetti, and U. Weiss, "Low-temperature properties of biased two-level systems: effects of frequency-dependent damping", Europhys. Lett. **10**:507 (1989).

17. S. Dattagupta, H. Grabert, and R. Jung, "The structure factor for neutron scattering from a two-state system in metals", J. Phys. Cond. Mat. **1**:1405 (1989).

18. H. Shiba, "The Korringa relation for the impurity nuclear spin-lattice relaxation in dilute Kondo alloys", Progr. Theor. Phys. **54**:967 (1975).

19. R. Görlich and U. Weiss, "Specific heat of the dissipative two-state system", Phys. Rev. B **38**:5245 (1988).

ANALYTICAL DYNAMICS OF MODULATED SYSTEMS

D. R. Westhead†, S. W. Lovesey‡ and G. Watson†

†Dept. of Theoretical Phys., Univ. of Oxford, 1-4 Keble Rd., Oxford, OX1 3NP
‡Rutherford Appleton Laboratory, Didcot, Oxfordshire, OX11 0QX

1 Introduction

In theoretical physics one faces the dilemma of whether to work with simple model systems which are analytically tractable or with more complicated and realistic systems which are only soluble, if at all, by numerical methods. Of course both approaches have their value. The purpose of this brief review is to deal with analytical methods applied to the calculation of dynamical properties of some simple models which exhibit two independent periodicities. These periodicities provide two independent length scales for the model. In principle the two periodicities could be incommensurable and much of the interest in this topic is the investigation of the dependence of calculated quantities on their ratio. We give results which show the general features of the dynamics of such models, which are probably present in more complicated systems. In addition, some results which are of relevance to experiments on particular systems are given.

In the main we will deal with the modulated spring model (de Lange and Janssen 1981). This is a one-dimensional model of a modulated crystal lattice, it consists of a chain of atoms joined by harmonic springs of modulated strength α_n, where

$$\alpha_n = m\{\alpha - \gamma \cos(nQ + \Delta)\}. \tag{1.1}$$

Here n labels the site, m the mass of a single atom, α is the average spring constant and the strength γ, modulation Q and phase Δ specify the spring modulation.

Many other systems have a mathematical structure which is very closely related to that of the modulated spring model. In this article we will look briefly at two other such systems, the Hofstadter problem and the problem of a longitudinally modulated magnet like Erbium. The Hofstadter problem (Hofstadter 1976) concerns the spectrum of a two-dimensional system of non-interacting electrons simultaneously in a two-dimensional periodic potential and perpendicular magnetic field. The two length scales associated with the system are the lattice spacing of the periodic potential and the magnetic length (i.e. the radius of a cyclotron orbit). This problem has seen much recent interest, particularly in the theory of the Quantum Hall Effect and in some theories of high T_c superconductivity. Rare earth magnets like Erbium exhibit phases in which the magnetic moment is modulated from site to site in the lattice. We are interested in the transverse spin excitation spectrum of such phases, observed in inelastic neutron scattering.

The paper is organised in the following way. In section two we look at properties of the spectrum (i.e. the set of energy eigenvalues of the system). For the above systems this takes the Hofstadter butterfly form. Section three reviews a method of calculation of the Green function associated with the problem, from which dynamical quantities can be calculated. In section four we survey some interesting results in lattice dynamics, particularly properties of defects in modulated lattices. It is possible to calculate the mean square velocity of a defect atom in the lattice, which is experimentally accessible as the second order Doppler shift of a Mössbauer peak. Also of experimental relevance, in section five we consider calculation of the dynamic structure factor for the modulated spring model. This is measured in an inelastic neutron scattering experiment. The structure factor is also of mathematical interest because, in a wider context, it is the dual spectrum of the model under investigation. Finally in section six we make some conclusions.

2 The Spectrum

The problems referred to in section one are all governed by second order linear difference equations, hence their identical mathematical structure. In the case of the modulated spring model this difference equation arises from the procedure which diagonalises the Hamiltonian,

$$H = \sum_n \left[\frac{p_n^2}{2m_n} + \frac{1}{2}\alpha_n(u_{n+1} - u_n)^2 \right]. \tag{2.1}$$

Here u_n is the displacement at site n, p_n is the conjugate momentum, α_n is the spring constant given in (1.1) and m_n is the mass of the atom at site n. A system which has $m_n = m$ for all n will be referred to as 'the pure system'. The most general case we consider has a single mass defect at site s whose mass is parameterised by λ, $m_n = m(1 - \lambda\delta_{n,s})$, this system is referred to as 'the perturbed system'.

We diagonalise (2.1) by introducing standard boson operators, b_σ, b_σ^+, defined by

$$u_n = \sum_\sigma f_\sigma(n)(b_{-\sigma}^+ - b_\sigma)/2\omega_\sigma, \tag{2.2}$$

where Planck's constant $\hbar = 1$. The quantities $f_\sigma(n)$ and ω_σ are the eigenfunction and the eigenvalue, respectively labelled by the index σ. Using (2.2), H takes the non-interacting form

$$H = \sum_\sigma (b_\sigma^+ b_\sigma + 1/2)\omega_\sigma, \tag{2.3}$$

provided that the eigenfunctions satisfy the difference equation

$$\Omega_n f_\sigma(n) + \alpha_{n+1} f_\sigma(n+1) + \alpha_n f_\sigma(n-1) = 0, \tag{2.4}$$

where $\Omega_n = \omega_\sigma^2 - \alpha_{n+1} - \alpha_n$. The functions $f_\sigma(n)$ have the orthonormality and closure relationships

$$\sum_\sigma f_\sigma(n) f_\sigma^*(m) = \delta_{n,m}/m_n, \tag{2.5a}$$

$$\sum_n m_n f_\sigma(n) f_{\sigma'}^*(n) = \delta_{\sigma,\sigma'}. \tag{2.5b}$$

For the case of the Hofstadter problem the difference equation arises by forming a site representation of the Schrödinger equation and incorporating the magnetic field by means of the Peierls substitution (Hofstadter 1976). The resulting eigenvalue problem is the Harper equation

$$\psi_{n+1} + \psi_{n-1} + e_n\psi_n = E\psi_n, \tag{2.6}$$

where $e_n = 2\gamma\cos(nQ + \Delta)$ and E is the eigenvalue. Here, γ is the ratio of tight binding overlap integrals and Q is proportional to the magnetic field.

The general solution of (2.4) can be written in terms of two particular solutions which we call $\{p_n\}$ and $\{q_n\}$, these are polynomials in ω_σ^2. If we choose the initial conditions to be $p_0 = q_1 = 0$ and $p_1 = q_0 = 1$ then the general solution for $f_\sigma(n)$ becomes

$$f_\sigma(n) = f_\sigma(0)q_n + f_\sigma(1)p_n. \tag{2.7}$$

At this point in the development we must consider the modulation wave-vector, Q. If Q is a rational multiple of 2π, i.e. $Q = 2\pi M/N$ where M and N are coprime integers, then the coefficients in the difference equation are periodic with period N and the system is commensurate. If $Q/2\pi$ is irrational then the coefficients are not periodic and the system is incommensurate. The relationship between these two cases is interesting. If we consider the symmetries of each system, they appear to be entirely different; the commensurate system has the symmetry of the discrete one dimensional translation group (translation through N sites) which is not present in the incommensurate case,while on the other hand the incommensurate system has the continuous SO(2) symmetry of change of phase of the modulation which is not present in the commensurate system. However, on physical grounds, we do not expect to be able to detect the difference between an irrational and a rational which approximates it. In view of this solutions for an arbitrary rational modulation are of great physical interest. The incommensurate system can then be viewed in terms of an infinite sequence of systems each having a rational modulation, $Q_i = 2\pi M_i/N_i$, where, as $i \to \infty$, $N_i \to \infty$ and M_i/N_i tends to a finite (irrational) limit.

We now proceed to survey some properties of a system with arbitrary period N. The translation group symmetry gives rise to many simplifications, the first of which is Floquet's theorem which states that for a given ω there exist solutions of (2.4) with the property that

$$f(n+N) = \xi f(n), \tag{2.8}$$

for some Floquet multiplier ξ. Using this along with (2.7) for $n = 0$ and $n = 1$ yields

$$\xi^2 - \xi(p_{N+1} + q_N) + 1 = 0, \tag{2.9}$$

where we have used the Wronskian relationship $p_{n+1}q_n - p_n q_{n+1} = \alpha_1/\alpha_n$. This equation yields two Floquet multipliers $\xi_1(\omega)$ and $\xi_2(\omega)$ which obey $\xi_1\xi_2 = 1$. Therefore, either $|\xi_1| < 1$ and $|\xi_2| > 1$, which gives solutions for which $|f(n)|$ is not a bounded function (unstable solutions), or, $\xi_1 = \exp(i\theta)$ and $\xi_2 = \xi_1^*$, which gives solutions for which $|f(n)|$ is a bounded function (stable solutions). The stable solutions can be viewed as Bloch states; if we write $\theta = kN$ then (2.9) is the dispersion relationship for the Bloch wave-vector k

$$p_{N+1} + q_N = 2\cos(kN). \tag{2.10}$$

Detailed inspection of (2.9) shows that there will be stable solutions of (2.4) if and only if the discriminant satisfies,

$$D_N \equiv 4 - (p_{N+1} + q_N)^2 \geq 0 \tag{2.11}$$

It is straightforward to show from (2.4) that D_N is a polynomial of order $2N$ in ω^2. The spectrum therefore consists of N bands of stable states where $D_N \geq 0$.

The zeros of D_N are the band edges. Many results concerning these zeros are reviewed by Lovesey (1988 b). In particular it is possible to prove that they are all real and that they are ordered in such a way that the bands of stable states have a finite width.

As the periodicity N becomes large the spectrum becomes increasingly fragmented, giving the Hofstadter butterfly. It has proved difficult to obtain results relating to the nature of the spectrum in the incommensurate limit.

3 Green Function

For all the problems considered in this article it is possible to define Green functions from which dynamical properties of interest can be calculated. For the modulated spring model we have the (retarded) lattice displacement Green function

$$G_{n,m}(\omega) = -i \int_0^{\infty} dt \ \exp(i\omega t - \eta t)\langle[u_n, u_m(t)]\rangle, \tag{3.1}$$

where $\eta \to 0^+$. Analogous definitions are obtained for the other systems by replacing the displacement operators with electron creation and annihilation operators for the Hofstadter problem and with spin operators for the magnetic problem (see Lovesey 1988a and 1988b).

Using (2.2) we find the alternative form

$$G_{n,m}(\omega) = \sum_{\sigma} \frac{f_\sigma(n) f_\sigma^*(m)}{(\omega + i\eta)^2 - \omega_\sigma^2}, \tag{3.2}$$

which is useful in certain calculations.

It is straightforward to obtain expressions for dynamical quantities in terms of this Green function. From (3.1), a fluctuation-dissipation is readily derived

$$\langle u_n u_m(t)\rangle = -\frac{1}{\pi} \int_{-\infty}^{\infty} d\omega[1 + n(\omega)] \exp(i\omega t) Im.[G_{,n}(\omega)m], \tag{3.3}$$

where $n(\omega) = [exp(\omega/t) - 1]^{-1}$. This correlation function can be used to calculate the internal energy and dynamic structure factor of the system (see sections four and five). Further, the normalised density of states,

$$Z(\omega) = \frac{1}{N_0} \sum_{\sigma} \delta(\omega - \omega_\sigma), \tag{3.4}$$

where N_0 is the number of atoms in the chain, is found to be

$$Z(\omega) = -\frac{2\omega}{\pi N_0} \sum m_n Im.G_{n,n}(\omega), \tag{3.5}$$

where the sum is taken over all sites.

An exact expression for the Green function of the pure system, denoted by $P_{n,m}(\omega)$, can be obtained by solution of the equation of motion (we report an expression for the Green function of the perturbed system in section four). The equation of motion can be obtained from either (3.1) or (3.2), in the pure system this is a forced version of (2.4), namely,

$$\Omega_n P_{n,m}(\omega) + \alpha_{n+1} P_{n+1,m}(\omega) + \alpha_n P_{n-1,m}(\omega) = \delta_{n,m} \ , \tag{3.6}$$

where $\Omega_n = m\omega^2 - \alpha_n - \alpha_{n+1}$. The algebraic solution of this equation is covered by Lovesey and Westhead (Lovesey and Westhead 1990a). In a short article of this type we feel that it is not appropriate to give exhaustive mathematical details, instead we prefer to use the available space for the more complete account of the results that we obtain. Below, we present an outline derivation of the solution to (3.6), the interested reader will be able to fill in the details from the references.

Iteration of (3.6) for $n > m$ gives

$$\alpha_{n+1} \frac{P_{n+1,n}(\omega)}{P_{n,n}(\omega)} = h_n, \tag{3.7}$$

where

$$h_n = \cfrac{-\alpha_{n+1}^2}{\Omega_{n+2} - \cfrac{\alpha_{n+2}^2}{\Omega_{n+3} - \cdots}},$$

is an infinite continued fraction. Iterating for $n < m$ in the same manner we obtain

$$\alpha_n \frac{P_{n-1,n}(\omega)}{P_{n,n}(\omega)} = g_n, \tag{3.8}$$

where

$$g_n = \cfrac{-\alpha_n^2}{\Omega_{n-1} - \cfrac{\alpha_{n-1}^2}{\Omega_{n-2} - \cdots}}.$$

Finally, using these expressions in (3.6) for $n = m$ we obtain for the diagonal elements of the Green function

$$mP_{n,n}(\omega) = \frac{1}{\Omega_n + g_n + h_n}. \tag{3.9}$$

Simplification arises at this stage in the analysis when the coefficients α_n are periodic with period N. With this condition it is possible to calculate the values of the continued fractions $\{h_n\}$ and $\{g_n\}$ (see Lovesey and Westhead 1990a, in appendix). In particular, it can be shown that $h \equiv h_0$ satisfies the quadratic equation

$$p_N h^2 + \alpha_1 (q_N - p_{N+1}) h - \alpha_1^2 q_{N+1} = 0 \tag{3.10}$$

A similar equation is obtained for g_0. Using these results in (3.9) yields a closed expression for the diagonal elements of the Green function,

$$mP_{n,n}(\omega) = \pm \frac{q_n^2 p_N + p_n q_n (p_{N+1} - q_N) - p_n^2 q_{N+1}}{\alpha_1 \sqrt{-D_N}}. \tag{3.11}$$

Here the discriminant D_N is defined by (2.11) and a prescription for choosing the sign is given in Lovesey and Westhead 1990a.

At this point we can consider the off-diagonal elements of the Green function, of which we have said little in previous papers. In section five we will require the result that $P_{n,0}(\omega)$ obeys Floquet's theorem in the region $n > 0$ and in the region $n < 0$ with the Floquet multiplier changing as we cross $n = 0$. We offer an outline proof of this fact below.

In the regions $n < 0$ and $n > 0$ the Green function $P_{n,0}(\omega)$ satisfies the homogeneous (un-forced) equation. For $n > 0$,

$$P_{n,0}(\omega) = P_{1,0}(\omega) p_n + P_{0,0}(\omega) q_n, \tag{3.12}$$

and using (3.7)

$$P_{n,0}(\omega) = (h p_n / \alpha_1 + q_n) P_{0,0}(\omega). \tag{3.13}$$

Furthermore, because the coefficients are periodic, the function $P_{n+N,0}(\omega)$ must satisfy the same recursion relationship as $P_{n,0}(\omega)$. Repeating the same analysis we have,

$$P_{n+N,0}(\omega) = (h p_n / \alpha_1 + q_n) P_{N,0}(\omega). \tag{3.14}$$

Eliminating p_n and q_n from (3.13) and (3.14), gives

$$P_{n+N,0}(\omega) = \beta P_{n,0}(\omega), \tag{3.15}$$

in which $\beta = P_{N,0}(\omega)/P_{0,0}(\omega)$ is not a function of n. From this result it follows by induction that the Green function satisfies Floquet's theorem for $n > 0$. By setting $n = N$ in (3.13) and eliminating h with (3.10) we obtain

$$\beta^2 - \beta(p_{N+1} + q_N) + 1 = 0, \tag{3.16}$$

which is identical with (2.9), so β is identified with the Floquet multiplier. The proof for $n < 0$ follows in exactly the same way except that the Floquet multiplier should be identified with the other root of (3.16).

4 Lattice Dynamics

4a Density of States and Localised Defect Modes

The density of states is of interest as the most fundamental dynamical property of the system. Again we are specifically interested in the modulated spring model, but the same calculation follows through for the Hofstadter problem. Our aim is to examine the density of states in the pure system, where all the masses on the chain are equal, then to consider the effect of a single isotopic mass defect. We will show that such a defect generates localised vibrational modes whose frequencies lie in the gaps between the stable bands of states.

A closed expression for the density of states in the pure system can be obtained by substituting (3.11) into (3.5) and using the Christoffel-Darboux formula (Lovesey 1988b). This gives

$$Z_p(\omega) = \begin{cases} \frac{2\omega|p'_{N+1}+q'_N|}{\pi N\sqrt{D_N}}, & \text{if } D_N \geq 0; \\ 0, & \text{otherwise,} \end{cases} \tag{4.1}$$

where the prime denotes differentiation with respect to ω^2. This formula is non-zero only in the bands of stable states, and it exhibits the inverse square root singularities characteristic of one-dimensional systems.

The system with a single mass defect can also be solved analytically. If the defect is situated at site s and has mass $m_s = m(1-\lambda)$, then a standard calculation (see Lovesey 1987) leads to a formula for the perturbed Green function, $G_{n,m}(\omega)$,

$$G_{n,m}(\omega) = P_{n,m}(\omega) + \frac{m\lambda\omega^2 P_{n,s}(\omega)P_{s,m}(\omega)}{1 - m\lambda\omega^2 P_{s,s}(\omega)}. \tag{4.2}$$

Substitution of this formula into (3.5) gives the density of states in the perturbed system

$$Z(\omega) = Z_p(\omega) - \frac{2\omega}{\pi N_0} Im. \frac{d}{d\omega^2}\ln[1 - m\lambda\omega^2 P_{s,s}(\omega)]. \tag{4.3}$$

If we consider this expression in regions where $Z_p(\omega)$ is zero, i.e. outside the stable bands of the pure system, then a straightforward manipulation shows that

$$Z(\omega) = \frac{1}{N_0}\sum_i \delta(\omega - \omega_i). \tag{4.4}$$

Here the frequencies $\{\omega_i\}$ are the roots of the equation

$$1 - m\lambda\omega^2 P_{s,s}(\omega) = 0. \tag{4.5}$$

The frequencies $\{\omega_i\}$ lie outside the stable bands of states of the pure system, they are the so-called localised mode frequencies. Detailed consideration of the form of (4.5) shows that for a light mass defect ($0 < \lambda < 1$) there are N localised modes, one in every band gap and one high frequency mode lying above all the bands. For a heavy defect ($\lambda < 1$) there are $N-1$ localised modes, one in every band gap. There is no high frequency localised mode for a heavy defect.

The localised modes may be suitable for experimental investigation, for instance by neutron scattering. The high frequency mode associated with a light defect may be particularly useful. In the limit of a very light defect its frequency approaches the value

$$m\omega_0^2 = \frac{\alpha_s + \alpha_{s+1}}{(1/\lambda) - 1}. \tag{4.6}$$

The position of this mode is clearly sensitive to the modulation parameters N, γ and Δ which appear in the numerator.

Finally we note that though the formula (4.4) does not reduce to zero in the limit $\lambda \to 0$, in fact the frequencies $\{\omega_i\}$ move onto the edges of the stable bands in this limit. It is possible to show that the density of states (4.3) preserves the normalisation to unity (Lovesey and Westhead 1990a), so there is a small change in the intensity in the stable bands to compensate for the intensity given by (4.4).

4b Thermodynamic Quantities

In this section we present results for the mean-square velocity of a defect atom and for the defect energy. Again we find it necessary to omit details in favour of results; the relevant manipulations can be found in Lovesey and Westhead 1990a. For the mean-square velocity, v_n, of the atom at site n, we obtain the result,

$$m_n v_n^2 = T \sum_{k=-\infty}^{\infty} \left[1 + r_k G_{n,n}(\omega^2 = -r_k)\right] , \tag{4.7}$$

where $r_k = (2\pi k T)^2$ and the Boltzmann constant is set to unity. For $n = s$ (4.7) gives the mean-square velocity of the defect atom relevant to a Mössbauer experiment.

In order to calculate the defect energy, defined as the change in the internal energy per particle upon the addition of a single defect, we use (4.7) along with the fact that for a harmonic system the total internal energy is given by $\sum_n m_n \langle v_n^2 \rangle$. This then yields for the defect energy

$$\epsilon(s) = -\frac{2T}{N_0} \sum_{k=1}^{\infty} \omega^2 \frac{d}{d\omega^2} \ln[1 - m\lambda\omega^2 P_{s,s}(\omega)]\bigg|_{\omega^2=-r_k} . \tag{4.8}$$

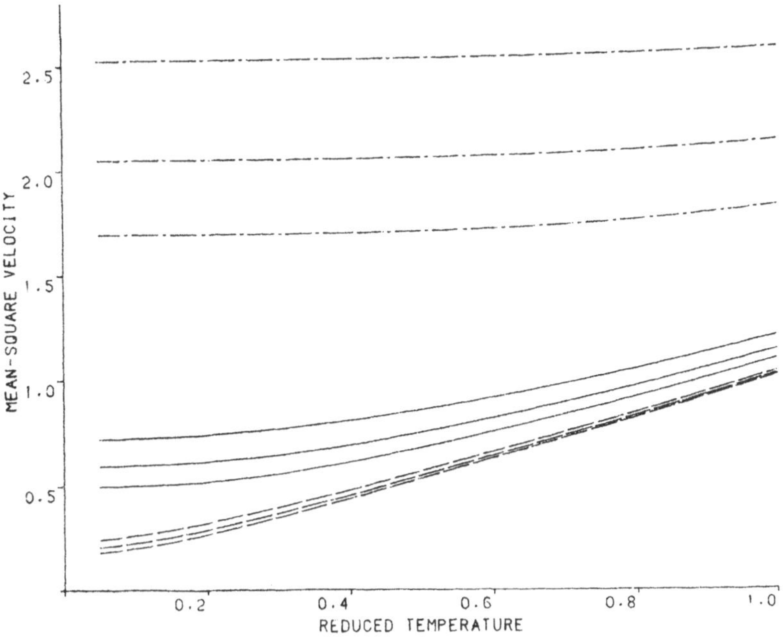

Figure 1. Mean-square velocity of a defect atom is shown as a function of reduced temperature for a system with $N = 5$, $M = 1$, $\gamma = 0.5\alpha$ and $\Delta = 2\pi/5$. We show a heavy defect ($\lambda = -5.0$, dash-dot curve), the pure system ($\lambda = 0.0$, unbroken curve) and a light defect ($\lambda = 0.9$, dashed curve). For each case there are three distinct sites for the defect.

The numerical implementation of (4.7) and (4.8) is straightforward. Figures 1 and 2 are a sample of our results. In figure 1 we show mean-square velocity of the defect atom as a function of reduced temperature for a system of periodicity $N = 5$. The phase, $\Delta = 2\pi/5$, is chosen so that only three of the five sites are distinct and the modulation is set to $\gamma = 0.5$. Results are shown for a heavy defect ($\lambda = -5.0$), the pure system ($\lambda = 0.0$) and for a light defect ($\lambda = 0.9$). Clearly the distinction between the two inequivalent sites is increased by making the mass defect lighter, more extensive numerical work shows this to be true in general. A light defect is a much more sensitive probe of modulation effects than a heavy one.

Figure 2 shows the defect energy as a function of the phase, Δ, of the modulation for a system with $N = 5$, $M = 1$, $\lambda = 0.9$ and a range of values of γ. The defect energy shows a pronounced minimum as a function of Δ, which increases in depth as the modulation γ increases. A defect will therefore lock in the phase of the modulation to the value at this stationary point in order to minimise energy. We should note that even though change of Δ is a continuous symmetry of the incommensurate system, the inclusion of a mass defect breaks this symmetry, implying that the lock-in effect will survive in the incommensurate limit.

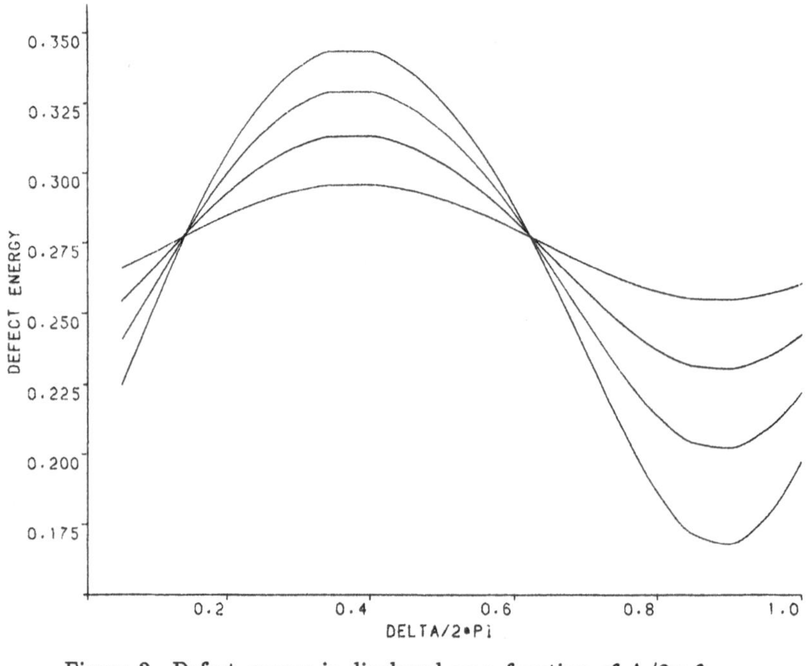

Figure 2. Defect energy is displayed as a function of $\Delta/2\pi$ for a system with $N = 5$, $M = 1$ and four equally spaced values of γ in the range $0.2\alpha \leq \gamma \leq 0.8\alpha$. We take a light defect, $\lambda = 0.9$, and a reduced temperature of $T = 0.2$.

5 Dynamic Structure Factor

The physical motivation for the examination of the modulated spring model in k-space is the calculation of the structure factor, proportional to the inelastic neutron cross-section. In this section we will consider calculation of the structure factor for a general modulation Q; for particular examples readers should see Lovesey and Westhead 1990b. We define a k-space

Green function

$$F(k,\omega) = -i\int_0^\infty dt\ \exp(i\omega t - \eta t)\langle[\tilde{u}_k, \tilde{u}_k^+(t)]\rangle, \tag{5.1}$$

where $\tilde{u}_k$ is the Fourier transform of the lattice displacement, $\sum_n \exp(-ikn)u_n$. The structure factor is then given by

$$S(k,\omega) = -\frac{1}{\pi}[1-\exp(-\omega/T)]^{-1} Im.F(k,\omega), \tag{5.2}$$

(see Lovesey 1987).

Again we attempt to calculate the Green function from an equation of motion. To this end, it is convenient to introduce a new Green function, $F_{n,m}(\omega)$, which will yield the correlation function $\langle\tilde{u}_{k+nQ}\tilde{u}^+_{k+mQ}(t)\rangle$. By construction $F(k,\omega)$ is then given by $F_{0,0}(\omega)$. With this notation, the equation of motion of the Green function, obtained by direct Fourier transformation of the real space equation (3.6), is found to be

$$E_n F_{n,m}(\omega) - T_n F_{n-1,m}(\omega) - T_{n+1}F_{n+1,m}(\omega) = \sum_s \exp[(isN(Q/2+\Delta)]\delta_{n+sN,m}\ , \tag{5.3}$$

where

$$E_n = \omega^2 - 4\alpha|\sin\{(k+nQ)/2\}|$$

and

$$T_n = -2\gamma\sin\{(k+nQ)/2\}\sin\{(k+(n-1)Q)/2\}.$$

It is worth pausing a while to consider the form of (5.3). It is set up for a modulation wavevector which is a rational multiple of 2π, $Q = 2\pi M/N$. The equation for the incommensurate system is obtained in the $N\to\infty$ limit

$$E_n F_{n,m} - T_n F_{n-1,m} - T_{n+1}F_{n+1,m} = \delta_{n,m} \tag{5.4}$$

Thus the phase cancels from the equation for the neutron cross-section in the incommensurate limit as expected by symmetry. Further, we note that (5.4) is of the same form as the real space equation (3.6), in fact they transform into each other under $E_n \leftrightarrow \Omega_n$ and $T_n \leftrightarrow -\alpha_n$. This is an example of the self duality exhibited by systems with truly incommensurate modulation. This duality can be used to obtain a solution of (5.4), *but only for the case of periodic coefficients*, ie. a commensurate system. For this reason we will refer to this solution as $\tilde{F}_{n,m}$. From (3.11) we obtain

$$\tilde{F}_{0,0}(\omega) = \frac{\pm v_N}{T_1\sqrt{(v_{N+1}+u_N)^2-4}}, \tag{5.5}$$

where we use v_n and u_n to denote the polynomials p_n and q_n when the above duality transformation has been made.

A few words about the interpretation of this formula are in order, after all, it has been obtained by using a rational number in an equation which is strictly only valid for an irrational. Substituting (5.5) in (5.2) we see that it yields a structure factor consisting of N bands of response (the function $\tilde{F}_{0,0}(\omega)$ is shown in figure 3 for $N=5$, $M=1$, $k=\pi/5$ and $\gamma=\alpha=0.25$). Numerically we find that the width of the bands decreases very rapidly with increasing N, we see this as justification for considering $\tilde{F}_{0,0}$ as an approximation to the solution for the incommensurate system which converges rather rapidly as N becomes large, i.e. $N>10$. Numerical work is easy for such values of N.

Of course, a solid state physicist will want to know how the solution for a truly commensurate system can be obtained. In particular, where is the delta function type of response, which conserves crystal momentum and energy and corresponds to scattering from normal modes (phonons)? In fact this response is obtained as the solution to equation (5.3), which applies to rational values of $Q/2\pi$. This solution can be expressed in terms of $\tilde{F}_{n,m}$, the solution to (5.4) for rational values of $Q/2\pi$, as follows:

$$F_{0,0}(\omega) = \sum_s \exp[iNs(Q/2 + \Delta)]\tilde{F}_{sN,0}(\omega). \tag{5.6}$$

This sum can be evaluated using equation (3.15). Focussing our attention on the imaginary part of $F_{0,0}$ we need only consider regions of frequency where $\tilde{F}_{0,0}$ is pure imaginary (i.e. $4 - (v_{N+1} + u_N)^2 > 0$). In these regions the Floquet multiplier β, which applies to the solutions of the k-space equation (5.4), can be written $\beta = \exp(i\theta)$. Using this Floquet multiplier in (5.6) we obtain

$$Im.F_{0,0}(\omega) = [\delta(NQ/2 + N\Delta + \theta) + \delta(NQ/2 + \Delta - \theta)]Im.\tilde{F}_{0,0}(\omega). \tag{5.7}$$

The value of θ can be obtained in terms of v_{N+1} and u_N from (3.16), this gives a delta function response (5.7) at frequencies defined by

$$\cos(\theta) = \cos[N(Q/2 + \Delta)] = (v_{N+1} + u_N)/2. \tag{5.8}$$

It is possible to show that these frequencies always lie in regions where $\tilde{F}_{0,0}(\omega)$ is pure imaginary.

A fairly straightforward though rather lengthy proof is required to show that the frequencies defined by (5.8) for external wave-vector k are the same as the normal mode frequencies for Bloch wave-vector k defined by (2.10). This confirms that the response conserves the crystal momentum. The weight associated with scattering from a normal mode of frequency of frequency ω is given, from (5.7), by $Im.\tilde{F}_{0,0}(\omega)$. Thus the k-space Green function $\tilde{F}_{0,0}(\omega)$ provides a way to calculate the strength of scattering from a given normal mode which is easier to apply than the conventional analysis that would entail numerical calculation of $f_\sigma(n)$ from (2.4).

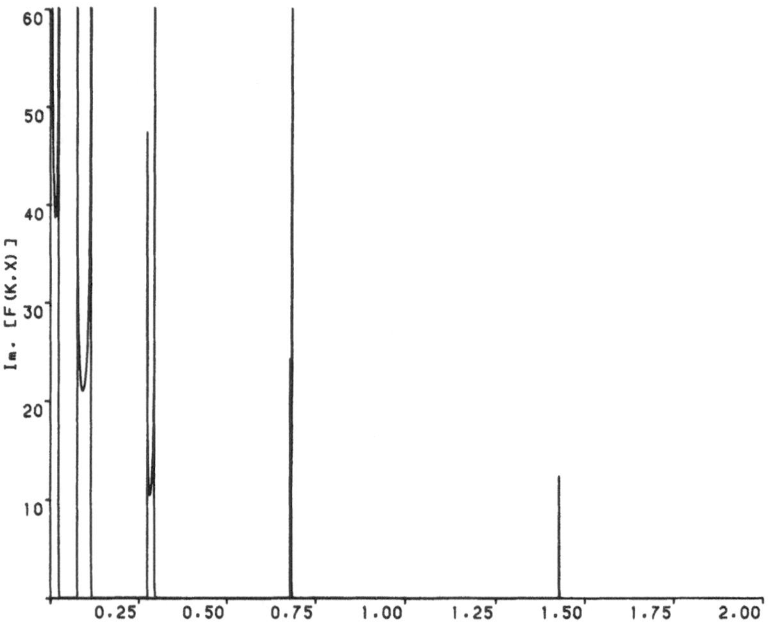

Figure 3. The function $Im.\tilde{F}(\pi/5,\omega)$ for a system with $N = 5$, $M = 1$ and $\gamma = \alpha = 0.25$.

6 Conclusions

We have reviewed the unusual nature of the spectrum of some modulated systems and presented a method for the calculation of there dynamic properties via a retarded Green function. The results are of interest in many areas of physics. Work presented in sections four and five is of experimental interest to those involved in Mössbauer and neutron scattering studies of modulated crystal structures. It is of theoretical interest in the fields of lattice dynamics and statistical physics, where it can be regarded as a non-trivial generalisation of the Rubin model. Section five also contains a fresh consideration of the mathematical difference between a commensurate system and an incommensurate, in particular the interesting self duality shown only for the incommensurate. It is shown how this self duality can be exploited to make a calculation of the structure factor and how this method must be adapted in order to solve the commensurate case.

We will do further work on the properties of defects in modulated lattices, particularly their effect on the structure factor. The Raman effect and other problems in light scattering are of immediate interest.

References

de Lange C. and Janssen T. (1981) J. Phys. C14, 5269

Hofstadter Phys. Rev. B14 (1976), 2239

Lovesey S. W. (1986) Condensed Matter Physics: Dynamic Correlations, Frontiers in Physics Vol.61(Menlo Park CA:Benjamin/Cummings)

Lovesey S. W. (1987) Theory of Neutron Scattering from Condensed Matter Vol 1 (Oxford: O.U.P.)

Lovesey S. W. (1988)a J. Phys. C21, 2805

Lovesey S. W. (1988)b J. Phys. C21, 4967

Lovesey S. W. (1989) J. Phys. Condens. Matter 1, 2731

Lovesey S. W. and Westhead D. R. (1990)a To be published in J. Phys. Condens. Matter

Lovesey S. W. and Westhead D. R. (1990)b To be published in J. Phys. Condens. Matter

STUDY OF THE CONTINUOUS SPECTRUM OF SEMICONDUCTOR HETEROSTRUCTURES

Witold Trzeciakowski*

Istituto di Fisica Teorica della Materia, FORUM INFM
Largo E.Fermi 2, 50125 Firenze, Italy

Massimo Gurioli

Laboratorio Europeo di Spettroscopia Non Lineare
Dipartimento di Fisica, Università di Firenze
Largo E.Fermi 2, 50125 Firenze, Italy

INTRODUCTION

Many optical and transport experiments on layered semiconductor structures involve the continuous energy spectrum. One important example are the quantum wells in a homogeneous electric field; the overall spectrum becomes continuous, bound states transform into resonances. Another example are the tunnelling structures (single barrier or double barrier). Very often the analysis of the continuous spectrum is performed in order to determine the dynamic characteristics of the structures, e.g. lifetimes of quasi-bound states[1,2], high-frequency properties of tunneling diodes[3,4] etc. Resonances are often treated with perturbative methods, i.e. the interaction between the discrete state and the continuum is assumed to be weak.

In the present paper we would like to demonstrate, that the direct determination of the density of states (DOS) is a simple and general method for characterizing the continuous spectrum. Apart from the resonances that can be broad (strongly coupled to the continuum) the study of the DOS reveals many interesting structures in the continuum, directly related to the behaviour of the wavefunctions. Furthermore we show that DOS is, in same cases, a better characteristic of structures in the continuous spectrum than transmission. First we briefly describe the method for determining the DOS, then we apply it to three important systems: tunneling structures, single quantum wells and the quantum wells in a uniform electric field.

METHOD

We put the considered structure in a big box so that the spectrum is dense but discrete. Suppose the condition for the energy levels E_n^o in an empty box is:

$$D_o(E_n^o) = 0 \ , \qquad (1)$$

and for a box with the structure:

$$D(E_n) = 0 \ . \qquad (2)$$

---

* on leave from "Unipress", Polish Academy of Sciences, Sokolowska 29, 01-142 Warsaw, Poland.

Microscopic Aspects of Nonlinearity in Condensed Matter
Edited by A.R. Bishop *et al.*, Plenum Press, New York, 1991

The DOS is given by the spacing between the levels:

$$\rho_o(E_n^o) = \frac{1}{\Delta_n^o}, \tag{3}$$

$$\rho(E_n) = \frac{1}{\Delta_n}, \tag{4}$$

where $\Delta_n^o = E_{n+1}^o - E_n^o$, $\Delta_n = E_{n+1} - E_n$. Both of these densities increase with the box size. However, the difference between them, which characterizes the considered structure, stays finite even for an infinite box. Thus we can write:

$$\Delta_n = \Delta_n^o + x_n \tag{5}$$

with $|x_n| << \Delta_n^o$. Therefore the change of the density of states $\Delta\rho$, from Eqs.(3)-(4), becomes:

$$\Delta\rho = \frac{1}{\Delta_n} - \frac{1}{\Delta_n^o} \cong -\frac{x_n}{(\Delta_n^o)^2} \tag{6}$$

The shift x_n can be obtained by expanding D(E) around $E_n+\Delta_n^o$, and we finally obtain:

$$\Delta\rho \cong \frac{1}{(\Delta_n^o)^2}\frac{D(E_n+\Delta_n^o)}{D'(E_n+\Delta_n^o)} \tag{7}$$

where D' indicates the first derivative of D.

Thus, if we know the eigenvalue condition (2) we can easily determine $\Delta\rho$. The spacing Δ_n^o in an empty box is often known exactly (e.g. $\Delta_n^o=(h/2L)^2(2n+1)/2m^*$ for a particle of mass m^* in a flat box of size L) but it can always be determined from Eq.(1).

APPLICATION TO TUNNELING STRUCTURES

This part of the work has already been published[5], therefore we only stress the main conclusions. The DOS seems to be a better characteristic of resonances than transmission, which is usually applied in this case. For narrow resonances both methods give the same positions and widths (Fig.(1a)), for broad structures the transmission is more obscure (Fig.(1b)).

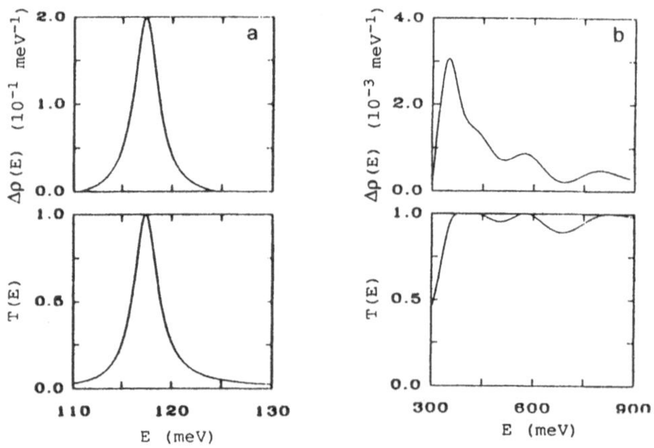

Fig.(1) Comparison between DOS and transmission for a double barrier structure. The barriers are 50 Å wide and 200 meV high, and the well is 100 Å wide; the effective mass m^* is $m^*=0.067m_o$ (electrons for GaAs)

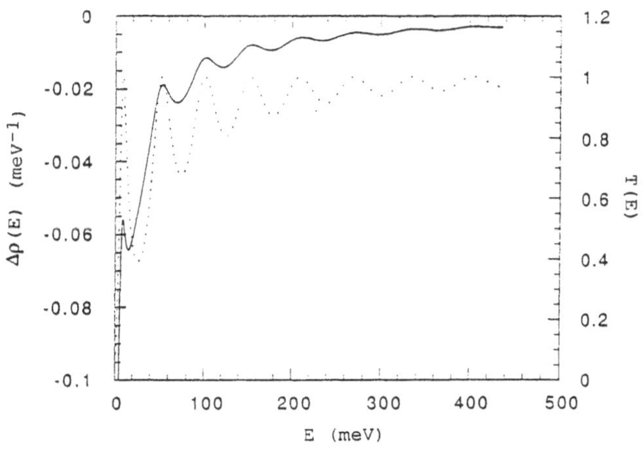

Fig.(2) Comparison between DOS (continuous line) and transmission (dashed line) for a single quantum well 200 Å wide and 200 meV deep; $m^*=0.45m_o$ (heavy holes for GaAs)

CONTINUUM STATES OF A SINGLE QUANTUM WELL

In Fig.(2) we compare DOS and transmission for the continuous energy spectrum of a 200 Å, 200 meV quantum well, showing that the resonances are present in both these quantities. Here it is important to remark that $\Delta\rho$ is negative because the continuum is depleted from states which were "sucked" into the well and became bound. It is easy to check that the positive (negative) $\Delta\rho$ indicates an increased (decreased) localization of the wavefunction in the region of the well. Transmission always changes from 0 to 1 and therefore does not yield this information.

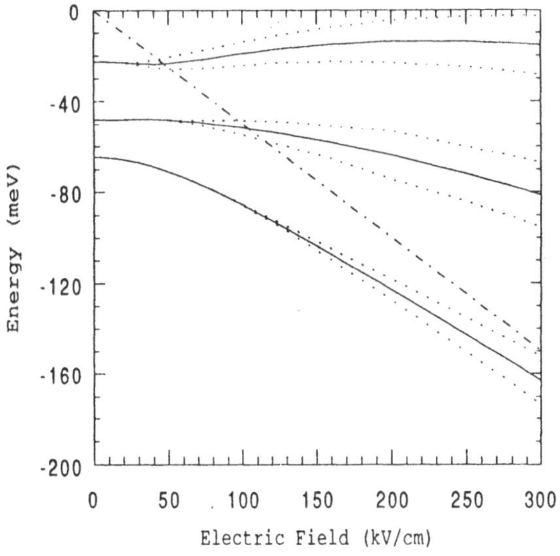

Fig.(3) Position (solid lines) and widths (dotted lines) of the first three resonant states for 100Å, 70meV well ($m^*=0.45m_o$). Dash-dotted line indicates the position of the lower edge of the well.

QUANTUM WELL IN A UNIFORM ELECTRIC FIELD

In the presence of an electric field the whole spectrum becomes continuous. Our method allows to determine the electric-field dependence of the positions and widths of the resonances originating from the bound states in absence of field (Fig.(3)); in this part our results agree with the calculation of E.J.Austin and M.Jaros[6]. Another interesting feature of the spectrum which could not be obtained perturbationally are the modulated oscillations of $\Delta\rho$ above the top of the well (Fig.(4)). The fast oscillation depends on the values of the electric field while the modulation depends only on the width of the well (and mass m^*). Again the maxima (minima) of $\Delta\rho$ correspond to characteristic changes of the wavefunction in the well region. This oscillatory structure of the continuous spectrum, which is due to the interference of the wave reflected from the well boundaries and from the slope of the electrostatic potential, should manifest itself in optical measurements; in particular the electroreflectance data of P.C.Klipstein et al[7]. shows a structure in the energy region above the top of the well that could be related to the oscillation in the density of states. A more detailed analysis of these features exceeds the aim of this communication and will be discussed in a separated paper[8].

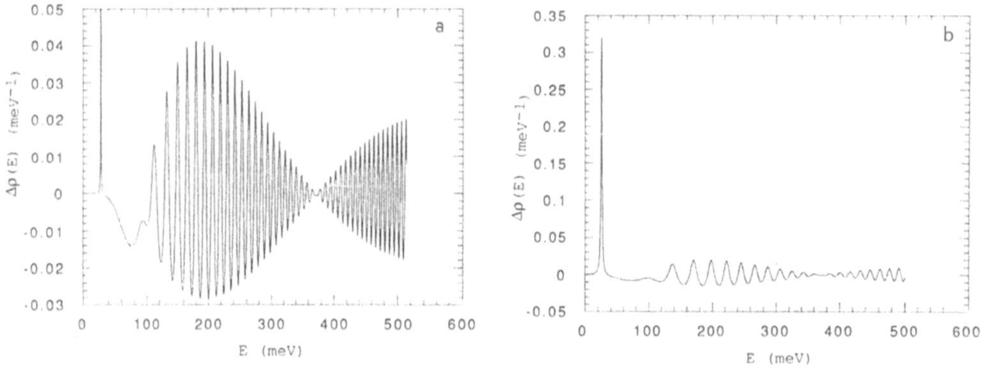

Fig.(4) Oscillations of $\Delta\rho$ for a single quantum well (30Å wide;70meV deep;$m^*=0.45m_0$) for two values of the electric field F. a) F=50 kV/cm ; b) F=100 kV/cm.

CONCLUSIONS

The study of the DOS is a simple and powerful method of investigating the continuous spectrum; it works for resonances, antiresonances, oscillatory structures etc. not susceptible to perturbative descriptions. It also yields some physical information on the localization of the wavefunctions.

REFERENCES

1. E.J.Austin and M.Jaros: Appl.Phys.Lett. **47**, 274 (1985)
2. T.B.Norris, X.Y.Song, W.J.Schaff, L.F.Eastman, G.Wicks, and G.A.Mourou: Appl.Phys.Lett. **54**, 60 (1989)
3. E.H.Hauge, J.P.Falck, T.A.Fjeldly: Phys.Rev.B **36**, 4203 (1987)
4. T.B.Bahder, C.A.Morrison, and J.D.Bruno: Appl.Phys.Lett. **51**, 1089 (1987)
5. W.Trzeciakowski, D.Sahu, and T.F.George: Phys.Rev.B **40**, 6058, (1989)
6. E.J.Austin, and M.Jaros: Phys.Rev.B **31**, 5569 (1985)
7. P.C.Klipstein, P.R.Tapster, N.Apsley, D.A.Anderson, M.S.Skolnick, T.M.Kerr, and K.Woodbridge: J.Phys.C, Solid StatePhys **19**, 857 (1986)
8. W.Trzeciakowski and M.Gurioli: to be published

RAMAN SCATTERING OF LIGHT BY ELECTRONS IN SUPERCONDUCTORS WITH A SMALL CORRELATION LENGHT

A.A. Abrikosov and L.A. Fal'kovskiĭ

L.D. Landau Institute of Theoretical Physics, Academy of Sciencies of the USSR
Kosygina, 2, GSP–1, Moscow, 117940, USSR

The reflection coefficient of pure anisotropic superconductors with a correlation lenght $\hbar v/\Delta$ much smaller than the penetration depth of light δ has been determined. The effect of anisotropy is discussed.

Abrikosov and Fal'kovskiĭ[1] and Abrikosov and Genkin[2] studied Raman-scattering spectra of pure superconductors with a large correlation length

$$\xi \sim \hbar v/\Delta \gg \delta, \tag{1}$$

where v is the velocity at the Fermi surface, Δ is the superconducting gap, and δ is the penetration depth of light. Dierker *et al.*[3] reported the observation of Raman scattering in a superconducting Nb_3Sn. Since condition (1) does not hold for this material, and since the inverse inequality probably applies, Dierker *et al.*[3] have made relevant theoretical predictions. They failed, however, to take into account a series of diagrams (which were singled out in Ref. 1) which in the limiting case

$$\xi \ll \delta \tag{2}$$

are of crucial importance. At the same time, the study of Raman electron scattering in superconductors in the case where condition (2) holds is of great current interest in view of the discovery of high-temperature superconductors with an extremely small correlation lenght. Experimental determination of Raman scattering of electrons[4] makes it possible to measure the energy gap of a contactless method.

We will derive here a theory in which all necessary diagrams are taken into account. Assuming that the temperature is zero, we will start with an isotropic model. As was pointed out in Ref. 1, in addition to the zeroth-order diagram for electron interaction (Fig. 2a in Ref. 1), it is necessary to take the sum over the entire series of chain diagrams (Fig. 3a in Ref. 1). While, at $\xi \gg \Delta$ these diagrams determine the reflection coefficient only near the threshold, $\omega_0 = \omega - \omega' = 2\Delta(\omega$ and ω' are the frequencies of the incident and scattered light), for $\xi \ll \delta$ they are important in a broad region, $\omega_0 - 2\Delta \sim \Delta$; we are using here a system of units with $\hbar = c = 1$.

The addition of new diagrams when the superconducting pairing is taken into account results in a replacement in the equation for the matrix element S_{j0} (see Ref.

1) of the expression $\psi_\alpha^+(x)\psi_\alpha(x)$ by the combination

$$R = A_1R_1 + A_2R_2 + A_3R_3, \tag{3}$$

where $R_1 = \psi_\alpha^+\psi_\alpha$, $R_2 = I_{\alpha\beta}\psi_\beta\psi_\alpha$, and $R_3 = I_{\alpha\beta}\psi_\alpha^+\psi_\beta^+$; here $I_{\alpha\beta}$ is a 2×2 antisymmetric matrix, $I_{\alpha\beta} = -I_{\beta\alpha}$, and $I^2 = -1$.

The equations for the coefficient A_i can be obtained as follows. By adding one link to the sum of an infinite series of diagrams in Fig. 3a of Ref. 1 this series transforms into itself. We thus obtain (in zeroth approximation)

$$A_i = \delta_{i_1} + a_{ki}A_k, \tag{4}$$

where the coefficients a_{ik} correspond to different types of links. The coefficients which are important below, for example, are described by the diagrams in Fig. 1. Analyzing the properties of the coefficients a_{ik} end Eq. (4), we find that $A_3 = -A_2$ and $A_1 \approx 1$ within the electron coupling constant g, which is assumed to be small. We then find

$$A_2 = a_{1\,2}/(1 - a_{2\,2} + a_{3\,2}). \tag{5}$$

We see from Fig. 1 that the diagram for $a_{2\,2}$ has a large logarithm. If the frequency transfer ω_0 and momentum transfer $\mathbf{q} = \mathbf{k} - \mathbf{k}'$ are moderately large ($\omega_0 \lesssim \Delta$, $vq \lesssim \Delta$), the coefficient $a_{2\,2}$ will contain the term $q\ln(2\omega_D/\Delta)$ which cancels out with 1 by virtue of the equation for the gap; the parameter ω_D limits the frequencies at which there is an attraction between electrons: this is the Debye frequency for the phonon mechanism. As a result, the terms of order g remain both in the numerator and denominator of A_2, i.e., $A_2 \sim 1$. In the case of a large energy transfer or large momentum transfer the interaction is unimportant, $A_2 \ll 1$, and only the first term is retained in (3).

Substituting into the matrix element S_{j0} the combination R instead of $\psi_\alpha^+\psi_\alpha$ (see Ref. 1) and performing the same transformations as in Ref. 1, we find (we assume, for

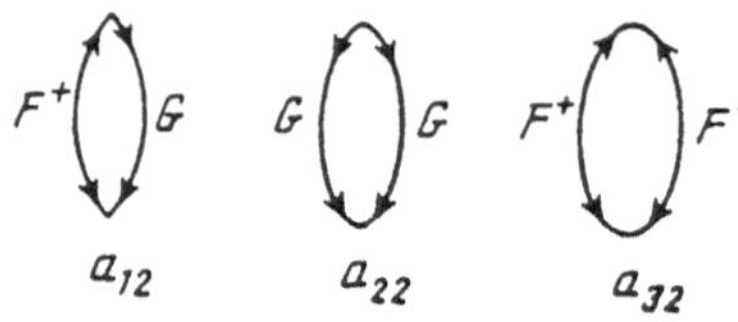

FIG. 1

simplicity, that the incidence and reflection are normal)

$$d\sigma = \frac{2^9 e^4 cos^2\theta d\omega' d\Omega'}{\pi m^2 \delta^2[(n+1)^2 + k^2][(n-1)^2 + k^2]} \int dqF(q)/(q^2 + 4\delta^{-2})^2, \tag{6}$$

$$F(q) = \int |uv' + vu' - 2A_2(uu' + vv')|^2 \delta(\omega_0 - \epsilon - \epsilon')d^3p/(2\pi)^3, \tag{7}$$

where n is the refractive index at the frequency ω; k is the absorption coefficient which determines the field penetration depth, $\delta = c/\omega k$; u and v are the Bogolyubov conversion factors; u^2, $v^2 = (1 \pm \xi/\epsilon)/2$; u' and v' correspond to the momentum p +

q; the vector **q** is directed normal to the surface, with an absolute q over which the integration is carried out; and θ is the angle between the polarization of incident light and the polarization of scattered light.

The coefficients appearing in A_2 are

$$a_{ik} = \frac{\lambda}{16} \int_{-\infty}^{+\infty} d\xi \int_{-1}^{1} d\mu[(\epsilon + \epsilon' + \omega_0 - i0)^{-1} + (\epsilon + \epsilon' + \omega_0 - i0)^{-1}]\rho_{ik},$$

where the weight function $\rho_{1\,2}$ for $a_{1\,2}$ is $-\Delta\omega_0/2\epsilon\epsilon'$, and for the difference $a_{2\,2} - a_{3\,2}$ we have $\rho = 1 + (\xi\xi' + \Delta^2)/\epsilon\epsilon'$, $\lambda = gp_0^2/\pi^2 v$ is a dimensionless coupling constant, and $\mu = \cos\mathbf{vq}$.

Since we are considering the case (2), and since the important terms in the integration over **q**, according to (6), are the terms $q \sim 1/\delta$, we must set $q = 0$ in the zeroth approximation. However, the expression inside the modulus in (7), with allowance for the δ function, will then vanish. This occurs precisely because of the incorporation of the chain diagrams. If these diagrams were ignored, as was the case in Ref. 3, then the first term, which is retained inside the modulus in (7), will, at $q = 0$, lead to a divergence near the threshold, $d\sigma \sim (\omega_0^2 - 4\Delta^2)^{-1/2}$.

Let us analyze the function $F(q)$ in the region $q \lesssim \Delta/v$. If we are near the threshold, i.e., if the condition $\omega_o - 2\Delta \ll \Delta$ holds, then the value $q_m = (\omega_0^2 - 4\Delta^2)^{1/2}/v$, falls within this region, beginning with which the function $F(q)$ reaches its asymptotic behavior $F(q) \sim (vq/\Delta)^3 \ln^{-2}(q/q_m)$. At $q \ll q_m$ the function $F(q) \sim (vq/\Delta)^3 q/q_m$. Clearly the principal contribution to the integral over q comes from $q \lesssim \max\{q_m, \delta^{-1}\}$. We thus find the following expression (in standard units), depending on the relationshio between q_m and $1/\delta$:

$$d\sigma = \frac{2^5}{\pi^4}\left(\frac{e^2 v}{\hbar c^2}\right)^2 \left(\frac{\hbar v}{\Delta\delta}\right)^2 \frac{cos^2\theta f d\omega' d\Omega'}{\Delta[(n+1)^2 + k^2][(n-1)^2 + k^2]}, \tag{8}$$

where

$$f \approx \begin{cases} 1 & \text{for} \quad q_m \gg 1/\delta & (8a) \\ \frac{\pi^2}{8}\ln(\delta/\xi)/\ln(1/q_m\delta)\ln(1/q_m\xi) & \text{for} \quad q_m \ll 1/\delta. & (8b) \end{cases}$$

Equation (8) can be used when $\omega_0 - 2\Delta \lesssim \Delta$. Far from the threshold, i.e., with the condition $\omega_0 - 2\Delta \gg \Delta$, satisfied, the result should correspond to the normal metal. A simple calculation leads to an expression which can be derived from (8) through the substitution

$$f = 4\pi^2(\Delta/\omega_0)^3 t^4[\ln(1 + t^{-2}) - (1 + t^2)^{-1}], \qquad t = \omega_0\delta/2v. \tag{8c}$$

This result describes Raman scattering by a normal metal for any relationship between the penetration depth and the frequency transfer. For case (2) which we are considering, Eqs. (8a) and (8c) are matched at $\omega_0 \approx 2.7\Delta$.

Equation (8) implies that near the threshold $d\sigma$ increases rapidly from zero to a value determined by (8) and (8a). A numerical estimate yields the value

$$d\sigma \approx 10^{-11}(\xi/\delta)^2(\hbar d\omega'/\Delta)d\Omega'. \tag{9}$$

Let us now consider qualitatively the consequences of anisotropy. The principal charge is that the function $F(q)$ in (7) does not vanish at $q = 0$. Allowance for the anisotropy of the mass tensor ($m^{-1}A^2 \to m_{ik}^{-1}A_iA_k$, where $\mathbf{n} = \mathbf{p}/p$) shows that the result is proportional to $\overline{(m_{ik} - \bar{m}_{ik})^2}$, where an average is weighted in a way that depends on the nature of electron interaction and on $\Delta(\mathbf{n})$. The final expression

$$d\sigma_0 \sim \pi^{-2}(e^2/\hbar c)^2(v/c)^2\delta d\omega' d\Omega'/v[(1+n)^2+k^2][(1-n)^2+k^2] \tag{10}$$

does not depend on ξ far from the threshold. Near the threshold it has the form

$$d\sigma \sim d\sigma_0[(\omega_o - 2\Delta_{min})/\Delta_{min}]^\alpha. \tag{11}$$

If the minimum of $\Delta(\mathbf{n})$ corresponds to a single point of the Fermi surface, we would have $\alpha = 1/2$. If the minimum falls on the line (in the strictly two-dimensional case, for example), we would have $\alpha = 0$. More detailed calculations give a factor $[\ln(\Delta/\omega_0 - 2\Delta_{min})]^{-2}$.

Our calculations show that integration involves the entire Fermi surface and that the result remains the same, regardless of whether we are dealing with a single crystal or a polycrystal comprised of crystallites with identical parameters but different orientations.

Substituting the numerical values ($\Delta \sim 10^2\mathbf{K}$) in (9), we find

$$d\sigma_0 \sim 10^{-11}\hbar d\Omega' d\omega'/\Delta.$$

This result corresponds to the result obtained experimentally in Ref. 4. Note that our result was obtained under the assumption that $\delta \gg \xi$, whereas the experiment of Ref. 4 apparently was carried out in accordance with the condition $\delta \sim \xi$.

REFERENCES

1 A.A. Abrikosov and L.A. Fal'kovskiĭ, Zh. Eksp. Teor. Fiz. **40**, 262 (1961) [Sov. Phys. JETP **13**, 179 (1961)].

2 A.A. Abrikosov and V.M. Genkin, Zh. Eksp. Teor. Fiz. **65**, 842 (1973) [Sov. Phys. JETP **38**, 417 (1974)].

3 B. Dierker, M.B. Klein, J.W. Webb, and Z. Fisk, Phys. Rev. Lett. **50**, 853 (1983).

4 A.V. Bazhenov, A.V. Gorbunov, N.V. Klassen, S.F. Kondakov, I.V. Kukushkin, V.D. Kulakovskiĭ, O.V. Misochko, V.B. Timofeev, L.I. Chernyshova, and B.N. Shepel', Pis'ma Zh. Eksp. Teor. Fiz. **46**, 35 (1987) JETP Lett. **46**, 42 (1987)].

QUANTUM SOLITONS ON QUANTUM CHAOS:

COHERENT STRUCTURES, ANYONS, AND STATISTICAL MECHANICS

R.K. Bullough

Department of Mathematics,
UMIST
P.O. Box 88
Manchester M60 1QD, UK

J. Timonen

Department of Physics
University of Jyväskylä
SF-40100 Jyväskylä
Finland

1. INTRODUCTION: SUMMARY OF THE STATISTICAL MECHANICS

This paper is concerned with the exact evaluation of functional integrals for the partition function Z (free energy $F = -\beta^{-1}\ell n\ Z$, β^{-1}= temperature) for integrable models like the quantum and classical sine-Gordon (s-G) models in 1+1 dimensions [1-12]. These models have wide applications in physics and are generic (and important) in that sense. The classical s-G model in 1+1 dimensions

$$\phi_{xx} - \phi_{tt} = m^2 \sin \phi \qquad (1)$$

(m > 0 is a "mass") has soliton (kink, anti-kink and breather) solutions. In Refs 1-12 we have reported a general theory of 'soliton statistical mechanics' (soliton SM) in which the particle description can be seen in terms of 'solitons' and 'phonons'. The situation concerning the phonons and breather solutions of models like the quantum and classical s-G models has proved unexpected and the latter part of this present report (the §4 on quantum and classical thermodynamic limits) is devoted to this problem and its actual solution.

The work so far [1-12] is concerned only with the 'one-body' quantities, namely Z or F. Recently [13] it has become clear that the methods used to evaluate Z or F can be extended to provide essentially exact solutions for the correlation functions. This aspect will be developed elsewhere. In this Introduction we sketch only the content of Refs 1-12 relevant to the present paper.

A new realisation [14,15] is that there is a natural anyon theory, with fractional quantum statistics, in the quantum SM. This paper sketches additional developments of that theory. It is well realised that a free anyon gas is almost certainly superfluid [16]. Such models may therefore prove important for understanding High-T_c superconductivity [16,17] — although (see below) no evidence of the breakdown of parity and time reversal invariance has been seen in experiments [18].

The new analytical methods developed in Refs 1-12 have wide ranging connections. Some of these are (literally!) sketched in the Fig. 1 attached.

Microscopic Aspects of Nonlinearity in Condensed Matter
Edited by A.R. Bishop *et al.*, Plenum Press, New York, 1991

"SOLITIONS"

U, V and Ψ are $N \times N$ Matrices $\left\{ \begin{matrix} sl(N,C) \\ gl(N,C) \end{matrix} \right\} = g \ni U, V$

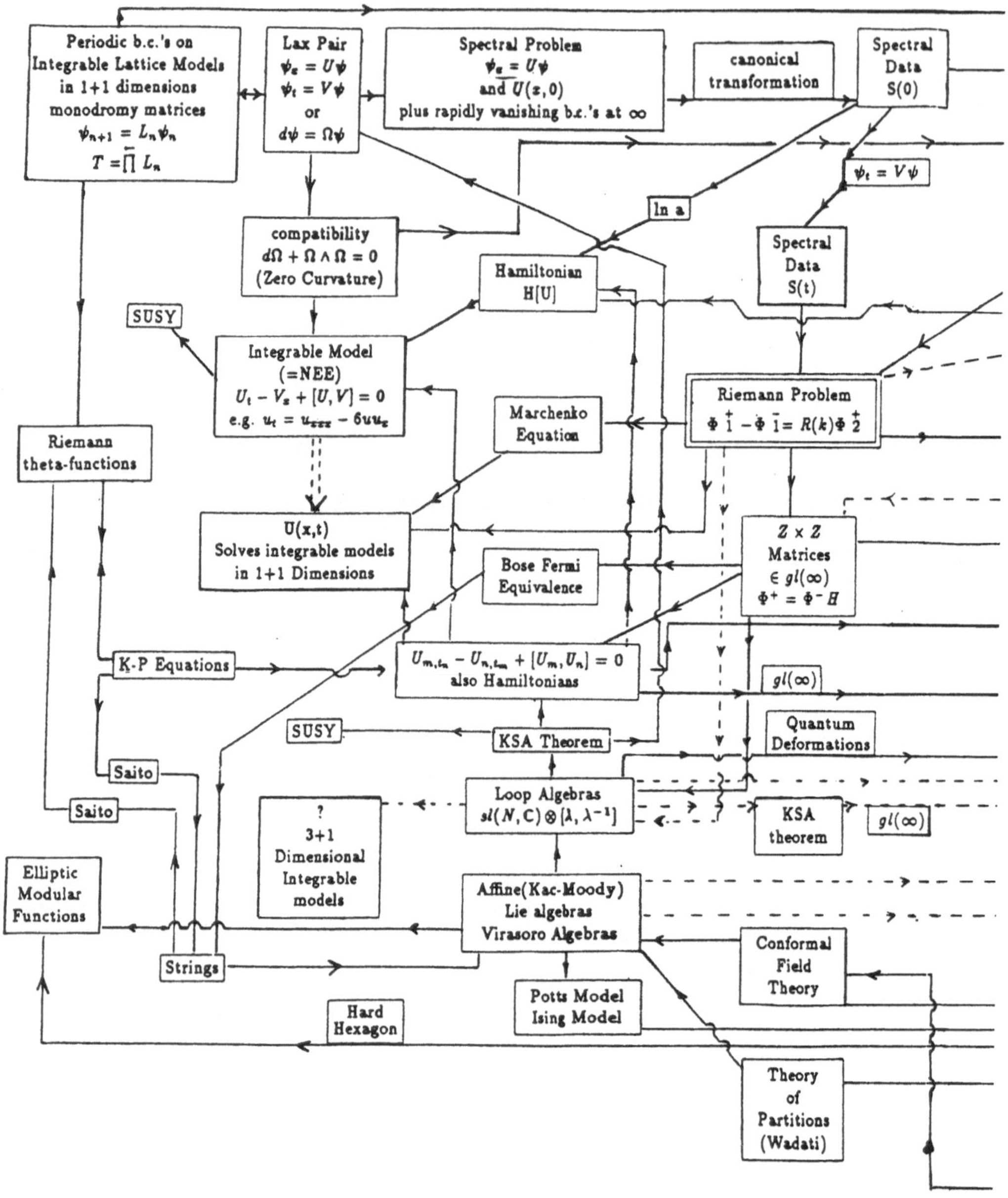

Figure 1. Overview of generalised 'Solition' theory as of June 1990. A hard arrow indicates minimal connection (at least) between 'boxes' is already established, and most hard arrows are actual mappings. Dashed arrows indicate expectation by the authors that some minimal connection can be achieved, or stronger - or indicate puzzles about connections already reported. Experimental outputs from the partition functions Z are shown at extreme right. High Tc superconductivity is in Refs. 16, 17 and (perhaps) in anyon theory (§ and Refs. 14, 15); Rydberg atoms are in Ref. 15 and references. For DNA see e.g. Ref. 7, for $CsNiF_3$ Ref. 36 and references. Free energies $F = -\beta^{-1} \ln Z$ yield specific heats e.g. Ref.1 and references. See also much work on spin chains etc., etc.

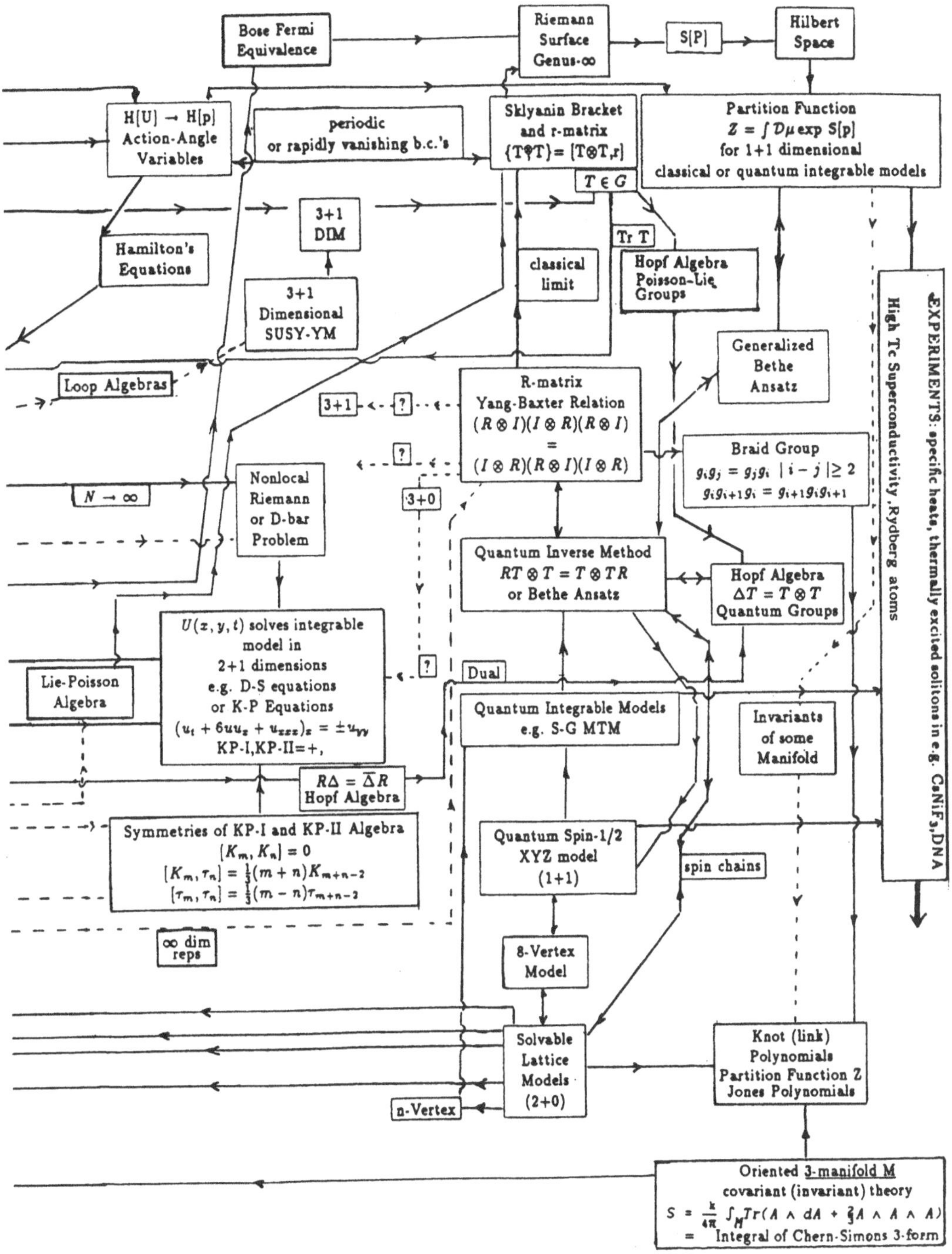

Bose Fermi Equivalence
Riemann Surface Genus-∞
S[P]
Hilbert Space
H[U] → H[p] Action-Angle Variables
periodic or rapidly vanishing b.c.'s
Sklyanin Bracket and r-matrix {T⊗T} = [T⊗T,r]
Partition Function $Z = \int \mathcal{D}\mu \exp S[p]$ for 1+1 dimensional classical or quantum integrable models
T ∈ G
3+1 DIM
Tr T
Hamilton's Equations
classical limit
Hopf Algebra Poisson-Lie Groups
3+1 Dimensional SUSY-YM
Generalized Bethe Ansatz
EXPERIMENTS: specific heats, thermally excited solitons in e.g. CsNiF3,DNA
High Tc Superconductivity, Rydberg atoms
Loop Algebras
R-matrix Yang-Baxter Relation $(R\otimes I)(I\otimes R)(R\otimes I) = (I\otimes R)(R\otimes I)(I\otimes R)$
3+1
?
Braid Group $g_i g_j = g_j g_i \ |i-j| \geq 2$ $g_i g_{i+1} g_i = g_{i+1} g_i g_{i+1}$
N → ∞
Nonlocal Riemann or D-bar Problem
3+0
Quantum Inverse Method $RT\otimes T = T\otimes TR$ or Bethe Ansatz
Hopf Algebra $\Delta T = T\otimes T$ Quantum Groups
$U(x,y,t)$ solves integrable model in 2+1 dimensions e.g. D-S equations or K-P Equations $(u_t + 6uu_x + u_{xxx})_x = \pm u_{yy}$ KP-I,KP-II=+,
Lie-Poisson Algebra
Dual
Quantum Integrable Models e.g. S-G MTM
Invariants of some Manifold
$R\Delta = \overline{\Delta}R$ Hopf Algebra
Symmetries of KP-I and KP-II Algebra $[K_m, K_n] = 0$ $[K_m, \tau_n] = \frac{1}{3}(m+n)K_{m+n-2}$ $[\tau_m, \tau_n] = \frac{1}{3}(m-n)\tau_{m+n-2}$
Quantum Spin-1/2 XYZ model (1+1)
spin chains
∞ dim reps
8-Vertex Model
Solvable Lattice Models (2+0)
Knot (link) Polynomials Partition Function Z Jones Polynomials
n-Vertex
Oriented 3-manifold M covariant (invariant) theory
$S = \frac{k}{4\pi}\int_M Tr(A\wedge dA + \frac{2}{3}A\wedge A\wedge A)$
= Integral of Chern-Simons 3-form

In this paper we are concerned with the expression for Z in the 'box' at top right (Fig. 1 assumes the trace Tr)

$$Z = \mathrm{Tr} \int \mathcal{D}\mu \exp S[p] \quad . \tag{2}$$

This form of the functional integral for Z needs explanation.

We have developed two alternative methods of analysis which are, in terms of results, wholly equivalent: one of these evaluates Z taken characteristically first of all in the form of the functional integral

$$Z = \mathrm{Tr} \int \mathcal{D}\Pi\mathcal{D}\phi \exp S[\phi] \tag{3}$$

where $S[\phi]$ is a classical action for a field $\phi(x,t)$

$$S[\phi] = \hbar^{-1}\int_0^{\beta\hbar} dt \; [i\int \Pi\phi_{,t}dx - H[\phi]] \quad . \tag{4}$$

Evaluation is by making a canonical transformation to the form (2) in which $\mathcal{D}\mu$ is a measure to be determined; then S[p] is (4) expressed in (appropriate) action and angle variables. This canonical transformation exists for integrable models in 1+1. This line of argument (sketched in Fig. 1 from top left to top right) we call the 'functional integral method' for evaluating $F = -\beta^{-1}\ell n\ Z$ [1,2,4,6,12]. The classical canonical transformation to action-angle variables involved introduces a natural description of the particles as soliton-like (and phonon-like) particles.

The second, alternative, method, somewhat easier to apply in practice, uses the same particle description and defines an entropy S, an energy E, and a free energy $F = (E-\beta^{-1}S)$ [1,2,4,5,12]. Minimisation of this F yields closed expressions for the actual free energy F which coincide with those derived from (2). Evidently this method generalises the Bethe Ansatz (BA) [19] and quantum inverse (QIM) methods [20,21]: it extends from the fermion descriptions used in the quantum BA and QIM to bose descriptions and to the classical statistical mechanics of Maxwell-Boltzmann particles [1-12] and we call it 'generalised BA' for that reason [1,2,4,5,12]. The boson descriptions are strictly equivalent to the fermion descriptions (bose-fermi equivalence). We have shown [2,4-6,9-12] by examples how such equivalence arises. The Fig. 1 shows where its mathematics may originate - in the classical infinite-dimensional Lie algebra gl(∞) which has at least two representations: representations as vertex operators on a bosonic space and as products of free fermion operators [22]. However quantisation of the integrable models in 1+1 can be traced through so-called *deformations* of Lie algebras to the 'quantum groups', and it may be that *deformation* of gl(∞) is the proper origin of bose-fermi equivalence [14,15]. Since the two descriptions (in terms of bosons or fermions) are particular cases of a more general anyon description we review here the bose-fermi equivalent theory for two particular integrable models — the sinh-Gordon model

$$\phi_{xx} - \phi_{tt} = m^2 \sinh \phi \quad , \tag{5}$$

which has a real field ϕ, and the *repulsive* NLS model

$$-i\phi_t = \phi_{xx} - 2c\phi^*\phi^2, \qquad c > 0 \quad , \tag{6}$$

in which ϕ is complex.

We need to explain that the form (4) for $S[\phi]$ is defined on a symplectic manifold (phase space) M co-ordinatised by Π,ϕ: M has a bracket, and $\{\Pi,\phi\}$ =

$\delta(x-x')$. Both the measure $\mathcal{D}\Pi\mathcal{D}\phi$ in (3) and $\mathcal{D}\mu$ in (2) are measures on M. The former can be found (§2): the latter is then to be calculated (see below).

Evidently (4) is not the usual Feynman form [23] of the action $S[\phi]$ and the measure in Z, eqn. (3) is not the Feynman measure. It is a much 'better' measure — see §2. Notice that if the classical Hamiltonian $H[\phi]$ of the model is quadratic in Π, then integration of (3) on Π yields Feynman's form (and measure). Evidently the s-G and sinh-G models are examples: for s-G

$$H[\phi] = \gamma_0^{-1}\int[\tfrac{1}{2}\Pi^2\gamma_0^{\,2} + \tfrac{1}{2}\phi_x^{\,2} + m^2(1 - \cos\phi)]dx \tag{7}$$

in which $\gamma_0 > 0$ is a (classical) coupling constant (in the quantum theory derived from (3) and (4) γ_0 is renormalised [12,24,25] to $\gamma_0'' = \gamma_0[1 - \gamma_0/8\pi]^{-1}$ used in §§3,4). By canonical transformation, analytical continuation in γ_0, and inverse canonical transformation, $H[\phi]$ (7) becomes [4,7] the classical H for the sinh-G model (5). The classical bracket for the repulsive NLS models is $\{\phi, \phi^*\} = i\delta(x-x')$; the coupling constant c is $c > 0$ (repulsive NLS) and $c < 0$ (attractive NLS) and these two models relate by a similar analytical continuation (in c). From the appropriate forms of (3) and (4) we still reach (2) by canonical transformation and (2) is generic.

The quantum statistics, fermi, bose or fractional, enters the theory through the action $S = S[p]$ expressed (generically) in terms of action-angle variables ($0 \le P(k) < \infty$, $0 \le Q(k) < 2\pi$; $\{P(k),Q(k')\} = \delta(k-k')$)

$$S[p] = \hbar^{-1}\int_0^{\beta\hbar} dt\ [i\int P(k)\ Q(k)_{,t}dk - H[p]] \ . \tag{8}$$

For sinh-G and repulsive NLS (8) is complete, but for s-G or attractive NLS extra terms arise from the solitons. For both of the 'repulsive' models $H[p] = \int_{-\infty}^{\infty} \omega(k)\ P(k)\ dk$ ($\omega(k) = k^2$, NLS; $\omega = (m^2 + k^2)^{\frac{1}{2}}$ sinh-G). These results are found for the classical models on the real line $-\infty < x < \infty$ with vanishing boundary conditions at $\pm\infty$. A critical feature in evaluating Z is to evaluate it in finite density thermodynamic limit. This is achieved through connection between action-angle variables under periodic b.c.s. of period L (say) and action-angle variables on the real line in the situation in which $L \to \infty$ in finite density: finite density means of course that, for N particles in L, $\lim_{L\to\infty} N/L = \bar{n} > 0$.

For $\hbar \to 0$ the action S[p] becomes an action for the classical statistical mechanics $S[p] = -\beta H[p]$. The P(k), Q(k) are real so

$$i\hbar^{-1}\int_0^{\beta\hbar} dt \int P(k)\ Q(k)_{,t}\ dk \equiv i\Phi[p] \tag{9}$$

is a pure phase. Evidently all of the quantum mechanics is in this phase.

Our strategy to evaluate Z in the form (2) is to replace the quantum mechanical phase $i\Phi$ by constraints. To find these constraints we discretize S[p] to

$$S[p_n] = \hbar^{-1}\int_0^{\beta\hbar} dt\ [i\sum_{n=1}^{N} P_n Q_{n,t} - H[p_n]] \ , \tag{10}$$

with $H[p_n] = \sum_{n=1}^{N} \omega(\tilde{k}_n)P_n$. We have introduced discretized modes $\tilde{k}_n$ and action-angle variables P_n, Q_n: $\{P_n, Q_n\} = \delta_{nm}$. For e.g. s-G or attractive NLS the classical models have soliton solutions and additional terms describing these[1,4,5,12,24] enter (9) and (10) as mentioned.

We can now *eliminate* the quantum mechanical phase $i\Phi[p_n] \equiv \hbar^{-1}\int_0^{\beta\hbar} dt\,[i\sum_{n=1}^{N} P_n Q_{n,t}]$ by setting $\hbar^{-1}\int_0^{\beta\hbar} dt\, P_n Q_{n,t} = 2\pi m_n$ where m_n is a non-negative *integer* for each label n. Since P_n is a constant, from $\oint P_n dQ_n = 2\pi P_n$

$$P_n = \hbar m_n > 0. \tag{11}$$

This is Bohr semiclassical quantisation! However it is also an exact set of constraints. So the deeper quantum structure is not prejudiced (see the phase shift equations (14) below). For the future we put $\hbar = 1$ so $h = 2\pi$: $P_n = m_n$ in (11).

In (11) m_n is *any* non-negative integer. There is a classification problem here which, at the time of going to press, is not yet solved (see §3). In Refs 1-12 we arbitrarily chose fermi statistics or bose statistics:-

$$\text{(fermions)} \qquad m_n = 0,1$$
$$\text{(bosons)} \qquad m_n = 0,1,2,\ldots\ , \tag{12}$$

and achieved consistency (and equivalence [2-5,7,11,12]). The constraints (11) replace the quantum mechanical phase Φ of (9) in (2). Note for the classical SM and Maxwell-Boltzmann particles there are no quantum mechanical constraints, there is no phase, and the P_m are the classical action variables. Thus Z is reduced to a classical form without constraints for the classical SM or to the same form with the quantum mechanical constraints for the quantum theories.

For sinh-G and repulsive NLS the measure $\mathcal{D}\mu$ in (2) is ($h = 2\pi$) the natural measure (see §2)

$$\mathcal{D}\mu = \prod_{n=1}^{N} dP_n dQ_n (2\pi)^{-1} \ . \tag{13}$$

However, because of the thermodynamic limit which is required, labels n are such that modes $\tilde{k}_n$ are admissible (labels n in (13) are admissible) if and only if they satisfy [2,4,6,7,11,12]

$$\tilde{k}_n = k_n - L^{-1}\sum_{m\neq n} \Delta\,(\tilde{k}_n, \tilde{k}_m)\; P_m \quad : \tag{14}$$

$\Delta(k,k')$ is a 2-body S-matrix phase shift for the model - there is Δ_f for fermions, Δ_b for bosons and Δ_c for classical Maxwell-Boltzmann particles (see §3). The constraints which are (14) we call "phase shift equations"[2,4,7,11,12,24]: if $P_m = 0,1$ (14) becomes the usual conditions for quantum BA (and QIM). The real generalisation here is therefore to bosons ($\Delta = \Delta_b$; $P_n = 0,1,2\ldots$) and classical M-B particles ($\Delta = \Delta_c$; P_n the classical action variable) - and to functional integral methods. It is the functional integral methods described here which illuminate the quantum constraints and which can (apparently) provide new physics (anyons, §3).

Since $P_m = O(1)$, in all three cases, the $L^{-1}\sum_{m\neq n}$ is O(1) as $L \to \infty$. The $k_n = 2\pi l_n L^{-1}$, l_n an integer labelled by n, and the $k_n \to k$, the $\tilde{k}_n \to \tilde{k}$ and both k, $\tilde{k}$ are dense as $L \to \infty$ in thermodynamic limit: the k and $\tilde{k}$ are then evidently the same only for free fields (for which the phase shifts Δ vanish). Notice now that $P_m \to P(\tilde{k})\, d\tilde{k}$ as $L \to \infty$ in which $d\tilde{k} = O(L^{-1})$: then $P(\tilde{k})$ is O(L) as $L \to \infty$ and $L^{-1} P(\tilde{k}) \to \rho(\tilde{k}) > 0$, a finite density of particles, as $L \to \infty$. Thus the action-angle variables P(k), Q(k) involved, though found on $-\infty < x < \infty$ with decaying b.c.s. at $\pm\infty$ (a zero density limit), are actually extended to describe the finite density limit. Similar considerations are involved in the soliton densities when there are classical soliton solutions.

If $\mathcal{D}\mu$ is given by (13) and the constraints (11) and (14) are imposed together the functional integral (2) with discretized action can be evaluated by iterating (14) through $H[p_n]$. This iterated expansion can be summed. The sums are (i.e. the series is the iterated series of)

$$\lim_{L\to\infty} FL^{-1} = \mu\bar{n} - (2\pi\beta)^{-1}\int_{-\infty}^{\infty} \ell n\,(1 + e^{-\beta\tilde{\varepsilon}(k)})\, dk \tag{15a}$$

$$\tilde{\varepsilon}(k) = \omega(k) - \mu - (2\pi\beta)^{-1}\int_{-\infty}^{\infty} [d\Delta_f(k,k')/dk]\ell n(1 + e^{-\beta\tilde{\varepsilon}(k')})dk' \tag{15b}$$

for fermions (in which μ is a chemical potential),

$$\lim FL^{-1} = \mu\bar{n} + (2\pi\beta)^{-1}\int_{-\infty}^{\infty} \ell n(1 - e^{-\beta\varepsilon(k)})\, dk \tag{16a}$$

$$\varepsilon(k) = \omega(k) - \mu + (2\pi\beta)^{-1}\int_{-\infty}^{\infty} [d\Delta_b(k,k')/dk]\, \ell n(1 - e^{-\beta\varepsilon(k')})dk' \tag{16b}$$

for bosons, and (with $\mu = 0$)

$$\ell n\, FL^{-1} = (2\pi\beta)^{-1}\int_{-\infty}^{\infty} \ell n\, \beta\varepsilon(k)\, dk \tag{17a}$$

$$\varepsilon(k) = \omega(k) + (2\pi\beta)^{-1}\int_{-\infty}^{\infty} [d\Delta_c(k,k')/dk]\, \ell n\, \beta\varepsilon(k')\, dk' \tag{17b}$$

for classical F and classical M-B particles [2,4,6,7,11,12].

For s-G (attractive NLS) both the quantum and classical results are more complicated because of the soliton contributions: (15a)-(17b) are evidently purely 'phonon' contributions.

Results (15) are exactly those found in Ref 26 by quantum BA for the bose gas (the bose gas is the quantum repulsive NLS model). We find all of (15a) - (17b) by our generalised BA also. We have now reported similar results, found by our two methods, functional integration and generalised BA, for some 11 different models [12,24]: the structure, 'oscillator' contributions to F with β-dependent excitation energies $\varepsilon(k)$ determined by (coupled in general) nonlinear integral equations containing phase shifts Δ, is typical of all of these results for other models [12,24].

In the §2 we give a derivation of the expressions (3) with (4) for $Z = \mathrm{Tr}\, e^{-\beta\hat{H}}$, $\hat{H}$ the quantum Hamiltonian operator: this argument enables us to focus on the usefulness of this form and, in its phase $i\Phi$, the apparent new physics. This new physics is developed to the point which we have reached, very recently, in the anyon theory given in §3. The §4 treats the problem of the quantum and classical thermodynamic limits for models which like s-G have particles with two degrees of freedom. A form of quantum chaos is apparently involved and it is this which motivates the first part of our title (§5).

2. THE FUNCTIONAL INTEGRAL Z

We first consider one degree of freedom and use the formalism for the corresponding quantum propagator in the text books already [27,28]. Take the simple operator $\hat{H}(p,q) = (\hat{p})^2(2m)^{-1} + V(\hat{q})$ acting on the Hilbert space $\mathcal{H}$ spanned by $|q\rangle$, $q \in \mathbb{R}$: $\hat{q}|q\rangle = q|q\rangle$; $\int dq\ |q\rangle\langle q| = \hat{I}$ (unit operator). $\mathcal{H}$ is also spanned by $|p\rangle$, $p \in \mathbb{R}$: $\hat{p}|p\rangle = p|p\rangle$, etc. Evidently $\langle q'|q\rangle = \delta(q'-q)$ $\langle p'|p\rangle = \delta(p'-p)$ (we can choose $|q\rangle$'s ($|p\rangle$'s) orthogonal with this normalisation. Since $\langle q|\hat{p}|p\rangle = p\langle q|p\rangle = -i\partial/\partial q\langle q|p\rangle$ (in coordinate q-representation), $\langle q|p\rangle = e^{ipq}(2\pi)^{-\frac{1}{2}}$. Also from $\hat{H}(p,q)$ in the form chosen $\langle p|\hat{H}|q\rangle = h(p,q)(2\pi)^{-\frac{1}{2}}e^{-ipq}$ where $h(p,q) \equiv p^2(2m)^{-1} + V(q)$. That this is the classical Hamiltonian *function* becomes plain from the analysis for the corresponding quantum propagator (see below) — the eigenvalues p,q coordinatise a symplectic manifold.

We have demonstrated so far that

$$\langle p|e^{-\Delta t\hat{H}}|q\rangle = e^{-\Delta t\ h(p,q)}(2\pi)^{-\frac{1}{2}}e^{-ipq} \quad . \tag{18}$$

Evidently

$$\begin{aligned} Z \equiv \mathrm{Tr}\ e^{-\beta\hat{H}} &= \int dq_M \langle q_M|e^{-\beta\hat{H}}||q_M\rangle \\ &= \int dq_M \int dp_M\langle q_M|p_M\rangle \int dq_{M-1}\langle p_M| \\ &\quad e^{-\beta\hat{H}M^{-1}}|q_{M-1}\rangle \int dp_{M-1}\langle q_{M-1}|p_{M-1}\rangle\langle p_{M-1}| \\ &\quad e^{-\beta\hat{H}M^{-1}}|q_{M-2}\rangle \dots \dots\langle p_1|e^{-\beta\hat{H}M^{-1}}|q_0\rangle \quad . \end{aligned} \tag{19}$$

For the trace we must impose periodicity so that $|q_0\rangle = |q_M\rangle$. There are M p_j's ($j = 1,\dots,M$) and M-1 q_j's ($j = 1,\dots,M-1$) for 'fixed end points' $q = q_0 = q_M$. The Tr traces over $q_0 = q_M$. Thus

$$Z = \lim_{M\to\infty} \int dq_M \int \prod_{j=1}^{M} dp_j \prod_{k=1}^{M-1} dq_k\ (2\pi)^{-M} \exp S[q] \tag{20}$$

where

$$\begin{aligned} S[q] = i\ \{&p_M(q_M - q_{M-1}) + \dots + p_1(q_1 - q_M) \\ &- \beta M^{-1}h(p_M,q_{M-1}) \dots -\beta M^{-1}h(p_1,q_M)\} \quad . \end{aligned} \tag{21}$$

Then as $M \to \infty$ ($\hbar = 1$)

$$S[q] = \int_0^\beta dt\left[ip(t)q_t(t) - h(p,q)\right] \tag{22}$$

and, for one degree of freedom with $h(p,q) = p^2(2m)^{-1} + V(q)$,

$$Z = \mathrm{Tr} \int \mathcal{D}p\mathcal{D}q \exp S[q]. \tag{23}$$

This form extends to a field theory $\phi(x,t)$ with an infinite number of degrees of freedom in the following way: the Tr means periodicity in "time" t, $0 \le t < \beta$ and the thermodynamic limit is reached by imposing periodicity in x, $-\frac{1}{2}L \le x < \frac{1}{2}L$. For $L < \infty$, $S[\phi]$ is defined on the space-time 'torus' $T(x,t) = \{0 \le t < \beta, -\frac{1}{2}L \le x < \frac{1}{2}L\}$. The 'paths' in the functional integral will be all paths parametrised by (x,t) in both $\Pi(x,t)$ and $\phi(x,t)$ just as (23) involves all paths in p and q parametrised by t alone in $0 \le t < \beta$. Discretize T(x,t): $0 \le t < \beta \to 0, t_1, t_2, \ldots, t_{N_0-1}, \beta$; $-\frac{1}{2}L \le x < \frac{1}{2}L \to -\frac{1}{2}L, x_{-N+1}, x_{-N+2}, \ldots, x_{N-1}, \frac{1}{2}L$ and there are $N_0 + 1$ points in t and $2N + 1$ points in x. Form

$$Z = (2\pi)^{-(2N+N_0)} \prod_{i=-N}^{N-1} \int_{-\infty}^{\infty} d\phi(x_i,0) \prod_{j=1}^{N_0-1} \int_{-\infty}^{\infty} d\phi(x_i,t_j) \int_{-\infty}^{\infty} d\Pi(x_i,t_j) \exp \sum_i \sum_j S_{ij} .$$
$$\phi(x_i,\beta) = \phi(x_i,0) \tag{24}$$

As the discretization tends to zero Z tends to (3) with (4). Notice that the Tr provides the product $\prod_{i=-N}^{N-1} \int_{-\infty}^{\infty} d\phi(x_i,0)$ part. The action $S = \sum_{i,j} S_{ij}$ and

$$S_{ij} = [i\Delta x_i \Pi(x_i,t_j)(\phi(x_i,t_j) - \phi(x_i,t_{j-1}) - \beta\Delta x_i \Delta t_j H[\Pi(x_i,t_j),\phi(x_{i-1},t_{j-1})] \tag{25}$$

with $\sum_{i=-N}^{N-1} \Delta x_i = L$, $\sum_{j=1}^{N_0-1} \beta\Delta t_j = \beta$. When the contribution to the quantum mechanical phase is eliminated (§1) there is no evolution in time t (H does not depend on t) and the paths for the functional integral for Z are parametrised by x alone namely paths for ϕ in x between $\phi = \phi(x,0)$ and $\phi = \phi(x,\beta) = \phi(x,0)$. Thus the elimination of the quantum mechanical phase, by replacing it by constraints, greatly simplifies the functional integral.

The measure $\mathcal{D}p\mathcal{D}q$ in (23) is the 'natural' measure $\mathcal{D}p\mathcal{D}q = \lim_{M\to\infty} \prod_{i=1}^{M} dp_j \prod_{k=1}^{M-1} dq_k (2\pi)^{-M}$ in which 2π is h. The measure in (23), and therefore in (3), is thus the natural measure which, after canonical transformation becomes the measure (13). This measure is formally the same in both the classical and quantum cases (for the sinh-G and repulsive NLS models) but the constraints (14) determine the acceptable labels n as explained (§1). For e.g. the s-G model which has solitons the measure is more complicated [1]. Moreover it is not the same in the quantum and classical cases (§4).

The functional integral for Z in its limiting form (3) is formally the Wick rotation of a quantum propagator taking $\phi(x,0)$ to $\phi(x,T)$. The classical action is

$$S = \frac{i}{\hbar} \int_0^T dt \, [\int \Pi\phi_t dx - H[\phi]] \quad . \tag{27}$$

Then $\delta S = 0$ is the classical Hamilton's equations of motion. Then we can define the usual bracket so that the manifold M coordinatised by Π,ϕ is a symplectic manifold (a phase space). The same argument applies to the *one* degree of freedom result (23) for which M is two dimensional. Notice that the action S, eq. (27) has no simple phase containing all of the $\hbar$ dependence. Thus the quantum mechanical phase Φ, §1, is a feature of the quantum partition function Z. In §3 next we derive (only formal at present) anyon theories from it.

Notice that the Wick rotation $t \to -it$ in (27) and $T = -i\beta$ regains formally the action (4). But $\Pi(x,t)$, $\phi(x,t)$ are *not* parametrised by $t \to -it$: Π,ϕ *are* the canonical variables parametrised by real time t. This is plain from the argument to (22) where h(p,q) is the usual classical Hamiltonian. The extension of the argument to (4) then shows $H[\phi]$ is the usual classical Hamiltonian for the field ϕ.

Notice finally that after canonical transformation to the form (2) and application of the constraints (14) with (11) , the constraints (11) in particular eliminate the quantum mechanical phase. Consequently the parameter t ('time') is eliminated from (2). This shows again how removal of the quantum mechanical phase as the constraints (11) simplifies both the expression of the functional integral *and* its evaluation.

3. ANYON THEORY

First of all we analyse bose-fermi equivalence for the sinh-G and repulsive NLS models.

The fermion description (15a,b) proves to be equivalent to the boson description (16a,b) in the following way. The 2-body S-matrix phase shifts are found to be [2,4]

$$\text{(repulsive NLS)} \qquad \Delta_f(k,k') = -2\tan^{-1}[c(k-k')^{-1}] \tag{28a}$$

$$\text{(sinh-G)} \qquad \Delta_f(k,k') = -2\tan^{-1}\left[\frac{m^2\sin(\frac{1}{8}\gamma_0'')}{k\omega(k')-k'\omega(k)}\right] . \tag{28b}$$

In both cases the *smooth* branch of the $\tan^{-1}$ functions $-2\pi < \Delta_f < 0$ is taken. This is consistent with fermion description exactly as it is in the quantum BA analysis [19]. However $\Delta_f(k=k') = -\pi$ in both cases and $e^{i\Delta_f} = e^{-i\pi} = -1$ corresponding to exchange of 2-fermions as required for fermi statistics.

The bose phase shift Δ_b is now defined as a 'true' phase shift such that $\Delta_b \to 0$ as $k \to \pm\infty$. Set $\Delta_b(k,k') = \Delta_f(k,k') + 2\pi\theta(k'-k)$ with θ the unit step. Then Δ_b has a -2π jump at $k = k'$ and $\Delta_b(k=k') = 0$ — so $e^{i\Delta_b} = +1$ corresponding to bose statistics. If in (15a,b) we define new bose excitation energies $\varepsilon(k)$ through

$$\ln(1 + e^{-\beta\tilde{\varepsilon}(k)}) = -\ln(1 - e^{-\beta\varepsilon(k)}) , \tag{29}$$

and we replace Δ_f by Δ_b, as just defined, in (15b), then the δ-function arising from $d\Delta_f/dk$ transfers terms to the left side of (15b) which are

$$\tilde{\varepsilon}(k) + \frac{1}{\beta}\ln(1 + e^{-\beta\tilde{\varepsilon}(k)}) = \varepsilon(k), \tag{30}$$

exactly equal to $\varepsilon(k)$ as stated by using (29). The right sides of (15a,b) are then exactly the right sides of (16a,b). Thus (15a,b) is *identically equivalent* to (16a,b) with these two phase shifts Δ_f, Δ_b.

We now investigate possible anyon theories. The quantisation conditions (11) are evidently not unique. Indeed even if m_n is a non-negative integer it is not necessarily the same integer for each label n: nor is it necessarily the fermion numbers or boson numbers of (12). There seems to be an arbitrary large number of alternative choices and corresponding quantum statistics. We shall see shortly by one example that if bose-fermi equivalence is imposed choices can become very restricted. We are still exploring the *classification* of admissible statistics. Natural statistics in *two* space dimensions or less are braid group statistics [16,29].

Introduce a winding number ν_n : then the analysis leading to (11) is $\int_0^\beta P_n Q_{m,t}\, dt = \oint P_n\, dQ_n = 2\pi m_n$ in which $\nu_n = 1$, namely one wind of 2π round the classical torus defined by Q_n $(h = 2\pi)$: $P_n = m_n$ for $\nu_n = 1$. Notice that we require periodic b.c.s. in t period β: thus $\beta\hbar \to 0$ (classical limit) requires $\nu_n = 0$ in the limit (zero times round the torus). For $\nu_n > 0$, $\oint P_n dQ_n = 2\pi m_n$ and

$$P_n = m_n \nu_n^{-1} \tag{31}$$

is the quantisation condition. The ν_n define homotopy classes m_ν in which all modes n have $\nu_n = \nu$. We can expect to develop a braid statistics for the m_ν [16].

We develop a *formalism* (which is all we have at present) for anyon models derived from the integrable models with classical $H[p] = \int_{-\infty}^{\infty} \omega(k)P(k)\, dk$ including the sinh-G and repulsive NLS models. We assume the constraints (14) are generic (they seem to be). Then the new quantisation conditions (31) mean

$$\tilde{k}_n = k_n - L^{-1} \sum_{\ell \neq n}^{N} \Delta_{ST}(\tilde{k}_n, \tilde{k}_\ell) m_\ell \nu_\ell^{-1} \quad . \tag{32}$$

The phase $\Delta_{ST}(k,k')$ is not yet known. However the phase shift $\Delta_S \equiv \Delta_{ST}\nu_n^{-1}$ is an anyon type phase shift in the sense of Zee [16]. We use Δ_{ST} to mean the actual anyon phase shift: we can argue at this stage whether $\Delta_{ST}(k=k')$ for true fermion anyons has the value $-\pi$ (compare Δ_f for fermions): certainly this then becomes $-\pi\nu_n^{-1}$ for the apparent anyon phase shift Δ_S and $e^{-i\Delta_S} = e^{-i\pi\nu_n^{-1}} = q_n$, a $2\nu_n$-th root of unity. For fermion semions [16] $e^{i\Delta_S} = e^{-i\pi/2} = -i$.

Next, for the type of model considered, which is still an integrable model because the classical P_m are still a sufficient number of constants, the action is

$$-\beta H[p_n] = -\beta \sum_{n=1}^{N} \omega(\tilde{k}_n)\nu_n^{-1} m_n \tag{33}$$

so that if we put, for simplest case, all $\nu_n = \nu$, and define $\beta' = \beta\nu^{-1}$, the functional integral we have to evaluate is exactly that treated in §1. With $\Delta_f(k,k') = \nu^{-1}\Delta_{ST}(k,k')$ for all k,k' (all labels n and ℓ) we evidently reach

$$\lim_{L\to\infty} FL^{-1} = \mu\bar{n} - (2\pi\beta)^{-1}\int_{-\infty}^{\infty} \ln(1 + e^{-\beta\nu^{-1}\tilde{\varepsilon}(k)})\, dk \tag{34}$$

$$\tilde{\varepsilon}(k) = \omega(k) - \mu - (2\pi\beta)^{-1}\int_{-\infty}^{\infty} [d\Delta_{ST}(k,k')/dk]\, \ln(1 + e^{-\beta\nu^{-1}\tilde{\varepsilon}(k')})dk'. \tag{35}$$

We have used $\Delta_S(\beta')^{-1} = \Delta_{ST}\beta^{-1}$ and $\beta' \equiv \beta\nu^{-1}$ to write down the result in this form. This result is a formal expression for the quantum free energy density of the anyon fluid. Its formal ground state energy can be calculated from this expression. Of interest to superfluidity is whether there is a regime with an effective *linear* dispersion relation [16]. For repulsive NLS type models $\omega(k) = k^2$ so the interest is whether the interaction described by the integral in $d\Delta_{ST}/dk$ can have such consequences. We are not in a position to say anything on that aspect yet. Compare nevertheless the bose gas (quantum repulsive NLS model) for which (15a,b) and (16a,b) both show [4] a free bose limit ($c \to 0$) *and* a free fermion limit ($c \to \infty$): $\omega(k) = k^2$ in both limits of course.

We can impose a fermi-bose equivalence on this fermion anyon theory: suppose Δ_{ST} for fermion anyons and boson anyons are related as were Δ_f and Δ_b but now with a jump $2\pi\lambda$ at $k = k'$ so that

$$\Delta_b = \Delta_f + 2\pi\lambda\theta(k'-k) \tag{36}$$

(in obvious notation for the Δ_{ST}). It seems indeed that it is only through a jump at $k = k'$ that Δ_{ST} for fermion anyons and boson anyons could be different from each other.

The condition for equivalence which is (30) becomes

$$\tilde{\varepsilon}(k) + \frac{\lambda}{\beta}\ln(1 + e^{-\beta\nu^{-1}\tilde{\varepsilon}(k)}) = \varepsilon(k) \ . \tag{37}$$

Put $x = e^{-\beta\nu^{-1}\tilde{\varepsilon}(k)}$, $y = e^{-\beta\nu^{-1}\varepsilon(k)}$. Then the condition corresponding to (29) means $y = x(1 + x)^{-1}$ and that for equivalence (37) then means $x^{\nu}(1 + x)^{-\lambda} = x^{\nu}(1 + x)^{-\nu}$; and this is possible for all x if and only if

$$\lambda = \nu > 0. \tag{38}$$

Thus there are (formal) bose-fermi equivalent anyon theories i.e. fermion anyon - boson anyon equivalent theories. This particular family just derived has $\nu = \nu_n$ for all labels n while the jump $2\pi\lambda = 2\pi\nu$ at $k = k'$. This combination of branches (if Δ_f for fermion anyons lies smoothly in $-2\pi < \Delta_f < 0$) is plainly exotic - unless $\nu = 1$.

If $\nu = 1$, $\lambda = 1$ and the bose-fermi equivalent theory is that for true bosons and true fermions resulting in (15a)-(16b). We might wish for bose-fermi-anyon equivalence. Anyon theories typically require spinless particles and breakdown of parity and time reversal invariance [16,18,29]. The repulsive NLS and sinh-G models are parity and time reversal invariant, and we find the true bose theory (ν_n = 1 all n) is equivalent to a true fermion theory (ν_n = 1 all n) only, and there is no anyon theory.

Evidently this approach to anyon theory and fractional statistics is little more than a collection of suggestions at this stage. It is not clear yet what quantisation conditions like (31) are actually possible and this is being investigated further. It is important too that we would prefer to be in *two* space dimensions (2+1 dimensions). Here similar considerations apply (there *is* a quantum mechanical phase Φ). Unfortunately no quantum integrable models in (2+1)-dimensions are available yet in which to develop comparable theory. No R-matrix theory ([4,11,12,20,21] and Fig. 1) is available yet in (2+1)-dimensions.

We complete this section on anyon theory by listing other possibilities for the quantisation conditions: m_n *any* non-negative integer and different in general for each label n is not eliminated so far; different winding numbers ν_n are not eliminated either. Moreover for any ν_n there are other possibilities: if $\nu_n = \nu_n'$ for labels n,n', then we could choose semion numbers $m_n = 0,\frac{1}{2}$; $m_n' = 0,\frac{1}{2}$ *together* (mode-mode *pairing*). We could choose $m_n = 0,\frac{1}{3}$, $m_n' = 0,\frac{2}{3}$ *together*. Moreover with $m_n = m_n' = m_n''$ etc. we could choose e.g. any integer ℓ and the numbers $\ell_1,\ell_2,\ldots$ relatively prime to ℓ and set $m_n = \ell_1\ell^{-1}$, $m_n' = \ell_2\ell^{-1}$, $m_n'' = \ell_3\ell^{-1}$, etc. together. Such mode-mode correlation (including mode pairing) seems to introduce new physics natural in the context of superfluidity.

4 QUANTUM AND CLASSICAL THERMODYNAMICS IN THE STATISTICAL MECHANICAL THEORY

The essential point made in this section is that when the classical integrable model has breather-like solutions the quantum and classical thermodynamic limits are *intrinsically* different. Examples are the s-G models, the Landau-Lifshitz models and spin-$\frac{1}{2}$ XYZ models (classical L-L' is the classical and quantum limit of quantum spin-$\frac{1}{2}$ XYZ), the attractive NLS models, the Heisenberg ferromagnets, the ferromagnets in a longitudinal field, the classical MTM model with commutative bracket [12,24]. The Toda lattice does not have breather-like solitons and the theory developed now does not apply (nor has it to) [8].

The quantum and classical s-G models are generic:- The classical s-G has breather solutions with two degrees of freedom. Nevertheless the discretized classical Hamiltonian density

$$H[p_n]L^{-1} = 2L^{-1}\sum_{n=1}^{N_s}(M^2 + p_n^2)^{\frac{1}{2}} + L^{-1}\sum_{n=-\frac{1}{2}N_{ph}}^{+\frac{1}{2}N_{ph}} \omega(\tilde{k}_n)P_n \qquad (39)$$

with $2N_s + N_{ph} = N$ the total number of degree of freedom on the period L, M the kink mass, yields [1,12] a low temperature asymptotic expansion for the classical free energy density lim FL^{-1} with $L \to \infty$ in thermodynamic limit which agrees *term by term* with that we find by the transfer integral method (TIM [1,3,4,12]). In (39) there are kink with antikink contributions and phonon contributions only: there are no classical breather contributions.

On the other hand [4,12,14,24] we show that quantum s-G has a quantum FL^{-1} described by N_b-2 quantum breather contributions and one kink-antikink pair contribution. The same results have been found for the fermi-bose equivalent massive Thirring model (MTM) by quantum BA [3,4,30,31] and from the quantum spin-$\frac{1}{2}$ XYZ model [32,33] (see the Fig.1). The number $N_b = [8\pi\gamma_0^{-1}]$ where [...] is integral part, and $\gamma_0 > 0$ is the bare coupling constant of quantum s-G (compare classical γ_0 in (7)). This renormalises to $\gamma_0'' = \gamma_0[1-\gamma_0/8\pi]^{-1}$ and $N_b = [8\pi\gamma_0''^{-1}] + 1$. For the MTM $\mu = \pi(1-\gamma_0/8\pi)$ and the coupling constant is g = $-\cot\mu$: Ref. 32 derives FL^{-1} and N_b-1 coupled integral equations for its excitation energies where $8\pi\gamma_0^{-1}$ = integer = N_b only: Ref. 33 does this for $8\pi\gamma_0^{-1}$ = integer + ε (ε infinitessimal > 0). Our methods [12,24] yield the same results but for all $0 \le \gamma_0 < 8\pi$ (see (44) below). For the functional integration for quantum Z the classical H is made up of breather contributions and kink-antikink pair contributions. The classical phase space is co-ordinatised in action-angle variables on the real line with vanishing b.c.s. at $\pm\infty$ by [5,28] kink-antikink co-ordinates $-\infty < p_\ell, q_\ell, \bar{p}_\ell, \bar{q}_\ell < +\infty$ with $\{p_\ell, q_m\} = \{\bar{p}_\ell, \bar{q}_m\} = \delta_{\ell m}$, together with classical breather co-ordinates $\hat{p}_\ell, \hat{q}_\ell$ ($-\infty < \hat{p}_\ell, \hat{q}_\ell < +\infty$; $\{\hat{p}_\ell, \hat{q}_\ell\} = \delta_{\ell m}$) describing translations and internal co-ordinates $4\gamma_0''^{-1}\Theta_\ell$, Φ_m ($0 \le \Theta_\ell < \frac{\pi}{2}$, $0 \le \Phi_m < 8\pi$; $\{4\gamma_0''^{-1}\Theta_\ell, \Phi_m\} = \delta_{\ell m}$ with $4\gamma_0''^{-1}\Theta_\ell$ the action). The contribution to the quantum mechanical phase angle Φ (Φ_ℓ is the breather's angle variable not the phase angle Φ!) is evidently $\sum_{\ell=1}^{N_k} p_\ell q_{\ell,t} + \sum_{\ell=1}^{N_{\bar{k}}} \bar{p}_\ell \bar{q}_{\ell,t}$ from $N_k = N_{\bar{k}} = N_s$ kinks and antikinks. The analogue of the quantisation conditions leading to (11) is $\int_0^\beta p_\ell q_{\ell,t} dt = \oint p_\ell dq_\ell = 2\pi m_\ell$ with m_ℓ any integer, and since p_ℓ = constant, $\oint p_\ell dq_\ell = p_\ell L$ (with $L \to \infty$) and

$$p_\ell = 2\pi m_\ell L^{-1} \quad . \tag{40}$$

The p_ℓ become dense as $L \to \infty$ in the usual way. Thus p_ℓ (and $\bar{p}_\ell$ likewise) gets the usual quantisation of translational momentum. Notice that in thermodynamic limit through periodic b.c.s. (40) also applies in the classical cases.

For the quantum breathers the translational momentum $\hat{p}_\ell$ is packed in the same way as (40) and this removes a contribution by breathers to the quantum mechanical phase. But internal breather co-ordinates contribute $\sum_\ell 4\gamma_0''^{-1}\Theta_\ell\Phi_{\ell,t}$, and the quantum constraints which remove this are

$$\int_0^\beta dt\ 4\gamma_0''^{-1}\Theta_\ell\Phi_{\ell,t} = 2\pi m_\ell \quad ; \tag{41}$$

m_ℓ is now a non-negative integer. For homotopy unity (single turn round the torus $0 \le \Phi < 8\pi$), $4\gamma_0''^{-1}\Theta_\ell \times 8\pi = 2\pi m_\ell$ and [12,24,25]

$$\Theta_\ell = m_\ell \gamma_0''/16 \quad . \tag{42}$$

For homotopy ν_ℓ

$$\Theta_\ell = m_\ell \gamma_0''/16\nu_\ell \quad . \tag{43}$$

Since the mass spectrum for breathers is $2M\sin\Theta_\ell$ (where M is the renormalised kink mass) (43) is a new mass spectrum depending on the homotopy. Evidently the quantisation constraints (31) describe new quantum models when $\nu_n \neq 1$ also. We shall not pursue this aspect of anyon theory further here (eqn. (43) defines new quantum soliton anyons) for we are concerned in this §4 only with thermodynamic limits. We analyse these only for the 'normal' models like 'normal' s-G.

If $\nu_\ell = 1$, Θ_ℓ is given by (42). Then the largest m_ℓ is $[\frac{1}{2}\pi \times 16\gamma_0''^{-1}] = N_b - 1$. There are indeed N_b-2 quantum breathers and the one 'breather' which is the kink-antikink pair. These quantum breathers are evidently *discrete* in the quantum thermodynamic limit $L \to \infty$: they take the discrete values (42) spaced by $\frac{1}{16}\gamma_0''$. In this limit the free energy proves to be [12,24]

$$\lim_{L\to\infty} FL^{-1} = \mu_s \bar{n}_s + \sum_{\ell=1}^{N_b-2} \mu_\ell \bar{n}_\ell - (2\pi\beta)^{-1} \sum_{\ell=1}^{N_b-2} \int d\alpha_\ell M_\ell \cosh\alpha_\ell \times$$
$$\times \ln(1 + e^{-\beta\varepsilon_\ell}) - 2(2\pi\beta)^{-1} \int_{-\infty}^{\infty} d\alpha_s M_s \cosh\alpha_s \ln(1 + e^{-\beta\varepsilon_s}) \tag{44}$$

in which $M_\ell = 2M\sin(\ell\gamma_0''/16)$, $M_s = M$, $\varepsilon_\ell \equiv \varepsilon_\ell(\alpha_\ell)$, $\varepsilon_s \equiv \varepsilon_s(\alpha_s)$; μ_ℓ, μ_s are chemical potentials for each particle and there is a renormalised field mass m $\equiv \frac{1}{8}\gamma_0''M$. Note that each of these particles is a *fermion*. The ε_ℓ and ε_s are determined from N_b-1 coupled integral equations generalising (15b) in fermion description.

In the *semi*-classical limit in which γ_0 and $m \to 0$ at fixed M [3], $N_b \to \infty$ and the N_b-2 quantum breathers become one collective *boson*. The mass spectrum with spacing m becomes dense. The bosons can now lower their energies by falling into the ground state as $L \to \infty$ e.g. as $O(L^{-1})$. In this description N_b-2 discrete quantum breathers for which the internal action variable Θ is given by (42) become dense with $\Theta_\ell = O(L^{-1})$ in classical limit. The actual free energy (with all $\mu_s, \mu_\ell = 0$) becomes [3]

$$\lim_{L\to\infty} FL^{-1} = (2\pi\beta)^{-1} \int_{-\infty}^{\infty} \omega(x)(1 - e^{-\beta\varepsilon(x)})dx - 2(2\pi\beta)^{-1} \int E(x)(1 + e^{-\beta\tilde{\varepsilon}(x)})dx \tag{45}$$

using rapidities $\omega(x) = m\cosh x$, $E(x) = M\cosh x$ ($m > 0$, is now the classical field mass). The excitation energies for *phonons* $\varepsilon(x)$ and for kink-antikink pairs $\tilde{\varepsilon}(x)$ are determined by *two* coupled integral equations [3]. The classical limit replaces $\ln(1 - e^{-\beta\varepsilon(x)})$ by $\ln\beta\varepsilon(x)$ and $\ln(1 + e^{-\beta\tilde{\varepsilon}(x)})$ by $e^{-\beta\tilde{\varepsilon}(x)}$. Iteration of the classical coupled integral equations for $\varepsilon(x)$ and $\tilde{\varepsilon}(x)$ then yields the low temperature asymptotic expansion for lim FL^{-1} in agreement with the TIM result [1,12].

In an *ab initio* classical calculation such as in Refs. 1,34 (Ref. 34 shows that the Heisenberg magnet in a longitudinal field is another generic example) one must therefore distinguish the classical from the quantum thermodynamic limits. Quantum breathers which have compact phase spaces for internal variables Θ,Φ must be discrete as $L \to \infty$: this is the result (42). But classical breathers become dense as $L \to \infty$, and for these it becomes necessary to *define* a classical thermodynamic limit for particles which have two degrees of freedom. We do this as follows: we pack translational momenta $\hat{p}_n$ (under periodic b.c.s.) as in (40) before. To each label n there are internal actions $\Theta_{n\ell}$, $\ell=1,\dots,N_b-2$ in the quantum case; $N_b\to\infty$ in the classical case. For this we make the $\Theta_{n\ell}$ for each ℓ dense in the same way as the classical $\hat{p}_n$ become dense i.e. neighbouring $\Theta_{n\ell}$'s are packed with spacings in ℓ of order $O(L^{-1})$. We then find the formal breather Hamiltonian density

$$\lim_{L\to\infty} H_{\text{breather}} L^{-1} = \lim_{L\to\infty}(L)\int_{-\infty}^{\infty}\int_{0}^{\frac{\pi}{2}}\rho_b(\hat{p},\Theta)[4M^2\sin^2\Theta + \hat{p}^2]^{\frac{1}{2}}d\hat{p}d\Theta \tag{46}$$

in which $\rho_b(\hat{p},\Theta)$ is a finite density classical breather particle density, and this contribution to classical H diverges as O(L): the corresponding phase shift equations are also inconsistent as $L \to \infty$.

The situation can be retrieved by restricting the $\Theta_{n\ell}$ to an angle $O(L^{-1})$ densely as the single collective boson described above suggested. Since for small $\Theta_{n\ell}$

$$H_{\text{breather}} \sim \sum_n 2M(1 - V_n^2)^{-\frac{1}{2}}\sum_{\ell}\Theta_{n\ell}\ , \tag{47}$$

let

$$16\gamma_0^{-1}\sum_{\ell}\Theta_{n\ell} \equiv \bar{P}_n \quad (= O(1))\ ,\ \tfrac{1}{4}L^{-1}\sum_{\ell'}\Phi_{n'\ell'} \equiv \bar{Q}_{n'} \quad (0 \le \bar{Q}_{n'} < 2\pi). \tag{48}$$

The Poisson brackets mean that

$$\{\bar{P}_n,\bar{Q}_{n'}\} = (L^{-1}\sum_{\ell} 1)\delta_{nn'} = \delta_{nn'}. \tag{49}$$

Thus the densely packed breathers act as (large amplitude) phonons!

On the other hand reference to a classical integrable lattice theory for an s-G lattice under periodic b.c.s. show that [2,12,35] the action variables P_n under periodic b.c.s. are $O(L^{-1})$ as $L \to \infty$. There are therefore no (small amplitude) phonon variables P_n, (with Q_n) in classical limit. Then these are replaced by the (large amplitude) phonon variables $\bar{P}_n$ (with $\bar{Q}_n$) : then $\bar{P}_n \to P(\tilde{k})d\tilde{k}$ as $L \to \infty$ and $P(\tilde{k})$ is O(L) and the finite phonon density $\rho(\tilde{k}) = P(\tilde{k})L^{-1}$ and this phonon contribution finally appears in the classical (semiclassical) free energy (45).

This situation is generic for classical s-G, classical MTM, classical L-L', classical attractive NLS, and the Heisenberg magnet. The TIM results for the s-G [1,12] and the Heisenberg model [34] in particular show that the analysis described in this §4 apparently completely solves the problem (see some references in Ref. 1) of classical breathers taken in thermodynamic limit.

5. QUANTUM SOLITONS ON QUANTUM CHAOS

The classical phonons are conceptually coupled to a heat bath in thermal equilibrium: we can argue that the heat bath provides large dimensional Hamiltonian chaos (believed to be ergodic). On the other hand the quantum 'phonons' are evidently the N_b-2 discrete quantum breathers. These are localised in k-space as described by (42). And (as described by quantum BA for the equivalent MTM [30,31]) their wave functions are 'strings' [3,19] which are localised. The quantum breathers become the phonons (supposed classical chaotic) in classical limit. In this picture the quantum breathers model a *quantum chaos*. This was the point of our title! Further work is needed to make this picture complete.

REFERENCES

1. J. Timonen, M. Stirland, D.J. Pilling, Yi Cheng and R.K. Bullough, Phys. Rev. Lett. 56:2233 (1986).
2. R.K. Bullough, D.J. Pilling and J. Timonen, J. Phys. A: Math. Gen. 19:L955 (1986).
3. J. Timonen, R.K. Bullough and D.J. Pilling, Phys. Rev. B34:6525 (1986).
4. R.K. Bullough, D.J. Pilling and J. Timonen, 'Soliton statistical mechanics' in: "Solitons", M. Lakshmanan ed. Springer-Verlag, Heidelberg (1988), pp. 250-281, and references.
5. J. Timonen, Yu-zhong Chen and R.K. Bullough, Nucl. Phys. B (Proc. Suppl.) 5A:58 (1988).
6. R.K. Bullough, Yu-zhong Chen, S. Olafsson and J. Timonen, 'Statistical Mechanics of the NLS Models and their Avatars' in: "Integrable Systems and Applications" Springer Lecture Notes in Physics 342, M. Balabane, P. Lochak, and C. Sulem eds., Springer-Verlag, Heidelberg (1989), pp. 12-26.
7. R.K. Bullough, D.J. Pilling and J. Timonen, 'Soliton statistical mechanics and the thermalisation of biological solitons', in: "Nonlinear Coherent Structures in Physics" J. Pouget, M. Remoissenet and R. Ribotta eds., Journal de Physique Colloque C3, suppl. au n° 3, Tome 50 (mars 1989) C3-41-C3-51.
8. R.K. Bullough, J. Timonen, Yu-zhong Chen, Yi Cheng and M. Stirland, 'Quantum and classical statistical mechanics of the integrable models', in: "Nonlinear evolution equations: integrablility and spectral methods" A. Degasperis, A.P. Fordy and M. Lakshmanan eds., Manchester University Press, Manchester (1990). pp. 605-617.
9. R.K. Bullough and S. Olafsson, 'Complete integrability of the integrable models: quick review', in: IXth Intl. Congress on Math. Phys. 17-27 July, 1988, Swansea, Wales, B. Simon, A. Truman and I.M. Davies eds., Adam Hilger, Bristol (1989), pp. 329-34.
10. R.K. Bullough and S. Olafsson, 'Algebra of Riemann-Hilbert Problems and the Integrable Models - A Sketch', in: "Proc. 17 Intl. Conf. on Diff. Geom. Methods in Theor. Phys.", Allan I. Solomon ed., World Scientific, Singapore (1989), pp. 295-309.
11. R.K. Bullough, S. Olafsson, Yu-zhong Chen and J. Timonen 'Integrability Conditions', in: "Proc. XVIIIth Intl. Conf. on Diff. Geom. Methods in Theor. Phys." (Lake Tahoe, July 2-9, 1989: NATO ARW Physics and Geometry) Ling-Lie Chau and Werner Nahm eds., Plenum, New York (1990) and references. To appear 1990.

12. R.K. Bullough, Yu-zhong Chen and J. Timonen, 'Soliton statistical mechanics: thermodynamics limits for the integrable models', in: "Proc. IV Intl. Workshop on Nonlinear and Turbulent Processes in Physics", V.E. Zakharov, A.G. Sitenko, N.S. Erokhin and V.M. Chernousenko eds., World Scientific, Singapore (1990) and references. To appear (1991), 46pp.
13. A.R. Its, A.G. Izergin and V.E. Korepin, Phys. Letts. A141:121 (1989).
14. R.K. Bullough and J. Timonen, 'Quantum and classical integrability: new approaches in Statistical Mechanics', in: "Nonlinear Science: the Next Decade" A.R. Bishop and D.K. Campbell eds., Physica D. Nonlinear Phenomena. To appear (1991).
15. R.K. Bullough and J. Timonen, 'Quantum groups and quantum complete integrability: theory and experiment', in: "Proc. XIXth Intl. Conf. on Diff. Geom. Methods in Theor. Phys." (Rapallo, Italy, June 19-24, 1990) U. Bruzzo and C. Bartocci eds., Lecture Notes in Physics, Springer-Verlag, Heidelberg. To appear (1991).
16. A. Zee 'Semionics: a theory of high temperature superconductivity' in: "High Temperature Superconductivity; Proc. of the Los Alamos Symp. 1989", K.S. Bedell, D. Coffey, D.E. Keltzer, D. Pines and J.R. Schrieffer eds., Addison Wesley Publ. Co., Redwood City, Calif. (1990), pp. 248-298.
17. P.W. Anderson, 'The Normal State of High T_c Superconductivity: A New Quantum Liquid', in: "High Temperature Superconductivity" Ref. 16, - and references.
18. Inf. discussion with Dr. E. Courtens, IBM Zurich Rüschlikon Laboratories.
19. H.B. Thacker, Rev. Mod. Phys. 53:253 (1981).
20. P.P. Kulish and E.K. Sklyanin; 'Quantum Inverse Scattering Method', in: "Proc. of the Tvärminne Symp. Finland, 1981", J. Hietarinta and C. Montonen eds. Springer-Verlag, Heidelberg (1982).
21. For example E.K. Sklyanin, L.A. Takatadhyan and L.D. Faddeev, Theor. Mat. Fiz. 40:194 (1979).
22. M. Jimbo and T. Miwa 'Infinite Dimensional Lie Algebras', in: "Integrable Systems in Statistical Mechanics", G.M. D'Ariano, A. Montorsi and M.G. Rasetti eds., World Scientific, Singapore (1985), pp. 105-125.
23. R.P. Feynman and A.R. Hibbs "Quantum Mechanics and Path Integrals" McGraw-Hill Book Co., New York (1965).
24. Yu-zhong Chen, 'Classical and quantum statistical mechanics of the 1+1 dimensional integrable models', Ph.D. Thesis, University of Manchester, July (1989).
25. R.F. Dashen, B. Hasslacher and A. Neveu, Phys. Rev. D11:3423 (1975).
26. C.N. Yang and C.P Yang, J. Math. Phys. 10:1115 (1969).
27. C. Itzykson and J.B. Zuber "Quantum Field Theory" McGraw-Hill Book Co., New York (1980).
28. R.K. Bullough 'Statistical Mechanics of the sine-Gordon Field: Part I', in: "Nonlinear Phenomena in Physics" F. Claro ed., Springer-Verlag, Heidelberg (1985), pp. 70-103.
29. D. Fröhlich, Lectures at XIXth Intl. Conf. on Diff. Geom. Methods in Theor. Phys. (Rapallo, Italy, June 19-24, 1990). To be published in the 'Proceedings' Ref. 15.
30. S.G. Chung and Yia-Chung Chang, J. Phys. A: Math. Gen. 20:2875 (1987).
31. J. Timonen and R.K. Bullough. To be published.
32. M. Fowler and X. Zotos, Phys. Rev. B 24:2634 (1981); 25:5806 (1982).
33. M. Imada, K. Hida and M. Ishikawa, J. Phys. C. 16:35 (1983).
34. R.K. Bullough, Y-z. Chen, J. Timonen, V. Tognetti and R. Vaia, Phys. Letts. A 145:54 (1990).
35. Yi Cheng 'Theory of Integrable Lattices', Ph.D. Thesis, University of Manchester, January (1987).
36. J. Timonen and R.K. Bullough, Phys. Letts. 82A:183 (1981).

STARK EFFECT FOR DIFFERENCE SCHRÖDINGER OPERATOR

E.I.Dinaburg

Shmidt Institute of Earth Physics
Academy of Sciences
Moscow, USSR

We consider a tight-binding model of the electron moving in an uniform electric field and study the spectrum of the corresponding Schrödinger operator and its small perturbations.

In the appropriate scale the problem can be reduced to the following.

Let $\mathbb{Z}^d$ be the d-dimensional usual lattice (d>1). Any site of $\mathbb{Z}^d$ is defined by vector $n=(n_1,\ldots,n_d)$ with integer components n_i $(1\leq i\leq d)$ and the length of any bond of $\mathbb{Z}^d$ is 1. Consider the operator

$$(H^{(V)}\psi)(n)=\sum_{i=1}^{d}\Delta_i\psi(n)+(E,n)\psi(n)+V(n)\psi(n) \tag{1}$$

acting in the space $l^{(2)}(\mathbb{Z}^d)$ of functions $\psi(n)$ on $\mathbb{Z}^d$, such that $\sum_n|\psi(n)|^2<\infty$. Here Δ_i is the second difference operator along the i-th coordinate direction of $\mathbb{Z}^d$, E is the vector of the electric field strain, V(n) is the potential of some perturbation.

We study for (1) the spectral problem

$$H^{(V)}\psi=\varepsilon\psi \tag{2}$$

At first consider the spectrum of the nonperturbated operator $H^{(0)}$ $(V\equiv0)$. The eigenvalues of $H^{(0)}$ can be found explicitly.

It is easy to see that $H^{(0)}$ has the following hidden symmetry: if $\psi(n)$ is an eigenfunction (e.f.) of $H^{(0)}$ with the eigenvalue (e.v.) ε then for any $m\in\mathbb{Z}^d$ $\psi(n-m)$ is also an e.f. of $H^{(0)}$ with e.v. $\varepsilon_1=\varepsilon+(E,m)$.

Let $\lambda=(\lambda_1,\ldots,\lambda_1)$ be a point of the torus $\mathbb{T}^d$ and $\hat{\psi}(\lambda)=\sum_n\psi(n)\exp\{2\pi i(\lambda,n)\}$ be the Fourier transform of $\psi(n)$. Then $\hat{\psi}(\lambda)$ satisfies the equation

$$\sum_{i=1}^{d}2(\cos2\pi\lambda_i-1)\hat{\psi}(\lambda)+(E,\mathrm{grad}\hat{\psi}(\lambda))(2\pi i)^{-1}=\varepsilon\hat{\psi}(\lambda). \tag{3}$$

The collection $\{\hat{\psi}_m(\lambda)\}$, $m\in\mathbb{Z}^d$ of e.f. of (3)

$$\hat{\psi}_m(\lambda)=\exp\{2\pi i(m,\lambda)\}\prod_{k=1}^{d}\exp\{-2iE_k^{-1}\sin2\pi\lambda_k\}$$

Microscopic Aspects of Nonlinearity in Condensed Matter
Edited by A.R. Bishop *et al.*, Plenum Press, New York, 1991

with e.v. $\varepsilon_m=(E,m)-2d$ is the orthonormed basis of the space $\mathcal{L}^2(\mathbb{T}^d)$.

The inverse Fourier transform of $\{\hat{\psi}_m(\lambda)\}$ gives the full collection $\{\psi_m(n)\}$ of e.f. for $H^{(0)}$. Remark that if the components E_i $(1\le i\le d)$ of the vector E are rationally independent the spectrum of $H^{(0)}$ is everywhere dense.

So the problem concerning the spectrum of perturbation of the operator with the purely point everywhere dense spectrum appears.

Now we formulate some results (proofs can be found in reference[1]).

Any vector E can be represented in the form $E=|E|\omega$ where $|E|$ is the length of E and ω is the vector defining the direction of E, $|\omega|=1$. Let μ be the Lebesque measure on the unit sphere S^d, $\mu(S^d)=1$.

The first result is connected with a periodic perturbation V. The last means that there are such d-linearly independent vectors $e_1,\dots,e_d$ that $V(n+e_i)=V(n)$, $1\le i\le d$.

Theorem 1. *Assume that V is a periodic function on* $\mathbb{Z}^d$ *and* $M=\sup|V|$. *Then for any given* $E_0>0$ *there is* $M_0=M_0(E_0)$ *such that for* $M<M_0$ *there exists a set* $\Omega_M\subset S^d$, $\mu(S^d)>0$ *such that for all* $E=E_0\omega$, $\omega\in\Omega_M$, *the operator* $H^{(V)}$ *has purely point and everywhere dense spectrum. E.v. of* $H^{(V)}$ *can be labeled by sites n of* $\mathbb{Z}^d$ *and have the following form:*

$$\lambda_n^{(V)}=(E,n)-2d+\delta(n),$$

where $\delta(n)$ *is a periodic function on* $\mathbb{Z}^d$ *with periods of the function V. Moreover* $\mu(\Omega_M)\longrightarrow 1$ *as* $M\longrightarrow 0$.

The second result is connected with a local perturbation V.

Theorem 2. *Assume that V(n) satisfies the inequality*

$$|V(n)|<M(|n|+1)^{-2d+1-\gamma},\quad \gamma>0.$$

Then for $H^{(V)}$ *all statements of the Theorem* 1 *are true except the periodicity of* $\delta(n)$.

The proofs of these theorems are based upon some version of the Kolmogorov-Arnold-Moser theory.

<u>References</u>

1. E.I.Dinaburg, Stark Effect of Difference Schrödinger Operator. *Teor. i Mat. Fizica*, V.78, N1:70 (1989) (in Russian).

DOUBLE PEAK STRUCTURE IN THE DENSITY OF STATES OF A KONDO LATTICE

Maria Marinaro, Canio Noce and Alfonso Romano

Dipartimento di Fisica Teorica e S.M.S.A.
Unità C.I.S.M e I.N.F.M. di Salerno
Università di Salerno, 84081 Baronissi (Salerno), Italy

INTRODUCTION

The aim of this paper is to present some results in the study of strongly correlated electron systems, which can be useful to clarify some open problems in the analysis of two large classes of physical systems, known as the intermediate valence (IV) and the heavy fermions (HF) systems[1].

It is generally assumed that the striking anomalies exhibited by these systems, specially in response functions such as resistivity, susceptibility and specific heat, are related to the behaviour of the electron density of states (DOS) near the Fermi level. In fact, while for normal metals the DOS can be considered constant in the relevant region around the Fermi level, for the above mentioned systems the DOS has in this region many features[2] which, depending on the value of the temperature, make them behave either as itinerant electron or as localized electron systems.

A problem not yet solved, both from a theoretical and experimental point of view, is the dependence of the DOS structure at the Fermi level on temperature. In particular, it is not clear if some special features which strongly affect the behaviour of the system, disappear at temperatures larger than the Kondo temperature.

Many works have been devoted to this problem, among which we recall the phenomenological model (see Sect.2) introduced some years ago by Marabelli and Wachter[3]. By using this model the authors were able to fit very well the experimental data of both IV and HF systems. Their basic assumption is that the DOS in the region around the Fermi level shows a double peak structure which does not depend on the temperature.

In Sect.3 we give a microscopic base to this phenomenological model. We start from the exact solution of the periodic version of the Anderson model[4] in the limit of zero width conduction band[5], then introducing in a perturbative way the effect of the kinetic term by means of an approximate Dyson equation[6]. We compute the DOSs for localized (f) and conduction (s, p or d) electrons and find out that they show near the Fermi level the same structure phenomenologically introduced in ref.3. In particular, the double peak structure exhibited by the f-DOS is found to be not vanishing even at temperatures larger than the Kondo temperature.

Microscopic Aspects of Nonlinearity in Condensed Matter
Edited by A.R. Bishop *et al.*, Plenum Press, New York, 1991

PHENOMENOLOGICAL MODEL FOR THE DENSITY OF STATES

Photoemission experiments as well as far infrared and point contact spectroscopy on some IV and HF systems are well understood in terms of a double peak structure in the f-DOS near the Fermi level. Based on this result, Marabelli and Wachter introduced in ref.3 a model DOS consisting in two gaussian peaks a few meV wide, separated by a gap of the same order of magnitude.

These authors assumed this shape of the f-DOS to hold for both IV and HF systems, the only difference being related to the position of the Fermi level. Their conclusion is that the Fermi level is in the gap for IV systems and in one of the peaks for HF systems.

Assuming that the model f-DOS does not change with temperature, the authors fit very well the resistivity, specific heat and susceptibility of several systems in a temperature range extending up to 300 K[2,3]. Using the width of each peak and the gap as free parameters, they found that the values of these parameters which gave the best fits were very close to the measured ones.

It is worth noting that, due to the assumption of a rigid DOS, the influence of the temperature in the fitting procedure is wholly contained in the Fermi distribution function $f(E)$.

In this kind of calculations, the most relevant quantity is the derivative with respect to the temperature of $f(E)$ at the Fermi level E_F. It passes from a delta function ($\delta(E-E_F)$) at very low temperatures to a peak whose width is much larger than the whole DOS structure near E_F at room temperature. As a consequence, IV and HF systems exhibit a behaviour which is more similar to the one of metals at low temperatures and to the one of systems with localized magnetic impurities at higher temperatures.

MICROSCOPIC CALCULATION OF THE DENSITY OF STATES

For a microscopic description of IV and HF systems, one usually refers to the Anderson model describing the interaction between localized electrons (coming from magnetic impurities) and conduction electrons (coming from the host metal).

This model has been exactly solved only in the case of single magnetic impurity, giving an f-DOS which shows at low temperatures a characteristic peak (Kondo resonance) in the region around the Fermi level[7]. On the other hand, it is well known that the existence in the same region of two narrow peaks separated by a small hybridization gap is a consequence of a coherent impurity scattering and therefore cannot be explained in terms of a single impurity model. For this reason it is necessary to consider the lattice version of the Anderson model, known as periodic Anderson model, whose Hamiltonian is

$$H = H_f + H_c + H_{fc}$$

where

$$H_f = \sum_{i,\sigma} \epsilon_f f_{i\sigma}^\dagger f_{i\sigma} + U \sum_i f_{i\uparrow}^\dagger f_{i\uparrow} f_{i\downarrow}^\dagger f_{i\downarrow}$$

$$H_c = \sum_{\vec{k},\sigma} \epsilon_{\vec{k}} c_{\vec{k}\sigma}^\dagger c_{\vec{k}\sigma}$$

$$H_{fc} = V \sum_{i,\sigma} (f_{i\sigma}^\dagger c_{i\sigma} + c_{i\sigma}^\dagger f_{i\sigma}) \quad .$$

Here H_f is the Hamiltonian of a set of localized electrons with a strong Coulomb repulsion U at the same site, H_c describes a system of uncorrelated itinerant electrons and H_{fc} is the term accounting for the hybridization between localized and conduction states. Standard second quantization notation is used, with the operators f and c referring to the localized and conduction electrons, respectively.

Even though this model cannot be exactly solved, nonetheless computations of the densities of states have been performed by various authors by means of different approximation techniques[8]. The results can be splitted in two groups. Some authors find the double peak at the Fermi level disappearing above the Kondo temperature[9], whereas others find no such a temperature effect on this structure[2,10].

In order to clarify this point, we have computed the DOSs for both localized and itinerant electrons, making use of a perturbative expansion in the conduction bandwidth. In our approach we start from the exact computation at finite temperature of the Green functions corresponding to the periodic Anderson Hamiltonian with no conduction band, i.e. with $\epsilon_{\vec{k}} = 0$. Let us denote these functions as G^0_{ff}, G^0_{cc} and G^0_{fc}. Then we introduce the effect of the kinetic term through the following Dyson equation in $(\vec{k},\omega_\nu)$-space[6] (ω_ν is the Matsubara frequency):

$$\widehat{\mathcal{G}}(\vec{k},\omega_\nu) = \widehat{\mathcal{G}}^0(\omega_\nu) + \widehat{\mathcal{G}}^0(\omega_\nu)\widehat{\Sigma}(\vec{k},\omega_\nu)\widehat{\mathcal{G}}(\vec{k},\omega_\nu) \quad .$$

Here

$$\widehat{\mathcal{G}}(\vec{k},\omega_\nu) = \begin{pmatrix} G_{ff}(\vec{k},\omega_\nu) & G_{fc}(\vec{k},\omega_\nu) \\ G_{cf}(\vec{k},\omega_\nu) & G_{cc}(\vec{k},\omega_\nu) \end{pmatrix}$$

is the exact matrix propagator, $G_{ab}(\vec{k},\omega_\nu)$ being the Fourier transform of

$$G^{ij}_{ab}(\tau-\tau') = -Tr\left[e^{-\beta H} a_i(\tau) b_j(\tau')\right] \equiv -\langle T[a_i(\tau)b_j(\tau')]\rangle$$

(i and j are site indices), and $\widehat{\mathcal{G}}^0(\omega_\nu)$ is the unperturbed one. Note that the latter shows no dependence on $\vec{k}$ because it refers to an Hamiltonian which does not contain bands of finite width.

The above Dyson equation is an approximation, because, due to the presence of the bilinear correlation energy U term in the unperturbed Hamiltonian, Wick theorem cannot be applied. Nonetheless, we believe that this approximation it not so drastic, because it consists in neglecting the fluctuations around mean values computed by taking exactly into account the effect of the most relevant terms of the Hamiltonian, i.e. the correlation energy U and the hybridization V.

The densities of states for conduction and localized electrons are computed by means of the standard formulas

$$\rho_f(\omega) = -\frac{1}{\pi N}\sum_{\vec{k}} Im\, G_{ff}(\vec{k}, i\omega_\nu \to \omega + i\eta)$$

$$\rho_c(\omega) = -\frac{1}{\pi N}\sum_{\vec{k}} Im\, G_{cc}(\vec{k}, i\omega_\nu \to \omega + i\eta) \quad .$$

Instead of specifying the form of the energy spectrum, we have computed ρ_f and ρ_c

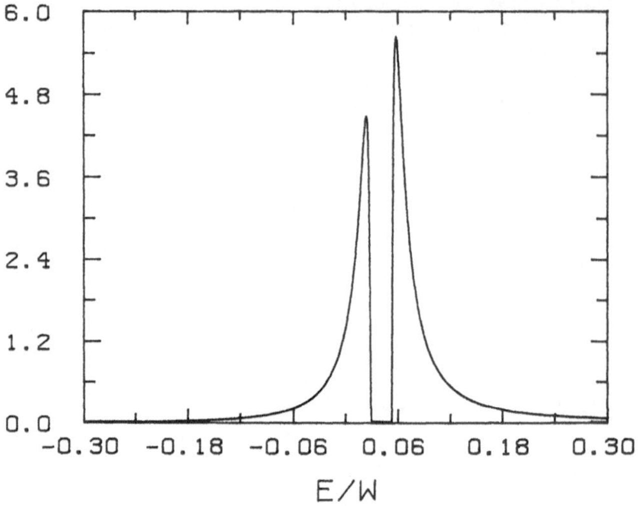

Fig. 1. f-DOS double peak structure at $k_B T/W = 0.43 \cdot 10^{-3}$.

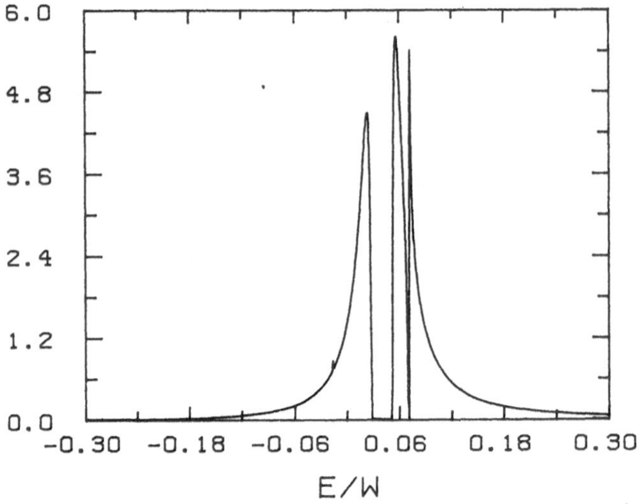

Fig. 2. f-DOS double peak structure at $k_B T/W = 1.3 \cdot 10^{-2}$.

assuming a parabolic density of states for the unperturbed conduction electrons[6]:

$$\rho_c^0(\omega) = \begin{cases} (3/4W)\left[1-(\omega/W)^2\right] & \text{if } |\omega| \leq W \\ 0 & \text{otherwise} \end{cases}$$

where W is the half-width of the conduction band.

In figs.1,2 we report the f-DOS in proximity of the Fermi level for two values of the temperature corresponding to $k_B T/W = 0.43 \cdot 10^{-3}$ and $k_B T/W = 1.3 \cdot 10^{-2}$. The Hamiltonian parameters have been chosen as $V/W = 0.2$, $U/W = 3$, $\epsilon_f/W = -0.5$. For reasonable choices of W, the first temperature value lies near the Kondo temperature and the second well above it. The figures show that a double peak structure with a small hybridization gap is present near the Fermi level. The latter always falls in the gap, because in our calculations we have fixed an even number of electrons per site[6].

For temperatures larger than the Kondo one, this feature does not disappear, the only effect of the temperature consisting in an increase of the number of the f-peaks near the Fermi level (see fig.2).

ACKNOWLEDGEMENTS

The authors acknowledge the financial support received from Progetto Finalizzato "Tecnologie Superconduttive e Criogeniche" of Consiglio Nazionale delle Ricerche (Italy).

REFERENCES

1. T. Kasuya and T. Saso, eds., "Theory of heavy fermions and valence fluctuations", Springer, Berlin (1985);
   G. Czycholl, Approximate treatments of intermediate valence and heavy fermion model systems, Phys. Rep. 143: 277 (1986);
   P. Schlottmann, Some exact results for dilute mixed-valent and heavy-fermion systems, Phys. Rep. 181: 1 (1989).
2. F. Marabelli and P. Wachter, Far-infrared properties of intermediate valence and heavy fermion materials, in: "Theoretical and experimental aspects of valence fluctuations and heavy fermions", L. C. Gupta and S. K. Malik, eds., Plenum, New York (1987).
3. F. Marabelli and P. Wachter, Electronic structure of UPt_3: a low energy optical study, J. Magn. Magn. Mat. 62: 287 (1986);
   F. Marabelli and P. Wachter, Electronic structure and magnetic properties of heavy fermions, J. Magn. Magn. Mat. 70: 364 (1987).
4. P. W. Anderson, Localized magnetic states in metals, Phys. Rev. 124: 41 (1961).
5. F. Mancini, M. Marinaro and Y. Nakano, Exact results for the Anderson model in the limit of zero-width conduction band, Physica B 159: 320 (1989);
   F. Mancini, M. Marinaro, Y. Nakano, C. Noce and A. Romano, A diagram method for the Anderson model. Limit of zero-width conduction band, Nuovo Cimento D 11: 1709 (1989).
6. M. Marinaro, C. Noce and A. Romano, Densities of states in the periodic Anderson model, submitted to J. Phys. Condensed Matter (1990).
7. A. M. Tsvelick and P. B. Wiegmann, Exact results in the theory of magnetic alloys, Advan. Phys. 32: 453 (1983).
8. P. A. Lee, T. M. Rice, J. W. Serene, L. J. Sham and J. W. Wilkins, Theories of heavy-electron systems, Comments Cond. Matt. Phys. 12: 99 (1986).

9. C. Lacroix, Coherence effects in the Kondo lattice, J. Magn. Magn. Mat. 60: 145 (1986);
N. Grewe, One particle excitation spectrum of the Kondo lattice, Solid State Commun. 50: 19 (1984);
H. Schweitzer and G. Czycholl, Second order U-perturbation approach to the Anderson lattice model in high dimensions, Solid State Commun. 69: 171 (1989).
10. T. Oguchi and A. J. Freeman, Local density band approach to f-electron systems - Heavy fermion superconductor UPt_3, J. Magn. Magn. Mat. 52: 174 (1985);
B. H. Brandow, Variational theory of valence fluctuations: ground states and quasiparticle excitations of the Anderson lattice model, Phys. Rev. B 33: 215 (1986).

LOW-TEMPERATURE ANALYSIS OF MICROEMULSION MODEL

A.E.Mazel

International Institute of Earthquake Prediction Theory and Mathematical Geophysics
Academy of Sciences
Moscow, USSR

The microemulsion model was proposed by B.Widom for the description of the microfilm structure in the "oil-water" mixture with some values of the mixture parameters. The model is a classical 3-D spin lattice model. The spin takes the values ± 1 and the interaction is described by the Hamiltonian:

$$H(\varphi(\mathbb{Z}^3))=-I\sum_{\mathrm{dist}(x,y)=1}\varphi(x)\varphi(y)-J\sum_{\mathrm{dist}(x,y)=\sqrt{2}}\varphi(x)\varphi(y)-K\sum_{\mathrm{dist}(x,y)=2}\varphi(x)\varphi(y), \tag{1}$$

where $\varphi(\mathbb{Z}^3)$ is a configuration on $\mathbb{Z}^3$, $x,y\in\mathbb{Z}^3$, $J=2K$. This Hamiltonian includes the interaction of nearest neighbours in the first sum, the interaction of diagonal neighbours in the second sum and the interaction of next nearest neighbours in the third sum.

We shall consider a more general model with arbitrary J,K and $I>0$. The model with $I<0$ is transformed into the previous one if the signs of all spins with even coordinates are changed.

Usually for models of this type a low-temperature phase diagram is a small perturbation of a ground states diagram. It is well known from the Pirogov-Sinai theory that this situation arises in particular in models with the finite number of the ground states satisfying the Peierls stability condition: the energy jump on the boundary between two ground states is proportional to the boundary area.

Therefore it is natural to begin the investigation of the Hamiltonian (1) by studying its ground states. With this purpose we can rewrite the initial Hamiltonian as the following double sum:

$$\sum_{x\in\mathbb{Z}^3}\varphi(x)\sum_{y\in N(x)}\left[-0.5I\sum_{\mathrm{dist}(x,y)=1}\varphi(y)-0.5J\sum_{\mathrm{dist}(x,y)=\sqrt{2}}\varphi(y)-K\sum_{\mathrm{dist}(x,y)=2}\varphi(y),\right] \tag{2}$$

where $N(x)$ is the set of the nearest neighbours of the lattice site x. The minimum of each item of the external sum is reached on some family $\Phi(I,J,K)$ of the configurations in the volume $N(x)$. Due to the symmetry of the Hamiltonian the family $\Phi(I,J,K)$ does not depend on the site x and is invariant with respect to the permutation of the coordinate axes and to changing of all spin signs.

Obviously if there exists a periodic configuration

$$\varphi(\mathbb{Z}^3):\ \varphi(N(x))\in\Phi(I,J,K) \text{ for all } x \qquad (3)$$

then it is a ground state. Moreover in this case all periodic ground states are the configurations of type (3) and are uniquely defined by the family $\Phi(I,J,K)$.

The diagram of the families $\Phi(I,J,K)$ is shown in Fig.1. It contains six domains with different $\Phi(I,J,K)$. Only one configuration is drawn for each $\Phi(I,J,K)$ meaning that all other configurations can be obtained from this one after the permutation of the axes and(or) after changing of all spins sign.

It can be verified that in the interior of each domain O_1-O_6 the corresponding family $\Phi(I,J,K)$ generates the finite number of the ground states satisfying the Peierls condition.

On the domains boundaries the family $\Phi(I,J,K)$ is the union of the families from the contiguous domains. This increasing of the $\Phi(I,J,K)$ leads to a violation of the Peierls condition. Hence the phase diagram analysis becomes much more complicated in the neighbourhood of this boundaries.

We can construct the low-temperature phase diagram of the model only in those parts of parameter space where all periodic ground states are layered configurations. Recall that a configuration is called layered if it takes the constant value on any lattice plane orthogonal to one coordinate axis. For such configuration φ a sequence of l adjacent planes is called the layer of thickness l if φ takes the same constant value on these planes and the opposite value on the planes which are nearest to these ones. The finite sequence $\langle l_1,\dots,l_k\rangle$ of integer positive numbers defines the class of layered periodic configurations. The class is generated by the sequence of adjoining layers of thicknesses $l_1, l_2,\dots,l_k$. We use a special notation $\langle\infty\rangle$ for the class containing two ferromagnetic configurations.

Inside the domain O_1 ground states are the ferromagnetic configurations , inside the domain O_2 - configurations $\langle 1\rangle$, inside the domain O_3 - configurations $\langle 2\rangle$. On the interior of the O_1 and O_2 common boundary there exists the infinite set of ground states consisting of any periodic layered configurations containing only the layers of thickness 1 or 2. On the interior of the O_1 and O_3 common boundary an arbitrary layered configuration is a ground state if it does not contain layers of thickness 1. In the triple point $I=-4J$, $K=0$ any layered configuration is a ground state. On the interior of the O_1 and O_2 common boundary only configuration $\langle\infty\rangle$ and $\langle 1\rangle$ are ground states and Peierls condition is satisfied.

The stability analysis of the ground states described has shown that the low-temperature phase diagram is not a small perturbation of the ground states diagram. The diagram contains only a finite number of the phases among which the first order phase transitions are occured. The exact picture of phase transitions is described by the following theorem.

Theorem. *For each fixed* I *and large enough inverse temperature* β $(I\beta \gg 1)$:

in a small neighbourhood of the triple point $J=I/4$, $K=0$ *there exists a full phase diagram for the phases* $\langle 1\rangle$, $\langle 2\rangle$, $\langle\infty\rangle$ *(see Fig. 2.a);*

in a small neighbourhood of the interval $-I/2<K<\beta^{-1}$ *on the line* $I+4J+2K=0$ *finite number of phases from the set:* $\langle 2\rangle$, $\langle 3\rangle$, $\langle 4\rangle$, $\langle\infty\rangle$ *survive (see Fig. 2.b);*

in a small neighbourhood of the interval $-I/6<K<\beta^{-1}$ *on the line* $I+4J-2K=0$ *finite number of phases from the set:* $\langle 1\rangle$, $\langle 2\rangle$, $\langle 2,1\rangle$ *survive (see Fig. 2.c).*

It is necessary to take into account the entropy effects to understand why at non zero temperature only a finite number of phases survive.

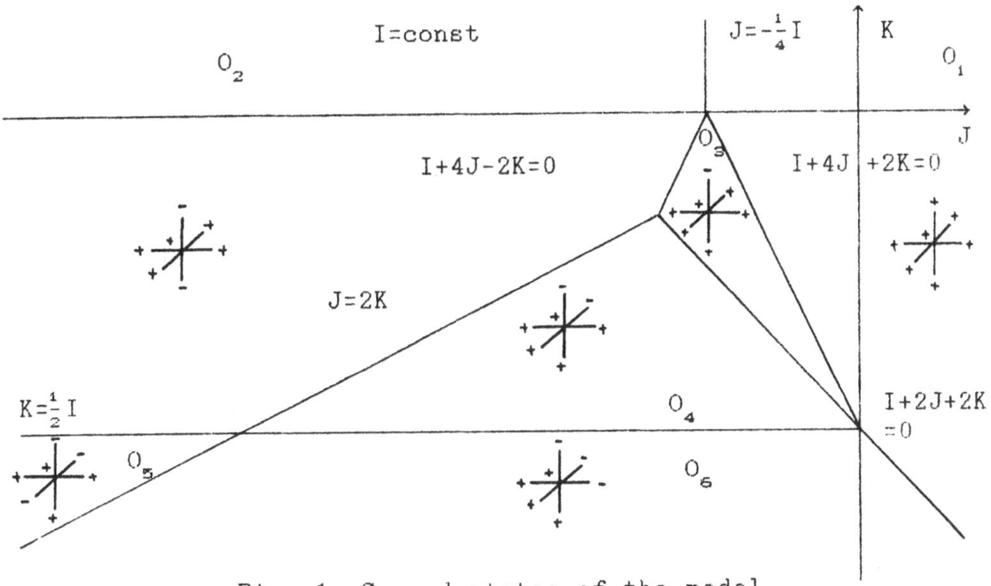

Fig. 1. Ground states of the model.

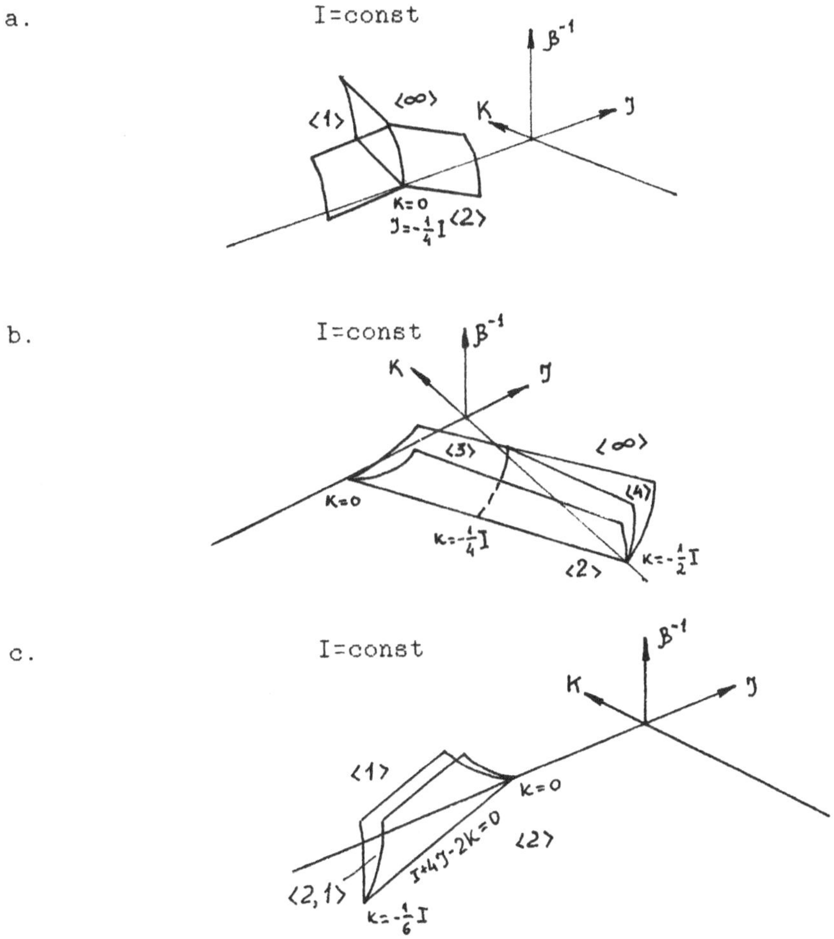

Fig. 2a-c. The phase diagram of the model

Different ground states generate the different free energy of the gas corresponding to the low-energy perturbations. Hence among all ground states we can fix the so called dominant ground states having the minimal sum of the specific energy and the free energy of the corresponding gas of the low-energy perturbations. If the number of the dominant ground states is finite and each dominant ground state satisfies some additional stability condition (the analog of the Peierls condition) then the extension of the Pirogov-Sinai theory can be constructed implying that the low-temperature phase diagram of the model is a small perturbation of the dominant ground states diagram.

All details of the proofs can be found in the reference[1].

References

1. E.I.Dinaburg and A.E.Mazel´, Analysis of Low-Temperature Phase Diagram of Microemulsion Model, *Commun. Math. Phys.*, V.125:27 (1989).

QUANTUM MONTE CARLO COMPUTATION OF STATIC AND TIME-DEPENDENT THERMODYNAMIC PROPERTIES OF LENNARD-JONES CRYSTALS

Arthur R. McGurn

Department of Physics
Western Michigan University
Kalamazoo, MI 49008 U.S.A.

Alexei A. Maradudin and Richard F. Wallis

Department of Physics
University of California
Irvine, CA 92717 U.S.A.

I. Introduction

Recently there has been much interest in the study of the thermodynamics of many-body quantum mechanical systems by means of Quantum Monte Carlo (QMC) computer simulation techniques[1-4]. Applications of QMC methods to the study of spin systems, the Hubbard model, and to systems of boson and fermion particles have shown them to be effective in the accurate computation of static thermodynamic properties[1-4]. However, in spite of these successful QMC treatments of static thermodynamics, the determination of time-dependent thermodynamic averages (spin correlation functions, atomic displacement correlation functions, etc.), related to the time-dependent response of quantum many-body systems, has been less forthcoming. Aside from the increased bookkeeping difficulties associated with having to determine averages which depend on the new variable of time, one finds that this variable enters time-dependent computed averages in such a way that analytic continuation techniques, commonly employed in Green's function treatments of such properties, are of little or no value in developing QMC methods for this problem.[3,5]

In this paper we shall address two problems. The first is the computation of the static thermodynamic properties of a crystalline system of atoms which interact with one another by Lennard-Jones pair potentials. We shall first discuss a one-dimensional chain of such atoms. This system is not truly crystalline in that it has no long range atomic order but, at very low temperatures, the system has a significant amount of short range order which renders it approximately crystalline over long segments of the chain.[6,7] Next, a fully three-dimensional system of atoms, which constitutes a model of an inert gas solid, will be studied. The energy, specific heat, pressure, and volume of both of these models will be determined.

The second problem we address is the development of a new method for calculating the time-dependent thermodynamic response functions of a many-body quantum system in the context of a QMC simulation. Specifically, we shall represent the frequency-dependent spectral density of these response functions in terms of a continued fraction expansion in the frequency moments of the spectral density.[8] This type of moment expansion is effective in reproducing spectral densities which are represented as sharply peaked functions about a single frequency or, in the case of spectral densities which are even functions of frequency, sharply peaked about a pair of oppositely signed frequencies.

As a specific example of the treatment of time-dependent responses in the context of the QMC, we present some preliminary results from our current work on the Lennard-Jones chain of atoms. We compute the spectral densities of the time-dependent atomic position correlation functions, and study the dependence of these functions on temperature. As these correlation functions are related to the cross section for inelastic neutron scattering with the creation or destruction of a single phonon in the chain, we can use them to study the change in frequency and lifetime of these excitations with temperature.

In all of the QMC computations of the static and time-dependent properties of the Lennard-Jones crystals presented in this paper, we take into account the full nonlinearity of the atomic interactions described by the Lennard-Jones pair potential. Comparisons of the QMC results to results in the quantum harmonic approximation and results for the corresponding system in the classical limit will be made. The QMC will be shown to be effective in determining the full range of thermodynamic properties of an anharmonic, quantum mechanical, vibrational system of atoms.

The organization of this paper is as follows: In Section II we shall present a discussion of the QMC method for determining static thermodynamic properties. This method will be applied to a one-dimensional chain and to a three-dimensional face-centered cubic crystal. In Section III we will describe our QMC treatment of the time-dependent correlation functions and will present some preliminary results of work on the Lennard-Jones chain of atoms.

II. Static Properties

In this section we shall briefly outline the development of the QMC method as applied to the one-dimensional chain of atoms, and will then present results for the static thermodynamic properties of one-dimensional[7] and three-dimensional[9] Lennard-Jones systems. The development of the QMC method for three-dimensional crystals[9], which is a simple generalization of that presented for the one-dimensional chain, will not be given in detail here.

We consider a one-dimensional chain of N distinguishable atoms of mass m occupying a segment of length L of the x-axis and subject to periodic boundary conditions in the length L of the segment. The atoms, which never pass one another on the chain, are constrained to move only parallel to the x-axis, and their positions $\{x_i\}$ are subject to the restrictions $0 \leq x_1 < x_2 \ldots < x_{N-1} < x_N \leq L$. The Hamiltonian of the system is given by

$$H = H_0 + H_1, \tag{1}$$

where

$$H_0 = -\frac{\hbar^2}{2m}\sum_{j=1}^{N}\frac{\partial^2}{\partial x_j^2} \tag{2a}$$

is the kinetic energy of the chain, while

$$H_1 = \sum_{j=1}^{N}\phi(x_{j+1}-x_j) \tag{2b}$$

with $x_{N+1} = x_1$ and

$$\phi(r) = 4\epsilon\left[\left(\frac{\sigma}{r}\right)^{12} - \left(\frac{\sigma}{r}\right)^{6}\right] \tag{3}$$

is the potential energy of the chain. We note that in Eq. (2b) only nearest neighbor interactions are considered and that the pair potential in Eq. (3) is the standard Lennard-Jones (6-12) potential used to model inert gas systems.

While, in the limit of classical mechanics, it is easy to calculate exactly the thermodynamics of the system described above[10], the quantum mechanical solution for this problem is more difficult owing to the noncommutivity of the operators H_0 and H_1. Although many approximate techniques have been developed to deal with this difficulty[11], we shall be specifically interested in presenting a computer simulation method, the QMC method, for determining these properties. The QMC method is limited only by the numerical capabilities of the computers employed and not, as are approximation methods, by specific approximating simplifications in the mathematical equations representing the thermodynamics. We shall now outline the development of the QMC method for the one-dimensional chain of atoms[7].

The QMC treatment of our many atom system is based on the application, to the expression for its quantum mechanical partition function, of an operator identity due to Trotter[12,13] which re-expresses the quantum mechanical partition function as a path-integral written in terms of classical (commuting) variables. The Trotter identity states that for $[H_0,H_1] \neq 0$

$$e^{-\beta(H_0+H_1)} = \lim_{M\to\infty}\left[e^{-\frac{\beta}{M}H_0}\, e^{-\frac{\beta}{M}H_1}\right]^M , \tag{4}$$

so that the partition function of the system described by Eq. (1) can be written as

$$\begin{aligned} Z &= \mathrm{Tr}\, e^{-\beta H} = \sum_{n_0} \langle n_0 | e^{-\beta H} | n_0 \rangle \\ &= \lim_{M\to\infty} \sum_{n_0} \langle n_0 | [e^{-\frac{\beta}{M}H_0}\, e^{-\frac{\beta}{M}H_1}]^M | n_0 \rangle , \end{aligned} \tag{5}$$

where we have expressed the trace in terms of a complete orthonormal set of basis states, $\{|n_0\rangle\}$. If we now insert representations of the identity operator between each product of exponentials on the right hand side of the second of Eqs. (5), we have

$$Z = \lim_{M\to\infty} \sum_{n_0}\sum_{n_1}\sum_{n_2}\cdots\sum_{n_{2M-1}} \langle n_0 | e^{-\frac{\beta}{M}H_0} | n_1 \rangle \langle n_1 | e^{-\frac{\beta}{M}H_1} | n_2 \rangle \times$$

$$\times \ldots \langle n_{2M-1}| e^{-\frac{\beta}{M} H_0} |n_{2M-1}\rangle\langle n_{2M-1}| e^{-\frac{\beta}{M} H_1} |n_0\rangle, \tag{6}$$

and the partition function is quite generally expressed as a trace over a matrix product. For the particular problem defined by Eqs. (1) and (2) it is convenient to choose the complete set of orthogonal basis states $\{|n_i\rangle\}$ in the position space representation of the N atoms (i.e., $|n_i\rangle = |x_1^{(i)}, x_2^{(i)}, \ldots, x_N^{(i)}\rangle$, where $x_j^{(i)}$ is the position of the jth atom in the segment). In this representation, Eq. (6) reduces in the $M \to \infty$ limit to a path integral given by[14] $(x_j^{M+1} = x_j^1)$

$$Z_M^{P-I} = \left[\frac{mM}{2\pi\beta\hbar^2}\right]^{\frac{NM}{2}} \int \prod_{i=1}^{M} \left[d^N x^{(i)} e^{-\frac{mM}{2\beta\hbar^2}\sum_{j=1}^{N}(x_j^{(i)} - x_j^{(i+1)})^2}\right] \times$$

$$\times\, e^{-\frac{\beta}{M}\sum_{i=1}^{M}\sum_{j=1}^{N}\phi(x_{j+1}^{(i)} - x_j^{(i)})}, \tag{7}$$

for $L\to\infty$ and $N\to\infty$, so that

$$Z = \lim_{M\to\infty} Z_M^{P-I}. \tag{8}$$

A faster convergence[14] in M to the $M \to \infty$ limit of Z_M^{P-I} can be achieved in Eqs. (7) and (8) if we include $0(1/M^2)$ corrections to the approximation

$$\exp\left[-\beta(H_o+H_1)\right] \cong \left[\exp(-\beta H_o/M)\exp(-\beta H_1/M)\right]^M$$

used in writing Z_M^{P-I}. Doing this we find

$$Z_M^{P-I} = \left[\frac{mM}{2\pi\beta\hbar^2}\right]^{\frac{NM}{2}} \int \prod_{i=1}^{M} \left[d^N x^{(i)} e^{-\frac{Mm}{2\beta\hbar^2}\sum_{j=1}^{N}(x_j^{(i)} - x_j^{(i+1)})^2}\right] \times$$

$$\times \exp\left[-\frac{\beta}{M}\sum_{i=1}^{M} v'(x_1^{(i)}, x_2^{(i)}, \ldots, x_N^{(i)})\right], \tag{9a}$$

where

$$v'(x_1, x_2 \ldots x_N) = H_1(x_1, x_2 \ldots, x_N) + \frac{1}{24}\frac{\hbar^2}{m}\left(\frac{\beta}{M}\right)^2 \sum_{j=1}^{N}\left[\frac{\partial H_1}{\partial x_j}\right]^2, \tag{9b}$$

which yields the same limit in Eq. (8) as does Eq. (7) but is an expression which provides the basis for a much more efficient computer simulation than does Eq. (7). The basis of our QMC averaging is then provided by Eqs. (8) and (9).

The energy, specific heat, and pressure, as functions of M, can be determined from the standard relations

$$E_M = -\frac{\partial}{\partial\beta}\ln Z_M^{P-I} \tag{10a}$$

$$C_M = \frac{\partial}{\partial T} E_M \tag{10b}$$

$$P_M = \beta \frac{\partial}{\partial L} \ln Z_M^{P\text{-}I} \ , \qquad (10c)$$

respectively, where the $M \to \infty$ limit of these expressions yields the exact results for the thermodynamic properties of the quantum mechanical system described by Eqs. (1) and (2). The averages in Eqs. (10) can be computed using standard classical Monte Carlo methods based on Metropolis sampling[15] applied to the classical (commuting) variables, $\{x_i\}$, in terms of which the path-integral representation in Eq. (9) is expressed.

In Fig. 1 we present QMC results for the energy and specific heat of the atomic chain[7,9] computed for a fixed average nearest-neighbor separation of $2^{1/6}\sigma$, for $\alpha = \hbar/(\sqrt{m\epsilon}\ \sigma) = 0.03$ (appropriate to solid Argon) and $\alpha = 0.1$. Comparison is made with results of the quantum mechanical harmonic approximation (dashed lines) and the exact classical limit

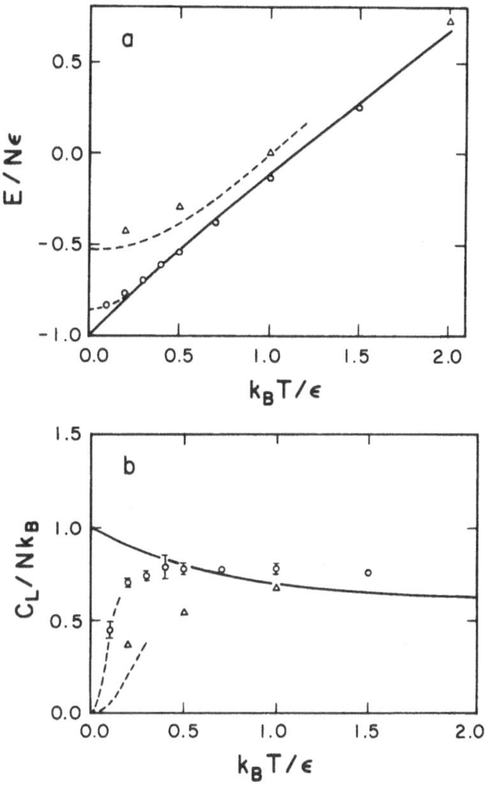

Fig. 1. The static thermodynamic properties of the Lennard-Jones chain: a) internal energy, $E/N\epsilon$ versus absolute temperature, T; b) specific heat at constant length, C_L/Nk_B, versus T. Quantum Monte Carlo results for $\alpha= 0.03$ (open circles) and $\alpha = 0.1$ (triangles) are shown, harmonic approximation results for $\alpha = 0.03$ and 0.1 are indicated by dashed curves, and the classical limit results are given by solid curves.

solution (solid line). Chains of N = 10 atoms and M = 10 and 15 were simulated, and the size of the simulation data points represent both the error and the spread in the data in going from M = 10 to 15. We surmise that the system with M = 10 appears to be well converged to the $M \to \infty$ limit, and note that for $\alpha = 0.1$ the anharmonicity and larger quantum fluctuations give rise to significant deviations of the QMC from the quantum harmonic and classical solutions. Due to limitations on space, the reader is referred to Refs. (7) and (9) for a presentation of the temperature dependence of the pressure of the atomic chain.

We next turn to the application of the QMC method to the determination of the thermodynamic properties of three-dimensional crystals[9]. Due to limitations of space, we present only a discussion of our results. In Fig. 2 results of QMC computations for an fcc system with nearest-neighbor Lennard-Jones interactions are presented for the temperature dependence of the energy and specific heat at constant volume. A path-integral formulation, similar to that made for the one-dimensional chain, was applied to a thirty-two atom cell subject to periodic boundary conditions to obtain simulation results for the average

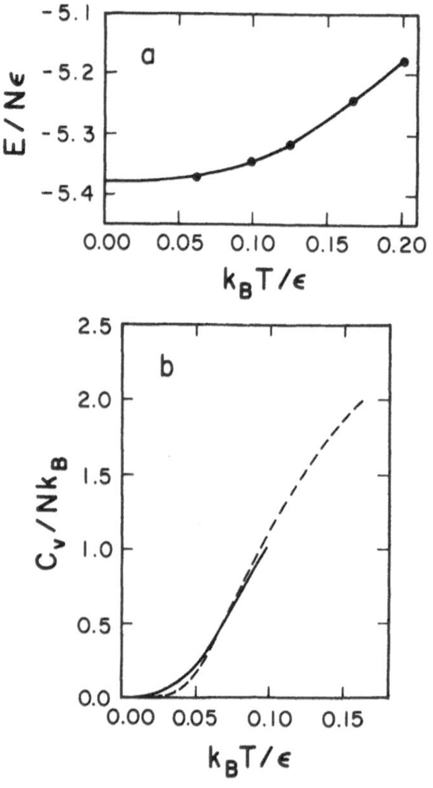

Fig. 2. QMC results (solid curves) for a) the internal energy, $E/N\epsilon$, and b) the specific heat at constant volume, C_v/Nk_B, versus temperature, T. The dots in a) are the QMC points, and the dashed line in b) is the quantum harmonic approximation.

energy versus temperature at zero pressure[9]. The simulation energy determined in this way was fitted for $k_BT/\epsilon < 0.20$ by the expression

$$E/N\epsilon = \epsilon_o + \epsilon_1 T^4 + \epsilon_2 T^6 + \epsilon_3 T^8, \tag{11}$$

and the specific heat at constant volume was determined from the temperature derivative of Eq. (11). The pressure in the system was set equal to zero at each temperature by allowing the lattice constant to vary with temperature. For the preliminary results in Fig. 2 only M = 10 data are shown, with the size of the points representing the error of the simulation data, and a comparison of these data with the quantum harmonic approximation is presented.

We see from the one- and three-dimensional results presented in Figs. 1 and 2 that the QMC method is effective in obtaining an accurate representation of the low temperature thermodynamic properties of anharmonic vibrational systems of atoms as well as in giving valuable interpolations between the very low and high temperature limits of such systems. More work is needed, however, to compare the preliminary results presented here to various analytical approximation techniques that have been applied to these problems in the past. Certainly the QMC method will prove valuable in obtaining quick, accurate, representations of the static thermodynamics of other types of less well understood quantum mechanical systems.

III. Dynamical Properties

In this section we develop QMC methods for computing the time-dependent response functions which describe the vibrational properties of a system of quantum mechanical atoms. The methods presented are based upon continued fraction representations of the spectral densities of the response functions in terms of their frequency moments[8], and are most effective in the determination of spectral densities which are peaked about a single frequency or a pair of oppositely signed frequencies. Although the methodology we propose is quite general in scope, we shall apply it specifically to study the time-dependent properties of the one-dimensional chain of atoms discussed in Section II.

For the one-dimensional system of atoms described by Eqs. (1) and (2) we consider quantum mechanical response functions of the form

$$C_k^o(\omega) = \int_{-\infty}^{\infty} \langle x_k(t) x_{-k}(0) \rangle e^{-i\omega t} dt \tag{12}$$

with

$$x_k = \frac{1}{\sqrt{N}} \sum_{j=1}^{N} x_j \; e^{-ik(ja)}, \tag{13}$$

where $a = 2^{1/6}\sigma$ is the average nearest neighbor atomic separation, $\langle \ldots \rangle$ indicates the thermodynamic average of the enclosed quantum mechanical operator, and k is specified by $k = 2\pi n/Na$ for $n = \pm 1, \pm 2, \ldots$. This response function is related to the cross section for the inelastic scattering of neutrons from the chain with the creation or destruction of single quasi-particle phonon excitations of wave number k or -k, respectively, and hence at low temperatures the spectral density of the response function in Eq. (12) should be peaked about an oppositely signed pair of frequencies.[16] The peak positions in the spectral density are the phonon frequencies of the chain, and the widths of the peaks are inversely proportional to the phonon lifetimes; both of these quantities are affected by the anharmonicity of the Lennard-Jones interactions and the quantum mechanical fluctuations present in the chains.

It is much easier to develop continued fraction representations for spectral densities which are even functions of the frequency because, in general, a smaller number of moments is required to approximate accurately a function over either the positive or negative frequency axis than over both the positive and negative frequency axes. In the following we shall, hence, deal with the computation of

$$C_k(\omega) = \tfrac{1}{2}\left[C_k^o(\omega)+C_k^o(-\omega)\right]. \tag{14}$$

By inserting complete sets of states in Eq. (12) and computing the spectral densities we find the simple relationship $C_k(\omega) = \tfrac{1}{2}\left[1+e^{\beta\hbar\omega}\right]C_k^o(\omega)$. Furthermore, the frequency moments of $C_k(\omega)$ are just the even frequency moments of $C_k^o(\omega)$; all odd moments of $C_k(\omega)$ are zero.

In the harmonic approximation for the Lennard-Jones chain, $C_k(\omega)$ can be easily shown to be given by

$$C_k(\omega) = \frac{\mu_0(k)}{2}\left[\delta(\omega-\omega_k)+\delta(\omega+\omega_k)\right], \tag{15}$$

where

$$\omega_k = 12(2)^{-1/6}\sqrt{\epsilon/m}\,\frac{1}{\sigma}\left[1-\cos(ka)\right]^{1/2} \tag{16}$$

$$\mu_0(k) = 2\pi\langle x_k(0)x_{-k}(0)\rangle = \frac{\pi\hbar}{m\omega_k}\coth\left[\frac{\beta\hbar\omega_k}{2}\right]. \tag{17}$$

When anharmonicity is included in the solution for $C_k(\omega)$ we expect the phonon peaks in $C_k(\omega)$ to develop widths due to the finite lifetimes of the excitations and the frequencies of the maxima of these peaks to be shifted away from their harmonic approximation values. If these effects are small, then a good quantitative approximation to the peak heights, frequencies of the peak maxima, and peak widths can be obtained from the zero, second, and fourth frequency moments of $C_k(\omega)$, and good approximations of the general functional form of $C_k(\omega)$ can be obtained from these moments in the context of truncated approximations to the full continued fraction expansion of $C_k(\omega)$.[17] We shall now discuss the continued fraction expansion of $C_k(\omega)$ in terms of its frequency moments.

From quite general thermodynamic considerations Mori[18,19] showed that response functions which are even functions of frequency, such as our function $C_k(\omega)$, can be expressed as

$$C_k(\omega) = 2\mathrm{Re}\left[\psi_k(\omega)\right]\int_{-\infty}^{\infty}\frac{d\omega'}{2\pi}\,C_k(\omega'), \tag{18}$$

where $\psi_k(\omega)$ is an infinite continued fraction

$$\psi_k(\omega) = \frac{1}{i\omega+}\,\frac{a_1^2(k)}{i\omega+}\,\frac{a_2^2(k)}{i\omega+}\,\cdots, \tag{19}$$

and the $\{a_n^2(k)\}$ are related to the frequency moments $\langle\omega^{2\ell}\rangle_k = \int_{-\infty}^{\infty}d\omega C_k(\omega)\omega^{2\ell}$ by

$$a_1^2(k) = \langle\omega^2\rangle_k/\langle\omega^0\rangle_k \tag{20a}$$

$$a_2^2(k) = \frac{\langle\omega^4\rangle_k}{\langle\omega^2\rangle_k} - \frac{\langle\omega^2\rangle_k}{\langle\omega^0\rangle_k} , \tag{20b}$$

etc. As a specific example of this continued fraction representation of $C_k(\omega)$, it is interesting to note that in the harmonic approximation to our Lennard-Jones chain of atoms $a_2^2(k) = 0$, so that Eq. (19) is terminated at the $a_2^2(k)$ terms and Eqs. (18) and (19) reduce to the result in Eq. (15). We therefore expect that for small chain anharmonicities and in the low temperature region in which quantum fluctuations are important, Eq. (19) can be terminated at $a_2^2(k)$ in some reasonable way to represent $C_k(\omega)$ accurately for our one-dimensional atomic chain.

One such termination of the continued fraction in Eq. (19) was proposed by Tomita _et al._(19) in their study of time-dependent response functions for the classical magnetic chain, and we shall use this termination in the preliminary results we present below for the spectral density of our Lennard-Jones chain of atoms. Tomita _et al._(19) proposed approximating $\psi_k(\omega)$ by

$$\psi_k(\omega) \simeq \frac{1}{i\omega + a_1^2(k) f(\omega, a_2^2(k))} , \tag{21}$$

where

$$f(\omega, a_2^2(k)) = \int_0^\infty dt\, e^{-\frac{1}{2} a_2^2(k) t^2} e^{-i\omega t}$$

$$= \frac{1}{a_2(k)} \left[\sqrt{\frac{\pi}{2}}\, e^{-\frac{y^2}{2}} - i\, e^{-\frac{y^2}{2}} \int_0^y ds\, e^{s^2/2}\right]\Bigg|_{y = \frac{\omega}{a_2(k)}} . \tag{22}$$

The termination given by $f(\omega, a_2^2(k))$, then, replaces the infinite continued fraction multiplying $a_1^2(k)$ in Eq. (19) by the Fourier transform of a Gaussian function whose spectral width gives the correct fourth frequency moment of the spectral density. Other methods of termination are available(16) but, as Tomita _et al._(19) point out, sharply peaked spectral densities are not sensitive to the differences among these methods.

To use Eqs. (21) and (22) as an approximate representation of $C_k(\omega)$, we must know the zeroth, second, and fourth frequency moments of $C_k(\omega)$ which are the same as those for $C_k^o(\omega)$. From Eq. (12) and the quantum mechanical equations of motion, we find

$$\langle\omega^{2n}\rangle = 2\pi(-1)^n \langle x_k^{(2n)}(0) x_{-k}(0)\rangle, \tag{23}$$

$$x_k^{(2n)}(0) = \frac{d^{2n} x_k(t)}{dt^{2n}}\Bigg|_{t=0} . \tag{24}$$

The frequency moments, expressed in Eqs. (23) and (24), are seen to be simple static averages which can be computed by using the standard QMC techniques described in Section II.

In particular, an example of the use of Eqs. (23) and (24) to determine the frequency moments of the spectral distribution can be seen in the computation of the second moment, $<\omega^2>$. From Eqs. (1), (2), and (24) we determine

$$x_k^{(2)} = -\frac{1}{h^2}\Big[[x_k,H],H\Big]$$

$$= -\frac{1}{m}\frac{1}{\sqrt{N}}\sum_n \frac{\partial}{\partial x_n}\Big\{\phi(x_n-x_{n-1}) + \phi(x_{n+1}-x_n)\Big\}e^{-ikna}, \qquad (25)$$

where a is the average nearest neighbor separation, so that from Eq. (23) we need to compute

$$<x_k^{(2)}(0)x_{-k}(0)> = -\frac{1}{m}(1-e^{ika})\frac{1}{\sqrt{N}}\sum_n <\Big(\frac{\partial}{\partial x_n}\phi(x_n-x_{n-1})\Big)x_{-k}>e^{-ikna}. \qquad (26)$$

The thermodynamic averages of the quantum mechanical variables in Eq. (26) can be reexpressed in terms of averages of path-integral operators (expressed in terms of the $\{x_j^{(i)}\}$ variables) based on the path-integral partition function of Eqs. (9) (see for example, Chapter 7 of Feynman and Hibbs[20]). The details of this mapping of thermodynamic averages of quantum mechanical operators onto the averages of path-integral operators in the path-integral formulation and a discussion of the evaluation of higher frequency moments will be presented in a future publication.

In Fig. 3 we present preliminary results for $C_k(\omega)$ calculated for k at the Brillouin zone boundary for a chain of ten atoms with M = 15 and k_BT/ϵ = 0.175, 0.2 and 0.3. Only the positive frequency parts of $C_k(\omega)$ are shown, and the full widths at half maximum at k_BT/ϵ = 0.175, 0.2, and 0.3 are $\Delta\omega$ = 0.0015 $\pm$ 0.0014, 0.66 $\pm$ 0.33, 4.6 $\pm$ 0.6, respectively, in units of $\sqrt{\epsilon/m}/\sigma$. In addition to the widths of these peaks, the anharmonicity of the system is seen to cause a gradual frequency shift in the phonon spectrum as the temperature is raised, so that for k_BT/ϵ = 0.175, 0.2 and 0.3, respectively, the maximum peak height occurs at ω = 17.25 $\pm$ 0.09, 18.13 $\pm$ 0.27, 19.37 $\pm$ 0.13, in units of $\sqrt{\epsilon/m}/\sigma$.

The results presented in Fig. 3 are as yet only preliminary. A study of the convergence properties in M of $C_k(\omega)$ and a detailed comparison with results obtained from standard approximation methods available in the study of phonons in anharmonic crystals will be presented elsewhere. Nevertheless, we feel that the proposed QMC methodology holds great promise in treating the low temperature properties of nonlinear quantum mechanical systems which have well defined single particle excitations. In both the static and dynamic QMC studies we have made on many-body vibrational systems, the QMC has been seen to be effective in:

1) computing accurate anharmonic quantum corrections to the harmonic approximations;

2) acting as an interpolation between the very low temperature region and the classical high temperature limit;

3) determining phonon frequency shifts and lifetimes which arise due to quantum fluctuations and system anharmonicities.

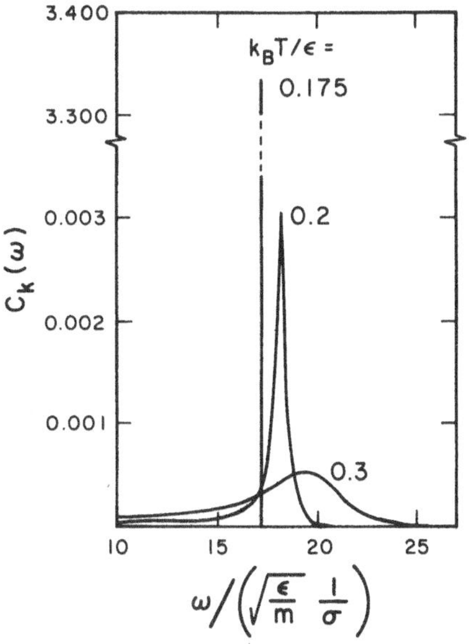

Fig. 3. Plot of $C_k(\omega)$ for the one-dimensional chain versus ω. The wavevector k is at the Brillouin zone boundary, the average nearest neighbor atomic separation is $a=2^{1/6}\sigma$, and results for temperatures $k_BT/\epsilon = 0.175$, 0.2, and 0.3 are shown. The plots presented at each k_BT/ϵ are made using the average values of $\mu_0(k)$, $a_1^2(k)$, and $a_2^2(k)$ from the simulation. The errors in peak frequency and in the full width at half maximum of these curves are given in the text.

For less well studied quantum systems than the present vibrational many-body models, we expect that the QMC method will be a valuable tool for quickly obtaining accurate representations of their static and dynamic properties.

Acknowledgement

This research was supported in part by NSF Grant No. DMR 89-18184.

References

1. NATO Advanced Research Workshop on Monte Carlo Methods in Quantum Problems, ed. Malvin H. Kalos, D. Reidel Pub. Co., Hingham, MA, USA (1984).
2. Quantum Monte Carlo Methods in Equilibrium and Non-Equilibrium Systems, ed. M. Suzuki, Springer-Verlag, Berlin and New York (1987).
3. Journal of Statistical Physics 43, pp. 729-1243 (1986).
4. International Workshop on Quantum Simulations of Condensed Matter Phenomena, eds. J. D. Doll and J. E. Gubernatis, World Scientific, Singapore (1990).
5. E. L. Pollock and D. M. Ceperley, Phys. Rev. B30, 2555-2568 (1984).
6. Statistical Physics, L. D. Landau and E. M. Lifshitz, Pergamon Press, Oxford (1980), p. 537.
7. A. R. McGurn, P. Ryan, A. A. Maradudin, and R. F. Wallis, Phys. Rev. B40, 2407-2413 (1989).
8. H. Mori, Prog. Theor. Phys. 34, 399-416 (1965).
9. A. A. Maradudin, R. F. Wallis, A. R. McGurn, M. S. Daw, and A.J.C. Ladd, in Lattice Dynamics and Semiconductor Physics, eds. J. Xia, Z. Gan, R. Han, G. Qin, G. Yang, H. Zheng, Z. Zhong, and B. Zhu, World Scientific, Singapore, (1990), pp. 103-157.

10. F. Gürsey, Proc. Camb. Phil. Soc. 46, 182 (1950).
11. A. A. Maradudin, in Physics of Phonons, ed. T. Paszkiewicz, Springer-Verlag, Berlin (1987), pp. 1-47.
12. H. F. Trotter, Proc. Am. Math. Soc. 110, 545 551 (1959)
13. M. Suzuki, Prog. Theor Phys. 56, 145-1469 (1976).
14. M. Takahashi and M. Imada, J. Phys. Soc. Jpn 53, 3765-3769 (1984).
15. Applications of Monte Carlo Methods in Statistical Physics, ed. K. Binder, Springer-Verlag, Berlin (1984).
16. S. W. Lovesey, Condensed Matter Physics: Dynamic Correlations, second edition, Benjamin/Cummings Pub. Co., Inc. (1986), Ch. 1.
17. S. W. Lovesey, in Physics in One Dimension, eds. J. Bernasconi and T. Schneider, Springer-Verlag, Berlin (1981) pp. 129-139.
18. H. Tomita and H. Mashiyama, Prog. Theor. Phys. 48, 1133-1149 (1972).
19. K. Tomita and H. Tomita, Prog. Theor. Phys. 45, 1407-1436 (1971).
20. R. P. Feynman and A. R. Hibbs, Quantum Mechanics and Path-Integrals, McGraw-Hill, New York (1965), Ch. 7.

EXACT FREE ENERGY OF A KSSH MODEL IN d-DIMENSIONS

Arianna Montorsi, and Mario Rasetti

Dipartimento di Fisica, and Unitá I.N.F.M.
Politecnico di Torino, 10129 Torino, Italy

Abstract

The *K.S.S.H.*-like model one obtains when taking into account spin-non-conserving spin-orbit interactions in the hamiltonian, has been shown to be exactly solvable in any number of dimensions for a particular choice of the coupling constant describing the hopping process amplitude. Here we review in details the solution and we discuss the behavior of the free energy.

1. Introduction

It has been recently shown[1] that a version of the model proposed by *Kievelson, Schrieffer, Su,* and *Heeger* (*K.S.S.H.*) [2] is exactly solvable for any number of dimensions d of the lattice Λ over which it is defined.

The model considered includes, besides the hopping term and the *Hubbard*-like local Coulomb repulsion term, an off-diagonal part intended to represent bond-charge repulsion interactions [3],[4]. Furthermore the hopping amplitudes are spin-dependent, so as to take into account spin-orbit coupling. The resulting hamiltonian is given by

$$\begin{aligned} H = & \sum_{\mathbf{i},\sigma} \varepsilon n_{\mathbf{i},\sigma} + U \sum_{\mathbf{i}} n_{\mathbf{i},\Uparrow} n_{\mathbf{i},\Downarrow} + \frac{1}{2} \sum_{<\mathbf{i},\mathbf{j}>} \sum_{\sigma,\sigma'} t_{\sigma,\sigma'} \left(a^\dagger_{\mathbf{i},\sigma} a_{\mathbf{j},\sigma'} + a^\dagger_{\mathbf{j},\sigma'} a_{\mathbf{i},\sigma} \right) \\ & - \frac{1}{2} \sum_{<\mathbf{i},\mathbf{j}>} \sum_{\sigma,\sigma'} \Delta t_{\sigma,\sigma'} \left(a^\dagger_{\mathbf{i},\sigma} a_{\mathbf{j},\sigma'} + a^\dagger_{\mathbf{j},\sigma'} a_{\mathbf{i},\sigma} \right) \left(n_{\mathbf{i},-\sigma} + n_{\mathbf{j},-\sigma'} \right) \quad ; \end{aligned} \tag{1}$$

where the last factor represents just the *bond-charge* repulsion term characteristic of the *K.S.S.H.* model. As customary, $a^\dagger_{\mathbf{i},\sigma}, a_{\mathbf{i},\sigma}$ are fermionic creation and annihilation operators ($\{a_{\mathbf{i},\sigma}, a_{\mathbf{j},\sigma'}\} = 0$, $\{a^\dagger_{\mathbf{i},\sigma}, a_{\mathbf{j},\sigma'}\} = \delta_{\mathbf{i},\mathbf{j}} \delta_{\sigma,\sigma'} \mathbb{I}$, $n_{\mathbf{i},\sigma} \doteq a^\dagger_{\mathbf{i},\sigma} a_{\mathbf{i},\sigma}$) over a d-dimensional lattice Λ ($\mathbf{i},\mathbf{j} \in \Lambda$, $\sigma \in \{\Uparrow,\Downarrow\}$), whereas the parameters ε, U, t, and Δt have the usual meaning [4], and $< \mathbf{i},\mathbf{j} >$ stands for nearest neighbours (*n.n.*) in Λ.

Notice that the extra terms in (1) due to spin-orbit coupling describe, in this formulation, spin-flip (spin non-conserving) hopping processes.

Microscopic Aspects of Nonlinearity in Condensed Matter
Edited by A.R. Bishop *et al.*, Plenum Press, New York, 1991

PROCESS

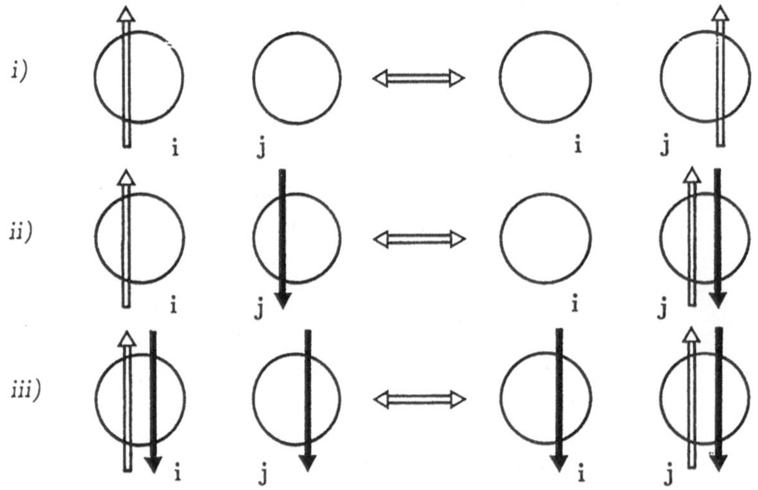

AMPLITUDE

$$\frac{1}{2}t$$

$$\frac{1}{2}(t - \Delta t)$$

$$\frac{1}{2}t - \Delta t$$

The model (1), with a special choice of the hopping parameters, is exactly solvable – in the sense of statistical mechanics – in any number of dimensions. Here we shall review the solution proposed in ref. [1], in order to evaluate explicitly the free energy.

Contrary to the Hubbard case, hamiltonian (1) assigns different amplitudes to the hopping processes, depending on the relative site occupation. This is schematically shown in Fig. 1., which in particular shows how the closer Δt is to t, the more the hopping processes between two singly occupied sites, and a doubly occupied and an empty site are inhibited.

In the particular case when $t_{\sigma,\sigma'} \equiv \Delta t_{\sigma,\sigma'}$ it is straightforward to verify the important feature that the hopping part in (1) commutes with the diagonal part of the hamiltonian, implying that both the number of electrons $N_e \doteq \sum_i n_{i,\Uparrow} + n_{i,\Downarrow}$ and the number of doubly occupied sites $\sum_i n_{i,\Uparrow} n_{i,\Downarrow}$ are conserved quantities. As this choice implies equal amplitudes (in absolute value) for the hopping processes from singly occupied to empty sites as from doubly occupied to singly occupied sites (see Fig. 1.), we add to the hamiltonian a term

$$\frac{1}{2}\gamma \sum_{<i,j>} \sum_{\sigma,\sigma'} t_{\sigma,\sigma'} \left(a^\dagger_{i,\sigma} a_{j,\sigma'} + a^\dagger_{j,\sigma'} a_{i,\sigma} \right) n_{i,-\sigma} n_{j,-\sigma'} \quad ,$$

which has the only effect to allow us changing the relative amplitudes of the two processes.

The resulting model is then exactly solvable if $t_{\sigma,\sigma} \equiv t_{\sigma,-\sigma} \doteq t$. Notice that the latter condition could be replaced by $t_{\sigma,-\sigma} = 0$ (*i.e.* spin conservation, no spin-flip processes allowed) [5]. As the model in this case turns out to have a much more complicated combinatorial structure, we confine here our attention to the former case.

2. Model and Dynamical Algebra

In the above hypothesis, it is convenient to introduce the new set of local operators :

$$\begin{aligned} A_i &\doteq \frac{1}{\sqrt{2}}\left(a_{i,\Uparrow} + a_{i,\Downarrow}\right) \quad ; \quad N_i \doteq A^\dagger_i A_i \quad ; \\ D_i &\doteq \frac{1}{2}\left(n_{i,\Uparrow} + n_{i,\Downarrow} + a_{i,\Uparrow} a^\dagger_{i,\Downarrow} + a_{i,\Downarrow} a^\dagger_{i,\Uparrow}\right) \quad ; \end{aligned} \tag{2}$$

such that $n_{i,\Uparrow} + n_{i,\Downarrow} = N_i + D_i$, and $n_{i,\Uparrow} n_{i,\Downarrow} = D_i N_i \; \forall i \in \Lambda$. The new spinless operators $A_i, A^\dagger_i, N_i$ still generate a fermionic algebra: $\{A_i, A_j\} = 0$; $\{A_i, A^\dagger_j\} = \delta_{i,j}\mathbb{I}$; $[A_i, N_i] = A_i \delta_{i,j}$, and all of them commute with the idempotent operators D_i:

$$D_i D_i = D_i \quad ; \quad [\bullet, D_i] = 0 \quad , \tag{3}$$

• being any of the operators in (2).

In terms of the latter the effective hamiltonian of our model, $\mathcal{H} \doteq H - \mu N_e$ (where μ denotes the chemical potential), which is required in the grand-canonical ensemble scheme, reads

$$\mathcal{H} = \sum_i \omega \left(N_i + D_i\right) + t \sum_{<i,j>} \kappa_{i,j} \left(A^\dagger_i A_j + A^\dagger_j A_i\right) + U \sum_i N_i D_i \quad ; \tag{4}$$

where $\omega \doteq \varepsilon - \mu$, and $\kappa_{i,j} \equiv \mathbb{I} - D_i - D_j + \gamma D_i D_j$.

The system described by (4) can be readily seen to have a dynamical algebra $\mathcal{L}$. Upon introducing the subindex α – a poly-index corresponding to a collection of binary variables, one for each lattice site – $\alpha \equiv \{s_i | s_i \in \{0,1\}; i \in \Lambda\}$ (the set of all α's has cardinality $2^{\mathcal{N}}$, $\mathcal{N}$ being the total number of sites in Λ) one can write

$$\mathcal{L} = \bigoplus_{\alpha} \mathcal{A}_\alpha \quad ; \tag{5}$$

where

$$\mathcal{A}_\alpha = \mathcal{P}_\alpha \mathcal{A} \quad . \tag{6}$$

Here $\mathcal{A}$ denotes the $\mathcal{N}^2$-dimensional $\mathbf{Z}_2$-graded algebra

$$\mathcal{A} = < N_i, A_i^\dagger A_j, \mathbb{I} | i,j \in \Lambda, i \neq j > \sim u(\mathcal{N}) \quad ; \tag{7}$$

$\mathcal{P}_\alpha$ is the projection operator $(\mathcal{P}_\alpha \mathcal{P}_{\alpha'} = \delta_{\alpha,\alpha'} \mathcal{P}_{\alpha'})$ onto the representation specified by a given collection of eigenvalues of the D_i's , each eigenvalue corresponding locally (*i.e.* at site $i \in \Lambda$) to the value of s_i entering the definiton of α. $\mathcal{P}_\alpha$ is straightforwardly constructed as $\mathcal{P}_\alpha = \prod_{i \in \Lambda} p_i^{(\alpha)}$, with

$$p_i^{(\alpha)} \doteq \frac{1}{2} e^{-i\frac{\pi}{2} s_i} \left(e^{i\frac{\pi}{2} D_i} + e^{i\pi s_i} e^{-i\frac{\pi}{2} D_i} \right) \quad , \tag{8}$$

namely,

$$p_i^{(\alpha)} = \begin{cases} \cos(\frac{\pi}{2} D_i) \equiv \mathbb{I} - D_i \,, & \text{if } s_i = 0 \;; \\ \sin(\frac{\pi}{2} D_i) \equiv D_i \,, & \text{if } s_i = 1 \,. \end{cases}$$

In view of (5), the hamiltonian $\mathcal{H}$ can be now written as

$$\mathcal{H} = \bigoplus_{\alpha} \mathcal{H}_\alpha \quad , \quad \mathcal{H}_\alpha \doteq \mathcal{H} \mathcal{P}_\alpha \quad . \tag{9}$$

3. Solution of the Model

Statistical mechanical solution of the model requires the evaluation of the grand-canonical partition function

$$Z_{\mathcal{N}} = \mathrm{Tr}_{\mathcal{L}} \{\exp(-\beta \mathcal{H})\} = \mathrm{Tr}_{\mathcal{L}} \{\exp(-\beta \mathcal{H}_0) \exp(-\beta \mathcal{H}_1)\} \quad , \tag{10}$$

where $\beta = 1/k_B T$, and use has been made of the property $[\mathcal{H}_0^{(\alpha)}, \mathcal{H}_1^{(\alpha)}] = 0$.

Due to (9), trace over the dynamical algebra $\mathcal{L}$, Tr, reduces to

$$\mathrm{Tr}_{\mathcal{L}} \equiv \sum_{\{\alpha\}} \mathrm{tr}_{\mathcal{A}_\alpha} \quad ; \quad \left(\sum_{\{\alpha\}} \equiv \sum_{\{s_i\}} \right) \quad . \tag{11}$$

Eq.(11) shows explicitly how the algebra direct sum structure (5) allowed us to deal with the operators D_i as with classical *Ising*-like dynamical variables (as one could of course expect from the commutation relations (3)). The main problem is therefore to evaluate $\mathrm{tr}_{\mathcal{A}_\alpha}$, namely the trace over the α-th copy of $\mathcal{A}$, corresponding to the eigenvalues s_i's defining α. In order to takle such problem, we refer the Hilbert space of states to the *Fock* basis $\{|\psi_k>\}$

$$|\psi_k> \doteq \bigotimes_i |\nu_i> \quad , \quad k = 1, \ldots, 2^{\mathcal{N}} \quad ,$$

where $|\nu_i>$ denotes the eigenvector of the operator $N_i : N_i|\nu_i>= \nu_i|\nu_i> \ \nu_i \in \{0,1\}$.

In the above basis,

$$\mathrm{tr}_{\mathcal{A}_\alpha} \{e^{-\beta\mathcal{H}_\alpha}\} = \sum_{\{\nu_i\}} \exp\left\{\sum_i [(\omega + Us_i)\nu_i + \omega s_i]\right\} < \psi_k|e^{-\beta\mathcal{H}_1^{(\alpha)}}|\psi_k> \quad . \tag{12}$$

The matrix elements $< \psi_k|\exp\{-\beta\mathcal{H}_1^{(\alpha)}\}|\psi_k>$ entering (12) can be expanded as a power series in the variable $\tau \doteq -2\beta t$ giving rise to

$$\begin{aligned} < \psi_k|e^{-\beta\mathcal{H}_1^{(\alpha)}}|\psi_k> &= \sum_{\ell=0}^{\infty} \frac{\tau^{2\ell}}{(2\ell)!} < \psi_k| \left\{ \sum_{<i,j>} \kappa_{i,j} A_i^\dagger A_j \right\}^{2\ell} |\psi_k> \\ &= 1 + \sum_{\ell=1}^{\infty} \frac{\tau^{2\ell}}{(2\ell)!} \sum_{\{\lambda_\ell\}} \Theta_{\lambda_\ell}(\{s_i\}) \Xi_{\lambda_\ell}(\{\nu_i\}) \quad . \end{aligned} \tag{13}$$

In the last factor at the *r.h.s.* of (13) $\{\lambda_\ell\}$ denotes the collection of all loops in Λ with ℓ edges (which can possibly multiply cover the same bonds in Λ). The equivalence between the two series in τ is based on the property that the matrix elements in the Fock state $|\psi_k>$ of the ℓ-th power of $\mathcal{H}_1^{(\alpha)}$ obviously have nonvanishing contributions only from those factors which are associated with bonds $< i, j >$ with non-dangling ends. One can easily check that such contributions are each the product of a term Θ_{λ_ℓ} - originated by the $\kappa_{i,j}$'s, hence depending only on the $\{s_i\}$'s – times a term Ξ_{λ_ℓ} – diagonal matrix element in the state $|\psi_k>$ of a sum of products of an equal number of operators A_i and $A_i^\dagger$ along the loop – naturally depending only on the $\{\nu_i\}$'s.

It is now convenient to decompose the set $\{\lambda_\ell\}$ into its disconnected components, the i-th of which is characterized by the number of edges $2\ell_i$ and the number of sites touched m_i. We denote the corresponding set by $\{\lambda_{\ell_i}^{(m_i)}|1 \le \ell_i \le \ell; 2 \le m_i \le L_i\}$ ($L_i \equiv \min(2\ell_i, \mathcal{N})$). The sum over $\{\lambda_\ell\}$ in (7) is then decomposed into

$$\sum_{j=1}^{J_\ell} \sum_{\{\ell_i\}}{}' (2\ell)! \prod_{n=1}^{j} \left(\frac{1}{(2\ell_n)!} \sum_{m_n=2}^{L_n} \sum_{\{\lambda_{\ell_n}^{(m_n)}\}} \Theta_{\lambda_{\ell_n}^{(m_n)}}(\{s_i\}) \Xi_{\lambda_{\ell_n}^{(m_n)}}(\{\nu_i\}) \right) \quad , \tag{14}$$

where $J_\ell \equiv \min(\ell, \frac{\mathcal{N}}{2})$, whereas $\sum_{\{\ell_i\}}'$ denotes $\sum_{\ell_1 \ge 1} \ldots \sum_{\ell_j \ge 1} \delta(\ell - \sum_{n=1}^{j} \ell_n)$.

Due to the particular form of $\kappa_{i,j}$ one has

$$\Theta_{\lambda_{\ell_n}^{(m_n)}}(\{s_i\}) \equiv \prod_{\substack{i=1 \\ \mathrm{mod}\ m}}^{m} \kappa_{i,i+1} = \prod_{i \in \lambda_{\ell_n}^{(m_n)}} \delta(s_i, 0) + \varrho^{2\ell_n} \prod_{i \in \lambda_{\ell_n}^{(m_n)}} \delta(s_i, 1) \quad ; \tag{15}$$

where i labels the sites of $\lambda_{\ell_n}^{(m_n)}$ numbered along the loop (so that $i+1$ is $n.n.$ of i). We shall see in the sequel that the explicit expression of $\Xi_{\lambda_{\ell_n}^{(m_n)}}(\{\nu_{\mathrm{i}}\})$ is not needed.

Eqs.(10) to (15) allow us to rewrite $Z_{\mathcal{N}}$ as the sum of two contributions, $Z_{\mathcal{N}} = Z_0 + Z_1$, the first of which – originated by the factor 1 in (13) – has the form

$$\begin{aligned} Z_0 &= \sum_{\{s_{\mathrm{i}}\}} \sum_{\{\nu_{\mathrm{i}}\}} \exp\left\{ \sum_{\mathrm{i}\in\Lambda} \left[(\omega + U s_{\mathrm{i}})\, \nu_{\mathrm{i}} + \omega s_{\mathrm{i}} \right] \right\} \\ &= \left[1 + 2e^{-\beta\omega} + e^{-\beta(U+2\omega)} \right]^{\mathcal{N}} \equiv z_0^{\mathcal{N}} \quad , \end{aligned} \tag{16}$$

whereas the second, containing the remaining part of the series, upon inserting the explicit expression (15) for $\Theta_{\lambda_{\ell_n}^{(m_n)}}(\{s_{\mathrm{i}}\})$ and performing the configuration sums over $\{s_{\mathrm{i}}\}$ and $\{\nu_{\mathrm{i}} | \mathrm{i} \notin \lambda_{\ell_n}^{(m_n)}\}$, turns out to be:

$$Z_1 = Z_0 \sum_{\ell=1}^{\infty} \tau^{2\ell} \sum_{j=1}^{J_\ell} \sum_{\{\ell_i\}}{}' \prod_{n=1}^{j} \left[\mathcal{Q}_{\ell_n}^{(n)}(\omega, z_0) + \varrho^{2\ell_n} \mathcal{Q}_{\ell_n}^{(n)}(\omega + U, e^{\frac{1}{2}\beta\omega} z_0) \right] \quad , \tag{17}$$

which contains the only remaining trace :

$$\mathcal{Q}_{\ell_n}^{(n)}(\Omega, \xi_0) \doteq \frac{1}{(2\ell_n)!} \sum_{m_n=2}^{L_n} \xi_0^{-2m_n} \sum_{\{\lambda_{\ell_n}^{(m_n)}\}} \sum_{\substack{\{\nu_{\mathrm{i}}\} \\ \mathrm{i}\in\lambda_{\ell_n}^{(m_n)}}} \Xi_{\lambda_{\ell_n}^{(m_n)}}(\{\nu_{\mathrm{i}}\}) \prod_{\mathrm{i}} e^{-\beta\Omega\nu_{\mathrm{i}}} \, . \tag{18}$$

(18) can be evaluated globally, by comparing the solution to an associated pure *tight-binding* model obtained by the same procedure adopted above with the conventional solution derived by transforming to wave-vector space [6].

Let us recall that a pure spinless tight-binding model can be characterized by the hamiltonian

$$H_{t.b.} = \omega \sum_{\mathrm{i}} N_{\mathrm{i}} + t \sum_{<\mathrm{i},\mathrm{j}>} \left(A_{\mathrm{i}}^{\dagger} A_{\mathrm{j}} + A_{\mathrm{j}}^{\dagger} A_{\mathrm{i}} \right) \quad . \tag{19}$$

Repeating the whole derivation which led from (4) to (18), one obtains the partition function

$$Z_{t.b.} = \tilde{z}_0^{\mathcal{N}} \left\{ 1 + \sum_{\ell=1}^{\infty} \tau^{2\ell} \sum_{j=1}^{J_\ell} \sum_{\{\ell_i\}}{}' \prod_{n=1}^{j} \mathcal{Q}_{\ell_n}^{(n)}(\omega, \tilde{z}_0) \right\} \quad , \tag{20}$$

where $\tilde{z}_0 \equiv 1 + e^{-\beta\omega}$.

On the other hand, the model (19) can be exactly solved in standard way for any number of dimensions d, as $H_{t.b.}$ is diagonal in the wave-vector space. One finds in this manner for $Z_{t.b.}$ an expression formally identical with (20), in which

$$\prod_{n=1}^{j} \mathcal{Q}_{\ell_n}^{(n)}(\omega, \tilde{z}_0) = \frac{1}{j!} \left(\frac{2e^{-\beta\omega}}{\tilde{z}_0^2} \right)^j \sum_{\mathbf{k}^{(1)},\ldots,\mathbf{k}^{(j)}}{}'' \prod_{n=1}^{j} \left(\frac{1}{(2\ell_n)!} \left[\varphi\left(\mathbf{k}^{(n)} \right) \right]^{2\ell_n} \right) \quad . \tag{21}$$

In (21), $\mathbf{k} = \{k_1, \ldots, k_d\}$ is a vector in reciprocal space, ranging over the domain D, which is a half of the first *Brillouin Zone* ($B.Z.$), the double prime in $\sum\limits_{\{\mathbf{k}^{(n)}\}}''$ is to remind the requirement that each term in the sum must have all indices $\mathbf{k}^{(n)}$, $n = 1, \ldots, m$ different from each other, and $\varphi(\mathbf{k}) = 2\sum\limits_{r=1}^{d} \cos(k_r)$.

(21) must hold for every j, and provides both the functional form $Q_{\ell_n}^{(n)}(\Omega, \xi_0) = 2\frac{e^{-\beta\Omega}}{\xi_0{}^2}\frac{Q_{\ell_n}^{(n)}}{n}$ and the recursive equations for the $Q_{\ell_n}^{(n)}$'s.

Insertion of the solutions for the latter in (17), leads then, after some manipulations, to

$$Z_1 = Z_0 \sum_{\ell=1}^{\infty} \tau^{2\ell} \sum_{j=1}^{J_\ell} \frac{1}{j!} \sum_{\{\ell_i\}}{}' \prod_{n=1}^{j} \frac{1}{(2\ell_n)!}\left(x + \varrho^{2\ell_n} y\right) \sum_{\mathbf{k}^{(1)},\ldots,\mathbf{k}^{(j)}}'' \prod_{m=1}^{j} \left[\varphi(\mathbf{k}^{(m)})\right]^{2\ell_m}, \tag{22}$$

where $x \doteq \frac{2}{z_0^2} e^{-\beta\omega}$, $y \doteq x e^{-\beta(\omega+U)}$.

Resorting to the integral representation for the δ-function, one can explicitly perform the sum $\sum\limits_{\{\ell_i\}}'$. Besides, making use of the equality

$$\sum_{\mathbf{k}^{(1)},\ldots,\mathbf{k}^{(j)}}'' \prod_{n=1}^{j} f(\mathbf{k}^{(n)}) = (-)^j j! \sum_{p_1,\ldots,p_j \in \pi_j} \prod_{n=1}^{j} \frac{1}{p_n!}\left(\frac{\mathcal{F}_n}{n}\right)^{p_n}, \tag{23}$$

where π_j denotes the set of integers $\{p_n | n = 1, \ldots, j\}$ such that $\sum\limits_{n=1}^{j} n p_n = j$ and $\mathcal{F}_n \doteq -\sum\limits_{\mathbf{k}} (f(\mathbf{k}))^n$, and interchanging the summations over j and ℓ one gets

$$Z_1 = Z_0 \sum_{j=1}^{\frac{1}{2}\mathcal{N}} (-)^j \sum_{\ell=j}^{\infty} \frac{1}{2\pi} \int_{-\pi}^{\pi} d\zeta \left(\tau e^{i\zeta}\right)^{2\ell} \sum_{p_1,\ldots,p_j \in \pi_j} \prod_{n=1}^{j} \frac{1}{p_n!}\left\{-\frac{1}{n}\sum_{\mathbf{k}}[G(\mathbf{k}, e^{-i\zeta})]^n\right\}^{p_n}, \tag{24}$$

with $G(\mathbf{k}, z) \equiv x(\cosh(z\varphi(\mathbf{k})) - 1) + y(\cosh(z\varrho\varphi(\mathbf{k})) - 1)$.

Expression (24) can be further elaborated into

$$Z_1 = Z_0 \sum_{\ell=0}^{\infty} \tau^{2\ell} \frac{1}{2\pi i} \oint_C \frac{dz}{z^{2\ell+1}} \left\{\sum_{j=1}^{\frac{1}{2}\mathcal{N}} (-\tau^2)^j \sum_{p_1,\ldots,p_j \in \pi_j} \prod_{n=1}^{j} \frac{1}{p_n!}\left(\frac{\mathcal{G}_n(z)}{n}\right)^{p_n}\right\}, \tag{25}$$

where $\mathcal{G}_n(z) \equiv -\frac{1}{z^2}\sum_{\mathbf{k}}[G(\mathbf{k},z)]^n$ is a regular function in $z = 0$ and the integration path C is the unit circle. The integral can be easily performed, and the partition function $Z_{\mathcal{N}}$ is then obtained as

$$Z_{\mathcal{N}} = 1 + \sum_{j=1}^{\frac{1}{2}\mathcal{N}}(-\tau^2)^j \sum_{p_1,\dots,p_j\in\pi_j} \prod_{n=1}^{j} \frac{1}{p_n!}\left(\frac{\mathcal{G}_n(\tau)}{n}\right)^{p_n} \quad . \tag{26}$$

In the thermodynamic limit $\mathcal{N} \to \infty$, expression (26) can be further simplified by use of the identity[7]

$$\sum_{m=0}^{\infty} \xi^m \sum_{p_1,\dots,p_j\in\pi_j} \prod_{n=1}^{j} \frac{1}{p_n!}\left(\frac{\mathcal{F}_n}{n}\right)^{p_n} = \exp\left(-\sum_{n=1}^{\infty}\frac{\xi^n}{n}\mathcal{F}_n\right) \quad , \tag{27}$$

and reads

$$Z_{\mathcal{N}} = Z_0 \exp(\mathcal{N}\mathcal{B}) \quad . \tag{28}$$

with $\mathcal{B} \doteq \frac{1}{\mathcal{N}}\sum_{\mathbf{k}} \ln\left(1 + G(\mathbf{k},\tau)\right)$.

We now turn to the study of the free energy. In the thermodynamic limit the free energy per site, namely $f = \lim_{\mathcal{N}\to\infty} -\frac{1}{\beta\mathcal{N}} \ln Z_{\mathcal{N}}$, is straightforwardly obtained from (28), and is given by

$$f = -\frac{1}{\beta}\left(\ln z_0 + \mathcal{B}\right) \quad . \tag{29}$$

In order to give meaning to such limit we have first to recall that it turns the sum over $\mathbf{k} \in \mathcal{D}$ into a convergent integral ($\lim_{\mathcal{N}\to\infty} \frac{1}{\mathcal{N}} \sum_{\mathbf{k}\in\mathcal{D}} \Rightarrow \frac{1}{(2\pi)^d}\int_0^{2\pi} d\vartheta_1 \dots \int_0^{2\pi} d\vartheta_{d-1} \int_0^{\pi} d\vartheta_d$). In [1] we evaluated explicitly this multiple integral for the particular choice $\varrho = \pm 1$, in which case $\mathcal{B}$ turns out to be a sum of modified Bessel functions I_0. This is infact still true for all ϱ, but the coefficients are themselves (finite) sums.

The behavior of the free energy per site is then given by equation (29) once ω is suitably chosen. Indeed, one has to fix the total electron occupation number, namely the ratio $n_0 \doteq \frac{< N_e >}{\mathcal{N}}$. This is done here following the customary convention of choosing the chemical potential (*i.e.* ω), as the solution of the equation

$$n_0 \equiv \frac{\partial f}{\partial \omega} \quad . \tag{30}$$

Equation (30) turns out to be highly non-linear, and to have different solutions. Among them, for each temperature, the one minimizing the free energy (29) has to be selected, which turns out to depend on the physical parameters. The detailed discussion of the phase space emerging from the non-trivial interplay of equations (29)-(30) is not given here, but preliminar analysis shows that the chemical potential plays infact a crucial role in it, and can be considered as the "order parameter" for describing the phase transition.

In conclusion, it is worth pointing out that the solution presented, besides constituting an interesting example of exactly solvable system, provides on one hand a promising starting point for working out perturbatively (*e.g.* in the parameter $t - \Delta t$) solutions of the generalized *K.S.S.H.* model (1); on the other hand, as the partition function $Z_{\mathcal{N}}$ is given exactly also for finite $\mathcal{N}$, it can provide a reference test for numerical calculations on a non-trivial quantum system. Finally, work is at present in progress on the problem of studying the solution of the same model within the fermi-linearization approximation scheme [8]. This can be done rather easily in that the underlying dynamical (super)algebras are the same. The exact solution discussed here can then become an excellent test-ground for the fermionic mean-field approach.

References

[1] A. Montorsi, and M. Rasetti, *submitted to Phys. Rev. Letters*

[2] S. Kivelson, W.P. Su, J.R. Schrieffer, and A.J. Heeger, *Phys. Rev. Lett.* **58**, 1899 (1987)

[3] For a discussion of the relevance of the bond-charge repulsion term, see:
D. Baeriswyl, P. Horsch, and K. Maki, *Phys. Rev. Lett.* **60**, 70 (1988)
J. T. Gammel, and D.K. Campbell, *ibid.*, page 71
S. Kivelson, W.P. Su, J.R. Schrieffer, and A.J. Heeger, *ibid.*, page 72

[4] J.E. Hirsch, *Physica* **C 158**, 326 (1989)

[5] A. Montorsi, and M. Rasetti, *in preparation*

[6] D.C. Mattis, *The theory of magnetism I*, Springer Verlag, Berlin, 1981, chapter 6

[7] J. Riordan, *Combinatorial Identities*, J. Wiley & Sons, New York, 1968, p. 183

[8] A. Montorsi, and M. Rasetti, *Mod. Phys. Letters* **B 4**, 613 (1990)

NONLINEAR PROPERTIES OF THE BCS GAP EQUATION

M.P. Sörensen, T. Schneider and M. Frick

IBM Research Division, IBM Research Laboratory
8803 Rüschlikon, Switzerland

The iterative solution of the BCS gap equation defines a nonlinear map, which can be studied with the tools developed for nonlinear dynamics. We demonstrate the usefulness of this approach in determining the phase diagram in the superconducting state, in the presence of a non-trivial pairing interaction, having mixed s and d-wave symmetry.

I. Introduction

In this paper we study the BCS equation [1] using techniques developed in the field of nonlinear dynamics [2-4]. The BCS equation is a nonlinear integral equation for the order parameter of the superconducting state. In Fourier space it reads [1,5].

$$\Delta(\mathbf{k}) = \frac{1}{N}\sum_{\mathbf{k}'} V(\mathbf{k},\mathbf{k}')F(\mathbf{k}') \quad , \tag{1}$$

where

$$F(\mathbf{k}) = \frac{\Delta(\mathbf{k})}{2E(\mathbf{k})}\tanh\left(\frac{1}{2}\beta E(\mathbf{k})\right) \quad , \tag{2}$$

and the quasiparticle excitation energy is given by

$$E(\mathbf{k}) = \sqrt{\varepsilon^2(\mathbf{k}) + \Delta^2(\mathbf{k})} \quad . \tag{3}$$

Here, $\mathbf{k}$ denotes the wave vector, $\Delta(\mathbf{k})$ the order parameter or gap, $V(\mathbf{k},\mathbf{k}')$ the pairing interaction, $\varepsilon(\mathbf{k})$ the quasiparticle energy in the normal state measured relative to the chemical potential μ, and $\beta = 1/k_B T$, where T denotes the temperature. Assuming a unit cell with lattice constants (a, a, s), then k_α adopts the values $k_\alpha = n_\alpha \pi/(a_\alpha N_\alpha)$, where

Microscopic Aspects of Nonlinearity in Condensed Matter
Edited by A.R. Bishop *et al.*, Plenum Press, New York, 1991

$-N_\alpha < n_\alpha \leq N_\alpha$, $\mathbf{a} = (a,a,s)$, and $N_x, N_y, N_z \in N$. Accordingly, Δ, V and E are periodic functions and invariant under the space group symmetry operations of the crystal in the superconducting state. In addition, only the solution yielding the lowest Gibbs free energy [5] i.e.

$$G = \frac{1}{N}\sum_{\mathbf{k}}\left\{\varepsilon(\mathbf{k}) - E(\mathbf{k}) + 2\Delta(\mathbf{k})F(\mathbf{k}) - \frac{2}{\beta}\ln\left(1 + \exp(-\beta E(\mathbf{k}))\right)\right\}$$
$$- \frac{1}{N^2}\sum_{\mathbf{k},\mathbf{k}'} V(\mathbf{k},\mathbf{k}')F(\mathbf{k})F(\mathbf{k}') \qquad (4)$$
$$= \frac{1}{N}\sum_{\mathbf{k}}\left\{\varepsilon(\mathbf{k}) - E(\mathbf{k}) + \Delta(\mathbf{k})F(\mathbf{k}) - \frac{2}{\beta}\ln\left(1 + \exp(-\beta E(\mathbf{k}))\right)\right\}$$

is a relevant physical solution. In the last expression of Eq. (4) we have used the gap equation (1). It is important to emphasize that the BCS gap equation defines only the extrema of the free energy. In the simplest limiting case, where $V(\mathbf{k},\mathbf{k}')$ is assumed to be a constant, there are only two extrema $\Delta^* = \pm\Delta \neq 0$ (other than the trivial one $\Delta^* = 0$), which are absolute minima. In anisotropic systems, however, the pairing interaction is more complicated and the gap equation has many solutions.

To illustrate the usefulness of concepts developed in nonlinear dynamics for finding solutions to Eq. (1), we consider the special case of on-site pairing, where $V(\mathbf{k},\mathbf{k}') = g_0$. The gap equation then simplifies to

$$\Delta(T) = \frac{g_0}{N}\sum_{\mathbf{k}'}\frac{\Delta(T)}{2E(\mathbf{k}')}\tanh\left(\frac{1}{2}\beta E(\mathbf{k}')\right) = f(\Delta,T)\Delta(T) = K(\Delta,T) \ , \qquad (5)$$
$$\beta = 1/k_B T \ .$$

As we know from BCS theory [1], there is the trivial solution $\Delta = 0$, stable above the transition temperature T_c, given by $f(\Delta = 0,T_c) = 1$, which becomes unstable for $T < T_c$. The nontrivial solutions $1 = f(\pm\Delta,T)$ are stable below T_c. In Fig. 1 we depict the stable solutions in terms of solid lines and the unstable ones as dashed lines. The Gibbs free energy, shown in Fig. 2, illustrates the stability at $T = 0$.

Alternatively, solving Eq. (5) starting from some initial value Δ_0 leads to the nonlinear map

$$\Delta_{i+1} = K(\Delta_i) \ , \qquad (6)$$

and a sequence of $\Delta_i(T)$, converging to a fixed point $\Delta^*(T)$. In the following we shall refer to the map in Eq. (6) as the BCS map. The fixed points are solutions of the gap equation. Since K depends parametrically on T, there are lines of fixed points as seen in the flow and bifurcation diagram shown in Fig. 1. The stability of the fixed points is determined by

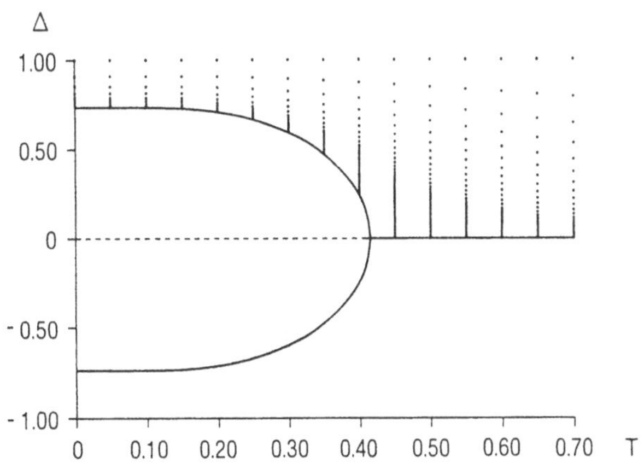

Fig. 1. Bifurcation and flow diagram for the one-dimensional BCS map, for $\varepsilon(\mathbf{k}) = -2(\cos(k_x a) + \cos(k_y a)) + 1.8\cos(k_x a)\cos(k_y a) - \mu$ and $\mu = 4.256$, $g_0 = 6$. The solid line corresponds to stable fixed points and the dashed one to unstable ones. The dots represent iterates Δ_i with initial value $\Delta = 1$ resulting from 100 iterations on a mesh with $N = 128^2$.

$$\lambda(T) = \left.\frac{dK(\Delta, T)}{d\Delta}\right|_{\Delta=\Delta^*} . \tag{7}$$

Above the transition temperature $\lambda(T) < 1$, the line of fixed points $\Delta^*(T) = 0$ $(T > T_c)$ is stable. At T_c, we have $\lambda(T_c) = 1$, because $\lambda(T_c) = f(0, T_c) = 1$, and a pitchfork bifurcation occurs, producing two lines of stable fixed points $\Delta(T) = \pm\Delta^* \neq 0$. Their stability follows from $\lambda(T) < 1$. In fact, for $T = 0$ we have

$$\lambda(T=0) = 1 - \frac{g_0\Delta^2}{2}\frac{1}{N}\sum_{\mathbf{k}}(\varepsilon^2(\mathbf{k}) + \Delta^2)^{-3/2} < 1 . \tag{8}$$

The stability of these fixed points can also be understood in terms of the Gibbs free energy shown in Fig. 2. The positions of the absolute minima correspond to the stable fixed points and for $T < T_c$, the extremum at $\Delta = 0$ is unstable. This simple example clearly reveals that the flow and bifurcation diagrams resulting from the BCS map are useful guides in identifying the relevant physical solutions.

In fact, for more complicated pairing interactions, the order parameter is **k**-dependent and can be decomposed into a sum of terms having different symmetry. In the superconducting phase, where $\Delta(\mathbf{k}) \neq \mathbf{0}$, not only one, but several phases with different symmetries, or mixed phases might appear and exchange their stability when parameters such as temperature and chemical potential are varied. Thus, even in the superconducting state, bifurcations or phase transitions might occur. In view of this, flow and bifurcation diagrams appear to be very useful to identify the lines of stable fixed points corresponding to phase boundaries and bifurcation points (critical points).

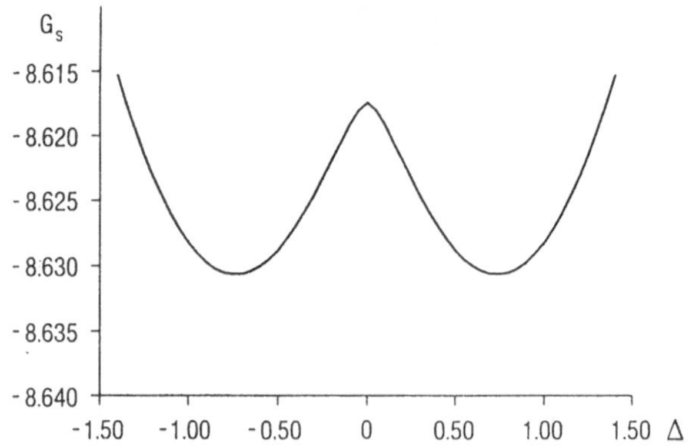

Fig. 2. Gibbs free energy versus Δ for $T < T_c$ for the parameters used in Fig. 1 and at $T = 0$.

In Section II we consider a model relevant to layered high-temperature superconductors having order parameters with s and d-wave symmetries. In particular, we study the two-dimensional BCS map, resulting from a nearest-neighbor pairing interaction with the aid of the flow and bifurcation diagram, the Gibbs free energy and the specific heat. By lowering the chemical potential at zero temperature, we also identify three superconducting phases, and the associated bifurcations correspond to transitions from s to d and mixed s, d-wave superconductivity. Similarly, a bifurcation is also found for fixed μ by increasing the temperature. It corresponds to a phase transition from mixed s, d-wave to d-wave superconductivity.

This nontrivial example clearly reveals the usefulness of the approach. The flow diagram of the BCS map provides an easily accessible overview of the location of the fixed points and the bifurcation diagram allows us to locate phase transitions.

II. Gap equation and the associated BCS map

For the special case $V(\mathbf{k}, \mathbf{k}') = g_0$ we have shown that an iterative solution of the gap equation leads to a nonlinear map. It is intuitively obvious that more realistic and complicated pairing interactions lead to a similar scenario. To substantiate this conjecture, we consider a model relevant to layered high-temperature superconductors [6,7]. Concentrating on the transition from the normal to the superconducting state, and assuming tight-binding Fermi liquid states, subject to an unretarded pairing interaction, the Hamiltonian reads

$$\mathcal{H} = \sum_{i,j,\sigma} t_{ij}\, a^{+}_{i,\sigma} a_{j,\sigma} - \sum_{i,j,\sigma} g_{ij}\, a^{+}_{i\uparrow} a^{+}_{j\downarrow} a_{j\downarrow} a_{i\uparrow} \ . \tag{9}$$

As depicted in Fig. 3, the t_{ij} term describes the hopping of the quasiparticles within and between the layers, by taking into account nearest-

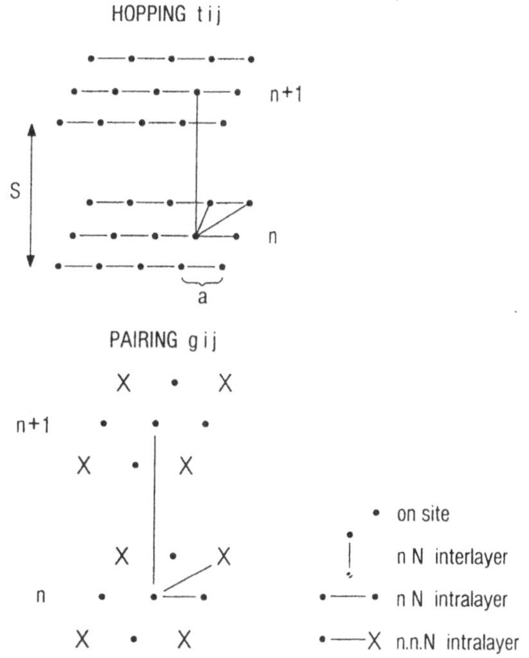

Fig. 3. (a) Schematic sketch of the lattice sites and the hopping of the quasiparticles. (b) Sketch of the pairing interactions.

neighbor (n.N.) hopping within the planes as $t_{ij} = A$, between the planes as $t_{ij} = AC$, and next-nearest-neighbor (n.n.N.) hopping within the sheets as $t_{ij} = AB$. Similarly, the pairing includes on-site interaction ($g_{ii} = g_0$), interaction between n.N.s within the layers ($g_{ij} = g_2$) and between the layers ($g_{ij} = g_1$) and n.n.N. pairing interaction within the sheets ($g_{ij} = g_3$) (Fig. 3). In this model, the pairing interaction is given by

$$\begin{aligned} V(\mathbf{k},\mathbf{k}') = g_0 &+ 2\,g_1 \cos((k_z - k'_z)s) \\ &+ 2\,g_2 \left[\cos((k_x - k'_x)a) + \cos((k_y - k'_y)a)\right] \\ &+ 4\,g_3 \cos((k_x - k'_x)a)\cos((k_y - k'_y)a) \quad , \end{aligned} \tag{10}$$

and the quasiparticle band reads

$$\varepsilon(\mathbf{k}) = A(-2(\cos(k_x a) + \cos(k_y a)) + 4B\cos(k_x a)\cos(k_y a) - 2C\cos(k_z s) - \mu) \quad , \tag{11}$$

where μ is the chemical potential. In the calculations below, both A and k_B are set to unity. This means that all energies are measured in units of A and the absolute temperature is measured in units of A/k_B. Restricting our study to singlet pairing, where

$$\Delta(\mathbf{k}) = \Delta(-\mathbf{k}) \quad , \tag{12}$$

the order parameter can be expressed in terms of

$$\Delta(\mathbf{k}) = \Delta_0 + 2\Delta_1 \cos(k_z s) + 2\Delta_{2x} \cos(k_x a) + 2\Delta_{2y} \cos(k_y a) + 4\Delta_{3s} \cos(k_x a)\cos(k_y a) + 4\Delta_{3d} \sin(k_x a)\sin(k_y a) \ , \tag{13}$$

by taking the periodicity and the crystal symmetry into account. The term $2\Delta_{2x}\cos(k_x a) + 2\Delta_{2y}\cos(k_y a)$ can also be written in the form $\Delta_{2s}(\cos(k_x a) + \cos(k_y a)) + \Delta_{2d}(\cos(k_y a) - \cos(k_y a))$, where $\Delta_{2s} = \Delta_{2x} + \Delta_{2y}$ and $\Delta_{2d} = \Delta_{2x} - \Delta_{2y}$. The classification of the order parameter (gap) is then given by Δ_0 describing s-wave pairing and by Δ_1, Δ_{2s} and Δ_{3s} describing extended s-wave pairing, while Δ_{2d} and Δ_{3d} correspond to d-wave pairing.

Here we consider the case $g_1 = 0$ such that $\Delta_1 = 0$. Insertion of Eqs. (10), (11) and (13) into (1) then gives

$$\Delta = \mathbf{B}(\Delta)\Delta = \mathbf{K}(\Delta) \ , \tag{14}$$

with $\mathbf{\Delta} = (\Delta_0, \Delta_{2x}, \Delta_{2y}, \Delta_{3s}, \Delta_{3d})$. $\mathbf{K}$ and $\mathbf{B}$ are nonlinear vector functions of Δ, and $\mathbf{B}$ is given by

$$\mathbf{B} = \begin{bmatrix} g_0 f_{11} & 2g_0 f_{12} & 2g_0 f_{13} & 4g_0 f_{14} & 0 \\ g_2 f_{21} & 2g_2 f_{22} & 2g_2 f_{23} & 4g_2 f_{24} & 0 \\ g_2 f_{31} & 2g_2 f_{32} & 2g_2 f_{33} & 4g_2 f_{34} & 0 \\ g_3 f_{41} & 2g_3 f_{42} & 4g_3 f_{43} & 4g_3 f_{44} & 0 \\ 0 & 0 & 0 & 0 & 4g_3 f_{55} \end{bmatrix} \tag{15}$$

The elements f_{ij} read

$$\begin{aligned} &f_{11} = \sum_{\mathbf{k}} \tilde{F}(\mathbf{k}), \qquad f_{12} = \sum_{\mathbf{k}} \cos(k_x a)\tilde{F}(\mathbf{k}), \\ &f_{13} = \sum_{\mathbf{k}} \cos(k_y a)\tilde{F}(\mathbf{k}), \qquad f_{14} = \sum_{\mathbf{k}} \cos(k_x a)\cos(k_y a)\tilde{F}(\mathbf{k}), \\ &f_{22} = \sum_{\mathbf{k}} \cos^2(k_x a)\tilde{F}(\mathbf{k}), \quad f_{23} = f_{14}, \quad f_{24} = \sum_{\mathbf{k}} \cos^2(k_x a)\cos(k_y a)\tilde{F}(\mathbf{k}), \\ &f_{33} = \sum_{\mathbf{k}} \cos^2(k_y a)\tilde{F}(\mathbf{k}), \qquad f_{34} = \sum_{\mathbf{k}} \cos(k_x a)\cos^2(k_y a)\tilde{F}(\mathbf{k}), \\ &f_{44} = \sum_{\mathbf{k}} \cos^2(k_x a)\cos^2(k_y a)\tilde{F}(\mathbf{k}), \\ &f_{55} = \sum_{\mathbf{k}} \sin^2(k_x a)\sin^2(k_y a)\tilde{F}(\mathbf{k}), \end{aligned} \tag{16}$$

where $\tilde{F}(\mathbf{k}) = F(\mathbf{k})/\Delta(\mathbf{k})$. The other f_{ij} follow from the symmetry relation $f_{ij} = f_{ji}$.

The gap equation can also be derived by minimizing the Gibbs free energy (4) with respect to $\Delta(\mathbf{k})$ for fixed chemical potential μ. The fixed μ corresponds to a grand canonical system. Accordingly, the gap equation is solved for a given μ and the solution with the lowest Gibbs free energy is the right one. When Δ is determined we can explicitly obtain the filling ρ from

$$\rho = \frac{1}{2N} \sum_{\mathbf{k}} \left[1 - \frac{\varepsilon(\mathbf{k})}{E(\mathbf{k})} \tanh\left(\frac{1}{2} \beta E(\mathbf{k}) \right) \right] . \tag{17}$$

A mathematically equivalent way of solving the gap equation is to fix ρ and determine Δ and μ from the Eqs. (1) and (17) together. Note that this procedure remains grand canonical because the gap equation (1) is derived from minimizing the Gibbs free energy with fixed μ and not from fixed ρ. The latter method is more difficult if more than one solution to the gap equation exists. In this case, the solution with the lowest Gibbs free energy should be chosen, but as μ in general will differ for different solutions of fixed ρ, we cannot decide which solution should be chosen by calculating the Gibbs free energy. The answer to this problem is to change ρ for each solution until all the different solutions have the same chemical potential μ, and then see which solution has the lowest Gibbs free energy.

In the following we solve the gap equation for fixed μ. Substitution of the order parameter (13) into the gap equation (1) reduces the task of solving N equations, with as many unknowns, to that of solving 5 equations (Eq. (14)). The solutions are the fixed points of the map

$$\Delta_{i+1} = \mathbf{K}(\Delta_i) \ , \tag{18}$$

which can be found numerically by iteration, or by estimating the zeros of $\Delta - \mathbf{K}(\Delta)$. The nonlinear map approach appears to be particularly useful, because many tools, developed for dynamical systems, are available and the flow and bifurcation diagrams provide an easy accessible overview, allowing us to identify the relevant solutions of the gap equation, when the parameters μ and T are varied.

Before investigating the BCS map (18) we should recall some crucial facts on the stability of fixed points and the flow around them [2,3]. The flow around fixed points Δ^* is determined by the eigenvalues λ_i and the associated eigenvectors $\mathbf{v}_i$ of the Jacobian $\mathbf{DK}(\Delta^*)$ of $\mathbf{K}$, taken at the fixed point. The stable manifold E^s, the unstable manifold E^u and the center manifold E^c are the spaces spanned by the eigenvectors whose eigenvalues have moduli less than 1, larger than 1 and equal to 1, respectively. Depending on the moduli of the eigenvalues λ_i, the nonlinear map has stable, unstable and center-invariant manifolds which, in the point Δ^*, are tangential to E^s, E^u and E^c, respectively. Sinks (asymptotically stable fixed points) are fixed points where all eigenvalues λ_i have a modulus less than 1 and similarly, sources (unstable fixed points) are fixed points where all eigenvalues have a modulus larger than 1. Hyperbolic fixed points appear when the moduli of some eigenvalues are less than 1 and the rest have moduli greater than 1. A necessary condition for the occurrence of bifurcation is that one of the eigenvalues be equal to one. This is equivalent to the condition [3]

$$\det[\mathbf{I} - \mathbf{DK}(\Delta^*)] = 0 \ , \tag{19}$$

where $\mathbf{I}$ is the identity matrix. For an asymptotically stable fixed point Δ^*, the determinant is positive while a negative value corresponds to an unstable fixed point.

As an example we will consider the two-dimensional BCS map resulting from nearest-neighbor intralayer pair interaction ($g_0 = g_1 = g_3 = 0$) next. Here the BCS map (18) reduces to

$$\Delta_{i+1} = \mathbf{K}(\Delta_i) = 2g_2 \begin{pmatrix} f_{22} & f_{23} \\ f_{32} & f_{33} \end{pmatrix} \Delta_i \ , \tag{20}$$

where $\Delta_i = (\Delta_{2x,i}, \Delta_{2y,i})$. The fixed points satisfy the following symmetry relation: if $\Delta^* = (\Delta_{2x}^*, \Delta_{2y}^*)$ is a fixed point, then $\Delta^* = \pm(\Delta_{2y}^*, \Delta_{2x}^*)$ is a fixed point as well. The Jacobian $\mathbf{DK}(\mathbf{0})$ taken at the trivial fixed point $\Delta^* = \mathbf{0}$ has the eigenvalues

$$\lambda_i = 2g_2(f_{22} \pm f_{23}) \ , \tag{21}$$

with eigenvectors $\mathbf{v}_1 = (\Delta_2, \Delta_2)$ and $\mathbf{v}_2 = (\Delta_2, -\Delta_2)$ corresponding to s-wave and d-wave pairing, respectively. For sufficiently high temperature T, both eigenvalues are less than 1 and the fixed point Δ^* is asymptotically stable. By lowering T the eigenvalues approach unity and, when the largest eigenvalue is 1, a pitchfork bifurcation to a solution $\Delta \neq \mathbf{0}$ occurs. The temperature T at which this bifurcation takes place is the superconducting transition temperature T_c.

The direction of the flow for the two-dimensional BCS map, Eq. (20), can easily be computed and provides an illuminating way of identifying all fixed points and their stabilities. Figure 4 depicts the flow diagram, using the parameter values $B = 0.45$, $C = 0.1$, $g_2 = 3.0$, $T = 0$ and $\mu = 1.5$, corresponding to $\rho = 0.7622$. The open squares are initial values of Δ and the filled squares indicate the flow direction after one iteration of Eq. (20). Points 1 and 2 are stable fixed points (sinks), and 7 and 8 are hyperbolic ones. The unstable fixed point $\Delta^* = \mathbf{0}$ appears as a source.

The μ-dependence of the fixed points is depicted in Fig. 5, where the numbering refers to different fixed points, and x and y label Δ_{2x}^*

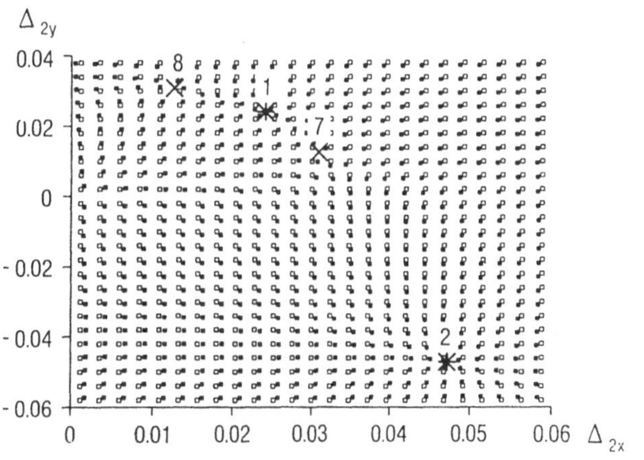

Fig. 4. Flow diagram of the map given in Eq. (20). Open squares are initial values of Δ and the solid squares indicate the direction of the flow after one iteration. Points 1 and 2 are stable fixed points, 7 and 8 are hyperbolic ones. Parameter values: $B = 0.45$, $C = 0.1$, $g_2 = 3.0$, $T = 0$ and $\mu = 1.5$, corresponding to $\rho = 0.7622$.

and Δ_{2y}^*, respectively. From the flow diagram (Fig. 4) we can readily identify points 1 and 2 as stable fixed points, and 7 and 8 as hyperbolic ones. Only half of the fixed points are shown, while the other half follows from the symmetry relation $\Delta(\mathbf{k}) = -\Delta(\mathbf{k})$. Thus full curves correspond to stable fixed points, while dashed curves represent unstable ones. Close to $\mu = 1.7$, a transition from s-wave to d-wave pairing takes place. The details of this transition are shown in the enlargement inserted in Fig. 5. Above $\mu = 2.8$, only s-wave pairing exists (lines $1x$

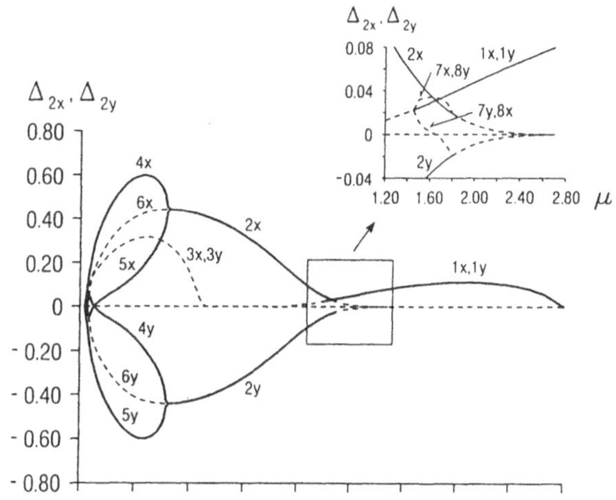

Fig. 5. Bifurcation diagram of $\Delta(\mu)$ at zero temperature of map (20) for the parameters $B = 0.45$, $C = 0.1$ and $g_2 = 3.0$. Solid curves mark stable fixed points and dashed curves unstable ones. The curves $1x = 1y$ and $3x = 3y$ correspond to s-wave pairing, $2x = -2y$ and $6x = -6y$ to d-wave pairing, and $4x = -5y$ and $4y = -5x$ to mixed s and d-wave pairing. The enlargement depicts the region around $\mu = 2$ in more detail. The curves $7x = 8y$ and $7y = 8x$ mark two distinct hyperbolic fixed points (mixed s and d-wave).

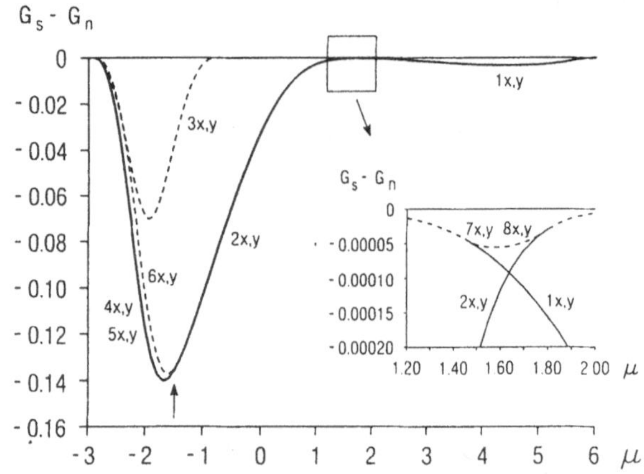

Fig. 6. Gibbs free energy at $T = 0$ for parameters used in Fig. 5. Bold curves represent the absolute minima, solid lines stable fixed points and dashed lines unstable ones. The labeling of the fixed points follows the notation in Fig. 5. The arrows mark the transition from s to d-wave ($\mu = 1.63$), and the one from d-wave to the mixed s, d-wave phase ($\mu = -1.4$). In the mixed phase the fixed points $4x, y$ and $5x, y$ are degenerate.

and $1y$), but close to $\mu = 2.8$ an unstable fixed point appears, having d-wave symmetry, $(2x, 2y)$. At $\mu = 1.8$, this fixed point bifurcates into a stable d-wave fixed point and two hyperbolic ones $(7x, 7y)$ and $(8x, 8y)$. Symmetry implies that the curves $7x$ and $8y$ are identical, i.e. $7x = 8y$, and similarly, $7y = 8x$. At the lower μ value, $\mu = 1.64$, two lines of stable fixed points, $1x = 1y$ and $2x = -2y$, cross. Thus, for fixed μ there can be several stable fixed points, and the one with the lowest Gibbs free energy is singled out as the physical one. From the Gibbs free energy shown in Fig. 6, and in particular from its enlargement in this region, it is clearly seen that the crossing leads to transition from s to

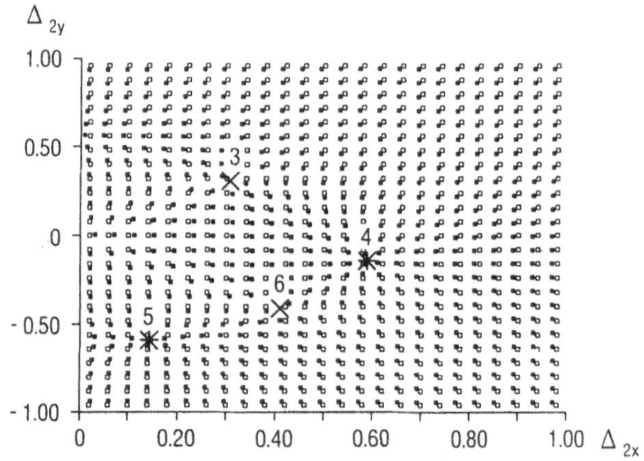

Fig. 7. Flow diagram of map in Eq. (20). Open squares are initial values of Δ and the solid squares indicate the direction of the flow after one iteration. Points 4 and 5 are stable fixed points, 3 and 6 are hyperbolic ones. Parameter values: $B = 0.45$, $C = 0.1$, $g_2 = 3.0$, $T = 0$ and $\mu = -2.0$, corresponding to $\rho = 0.1918$.

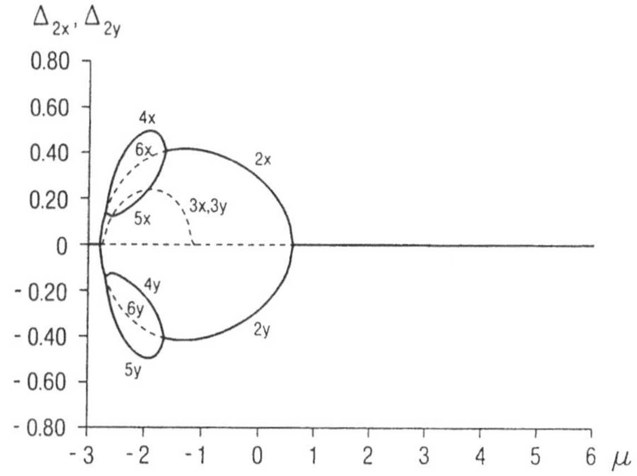

Fig. 8. Bifurcation diagram of $\Delta(\mu)$ at $T = 0.3$. Parameters are as in Fig. 5. Solid curves mark stable fixed points and dashed curves unstable ones. The curve $3x = 3y$ correspond to s-wave pairing, $2x = -2y$ and $6x = -6y$ to d-wave pairing, and $4x = -5y$ and $4y = -5x$ to mixed s and d-wave pairing.

d-wave pairing. Finally, at $\mu = 1.46$ the two hyperbolic fixed points 7 and 8 collide with the stable fixed point 1, which becomes unstable so that 7 and 8 disappear (pitchfork bifurcation).

From Fig. 7, which shows the flow diagram at $\mu = -2$, it is seen that new fixed points appear in this parameter range. We can identify points 4 and 5 as stable fixed points, and 3 and 6 as hyperbolic ones. From the bifurcation diagram (Fig. 5) we see that the stable fixed points result from a pitchfork bifurcation at $\mu = -1.4$. Here the stable d-wave state $2x = -2y$ ($\mu > -1.4$) becomes unstable and bifurcates into two mixed s, d-wave states, $4x = -5y$ and $4y = -5x$. From Fig. 6, it is seen that the mixed s, d-wave state possesses the lowest Gibbs free energy below $\mu = -1.4$. Thus by lowering μ, a transition from d to a mixed s, d-wave state occurs. Such a transition can also occur at fixed μ by increasing the temperature T. Clearly, a necessary condition is that several fixed points exist for this μ-value. As a function of temperature they might then exchange their stability. Indeed, by increasing T, the stable fixed points 4 and 5 move towards the unstable one, 6, as depicted in Fig. 8. When they collide a pitchfork bifurcation is expected. As shown in Fig. 9, the bifurcation from the mixed s, d-wave $(4x, 4y)$ and $(5x, 5y)$ to the d-state $(9x, 9y)$ occurs at $T = 0.66T_c$. This bifurcation will affect the temperature dependence of the specific heat:

$$C = T\frac{dS}{dT} = \frac{1}{2}\beta k_B^2 \sum_{\mathbf{k}} \left[E(\mathbf{k}) + \beta \frac{dE(\mathbf{k})}{d\beta} \right] E(\mathbf{k}) \mathrm{sech}^2\left(\frac{1}{2}\beta E(\mathbf{k}) \right) , \qquad (22)$$

where S is the entropy. In fact, for $\mu = -2.3$, the specific heat depicted in Fig. 10 clearly reveals two discontinuities in the temperature dependence at the bifurcation points. The jump at $T = 0.66T_c$ signals the transition from mixed s, d-wave to d-wave superconductivity, and the one at $T = T_c$ is due to the transition from the d-wave superconducting state to the normal phase.

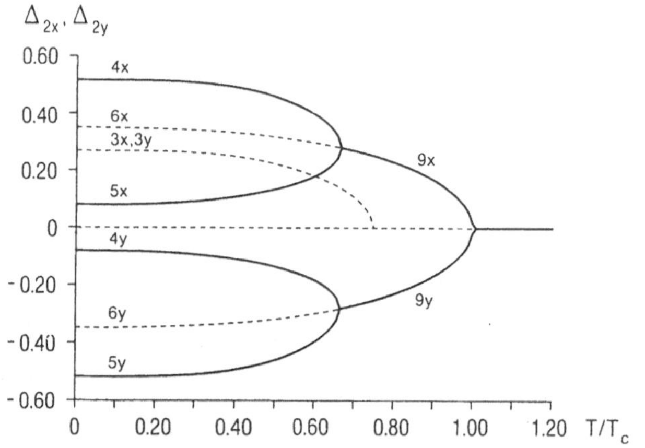

Fig. 9. Bifurcation diagram showing Δ_{2x} and Δ_{2y} versus the reduced temperature T/T_c at $\mu = -2.3$. Solid curves show stable fixed points and dashed curves show unstable ones. The curve $3x = 3y$ corresponds to s-wave pairing, $6x = -6y$ and $9x = -9y$ to d-wave pairing, and $4x = -5y$ and $4y = -5x$ to mixed s and d-wave pairing.

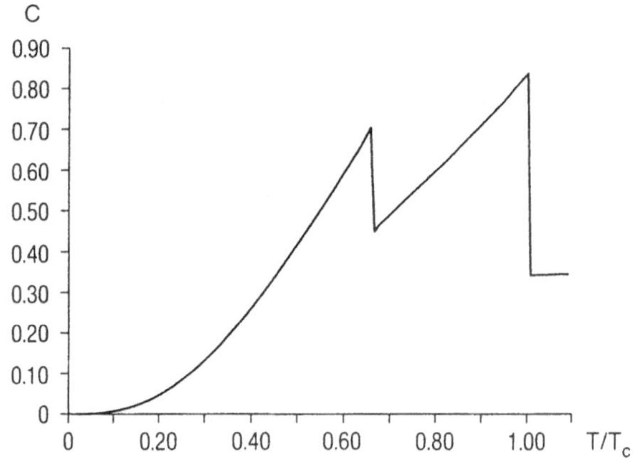

Fig. 10. The temperature dependence of the specific heat at $\mu = -2.3$ with parameters as in Fig. 4. The discontinuities signal the pitchfork bifurcations from a mixed s, d-wave to a d-wave and from d-wave to the normal state.

III. Summary and conclusions

We have established the relationship between the BCS gap equation for the order parameter in the superconducting state and nonlinear dynamics. Below the transition temperature, the stable fixed points with lowest Gibbs free energy represent the relevant physical solutions. The nonlinear dynamics approach appears to be particularly useful for anisotropic order parameters consisting of terms with different symmetries.

Here, we have explored a pairing interaction, which can be decomposed into terms with s and d-wave symmetries. Thus, the nonlinear map becomes two-dimensional (Eq. (18)) and superconducting phases with s, d and mixed s, d-wave symmetries can be expected. Here the nonlinear dynamics tools proved to be very useful. The flow diagram of the BCS map provides an easily accessible overview of the location of all fixed points and their stabilities. The bifurcation diagram allows us to locate the phase transitions within the superconducting state, between order parameters of different symmetry and between the superconducting and normal states. Considering a pairing interaction with s and d-wave coupling, we have illustrated the usefulness of the approach in determining the phase diagram in the μ, T-plane with the aid of the flow and bifurcation diagram, the Gibbs free energy and the specific heat.

References

1. J.R. Schrieffer, Theory of Superconductivity, New York: Addison-Wesley (1964).
2. J. Guckenheimer and P. Holmes, Nonlinear Oscillations, Dynamical Systems, and Bifurcations of Vector Fields, New York: Springer-Verlag (1983).
3. C.S. Hsu, Cell-to-Cell Mapping, New York: Springer-Verlag (1987), and references therein.
4. G. Gaeta, Physics Reports **189(1)** 1 (1990).
5. A.J. Leggett, Rev. Mod. Phys. **47(2),** 331 (1975).
6. T. Schneider, H. De Raedt, and M. Frick, Z. Phys. B - Condensed Matter **76,** 3 (1989).
7. T. Schneider and M.P. Sörensen, "Correlation between the Hall Coefficient, Penetration Depth, Transition Temperature, Gap Anisotropy and Hole Concentration in Layered High-Temperature Superconductors," submitted to Z. Phys. B - Condensed Matter (1990).

RESULTS OF MODE COUPLING THEORY FOR THE PARAMAGNETIC AND CRITICAL SPIN FLUCTUATIONS IN HEISENBERG MAGNETS

Alessandro Cuccoli*, Stephen W. Lovesey+ and Valerio Tognetti*

* Dipartimento di Fisica, Universitá di Firenze
Largo E. Fermi 2, I-50125 Firenze, Italy

+ Rutherford Appleton Laboratory
Chilton, Oxfordshire, 0X11 OQX, UK

INTRODUCTION

The progress in the understanding of the general features of static and dynamic critical phenomena, allowed by the use of renormalization group methods [1-2], raised in recent years the interest in experiments on the dynamics of spin fluctuations in simple paramagnets. The interpretation of the experimental data of magnetic neutron scattering on simple ferromagnets, however, revealed to be a tough challenge to available theoretical methods. In fact the performed experiments gave us not only the proof of the validity of the predictions of renormalization group concerning dynamic scaling and associated critical exponents [3], but provided us also with new more detailed informations on the shape of the magnetic response function which cannot be interpreted by renormalization group theory. In fact the latter gives at most an asymptotic form for the scaling function, but does not provide a means of calculation. The magnetic correlation functions can be obtained only within other approximation schemes, which can be divided basically in two classes: mode-coupling theories [4] and approximations developed starting from the consistency with low order frequency sum-rules [5]. The latter approaches give parametrized functional expressions for the response function which revealed very useful to fit experimental data when convoluted with instrumental resolution function, but they are not consistent with dynamic scaling. On the other side mode-coupling schemes not only provide explicit equations for response function, but these give also dynamic scaling with the right critical exponent; moreover mode coupling equations for Heisenberg paramagnets are consistent with the spherical model of static spin correlation function. Thus we have in mode coupling theory together with spherical model a complete and consistent description of static and dynamic correlations of spin fluctuations, which seems promising to study spin dynamics for arbitrary wave vector in the critical and paramagnetic phase.

BRIEF REVIEW OF MODE COUPLING THEORY

The mode-coupling theory is able to give a more satisfactory approach to the study of spin fluctuations in all the paramagnetic phase because it allows, also if approximately, to take into account the nonlinear dynamical coupling among fluctuation

Microscopic Aspects of Nonlinearity in Condensed Matter
Edited by A.R. Bishop *et al.*, Plenum Press, New York, 1991

modes. For an isotropic Heisenberg ferromagnet described by the hamiltonian:

$$H = -\frac{1}{2}\sum_{ij} J_{ij}\vec{s}_i \cdot \vec{s}_j - h\sum_i s_i^z \ , \tag{1}$$

where J_{ij} is the exchange interaction between spin operators $\vec{s}_i$ and $\vec{s}_j$, it leads to the following nonlinear equation for the spin relaxation function:

$$\partial_t F_q(t) = -2T\sum_k (J_k - J_{q-k})(J_k - J_q)\chi_k \int_0^t dt' F_k(t-t')F_{q-k}(t-t')F_q(t') \ . \tag{2}$$

$F_q(t)$ is connected to the scattering function $S(q,\omega)$ - observed in inelastic magnetic neutron scattering experiments - by the fluctuation-dissipation theorem:

$$S(q,\omega) = \frac{\omega}{1 - exp(-\beta\omega)}\chi_q F_q(\omega) \ . \tag{3}$$

From eq. (2) the dynamical scaling for isotropic ferromagnet is readily recovered; by adding the appropriate terms to the hamiltonian and consequently modifying eq. (3) [6], the mode-coupling approximation has proved to be useful also to describe the dynamical critical behaviour of ferromagnetic systems when the dipolar interaction is not negligible, and this allowed to explain also the most intriguing experimental findings of spin-echo experiments on europium oxide [7-8].

The use of the mode-coupling theory to interpret neutron scattering data collected out of the critical region, i.e. for high wave-vectors and temperatures, was up to now prevented by the fact that the known solutions of eq. (2) were obtained either analytically, for asymptotically small q [9], or numerically, but only for a simple cubic lattice with magnetic interactions restricted to nearest neighbours [10]. In a recent paper [11], in order to interpret experimental data, we have described in details a method to obtain a numerical solution of eq. (2) for the fcc magnetic lattice of europium compounds, which are considered the better physical realization of a simple Heisenberg ferromagnet with only *nn* and *nnn* interactions. Later work [12] extended the calculation to all three cubic lattices, and interactions up to ten shells of neighbouring magnetic ions. Results of such calculations are shown in the pictures and briefly described, with some comments, in the following section.

NUMERICAL RESULTS

Europium oxide is an fcc insulating ferromagnet with exchange interactions restricted to nearest and next nearest neighbours, and both interactions are ferromagnetic. The comparison of our numerical results with spin dynamics calculations [13] and neutron scattering data [14] at $T = 1.68T_c$ is shown in fig. 1. Three pole approximation, which is also reported in the picture, using for the parameters the value obtained in [14] fitting the experimental data, has been used to normalize the experimental data and it consents also to take into account the effect of resolution correction which is significant at small q. The experimental data were obtained from a powdered sample, but the average among the various direction is only a minor effect, because also our numerical calculation shows that the relaxation function in EuO comes out to be highly isotropic. The comparison of the results of mode coupling calculation with neutron scattering experimental data, also if a little bit worst, is favorable also for europium sulphide whose, major difference in respect to the europium oxide, is that the nnn interaction is antiferromagnetic; this makes the lineshapes for the various directions of the wave vectors near the zone boundary highly anisotropic [11].

Pd_2MnSn is similar in some respects to the europium compounds, in as much that it is an fcc ferromagnets with a relatively low critical temperature, but it is metallic. Nevertheless, in the ordered phase, the temperature dependence of the spin waves

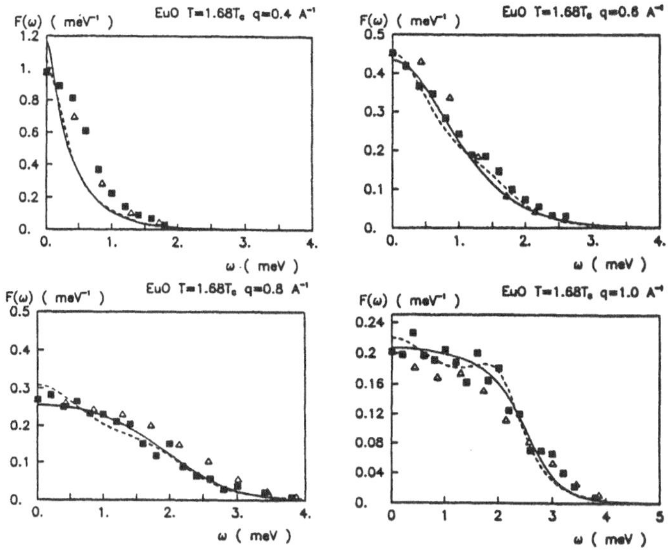

Fig. 1. Relaxation function of EuO at T = 1.68T$_c$ Full line: our numerical results; Open triangle: spin Dynamics simulation; Filled quad: neutron data; Dashed line: three pole approximation (see text).

energies and damping is well described by the interacting spin wave theory for a Heisenberg magnets, if allowance is made for interactions up to the eight shells of neighbours, which show an oscillating character at large distances. Therefore it has been suggested that Pd_2MnSn can be considered an ideal Heisenberg isotropic ferromagnet with long range interactions. Experimental data [17] for the Heusler compound Pd_2MnSn at $T = 1.1\ T_c$ and three wave vectors in the (111) direction are shown in Fig. 2 together with theoretical results obtained by mode coupling calculations and three pole approximation. Calculated spectra are convoluted with a lorentzian resolution function having as half width that given in the experimental paper.

Mode coupling calculation have also been carried out for a bcc lattice [12] with various sets of exchange constants proposed to model magnetic interactions in iron by a Heisenberg hamiltonian. Very different lineshapes are obtained for sets which give reasonable value for the critical temperature and reproduce the low wave vector behaviour of the spin waves dispersion curves. No final conclusion can however be drawn about the capability of the Heisenberg model and/or of the mode coupling theory to describe the dynamical behaviour of spin fluctuations in iron, because high resolution neutron scattering data, at constant wave vector and up to the zone boundary, are actually unavailable for this material.

CONCLUSIONS

The mode coupling approximation, for the dynamic properties of paramagnetic spin fluctuations in Heisenberg magnets, appears to provide a good interpretation of experimental data in insulating and metallic magnets. Good agreement between calculations and data requires the use of the correct lattice structure, and exchange constants. Spectra are found to be sensitive to the precise form of the exchange interaction, and this sensitivity can be exploited to distinguish between interactions that give tolerable spin wave dispersions, critical temperatures and other measurable quantities. New high resolution and constant-q data for paramagnetic Fe would be very helpful toward establishing the nature of the magnetic interactions and the reliability of coupled mode theory in metallic compounds. However another interesting test for mode coupling theory will be the interpretation of recently published neutron scattering data on Gadolinium, a metallic ferromagnet with a hcp magnetic lattice; work on this compound is in progress.

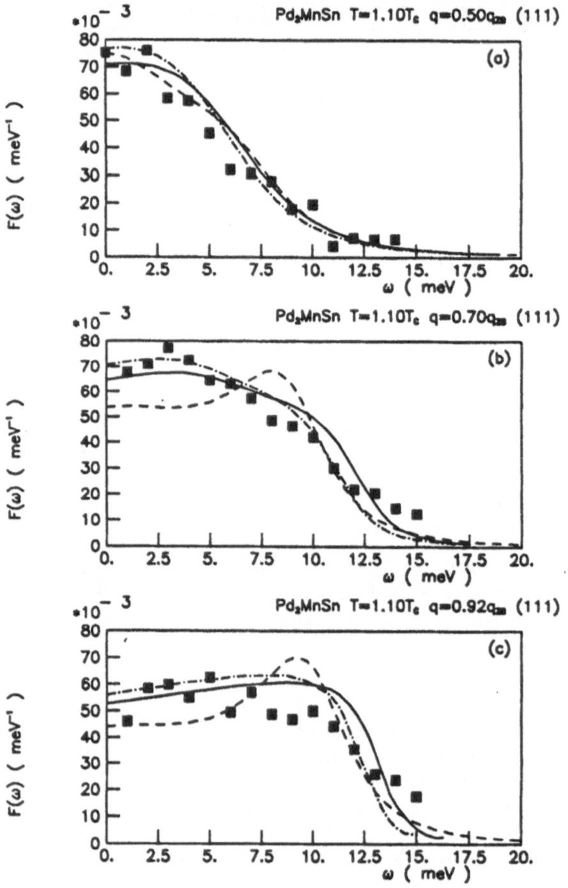

Fig. 2. Scattering function of Pd_2MnSn at $T = 1.1T_c$ Filled quad: neutron data; Solid and dash-dot curves: mode coupling calculations with six and eight shells of neighbours, respectively; Dashed curve: theoretical three pole approximation.

REFERENCES

[1] P.C. Hohenberg and B.I. Halperin, Rev. Mod. Phys. **49**, 1 (1977)
[2] R. Folk and H. Iro, Phys. Rev. B **32**, 1880 (1985)
[3] P. Böni, M.E. Chen and G. Shirane, Phys. Rev. B **35**, 8449 (1987)
[4] S.W. Lovesey, Condensed Matter Physic: Dynamic Correlations, Frontiers in Physics vol. 61, Benjamin Cummings (1986)
[5] S.W. Lovesey and R.A. Meserve, Jour. Phys. C **6**, 79 (1972)
[6] E. Frey and F. Schwabl, Phys. Lett. **123A**, 49 (1987)
[7] E. Frey, F. Schwabl and S. Thoma, Phys. Lett. **129A**, 343 (1988)
[8] F. Mezei, Physica **136B**, 417 (1986)
[9] S.W. Lovesey and R.D. Williams, Jour. Phys. C **19**, L253 (1986)
[10] J. Hubbard, Jour. Appl. Phys. **42**, 1390 (1971)
[11] A. Cuccoli, S.W. Lovesey and V. Tognetti, Phys. Rev. B **39**, 2619 (1989)
[12] A. Cuccoli, S.W. Lovesey and V. Tognetti, J. Phys. Cond. Matt. **2**, 3339 (1990)
[13] R. Chaudury and B.S. Shastry, Phys. Rev. B **37**, 5216 (1988)
[14] P. Böni and G. Shirane, Phys. Rev. B **33**, 3012 (1986)
[15] Mezei F., Farago B., Hayden S.M. and Stirling W.G., Physica **B156**, 226 (1989)
[16] H.G. Bohn, A. Kollmar and W. Zinn, Phys. Rev. B **11**, 6504 (1984)
[17] Graf H.A., Böni P., Shirane G., Kohgi M. and Endoh Y., Phys. Rev. B **40**, 4243 (1989)

QUANTUM THERMODYNAMICS OF A ANHARMONIC N-N CHAIN

Alessandro Cuccoli [a] , Valerio Tognetti [a] and Ruggero Vaia [b]

[a] Dipartimento di Fisica, Universitá di Firenze
Largo E. Fermi 2, I-50125 Firenze, Italy

[b] Istituto di Elettronica Quantistica - CNR
Via Panciatichi 56/30, I-50127 Firenze, Italy

INTRODUCTION AND SUMMARY

The thermodynamics of anharmonic crystals is an old subject[1] . Nevertheless, its study is strongly limited by the complexity of the available methods. For instance, direct quantum mechanical calculations involve the evaluation of very complicated integrals over wavevectors, which are of practical use only in particular limits[1] (low dimensionality and low-, high-temperature). Here we put forward a different approach.

The path-integral formulation of equilibrium statistical mechanics[2] provides an ideal starting point for attempting to account for the quantum character of a physical system in terms of corrections to the classical phase-space integral. Indeed, the latter is recovered by considering the only closed paths with minimal action, i.e. the pointlike ones. An expansion which locally accounts for the nontrivial paths permits to express the modifications due to quanticity by means of an "effective classical" hamiltonian. The calculations are then strongly simplified, since any classical method can be used. The idea is to devise an approximate "trial" action functional, which allows one to calculate the path-integral analytically. The advantage lies in that the choice of a trial functional is much more general than the choice of an approximating hamiltonian. With this purpose, Feynman introduced a variational approach[2] , defining an effective potential. However, it lacks the capability of exactly describing the quantum behavior of a linear system, and is not useful at low temperature. However, in recent years, Feynman's method has been improved[3] , simply by using a quadratic trial action and the solution of the quantum harmonic oscillator problem within the path-integral. The most interesting feature of this method is in that it accounts exactly for the quantum harmonic effects, without loosing the description of the nonlinear ones. The effective potential obtained in this way has been successfully used for many physical systems[3–5] , including field models[3,5] .

In the following we outline the variational method[3,5] and the calculation[6] of the quantum free energy of a linear array of atoms, interacting through a nearest-neighbor anharmonic potential. This chain is made "solid" by constraining it to a fixed total length, in such a way that fluctuations do not destroy it. The knowledge of the properties of such a system is the first step to study higher dimensional assemblies of interacting atoms, i.e. anharmonic crystals. Finally, we show an explicit application to the Lennard-Jones potential, and a comparison of the results with recent quantum Monte Carlo data.

Microscopic Aspects of Nonlinearity in Condensed Matter
Edited by A.R. Bishop *et al.*, Plenum Press, New York, 1991

THE EFFECTIVE POTENTIAL OF A N-N CHAIN.

We consider a chain of N atoms of mass m , described by the Lagrangian:

$$\mathcal{L} = \sum_{i=1}^{N} \frac{m}{2}\dot{x}_i^2 - V(x) \quad , \qquad V(x) = \sum_{i=1}^{N} \varepsilon \; u(x_i - x_{i-1}) \quad , \tag{1}$$

The constant ε sets the energy scale of the n-n interaction. Periodic boundary conditions are used, $x = (x_1, x_2, ..., x_N \equiv x_0)$, and the chain is constrained to a fixed length $L \equiv N \cdot d$.

As it is shown in ref. 3, the quantum partition function at the temperature $T \equiv \beta^{-1}$ can be approximated by its classical analogous, where $V(x)$ is replaced by the effective potential $V_{\text{eff}}(x)$ given by:

$$V_{\text{eff}}(x) = \widetilde{V}(x) - (m/4)\sum_k \alpha_k(x)\omega_k^2(x) + \beta^{-1}\sum_k \ln\left[\sinh f_k(x)/f_k(x)\right] \quad , \tag{2}$$

$$\widetilde{V}(x) = \int V(x + U^T\eta) \prod_k \left(\left[\pi\alpha_k(x)\right]^{-1/2} e^{-\eta_k^2/\alpha_k(x)} \, d\eta_k \right) \quad , \tag{3}$$

$$\alpha_k(x) = \hbar (m\omega_k)^{-1} \left(\coth f_k - f_k^{-1}\right) \quad , \qquad f_k(x) = \beta\hbar\omega_k/2 \quad , \tag{4}$$

and $\eta = \{\eta_k\}$, the index k assuming N values. The orthogonal matrix $U = \{U_{ki}(x)\}$ and the frequencies $\omega_k(x)$ are solution of the self consistent equations:

$$m \, \omega_k^2(x) \, \delta_{kl} = U_{ki} \, \widetilde{V_{ij}}(x) \, U_{lj} \quad , \tag{5}$$

where $V_{ij}(x) = \partial_{x_i}\partial_{x_j}V(x)$ and $\widetilde{V_{ij}}(x)$ is defined as in (3), which can be expanded as

$$\widetilde{V}(x) = \varepsilon \sum_{\ell=0}^{\infty} \frac{1}{\ell!} \sum_i \left(\frac{D_i}{2}\right)^{\ell} u^{(2\ell)}(x_i - x_{i-1}) \quad , \quad D_i(x) = \sum_k \left(U_{ki} - U_{k,i-1}\right)^2 \frac{\alpha_k}{2} \tag{6}$$

where $u^{(2l)}$ denotes the $2l$-th derivative of u and $D_i(x)$ is a renormalization factor.

The self consistent equations (4-5) are rather involved, due to the configuration dependence of U_{ki} , ω_k and D_i . For this reason we introduce a "low coupling" approximation, taking advantage of the observation that D_i is nothing but the pure quantum part of the average fluctuation of the variable $(x_i - x_{i-1})^2$, in gaussian approximation. As long as these fluctuations are very small, we can expand the effective potential starting from the most likely configuration, i.e. the equally spaced one. Since the latter is translation invariant, the matrix U_{ki} is nothing but a standard real Fourier transformation ($k = 2\pi n/N$), and it follows that the frequencies $\omega_k(x)$ are to be approximated with the renormalized frequencies $\Omega_k = 2\,\omega_d\kappa(T)\,\sin(k/2)$, where $\omega_d \equiv \sqrt{\varepsilon u''(d)/m}$ and $\kappa^2(T)$ is the (pure) quantum Hartree-Fock factor

$$\kappa^2(T) = \frac{1}{u''(d)} \sum_{\ell=0}^{\infty} \frac{u^{(2\ell+2)}(d)}{\ell!} \left(\frac{D}{2}\right)^{\ell} \quad . \tag{7}$$

Now $D = D(T)$ is self-consistently calculated with the frequencies Ω_k ,

$$D = N^{-1} \sum_k 4\sin^2(k/2) \; \hbar(2m\Omega_k)^{-1} \left(\coth F_k - F_k^{-1}\right) \quad , \tag{8}$$

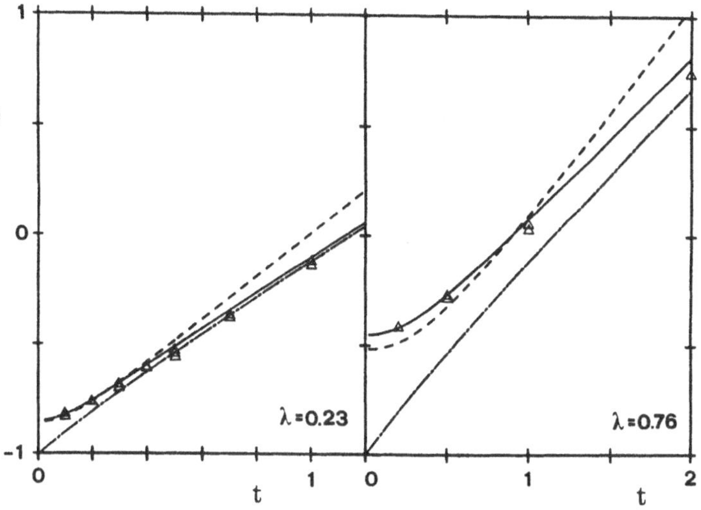

Fig. 1. Internal energy e per atom (units of ε) for the infinite Lennard-Jones chain with $d = d_0$. Full lines: effective potential; dashed lines: harmonic approximation; dash-dotted lines: classical result; triangles: quantum Monte Carlo, from ref. 9.

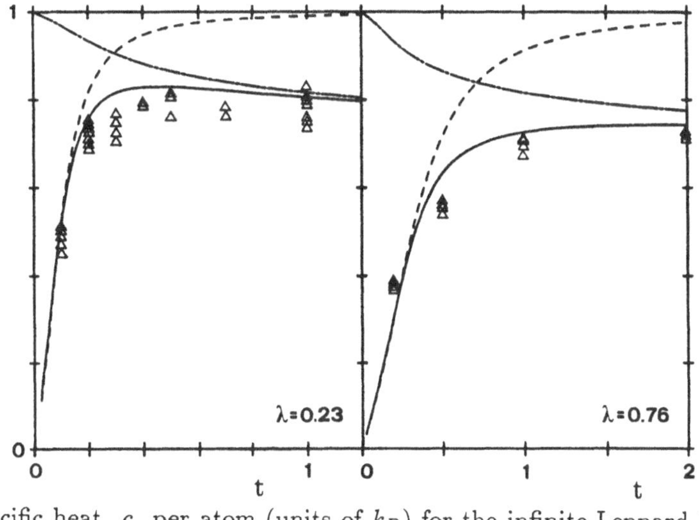

Fig. 2. Specific heat c per atom (units of k_B) for the infinite Lennard-Jones chain with $d = d_0$. Lines and symbols as in figure 1.

where $F_k = \beta\hbar\Omega_k/2$. The logarithmic term of (2) is expanded to first order in $\delta\omega_k^2(x) \equiv \omega_k^2(x) - \Omega_k^2$, yielding $V_{\rm eff} \simeq \sum_i u_{\rm eff}(x_i - x_{i-1})$, with

$$u_{\rm eff}(x) = \sum_{\ell=0}^{\infty} \frac{1}{\ell!}\left[u^{(2\ell)}(x) - \ell u^{(2\ell)}(d)\right]\left(\frac{D}{2}\right)^{\ell} + \frac{1}{\beta\varepsilon}\cdot\frac{1}{N}\sum_k \ln\left(\frac{\sinh F_k}{F_k}\right) . \qquad (9)$$

This pair potential can be shown to include all the one-loop quantum renormalization diagrams and is fully correct within terms of order D^2 . Indeed, in the limit of low T , it accounts for the full Hartree-Fock approximation, which is recovered through a self-consistent gaussian approximation of the configurational integral. The important feature is that the variational method produces this approximation on the bare quantum part. Of course, the low-T results are reliable as long as the Hartree-Fock approximation holds: the Ginzburg condition $u^{(4)}(d)\ D(T)\big[2u''(d)\big]^{-1} \ll 1$ should be satisfied. In cases with strong "coupling" this can lead to a lower temperature limit for the use of the effective potential. Furthermore, when T raises $D \simeq \hbar^2\beta/(12m) \to 0$, that is the renormalization becomes frequency-independent and the system becomes more and more "classical", in agreement with the results of the Wigner expansion[7] . Remarkably, the classical nonlinear features are fully accounted for by $u_{\rm eff}$ (this is not the case in the usual semiclassical theories). Moreover, the calculation of the configurational integral with the renormalized potential (9) also accounts for their quantum corrections.

THE LENNARD-JONES CHAIN

Let us apply the above framework to the case of a Lennard-Jones potential:

$$u(y/\sigma) = 4\left[(\sigma/y)^{12} - (\sigma/y)^6\right] \quad , \qquad (10)$$

where the constant σ sets the length scale of the interaction. We use the reduced variables $t = T/\varepsilon$, and $x = y/\sigma$. The (dimensionless) "coupling" parameter λ is defined as the ratio between the energy of the would-be quasi-harmonic excitations of the two-particle system and the well depth ε : $\lambda^2 = (\hbar^2\gamma^2)/(m\varepsilon\sigma^2)$, where $\gamma^2 = u''(d_0) = 72/\sqrt[3]{2}$ is the second derivative of $u(x)$ in its minimum $x = d_0 = \sqrt[6]{2}$. λ is related to the quantum fluctuation parameter α defined in ref. 9 by $\lambda = \sqrt{2}\gamma\alpha$.

In our computations we have retained from (9) terms up to $u^{(4)}(x)$, also according to the consistency of $u_{\rm eff}$ up to order D^2 . The free energy and the related thermodynamic quantities have been numerically evaluated by means of the transfer matrix method[8] . The calculation has been made for $\lambda = 0.23$ and $\lambda = 0.76$ (comparable to the characteristic parameters of the interaction of argon and neon atoms respectively), corresponding to the cases considered in the quantum Monte Carlo calculation of ref. 9, and the results for internal energy and specific heat per site are reported in the figures. The value $\lambda = 0.76$ actually gives rise to a strong coupling regime, since the frequency renormalization turns out to be $\kappa(t=0) = 1.38$. In this case the Ginzburg criterion is well satisfied for temperatures higher than $t \simeq 0.4$. It is apparent from the figures that our results interpolate very well from the quantum harmonic behavior at low t to the classical behavior at high t . The classical specific heat strongly deviates from the harmonic value $c = 1$, as a consequence of the strong anharmonicity of the interaction. Nevertheless, our results account also for the quantum correction to these deviations.

REFERENCES

1. A.A. Maradudin, P.A. Flinn and R.A. Coldwell-Horsfall, Ann. of Phys. **15**, 337 (1961).
2. R.P. Feynman, *Statistical Mechanics* (Benjamin, Reading MA, 1972)
3. R. Giachetti and V. Tognetti, Phys. Rev. Lett. **55**, 912 (1985); Phys. Rev. **B33**, 7647 (1986).

4. R.P. Feynman and H. Kleinert, Phys. Rev. **A34**, 5080 (1986).
5. R. Giachetti, V. Tognetti and R. Vaia, Phys. Rev. **A37**, 2165 (1988); **A38**, 1521 and 1638 (1988).
6. A. Cuccoli, V. Tognetti and R. Vaia, Phys. Rev. **B41**, 9588 (1990).
7. E.P. Wigner, Phys. Rev. **40**, 749 (1932).
8. F. Gürsey, Proc. Cambridge Phil. Soc. **46**, 182 (1950).
9. A.R. Mc Gurn, P. Ryan, A.A. Maradudin and R.F. Wallis, Phys. Rev. **B40**, 2407 (1989); see also A.R. Mc Gurn et al., in this volume.

SPECTRUM OF THE FERMI-LINEARIZED EXTENDED HUBBARD MODEL ON A DIMER

Luiz R. Evangelista♮, *Arianna Montorsi, and Mario Rasetti*

Dipartimento di Fisica, Politecnico di Torino, 10129 Torino, Italy

The extension of the Hubbard model proposed in refs. [1], [2] includes a diagonal term designed to account for Coulomb interaction between nearest neighbour sites. In the present note we discuss the spectrum of the fermi-linearized [3] version of such model on a dimer, assuming spin-independent hopping amplitudes.

The hamiltonian for the extended model is given by:

$$\begin{aligned} H = \sum_{\mathbf{i},\sigma} \varepsilon\, n_{\mathbf{i},\sigma} + U \sum_{\mathbf{i}} n_{\mathbf{i},\Uparrow} n_{\mathbf{i},\Downarrow} + \frac{1}{2} V \sum_{<\mathbf{i},\mathbf{j}>} \sum_{\sigma,\sigma'} n_{\mathbf{i},\sigma} n_{\mathbf{j},\sigma'} \\ + \frac{1}{2} \sum_{<\mathbf{i},\mathbf{j}>} \sum_{\sigma,\sigma'} t^{(\sigma,\sigma')} \left(a^\dagger_{\mathbf{i},\sigma} a_{\mathbf{j},\sigma'} + a^\dagger_{\mathbf{j},\sigma'} a_{\mathbf{i},\sigma} \right) \quad . \end{aligned} \tag{1}$$

$a^\dagger_{\mathbf{i},\sigma}, a_{\mathbf{i},\sigma}$ are fermionic creation and annihilation operators ($\{a_{\mathbf{i},\sigma}, a_{\mathbf{j},\sigma'}\} = 0$, $\{a^\dagger_{\mathbf{i},\sigma}, a_{\mathbf{j},\sigma'}\} = \delta_{\mathbf{i},\mathbf{j}}\, \delta_{\sigma,\sigma'} \mathbf{1}$, $n_{\mathbf{i},\sigma} \doteq a^\dagger_{\mathbf{i},\sigma} a_{\mathbf{i},\sigma}$) over a d-dimensional lattice Λ ($\mathbf{i}, \mathbf{j} \in \Lambda$, $\sigma \in \{\Uparrow, \Downarrow\}$), whereas the parameters ε, U, V, and t, have the usual meaning [2], and $< \mathbf{i}, \mathbf{j} >$ stands, as customary, for nearest neighbours ($n.n.$) in Λ.

Here we consider the special case when $t^{(\sigma,\sigma')} = t$, $\forall \sigma, \sigma'$. This naturally leads to introducing the new set of local operators :

$$\begin{aligned} A_{\mathbf{i}} \doteq \frac{1}{\sqrt{2}} \left(a_{\mathbf{i},\Uparrow} + a_{\mathbf{i},\Downarrow} \right) \quad ; \quad N_{\mathbf{i}} \doteq A^\dagger_{\mathbf{i}} A_{\mathbf{i}} \quad ; \\ D_{\mathbf{i}} \doteq \frac{1}{2} \left(n_{\mathbf{i},\Uparrow} + n_{\mathbf{i},\Downarrow} + a_{\mathbf{i},\Uparrow} a^\dagger_{\mathbf{i},\Downarrow} + a_{\mathbf{i},\Downarrow} a^\dagger_{\mathbf{i},\Uparrow} \right) \quad . \end{aligned} \tag{2}$$

It is important to notice that the operators $A_{\mathbf{i}}$, $A^\dagger_{\mathbf{i}}$, and $N_{\mathbf{i}}$ satisfy the same commutation and anticommutation relations as $a_{\mathbf{i}}$, $a^\dagger_{\mathbf{i}}$, and $n_{\mathbf{i}}$, and that the idempotent operators $D_{\mathbf{i}}$ commute with both $A_{\mathbf{j}}$, $A^\dagger_{\mathbf{j}}$, and $N_{\mathbf{j}}$.

Upon rewriting the hamiltonian (1) in terms of the above operators, we perform fermionic cluster-linearization over the hopping terms, according to the scheme discussed in refs. [3], [4], and standard bose linearization over the intersite Coulomb

♮ Permanent address : Departamento de Fisica, Universidade Estadual de Maringá, 87020 Maringá, Paraná, Brazil.

Microscopic Aspects of Nonlinearity in Condensed Matter
Edited by A.R. Bishop *et al.*, Plenum Press, New York, 1991

interaction terms[3], for a cluster (*dimer*) of two neighbouring sites. H reduces thus to a sum over an arbitrary set of dimers covering Λ of commuting dimer hamiltonians

$$H_d = \sum_{\alpha=1}^{2} \left[\varepsilon_\alpha N_\alpha + (q-1)\left(\vartheta_\alpha A_\alpha - \bar{\vartheta}_\alpha A_\alpha^\dagger\right)\right] + V N_1 N_2 + t\left(A_1^\dagger A_2 + A_2^\dagger A_1\right) + C \quad , \quad (3)$$

where $\varepsilon_\alpha \doteq \varepsilon + U D_\alpha + V[(q-1) < N > + \sum_{\mathrm{j=n.n.}\alpha} D_{\mathrm{j}}]$, $\vartheta_\alpha \equiv < A_\alpha^\dagger >$, $< \bullet >$ denoting the expectation value of $\bullet$ in the appropriate state (ground or Gibbs), and q the number of nearest neighbours per site in Λ. ϑ_α's are anticommuting variables belonging to the odd sector of the $\mathbb{Z}_2$-graded algebra $\mathcal{G} = \mathcal{G}_e \oplus \mathcal{G}_o$ defined in ref. [4]. Sub-index α labels the dimer sites.

Finally, $C \doteq \sum_{\alpha=1}^{2} [(q-1)\vartheta_\alpha \bar{\vartheta}_\alpha + \frac{1}{2} V \sum_{\mathrm{j=n.n.}\alpha} D_\alpha D_{\mathrm{j}}] - (q-1)V < N >^2$.

H_d as given in (3) has a dynamical algebra $\mathcal{A}$ which is the direct sum of 2^{2q} copies of a superalgebra $\mathcal{S}$ isomorphic with the Cartan extension by $B_0 \doteq N_1 N_2$ of $su(2|2)$[5]. The latter has a bosonic subalgebra $\mathcal{B}$ isomorphic with $su(2) \oplus su(2)$, generated by $\{B_\mu \,|\, \mu = 1, \ldots, 6\} \doteq \{A_1 A_2 \pm A_2^\dagger A_1^\dagger, \mathbb{I} - (N_1 + N_2) \; ; \; A_1^\dagger A_2 \pm A_2^\dagger A_1, N_1 - N_2\}$, and a fermionic sector $\mathcal{F} = \{F_\nu \,|\, \nu = 1, \ldots, 8\} \doteq \{A_\alpha N_\beta ; N_\beta A_\alpha^\dagger \,|\, \alpha, \beta = 1, 2\}$ (where the labeling in ν is assumed such that $F_{\kappa+4} = F_\kappa^\dagger \; ; \; \kappa = 1, \ldots, 4$). B_3 and B_6 are the two Cartan elements of $\mathcal{B}$. We denote by $\{c_{\lambda,\lambda'}^{(\mu)}, d_{\nu,\lambda}^{(\nu')}, e_{\nu,\nu'}^{(\mu)}\}$ – where ν, ν' range from 1 to 8 and μ from 1 to 6, while λ, λ' range between 0 and 6 – the structure constants of $\mathcal{S}$:

$$[B_\lambda, B_{\lambda'}] = \sum_{\mu=1}^{6} c_{\lambda,\lambda'}^{(\mu)} B_\mu \; ; \; [F_\nu, B_\lambda] = \sum_{\nu'=1}^{8} d_{\nu,\lambda}^{(\nu')} F_{\nu'} \; ; \; \{F_\nu, F_{\nu'}\} = \sum_{\mu=1}^{6} e_{\nu,\nu'}^{(\mu)} B_\mu \, .$$

Each copy of $\mathcal{S}$ is characterized by a different distribution of eigenvalues (0 , 1) for the 2^{2q} operators D_{j} entering (3).

As the aim of this paper is to analyze the spectrum of H (and hence of H_d in the linearized approach), we shall henceforth consider ε_α, C as fixed constants, and shall not handle the problem – straightforward once the spectrum is given[3] – of self-consistently determining $< N >$, and ϑ_α.

The diagonalized hamiltonian is obtained by implementing a generalized Bogolubov automorphism in $\mathcal{S}$ rotating H_d into the Cartan sector of $\mathcal{B}$. This can be done in two successive steps, first acting on hamiltonian (3) by the adjoint action of an antihermitian rotation operator $\mathcal{Z}_F \in \mathcal{F}$, $H' = \exp(\mathrm{ad}\mathcal{Z}_F)(H_d) \in \mathcal{B}$, with $\mathcal{Z}_F = \sum_{\kappa=1}^{4}(\varphi_\kappa F_\kappa + \bar{\varphi}_\kappa F_{\kappa+4})$, then transforming H' into $\mathcal{H} \doteq \exp(\mathrm{ad}\mathcal{Z}_B)(H')$ with $\mathcal{Z}_B = r\, B_5 + s\, B_2 \in \mathcal{B}$. The coefficients $\varphi_\kappa \in \mathcal{G}_o$ are to be chosen in such a way that H' turns out to be an element of $\mathcal{B}$, whereas $r, s \in \mathbb{R}$ are determined by imposing that $\mathcal{H}$, as required, is Cartan. Notice that $\mathcal{Z}_B$ does not need to contain all the elements of $\mathcal{B}$, but only those which are not Cartan.

Explicit evaluation of H' for generic φ_ν's gives

$$H' = V B_0 + \sum_{\mu=1}^{6} b_\mu B_\mu + \sum_{\kappa=1}^{4} \left(f_\kappa F_\kappa - \bar{f}_\kappa F_{\kappa+4}\right) + C \quad , \quad f_\kappa \in \mathcal{G}_o \quad , \quad (4)$$

where, as H' should be hermitian, $b_2 = 0$, and $b_5 = 0$. Upon denoting by $X_\mu^{(m)} \in \mathbb{R}$, and $Y_\nu^{(m)} \in \mathcal{G}_o$, the coefficients defined by the recursive relation

$$(\mathcal{Z}_F)_m \circ H_d \doteq [\,\mathcal{Z}_F, (\mathcal{Z}_F)_{m-1} \circ H_d\,] \equiv \sum_{\mu=1}^{6} X_\mu^{(m)} B_\mu + \sum_{\kappa=1}^{4} \left(Y_\kappa^{(m)} F_\kappa - \bar{Y}_\kappa^{(m)} F_{\kappa+4} \right) , \quad (5)$$

with m integer ≥ 1, and $(\mathcal{Z}_F)_0 \circ H_d \equiv H_d$, one can write b_μ in (4) as :

$$b_\mu = \frac{1}{2} \sum_{\mu'=1}^{6} \sum_{\ell=0}^{3} \mathcal{L}_{\mu,\mu'}^{(\ell)} X_{\mu'}^{(\ell)} \quad , \quad (6)$$

with $\mathcal{L}_{\mu,\mu'}^{(\ell)} \doteq \sum_{\alpha=0}^{4} \kappa_{\mu,\mu'}^{(\alpha)} z_\alpha^{-\ell} \frac{d^\ell}{d z_\alpha^\ell} \left[\cosh(z_\alpha) + (-)^\ell \cos(z_\alpha)\right]$. As for $\kappa_{\mu,\mu'}^{(\alpha)}$, one should first iterate the recursive relation (5), in order to remove the $Y_\nu^{(m)}$, into $|X^{(m+2)}> = \mathbf{R}|X^{(m)}>$, where $|X^{(m)}>$ is a 6-vector of components $X_\mu^{(m)}$ and $\mathbf{R}$ is a 6×6 matrix whose elements ($\in \mathcal{G}_e$) are expressed in terms of the structure constants as $(\mathbf{R})_{\mu,\mu'} = \sum_{\nu,\nu',\nu''=1}^{8} \varphi_\nu e_{\nu,\nu'}^{(\mu)} d_{\nu'',\mu'}^{(\nu')} \varphi_{\nu''}$. The matrix $\mathbf{R}^2$ turns then out to be block diagonal (two degenerate 1×1 and two 2×2 blocks). Denoting by $\mathbf{T}$ the rotation matrix diagonalizing $\mathbf{R}^2$ ($\mathbf{T}\mathbf{R}^2\mathbf{T}^{-1} = \mathrm{diag}(z_0^4, z_\alpha^4; \alpha = 0, \ldots, 4)$), the $k_{\mu,\mu'}^{(\alpha)}$ are finally given by $k_{\mu,\mu'}^{(\alpha)} = (\mathbf{T}^{-1})_{\mu,\alpha}(\mathbf{T})_{\alpha,\mu'}$.

By implementing the recursion relation (5), one also obtains

$$f_\nu = Y_\nu^{(0)} + \Gamma_\nu + \frac{1}{2} \sum_{\mu=1}^{6} \Theta_{\nu,\mu} \sum_{\mu'=1}^{6} \left\{ \sum_{\ell=1}^{4} \mathcal{L}_{\mu,\mu'}^{(\ell)} X_{\mu'}^{(\ell-1)} - X_{\mu'}^{(3)} \sum_{\alpha=0}^{4} \kappa_{\mu,\mu'}^{(\alpha)} z_\alpha^{-4} \right\} \quad . \quad (7)$$

Here the coefficients $\Theta_{\nu,\mu}$, defined by the recursion relation $Y_\nu^{(m+1)} = \sum_{\mu=1}^{6} \Theta_{\nu,\mu} X_\mu^{(m)}$, and Γ_ν are elements of $\mathcal{G}_o$ simply given in terms of the structure constants and the fermionic rotation parameters by :

$$\Theta_{\nu,\mu} = \sum_{\nu'=1}^{8} \varphi_{\nu'} d_{\nu',\mu}^{(\nu)} \quad , \quad \Gamma_\nu = \sum_{\nu'=1}^{8} \varphi_{\nu'} d_{\nu',0}^{(\nu)} \quad .$$

H' as given by (4) is an element of $\mathcal{B}$, if the coefficients f_ν given by (7) equal zero. We shall denote by $\tilde{\varphi}_\nu$ the rotation parameters in $\mathcal{Z}_F$ which are solutions of the system of non-linear equations one obtains imposing the latter requirement, and by $\tilde{b}_\mu$ the expressions (6) when $\tilde{\varphi}_\nu$ is inserted. We still denote as H' the rotated bosonic hamiltonian given now by $H' = VB_0 + \sum_{\mu=1}^{6} \tilde{b}_\mu B_\mu + C$. It is worth noticing how H' contains the off-diagonal pairing operator B_1, even though H_d doesn't. This makes the model considered particularly interesting, *e.g.* in view of high T_c superconductivity, as such operators were intrinsically generated by the dynamical algebra of the system.

We can finally focus our attention on the rotation of H' into $\mathcal{H}$, by means of the adjoint action of $\mathcal{Z}_B$. Of course, as the rotation is an automorphism in $\mathcal{B}$, $\mathcal{H}$ will still be of the form $(VB_0 + \sum_{\mu=1}^{6} h_\mu B_\mu)$ (we temporaryly neglected terms proportional to $\mathbb{I}$, as they are not affected by $\mathcal{Z}_B$). Hermiticity of $\mathcal{H}$ implies that $h_2 = 0 = h_5$. Explicit calculation for generic r, s shows then that

$$h_1 = \tilde{b}_1 c_s - \frac{1}{2}(1 - \tilde{b}_3 + V)s_s \quad , \quad h_3 = \frac{1}{2}(1 - \tilde{b}_3 - V)c_s - \tilde{b}_1 s_s + \frac{1}{2}V \quad ,$$

$$h_4 = \tilde{b}_4 c_r + \frac{1}{2}\tilde{b}_6 s_r \quad , \quad h_6 = \frac{1}{2}\tilde{b}_6 c_r - \tilde{b}_4 s_r \quad ,$$

where $c_t \doteq \cos(2t)$, $s_t \doteq \sin(2t)$, $t = r, s$.

As we want $\mathcal{H}$ to be diagonal, the equations $h_1 = 0 = h_4$ must be satisfied. The latter can be easily solved in r, s. Denoting the solution by $\tilde{r}, \tilde{s}$, one has

$$\tilde{r} = -\frac{1}{2}\arctan\left(2\frac{\tilde{b}_4}{\tilde{b}_6}\right) \quad , \quad \tilde{s} = \frac{1}{2}\arctan\left(2\frac{\tilde{b}_1}{1 - \tilde{b}_3 + V}\right) \quad . \tag{8}$$

With the above choices $\mathcal{H}$ reduces to $\mathcal{H} = VB_0 + \tilde{h}_3 B_3 + \tilde{h}_6 B_6$. A straightforward calculation provides then finally the spectrum in the form :

$$\text{spec}(\mathcal{H}) = (\tilde{h}_6 - \tilde{h}_3)N_1 - (\tilde{h}_6 + \tilde{h}_3)N_2 + VN_1N_2 + C' \quad ,$$

where $N_\alpha = 0, 1$ and

$$\tilde{h}_3 \doteq \frac{1}{2}\left(V - \sqrt{(1 - \tilde{b}_3 + V)^2 + 4\tilde{b}_1^2}\right) \quad , \quad \tilde{h}_6 \doteq \frac{1}{2}\sqrt{\tilde{b}_6^2 + 4\tilde{b}_1^2} \quad ;$$

$$C' \doteq C + \frac{1}{2}(1 - \tilde{b}_3) + \tilde{h}_3 \quad .$$

The diagonalization achieved is particularly convenient from the point of view of statistical mechanics. In fact in this sense the model considered, when one wants to compute *e.g.* the partition function, simply leads on the one hand to summing over a set of classical Ising-like configuration variables (the $D_{\mathbf{i}}$'s), on the other hand is reconductible to a well known combinatorial problem (the covering of lattice Λ by dimers). Work is in progress along these lines.

One of the authors (L. E.) thanks the Brazilian Agency CNPq for finantial support.

References

[1] K.B. Efetov, and A.I. Larkin, *Zh. Exp. Teor. Fiz.* **69**, 764 (1975) [*Sov. Phys. JETP* **42**, 390 (1976)]

[2] V.J. Emery, *Phys. Rev.* **B 14**, 2989 (1976); and *Phys. Rev. Lett.* **58**, 2794 (1987)

[3] A. Montorsi, M. Rasetti, and A.I. Solomon, *Int. J. Mod. Phys.* **3**, 247 (1989)

[4] A. Montorsi, and M. Rasetti, *Mod. Phys. Lett.* **B 4**, 613 (1990)

[5] V.G. Kac, *Adv. Math.* **26**, 8 (1977); and *Comm. Math. Phys.* **53**, 31 (1977)

LOCALIZED EXCITATIONS IN 1-D PEIERLS-HUBBARD MODELS

OF POLYACETYLENE AND MX CHAINS

J. Tinka Gammel

Theoretical Division and Center for Nonlinear Studies
Los Alamos National Laboratory, Los Alamos, NM 87545, USA
and Physikalisches Institut, Universität Bayreuth, D-8580 Bayreuth, F.R.G.

INTRODUCTION

Recent interest in novel low-D materials – *e.g.*, high-temperature superconducting copper oxides, "heavy-fermion" and charge-density wave systems, halogen-bridged transition-metal linear chain complexes (MX chains), and organic synthetic metals and superconductors – and their pressure, photoexcitation, and doping induced intragap states, has stimulated the theoretical study of competing electron-electron (e-e) and electron-phonon (e-p) interactions in reduced dimensions. For these new materials it is important in microscopic models to capture the essence of both e-p and e-e interactions and to represent faithfully their synergetic, or competing, effects. To this end, variants of the extended Peierls-Hubbard Hamiltonian (ePHH) have been widely used. In this article, intrinsic localized non-linear defects (solitons, polarons, and bipolarons) in 1- and 2-band versions of the 1-D ePHH are discussed and compared to experiments on the conducting polymer *trans*-polyacetylene, $(CH)_x$, and on the MX chain $[Pt(en)_2][Pt(en)_2Cl_2](ClO_4)_4$ ($en{=}N_2C_2H_8$), hereafter referred to as PtCl. Obtaining fits to many different observables with a single set of physically plausible parameters within a given model is *essential* for a true microscopic understanding of non-linear excitations in any specific material, and in general for an understanding of the whole class of novel low-D materials.

POLYACETYLENE: PARAMETERS AND SOLITONS

$(CH)_x$ has the simplest idealized structure of all the novel, low-D materials. Thus the hope of modeling (and understanding) it within the geometrically and quantum chemically simple (single chain, single molecular orbital, nearest neighbor, tight-binding) ePHH is greatest. In spite of this, the correct parameters for modeling $(CH)_x$ remain a matter of much debate.[1] We have determined[2] for a broad range of parameters, (*i*) the self-consistent lattice distortion, Δ, (*ii*) the longitudinal optical phonon mode, ω_{LO}^2, and (*iii*) the optical absorption properties (optical gap, E_g, and bandwidth, W), all as functions of the strength of the e-e and e-p interactions. Comparing the results of this parameter space search with the appropriate experimental values allows us to infer internally consistent parameters, which are then used to determine whether other observables – such as the soliton absorptions and the triplet gap – are consistent with experiment. This work, still in progress, is summarized below.

In the context appropriate to $(CH)_x$, the ePHH takes the form[1]

$$H=\sum_{\ell}(-t_0+\alpha\Delta_\ell)B_{\ell,\ell+1}+\frac{1}{2}K\sum_{\ell}\Delta_\ell^2+U\sum_{\ell}n_{\ell\uparrow}n_{\ell\downarrow}+V\sum_{\ell}n_\ell n_{\ell+1}\ . \qquad (1)$$

Here $c_{\ell\sigma}^\dagger$ creates an electron in the Wannier orbital at site ℓ with spin σ; $B_{\ell,\ell+1}=\sum_\sigma(c_{\ell\sigma}^\dagger c_{\ell+1\sigma}+c_{\ell+1\sigma}^\dagger c_{\ell\sigma})$; $n_{\ell,\sigma}=c_{\ell\sigma}^\dagger c_{\ell\sigma}$; t_0 is the hopping integral with α the e-p coupling describing its distance dependence; $\Delta_\ell=y_{\ell+1}-y_\ell$ where y_ℓ is the (adiabatic[3]) displacement of the ℓ-th atom along the axis; and K represents all other costs of distorting the lattice. Coulomb repulsions among electrons are parameterized with the conventional Hubbard U and V.

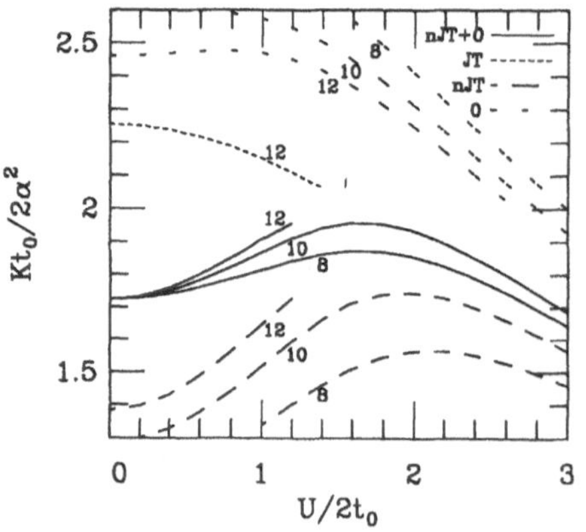

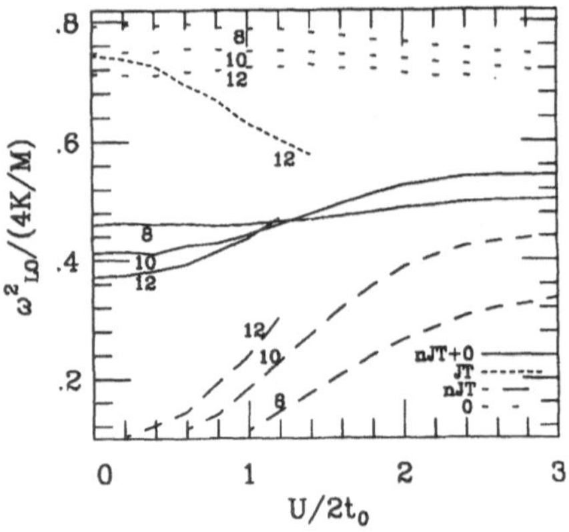

Fig. 1 (left). Self-consistent K as a function of Hubbard U for fixed Δ=0.1, t_0=0.5, α=0.5, V=0, and several N with nJT and chain BC averaged as discussed in the text. The counter-intuitive *increase* of K with U can be seen on systems as small as 8-sites. The results for the 12-site ring and JT, nJT and chain (0) BCs are also shown.

Fig. 2 (right). ω^2_{LO} for the same cases as in Fig. 1. Note the LO mode frequency tracks K.

For the ePHH at 1/2-filling, it is well known that Hartree-Fock (HF) results are unreliable. Thus we use the Lanczos exact diagonalization method (LEDM),[4] discussed in this context elsewhere,[2] to investigate the effects of competing e-p and e-e interactions small system size N and extrapolate to $N=\infty$. However, for small N, the size and boundary condition (BC) dependence is large. The importance of the Jahn-Teller/non-Jahn-Teller (JT/nJT) distinction for small N is familiar from the finite cyclic polyenes, where $4N+2$ rings do not dimerize; benzene, *e.g.*, has equal bond lengths. This JT/nJT dichotomy becomes a problem when extrapolating to the infinite $(CH)_x$ limit. For $U=V=0$, one can show that "complex-phase-averaging"[2] – Bloch's theorem – *exactly* reproduces the long chain behavior from studying short chains with different BCs, but it becomes ineffective in the strongly correlated ($U\rightarrow\infty$) limit for the 1/2-filled band systems currently considered. In this limit, an *amplitude* BC averaging technique has been found to be effective in reducing finite size effects and improving extrapolation to the infinite limit.[2]

The value of Δ_ℓ is found by minimizing the total energy. At fixed K, the LEDM is used iteratively to calculate Δ_ℓ from the self-consistency condition (SCC) $K\Delta_\ell=-\alpha\langle B_{\ell,\ell+1}\rangle$. For $\Delta_\ell=a_0+(-1)^\ell\Delta$, one can reverse the question and consider instead Δ as the independent variable, K (and a_0) then being determined from the SCC. While for strong e-p coupling Δ is reduced by U, it is well known that the Hubbard U counterintuitively initially increases Δ for weak e-p coupling.[5] Equivalently, for small fixed Δ, U initially increases K.[6] Fig. 1 shows K as a function of U for various lattice sizes with JT, nJT, chain, and BC averaged (nJT plus chain, weighted to reproduce the exact answer at U=0) BCs. Fig. 1 illustrates the effects of different BCs on the finite systems, as well as the convergence of the LEDM with system size. Phase-averaging yields a behavior that is nearly converged to the valued inferred for the infinite case, whereas any single BC yields this behavior only after (considerable) extrapolation. Thus BC averaging eliminates many of the problems in applying the LEDM to extract infinite system behavior. To calculate the LO phonon mode frequency, ω^2_{LO}, one needs only evaluate the total energy $E[\Delta]$ at three different Δ near the equilibrium geometry, requiring minimal additional computation. Fig. 2 shows ω^2_{LO} as a function of U and system size for the same cases as in Fig. 1. The studies at fixed K show that for the finite rings ω^2_{LO} goes soft at finite U, confirming the fact that Δ vanishes at this U, and that finite systems do have a "phase transition" to a SDW state. Using the Golden Rule, the problem of calculating the optical absorption coefficient within the LEDM reduces to finding the spectral weight of $J|\psi_0\rangle$, where J is the current operator.[2,7] For small systems, the optical spectra will be sparse. Indeed, benzene has only a single $\pi\rightarrow\pi^*$ absorption. To illustrate how BC averaging yields non-sparse spectra as in the infinite system, which agree with expectations based on strong- and weak-coupling arguments,[2,7] Fig. 3 shows a spectrum obtained using several BCs in the range $-1\leq x\leq 1$.

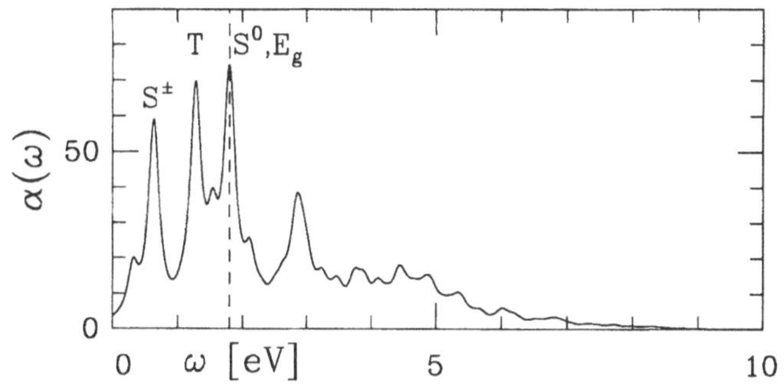

Fig. 3. Optical absorption spectra obtained using the BC averaging approach on a 10-site rings showing neutral (S^0) and charged ($S^\pm$) soliton and triplet (T) absorptions along with the ground state (E_g) for one set of parameters meeting all the sorting criterion, yet far from the "conventional wisdom": t_0=1.0, α=3.2, K=43.9, U=3.2, and V=0. Note that, to the extent that they can be determined within the LEDM, the soliton and triplet absorptions also do not rule out this set.

Table 1. Representative parameters scaled to Δ=0.03Å, W=10eV, E_g=1.8eV, and $\bar{\nu}_{LO}$=1450cm^{-1}.

N	t_0(eV)	α(eV/Å)	U(eV)	V(eV)	K(eV)	W(eV)
8	1.7	4.5	4.8	1.4	52.4	11.3
	1.4	4.3	4.0	0.6	53.1	10.4
	1.3	4.4	3.7	0.0	53.4	10.7
	1.3	4.0	4.3	1.1	51.8	11.3
	1.2	3.4	5.8	2.4	47.6	10.5
	1.0	3.1	6.0	2.4	44.9	10.5
10	1.5	4.3	4.3	0.6	53.6	11.4
	1.4	3.9	4.5	1.1	51.4	10.3
	1.4	3.6	5.8	2.3	49.1	11.0
	1.2	4.0	4.0	0.0	50.9	10.6
	1.1	2.9	5.9	2.5	43.3	10.6

It is useful first to recall how the parameters are determined in the conventional SSH model[8] of $(CH)_x$, which corresponds to Eq. 1 with U=V=0. Since this is a single-electron theory, it can be solved analytically for the ground state. One finds that the bandwidth is given by W=4t_0, the optical gap by E_g=4$\alpha\Delta$, and that the dimerization Δ is determined by the SCC πk=2(K$'(\delta)$ − E$'(\delta)$)/$(1-\delta^2)$, where δ=$\alpha\Delta/t_0$, k=$Kt_0/2\alpha^2$, and E$'$ and K$'$ are elliptic integrals. Fitting to W=10eV, E_g=1.4eV, and Δ=0.086Å yields t_0=2.5eV, α=4.1eV/Å, and K=21eV/Å^2. However, U=0 parameters can never reproduce the experimental fact that the 2^1A_g state lies below the 1^1B_u state,[9] and finite α is required if dimerization is to occur at all. Thus the observed optical gap must come from both e-e and e-p interactions, and hence this investigation using LEDM. The above LEDM results suggest that BC averaging on small systems provides reasonable estimates of the infinite limit. Hence one can use manageable size systems to carry out an exhaustive parameter search. Our procedure[2] is to span the dimensionless parameter space: t_d≡0.5, α_d≡0.5, δ, u=$U/2t_0$, and v=$V/2t_0$; determining k, e_g=$E_g/2t_0$, w=$W/2t_0$, and Ω^2_{LO}=$\omega^2_{LO}/(4K/M)$. To date, for N=6,8,10,12 and x=−1, 0, 1, a search over δ=.02, .04, ..., .20, v=0, .2, ..., 2., and u=v, v + .2, ..., 3., determining ground state properties only, is nearly completed. The same BC averaging scheme as in Fig. 1 was used to determine K. The optical gap was defined to be the lowest value of e_g for the 3 BCs (usually JT), and appeared to scale with U−V, as predicted by strong-coupling.[10] For U≃2V, the gap appeared to depend on U−2V as expected from decoupled dimer arguments.[2] The LEDM does not reliably give outer band edges (nor does experiment); thus w was determined by where the absorption had fallen to 1% of its maximum. Note this can differ considerably from the actual band edge. One also finds w/e_g and Ω^2_{LO} are remarkably insensitive to parameters. To determine the parameters describing $(CH)_x$, one then scales each of these dimension*less* data points to the actual gap and dimerization to determine the dimensional parameters. If the bandwidth and ω_{LO}^2 then agree with their actual values, one accepts these as possible parameters. Though for $(CH)_x$even some relatively basic information is not known precisely, it is "generally agreed" that Δ=0.03Å, E_g=1.8eV (to correct for 3D effects[9]), W=10eV, and $\bar{\nu}_{LO}=\omega_{LO}/(2\pi c)=1450\text{cm}^{-1}$ are the values to use. Listed in Table 1 are the results to date for the possible parameters describing $(CH)_x$. One finds $t_0<2.5$, as part of the bandwidth comes from e-e correlations.

The triplet (T=1.4eV), neutral ($S^0 \simeq E_g$), and charged ($S^{\pm}$=.6eV) soliton optical and infrared absorptions have also been identified. The studies of how well the inferred parameters match these data on non-uniform geometries – *i.e.*, whether the same parameters can be used for a microscopic description of the non-linear excitations – are currently underway. Fig. 3 shows, however, that even for a parameter set far from the "conventional wisdom", to the extent that they can correctly be determined by the LEDM for non-uniform geometries, predicted optical absorptions are in agreement with experiment.

MX CHAINS: ELECTRON/HOLE ASYMMETRY AND POLARONS

MX=PtCl compounds typically[11] have a ground state in which a large-amplitude dimerization of the Cl sublattice displacements is stabilized by an equally strong valence dimerization on the Pt sublattice. The strength of this charge-density-wave (CDW) distortion decreases as X is replaced with Br and I toward the valence delocalized (weak distortion) regime.[12] Indeed Pt_2Br (based on the MMX monomer) under pressure[13] and NiBr[14] are undimerized, and a spin-density-wave (SDW) ground state probably occurs. Tuning through the transition regions between ground states is especially interesting, as seen in recent studies of high-temperature superconducting oxides.[15] In fact the MX class has much in common with current modeling of oxide superconductors. At stoichiometry, both are nominally 3/4-filled, hybridized 2-band materials in which both e-e and e-p interactions are important, described by similar model Hamiltonians – the similarity with $Ba_{1-x}Pb_xBiO_3$ is especially direct.[16] This analogy extends to electron or hole *doping* of the broken symmetry ground states leading to discussion of polarons, bipolarons, excitons, and domain walls as self-trapped local defect states (bags).

Considering an isolated MX chain, taking into account a single orbital per site, and including only nearest neighbor interactions, the following ePHH is used[17]:

$$\begin{aligned} H &= \sum_{\ell}\{(-t_0+\alpha\Delta_\ell)B_{\ell,\ell+1} + [(-1)^\ell e_0 - \beta_\ell(\Delta_\ell+\Delta_{\ell-1})]n_\ell\} \\ &\quad + \sum_\ell U_\ell n_{\ell\uparrow}n_{\ell\downarrow} + \frac{1}{2}K\sum_\ell \Delta_\ell^2 + \frac{1}{2}K_{MM}\sum_\ell(\Delta_{2\ell}+\Delta_{2\ell+1})^2 \ . \end{aligned} \tag{2}$$

At stochiometry there are 3 electrons per MX unit – 3/4-filling. Here $c_{\ell,\sigma}$, $n_{\ell,\sigma}$, and $B_{\ell,\ell+1}$, as well as t_0, α, Δ_ℓ, and K are as in Eq. 1. The difference in on-site energies of M d_{z^2} and X p_z orbitals is modeled by $-e_0$ on X (odd) sites and $+e_0$ on M (even) sites. A second e-p coupling is included: the coupling of the d_{z^2} and p_z on-site energies to the breathing mode of the neighboring atoms along the chain (β_M and β_X, respectively). Again, Hubbard terms model the e-e interactions (U_M, U_X). A M-M spring, K_{MM}, is also added to model the ligand-ligand interaction. For PtCl, a combination of ground state experimental data [the Cl sublattice distortion amplitude (0.38Å), the $\sigma(Cl) \rightarrow d\sigma^*$ absorption for the oxidized monomer (4.7eV), and the IVCT band edge (2.5eV)] and quantum chemical calculations[18] on small PtCl clusters (including ligands) appropriate to the highly valence localized limit, leads us[17] to the following HF effective (U=0) parameter set for the Hamiltonian (Eq. 2): t_0=1.6eV, α=2.4eV/Å, e_0=0.96eV, β_M=0.16eV/Å, K=3.9eV/Å^2, and $K_{MM}=\infty$. For $K_{MM} \gtrsim K$ the results are insensitive to K_{MM} and β_X. For small to intermediate U, agreement of LEDM and HF calculations is excellent (for both ground and defect state structure and optical absorption). This is a consequence of the 3/4-filling in a CDW regime and is to be contrasted with the above discussion of a 1/2-filled 1-band model, where true e-e correlation effects dominate. It is anticipated that $U_M \sim 1-2$eV for PtCl. More detailed parameter determination is underway.

Fig. 4b shows the optical absorptions predicted for hole and electron polarons and bipolarons, calculated from the Golden Rule – the results were Lorentzian smoothed, but no inhomogeneous broadening was included.[17] One finds three distinct intra-gap absorptions features for each polaron species and one and two intragap features for the hole and electron bipolarons, respectively. Corresponding results have been obtained for negatively charged, positively charged and neutral kinks, and for excitons. Note the defects are accompanied (for $K_{MM} \neq \infty$) by a small local BOW (M sublattice) distortion even though the ground state is purely CDW. Early experimental studies of polaron states in MX materials

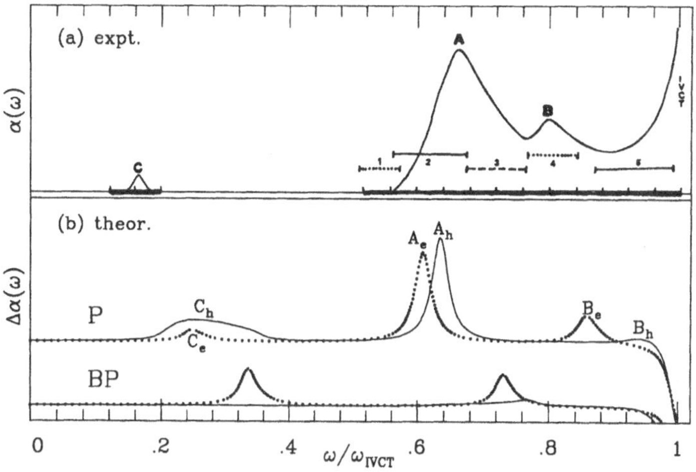

Fig. 4. Intragap optical absorption spectra, with the IVCT (2.5eV) scaled to unity. (a) The curve labeled EXPT show the absorption features for $[Pt(en)_2][Pt(en)_2Cl_2]\cdot(ClO_4)_4$ with RR enhancement regions (1-5) superimposed. The RR data separates the optical data into distinct features for electron (regions 1 and 4 – dotted – enhancing a phonon at 263 cm^{-1}) and hole (regions 2 and 5 – solid – enhancing phonon at 287 cm^{-1}) polarons. Region 3 – dashed – enhances a phonon at 272 cm^{-1}, and is probably due to an electron bipolaron (see text). (b) The curves labeled P and B are the electron (dotted) and hole (solid) polaron and bipolaron, using the parameters in the text, with uniform absorption subtracted to show the B features. Note the electron-hole asymmetry.

focused on high energy intragap absorption features ("A" and "B" bands).[19] These were then found to be accompanied by a much lower energy ("C") band,[20] as shown in Fig. 4a. Resonant Raman (RR) experiments performed using excitation to the red of the A band (1.3-1.5eV) before and after photo-excitation into the IVCT transition show that features at 287 and 263 cm^{-1} show the same growth and decay rates as the A, B and C bands.[21] Accordingly, they are attributed to local modes of the associated defects. On the basis of expected changes in the Pt-Cl force constants upon oxidation of Pt^{II} (hole) versus reduction of Pt^{IV} (electron) sites, the vibrations at 287 and 263 cm^{-1} are attributed to hole and electron polarons, respectively. The feature at 287 cm^{-1} exhibits two distinct regions of enhancement centered slightly to the red of the A band (1.4-1.7eV) and slightly to the red of the IVCT band edge (2.17-2.5eV). In contrast, the broader RR feature at 263 cm^{-1} is enhanced in a region further to the red of the A band (1.28-1.43eV) and in the region corresponding to the B band (1.92-2.1eV). These distinct enhancement regions are also depicted in Fig. 4a. Comparing Figs. 4a and b, the agreement is excellent and supports the interpretation of hole-polaron absorptions at $A_h \simeq 1.5$eV and $B_h \simeq 2.3$eV (RR profiles 2 and 5, enhancing the vibration at 287 cm^{-1}), with electron-polaron absorption at $A_e \simeq 1.3$eV, $B_e \simeq 2.0$eV and $C_e \simeq 0.4$eV (RR profiles 1 and 4, enhancing the *softer* vibration at 263 cm^{-1}). The predicted C_h structure is too broad to be detected by the experimental technique used here and will need an alternative experimental probe. The A band observed to grow in upon photo-excitation corresponds to hole and electron polarons with the electron polarons contributing to the red edge of this intragap transition. Further, the B band is dominated by the electron polaron while the expected B_h absorption is within, and unresolved from, the IVCT band edge. Single crystal transmission experiments[19,20] have shown no evidence for a resolved A band feature for the electron polaron, although this is a difficult region owing to poor instrument response. The RR results are important in that they demonstrate that electron and hole polarons have distinct vibrational modes which, in turn, have distinct excitation profiles. In this interpretation the resonance enhancement observed near 1.8eV (enhancing a phonon mode at 272 cm^{-1}) is attributed to *bipolarons*. Initial studies of the temperature dependence of the intensities of the 263 and 272 cm^{-1} modes confirm that as the temperature is raised from 13 to 270 K intensity shifts from the 272 (electron bipolaron) to the 263 (electron polaron) cm^{-1} mode.

The above shows that calculations for localized electron and hole polarons with parameters determined from ab initio and band structure input are in *quantitative* agreement with experimental observation of the photoexcitation of PtCl above the intervalence charge transfer (IVCT) band gap. Further, we find *asymmetry* exists between electron and hole polarons in terms of both their characteristic vibrational modes and their intragap absorption transitions. The importance of electron-hole asymmetry has also been suggested in recent studies of quasi-2-D oxide superconductors.[22]

CONCLUSION

Here, I have attempted to give a brief survey of the research I am involved in which focuses on trying to determine to what extent a detailed microscopic understanding can be obtained of the spectral signatures of intrinsic localized non-linear defects for several specific low-dimensional materials within the framework of the "simple" tight-binding ePHH. Although this research is in progress, it seems that for both $(CH)_x$ and PtCl, the above results confirm that the ePHH captures the essential behavior of the competing e-e and e-p interactions. We feel that the lessons and techniques reported here will lead to a more general microscopic understanding of non-linear excitations in novel low-D materials with strong competitions for broken-symmetry ground states.

Acknowledgements – While there is not space here to list them individually, I would like to thank the many people who have collaborated in this research effort. This work was supported by the U.S. D.O.E.-O.B.E.S., and by the D.F.G. under SFB213.

REFERENCES

1. For a recent review of the ePHH applied to $(CH)_x$, see: D. Baeriswyl, D.K. Campbell, and S. Mazumdar, to be published in *Conducting Polymers*, ed. by H. Kiess (Springer, New York, 1990).
2. (a) D.K. Campbell, J.T. Gammel, and E.Y. Loh, Jr., to be pub. in the *Proc. of the Ann. Adriatico Res. Conf. on Strongly Correl. Electrons*, Trieste, Italy, July 18-21, 1989. (b) J.T. Gammel, D.K. Campbell, E.Y. Loh, Jr., S. Mazumdar, and S.N. Dixit, to be pub. in the *Proc. of the MRS Symp. on Elec., Opt., and Mag. Prop. of Org. Solid State Mater.*, Boston, MA, Nov. 27- Dec. 2, 1989.
3. A brief summary, including several references, of expected quantum lattice effects in $(CH)_x$, ignored here, may be found in Ref. 1.
4. S. Pissanetsky, *Sparse Matrix Technology* (Academic, London, 1984).
5. *e.g.*, V. Waas, H. Büttner, and J. Voit, *Phys. Rev. B* **41**, 9366 (1990).
6. G.W. Hayden and Z.G. Soos, *Phys. Rev. B* **38**, 6075 (1988).
7. P. F. Maldague, *Phys. Rev. B* **16**, 2437 (1977).
8. W.P. Su, J.R. Schrieffer, and A.J. Heeger, *Phys. Rev. Lett.* **42**, 1698 (1979); *Phys. Rev. B* **22**, 2099 (1980).
9. B.E. Kohler, C. Spangler, and C. Westerfield, *J. Chem. Phys.* **133**, A171 (1964).
10. J. Bernasconi, M.J. Rice, W.R. Schneider, and S. Strässler, *Phys. Rev. B* **12**, 1090 (1975).
11. H.J. Keller, in *Extended Linear Chain Compounds*, ed. by J.S. Miller (Plenum Press, New York, 1982), Vol. 1, p. 357.
12. R.J.H. Clark, in *Advances in Infrared and Raman Spectroscopy*, ed. by R.J.H. Clark and R.E. Hester (Wiley Heyden, New York, 1984), Vol. 11, p. 95
13. B. I. Swanson, M. A. Stroud, S. D. Conradson, and M. H. Zeitlow, *Solid State Commun.* **65**, 1405 (1988).
14. H. Toftlund and O. Simonsen, *Inorg. Chem.* **23**, 4261 (1984); K. Toriumi, Y. Wada, T. Mitani, and S. Bandow, *J. Amer. Chem. Soc.* **111**, 2341 (1989).
15. *e.g.*, *Proceedings of the IBM Europe Institute 1988 Workshop on High Temperature Superconductivity, Oberlech, Austria (August 1988)*, ed. by K.A. Mueller: *IBM J. Res. Dev.* **33**, (1989).
16. D. Baeriswyl and A. R. Bishop, *Physica Scripta* **T19**, 239 (1987); *J. Phys. C: Solid State Phys.* **21**, 339 (1988); S. D. Conradson, M. A. Stroud, M. H. Zeitlow, B. I. Swanson, D. Baeriswyl, and A. R. Bishop. *Solid State Commun.* **65**, 723 (1988).
17. J.T. Gammel, R.J. Donohoe, A.R. Bishop, and B.I. Swanson, *Phys. Rev. B*, to be published.
18. C.A. Boyle, P.J. Hay, and R.L. Martin, unpublished.
19. S. Kurita, M. Haruki, and K. Miyagawa, *J. Phys. Soc. Japan* **57**, 1789 (1988); L. Degiorgi, P. Wachter, M. Haruki, and S. Kurita, *Phys. Rev. B* **40**, 3285 (1989).
20. R. J. Donohoe, S. A. Ekberg, C. D. Tait and B. I. Swanson, *Solid State Commun.* **71**, 49 (1989).
21. R. J. Donohoe, R. B. Dyer, and B. I. Swanson, *Solid State Commun.*, in press.
22. J. E. Hirsch and F. Marsigilio, *Phys. Rev. B* **39**, 11515 (1989).

MICROSCOPIC SPIN WAVE STUDY OF THE Co/Cu(100) EPITAXIAL MONOLAYER

M.G. Pini[a] , A. Rettori[b], D. Pescia[c], N. Majlis[d] (*) , S. Selzer[e] (*)

a-Istituto di Elettronica Quantistica, Consiglio Nazionale delle Ricerche, I-50127 Firenze, Italy
b-Istituto di Fisica dell'Universita' di Siena, I-53100 Siena, Italy and Unita' di Firenze GNSM CISM, I-50125 Firenze, Italy
c-Institut für Festkörperforschung der KFA Jülich, 5170 Jülich, Federal Republic of Germany
d-Institute of Theoretical Matter Physics, Consorzio di Fisica della Materia-INFM (forum), I-50125 Firenze, Italy and Scuola Normale Superiore, I-56100 Pisa, Italy
e-Istituto di Fisica dell'Universita' di Pisa, I-56100 Pisa, Italy

(*) Permanent address : Instituto de Fisica, Universidade Federal Fluminense, Caixa Postal 296, 24.210 Niteroi, RJ, Brazil

Recent progress[1] in the preparation and investigation of high quality epitaxial monolayers (ML) has renewed the interest for two-dimensional (2D) magnetism. In particular, the temperature dependence of the magnetization of the Co/Cu(100) monolayer was measured with great accuracy using spin-polarized low-energy electron diffraction (SPLEED), and magneto-optical Kerr effect.[2] The field dependence of the magnon energy gap was detected by means of light scattering.[2]

In the framework of a microscopic free spin wave approach, we discuss in detail the role of the different anisotropies in the stabilization of such a 2D system at finite temperatures, in particular that due to the dipolar interaction. From the comparison of our calculations with the experimental data we are able to settle the appropriate model and fix the values of the anisotropies.

We assume the following spin Hamiltonian:

$$\mathcal{H} = -J \sum_{<\ell,m>} \vec{S}_\ell \cdot \vec{S}_m - g\mu_B H \sum_\ell S_\ell^y$$

$$+ \frac{1}{2} \frac{g^2\mu_B^2}{a^3} \sum_{\ell \neq m} \left|\frac{a}{\vec{r}_{\ell m}}\right|^3 \left[\vec{S}_\ell \cdot \vec{S}_m - 3 \frac{(\vec{S}_\ell \cdot \vec{r}_{\ell m})(\vec{S}_m \cdot \vec{r}_{\ell m})}{(\vec{r}_{\ell m})^2} \right]$$

$$+ D\sum_\ell (S_\ell^x)^2 + \frac{K_4}{2} \sum_\ell [(S_\ell^y)^4 + (S_\ell^z)^4] \qquad (1)$$

where we take yz as the monolayer plane and the spins are located on the sites of a square lattice. With <,> we denote the set of nearest neighbour site pairs.

Microscopic Aspects of Nonlinearity in Condensed Matter
Edited by A.R. Bishop *et al.*, Plenum Press, New York, 1991

In addition to the usual isotropic ferromagnetic ($J>0$) exchange, we take into account the dipolar interaction, which forces the spins to lie within the yz plane. The dipolar interaction is of long range, and though very small compared to the exchange interaction, it is capable of stabilizing long-range order in a 2D ferromagnet. Its main effect is to change the small k behavior of the dispersion relation from $E_k \alpha k^2$ into $E_k \alpha k^{1/2}$.[3-5] The subsistence of long-range order at finite T in a 2D system does not violate in this case the Mermin-Wagner theorem,[6] which is based on the assumption that the interactions are of finite range.

We also assume : an easy-plane single-ion anisotropy ($D > 0$) and an in-plane anisotropy, favouring four equivalent directions along the diagonals of the square ($K_4 > 0$), as suggested by the experiment. According to the experimental results,[2] which show only in-plane remanence, we take as quantization axis for the magnetization a direction $\tilde{z}$ in the yz plane which forms an angle ϕ with the y=[010] axis, along which the magnetic field is applied. Using the standard Holstein-Primakoff transformation for the spin operators, the ground state configuration of the spins, is found by imposing the vanishing of the terms linear in a and a^+ in the Boson hamiltonian:

$$\sin\phi\ [g\mu_B H - 2K_4S^3\ \cos\phi\ \cos2\phi] = 0 \tag{2}$$

The quadratic Hamiltonian takes the following form :

$$\mathcal{H}_2 = \sum_k \left[A_k\, a_k^+ a_k + \frac{1}{2} B_k (a_k^+ a_{-k}^+ + a_k a_{-k})\right] \tag{3}$$

$$A_k = 2JSz(1-\gamma_k) + g\mu_B H \cos\phi + 3wy_0 - w(c_0 - c_k) - \frac{3}{2} w\ [y_k \sin^2\phi + z_k \cos^2\phi$$

$$-2\tau_k \sin\phi\cos\phi] + DS - 2K_4S^3(\sin^4\phi + \cos^4\phi) + \frac{3}{4} K_4S^3 (1-\cos4\phi) \tag{3a}$$

$$B_k = \frac{3}{2} w\ [y_k \sin^2\phi + z_k \cos^2\phi - 2\tau_k \sin\phi\cos\phi] + DS - \frac{3}{4} K_4S^3 (1-\cos4\phi) \tag{3b}$$

In terms of A_k , B_k the dispersion relation of the spin wave excitations and the T-dependent spin wave magnetization are expressed as:

$$E_k = \sqrt{A_k^2 - B_k^2} \tag{4}$$

$$\frac{\langle S^{\tilde{z}}(T)\rangle}{S} = 1 - \frac{1}{NS} \sum_k \frac{A_k - E_k}{2E_k} - \frac{1}{NS} \sum_k \frac{A_k}{E_k} \frac{1}{e^{E_k/K_BT} - 1} \tag{5}$$

For a square lattice, the number of n.n. is z=4 and $\gamma_k = \frac{1}{2}[\cos(k_y a) + \cos(k_z a)]$ with $-\pi \leq k_y a\ ,\ k_z a \leq \pi$. The factor $w = g^2\mu_B^2 S/a^3$ measures the strength of the dipolar interaction while τ_k, y_k, z_k denote 2D dipolar sums which can be calculated either numerically or analytically in the continuum limit $ka \ll 1$:

$$\tau_k = \sum_\ell \left|\frac{a}{\vec{r}_{\ell m}}\right|^3 \frac{y_{\ell m} z_{\ell m}}{r_{\ell m}^2} e^{i\vec{k}\cdot\vec{r}_{\ell m}} \sim -\pi ka \frac{1}{3} \sin(2\theta_k)$$

$$y_0 - y_k = \sum_\ell \left|\frac{a}{\vec{r}_{\ell m}}\right|^3 \left(\frac{y_{\ell m}}{r_{\ell m}}\right)^2 \left(1 - e^{i\vec{k}\cdot\vec{r}_{\ell m}}\right) \sim \pi ka\ [1 + \frac{1}{3} \cos(2\theta_k)]$$

$$z_0 - z_k = \sum_\ell \left|\frac{a}{\vec{r}_{\ell m}}\right|^3 \left(\frac{z_{\ell m}}{r_{\ell m}}\right)^2 \left(1 - e^{i\vec{k}\cdot\vec{r}_{\ell m}}\right) \sim \pi ka\ [1 - \frac{1}{3} \cos(2\theta_k)] \tag{6}$$

where θ_k denotes the angle between the direction of the wave vector and the y=[010] axis. One has $c_k = y_k + z_k$, and $y_0=z_0=c_0/2 = \frac{4\pi\bar{e}}{3}$ where the numeric factor $\bar{e}=1.0783$ for the square lattice.[5]

The explicit expression for the field dependence of the magnon gap is ($ka \ll 1$) :

$$E_k \sim [g\mu_B H \cos\phi - 2K_4S^3(\sin^4\phi+\cos^4\phi) + 2DS + 4\pi w\bar{e} - 2\pi wka + 2JS(ka)^2]^{1/2} \cdot$$
$$\cdot [g\mu_B H \cos\phi - 2K_4S^3 \cos(4\phi) + 2\pi wka \sin^2(\theta_k - \phi) + 2JS(ka)^2]^{1/2} \quad (7)$$

where ϕ is solution of eq. (2) i.e. it decreases monotonically with increasing field from the value $\pi/4$ for H=0 to 0 for $g\mu_B H \geq 2K_4S^3$. The expression (7) has a minimum for $g\mu_B H_0 = 2K_4S^3$ i.e. in correspondence to the lowest value of the field capable to align the magnetization along its direction. The cusp-like shape of the minimum turns out to be rounded assuming that the field direction forms a very small angle α with the y=[010] axis.

From the experimental value of H_0 =400 Oe,[2] with g=2 and S=1,[7] we estimate the strength of the in-plane anisotropy to be K_4= 0.027 K. In order to explain the high value of the energy gap for H=0, it is necessary to invoke the presence of a single-ion easy-plane anisotropy D=1.8 K, of the same order as the dipolar interaction $4\pi w$=1.88K (with lattice parameter $a=3.61/\sqrt{2}\ 10^{-8}$ cm). Using for the exchange constant the bulk value J=160 K,[2] the field dependence of the magnon gap, as measured by light scattering in Co/Cu(100) ,[2] is well reproduced in the whole range of fields, as can be seen in fig.1.[8]

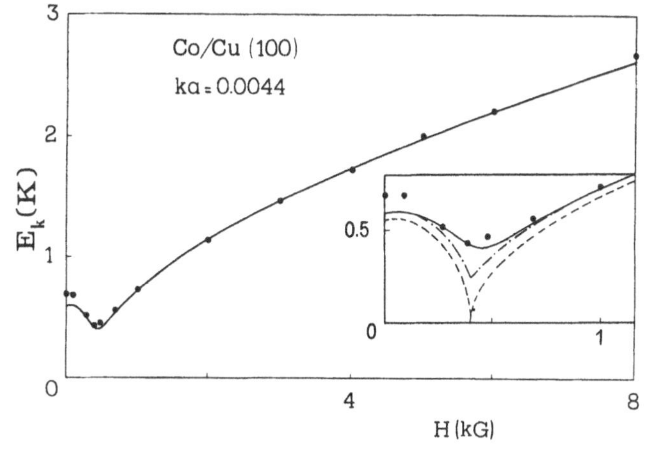

Fig.1- Energy of the magnons with λ=363 nm (ka=0.0044), propagating perpendicularly to the y=[010] axis ($\theta_k=\pi/2$), versus magnetic field for a monolayer of Co/Cu(100). Full line: spin-wave calculation for the field applied at an angle $\alpha=0.01\ \pi$ with the y axis. [Insert: the full line refers to ka=0.0044, $\alpha=0.01\ \pi$; dashed-dotted line: ka=0.0044, $\alpha=0$; dashed line: ka=0, $\alpha=0$]. Dots: experimental data .[2]

In fig.2 we report the temperature dependence of the zero field spin-wave magnetization of the Co/Cu(100) monolayer, compared with the experimental results[9] for a sample of 0.8 ML thickness. The zero-point deviation of the magnetization is very small, of the order of the ratio between the effective easy-plane anisotropy and the exchange. A temperature region follows, extremely narrow owing to the smallness of the energy gap, where the spin reduction is exponentially small. At higher temperatures, our numerical calculation gives a nearly linear temperature dependence of the magnetization, in good agreement with previous analytical estimates[3,5] and with the experimental data for 0.8 ML.[9]

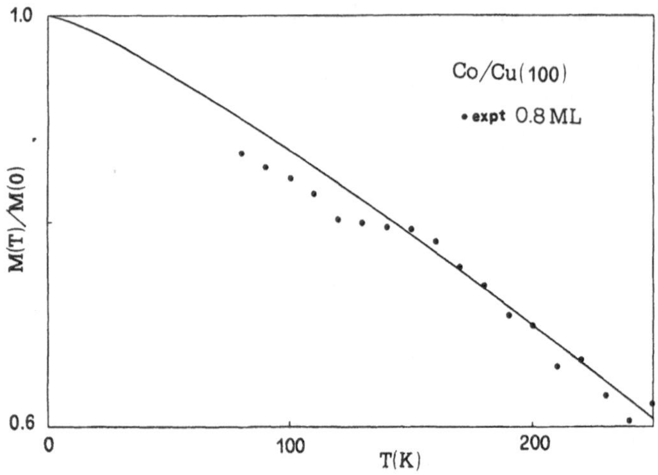

Fig.2- Temperature dependence of the spin-wave magnetization of a Co/Cu(100) monolayer (full line) compared with the experimental results (dots) for 0.8 ML .[9]

In fig.2 we have not reported the experimental magnetization of a 1.3 ML sample, which remains almost constant up to 400 K.[2]

It seems not possible to explain such a great thermal stability in terms of the small measured[2] value of the magnon gap in the same system (even taking into account the ordering effect of the dipolar interaction), without assuming an enhancement of the exchange in the monolayer with respect to the bulk value. This point requires further experimental investigation.

In conclusion, a simple spin-wave microscopic approach was used to obtain detailed information about the magnetic anisotropies which determine the magnetic properties of the epitaxial monolayer Co/Cu(100). In this regard, we took advantage of the light scattering data[2] for the magnon energy gap. The role of the dipolar interaction in the stabilization of long-range order in such 2D systems was found to be important.

REFERENCES

1. Applied Physics A 49/5+6 (1989): special issue on "Magnetism in Ultrathin Films" (guest editor D. Pescia) and references therein.
2. D. Kerkmann, J.A. Wolf, D. Pescia, Th. Woike and P. Grünberg, Solid State Comm. 72, 963 (1989); D. Kerkmann, Appl. Phys.A 49, 523 (1989).
3. S.V. Maleev, Sov.Phys.JETP 43, 1240 (1976).
4. V.L. Pokrovski, and M.V. Feigelman, Sov. Phys. J.E.T.P. 45 : 291 (1977).
5. Y. Yafet, J. Kwo , and E.M. Gyorgy, Phys. Rev. B 33 : 6519 (1988).
6. M.D. Mermin, and H. Wagner, Phys. Rev. Lett. 17 : 1133 (1966).
7. J.A.C. Bland, D. Pescia, and R.L. Willis, Phys.Rev.Lett. 58 : 1244 (1987).
8. Since the light scattering experiment[2] was made at room temperature, a more accurate determination of the values of the magnetic anisotropies would require the renormalization of the Hamiltonian parameters. This is deferred to future work.
9. D. Pescia et al. , private communication (1990).

DYNAMIC SCALING IN THE TWO-DIMENSIONAL HEISENBERG ANTIFERROMAGNET

Gary M. Wysin† and Alan R. Bishop*

†Department of Physics, Kansas State University, Manhattan, KS 66506-2601 USA
*Los Alamos National Laboratory, Los Alamos, NM 87545 USA

INTRODUCTION

The two-dimensional Heisenberg antiferromagnet has generated considerable interest with respect to its applications for understanding copper-oxide-based high temperature superconductors.[1] The model is described by a Hamiltonian,

$$H = J \sum_{(n,m)} \vec{S}_n \cdot \vec{S}_m \tag{1}$$

where $J > 0$ and the sum is over nearest neighbor spin variables $\vec{S}_n$. While analytic calculations of either ground state or dynamic properties are very difficult for this and related nonlinear spin models, it is sometimes possible to extract important information from numerical calculations. Some success in obtaining ground state properties for the spin-1/2 model has resulted from quantum Monte Carlo calculations.[2] However, the principle interest here is in dynamics, for which quantum Monte Carlo calculations are emerging but not yet well-developed. Nevertheless, progress in obtaining quantities such as the dynamic structure function $S(q,\omega)$ is occuring.[3]

One important approach has been the dynamic scaling theory proposed by Chakravarty, Halperin and Nelson (CHN).[4] The CHN theory is based on a mapping (following that of Haldane[5]) of the quantum Heisenberg antiferromagnet onto an equivalent quantum O(3) nonlinear sigma model, whose properties are studied by including quantum fluctuations into a classical O(3) model. This equivalent classical model is a ferromagnetically coupled rotor model, whose Hamiltonian is

$$H = I \sum_n \frac{1}{2} \dot{\vec{\Omega}}_n^2 - K \sum_{(n,m)} \vec{\Omega}_n \cdot \vec{\Omega}_m \tag{2}$$

where $\vec{\Omega}_n$ are the rotor variables with a kinetic energy in addition to the near-neighbor potential energy. The mapping is based on an assumption of low-energy perturbations about the classical antiferromagnetic ground state, and therefore is plausible at adequately low temperature, low frequency, and long wavelengths. The utility of the mapping is that it makes it possible to employ a numerical calculation of classical dynamics of the rotor model and then infer the corresponding quantum properties. This approach was followed by Tyc, Halperin and Chakravarty,[6] who made a simulation of the

Microscopic Aspects of Nonlinearity in Condensed Matter
Edited by A.R. Bishop *et al.*, Plenum Press, New York, 1991

classical rotor model using Langevin dynamics, in an attempt to verify the dynamic scaling assumptions.

A related calculation[7] of dynamics is summarized here, but for the direct evaluation of spin dynamics of the two-dimensional classical Heisenberg antiferromagnet (Eq. 1). This excludes the quantum fluctuations but at the same time avoids the mapping to the rotor model; in the continuum limit the two models should be equivalent. Here we present a finite temperature simulation of the spin dynamics to produce $S(q,\omega)$, from which the wavevector and temperature dependence of the spinwave frequency ω_q and damping γ_q are estimated. These estimates are compared with the dynamic scaling theory.

DYNAMIC SCALING IN THE HEISENBERG ANTIFEROMAGNET

The principle quantity of interest is the dynamic structure function $S(q,\omega)$, defined in terms of the space- and time-displaced correlation function,

$$S(q,\omega) \equiv \frac{1}{2\pi} \int_{-\infty}^{\infty} dt\, e^{-i\omega t} \frac{1}{3} \langle \vec{S}_q(0) \cdot \vec{S}_{-q}(t) \rangle \tag{3}$$

where

$$\vec{S}_k(t) = \frac{1}{N} \sum_n e^{i\vec{k}\cdot\vec{r}_n} \vec{S}_n(t), \quad \vec{q} \equiv (\pi,\pi) - \vec{k} . \tag{4}$$

The wavevector $\vec{q}$ is measured from the antiferromagnet Bragg point. In a scaling regime, (long wavelengths, low temperatures and frequencies) the scaling hypothesis assumes that there is an important physical length scale, which is the correlation length ξ, and a corresponding important time scale τ_s, or, correlation time, and these depend only on the temperature. Dynamical quantities measured at different temperatures will have their frequency and wavelength dependencies determined by the scaled wavevector, $Q \equiv q\xi$, and the scaled frequency, $\Omega \equiv \omega\tau_s$, rather than on q, ω and T separately. More specifically, the CHN scaling assumption for $S(q,\omega)$ is to write it in the form,

$$S(q,\omega) = \tau_s\, S(\vec{q})\, \Phi(\vec{q}\xi, \omega\tau_s) \tag{5}$$

where $\Phi(Q,\Omega)$ is an undetermined scaling function. The amplitude is determined by the static structure function,

$$S(\vec{q}) = \frac{1}{3} \langle \vec{S}_q \cdot \vec{S}_{-q} \rangle \tag{6}$$

which itself displays scaling in the wavevector. Furthermore, the correlation length and time were shown to be related by[4]

$$\frac{\xi}{\tau_s} = \frac{c}{2\pi} \left(\frac{T}{2\pi J S^2} \right)^{1/2} \tag{7}$$

where c is the long-wavelength spinwave velocity.

The form of the scaling function is assumed here to be a product of Lorentzians,[8]

$$S(q,\omega) = \frac{A_q}{[(\omega+\omega_q)^2+\gamma_q^2]\,[(\omega-\omega_q)^2+\gamma_q^2]} \tag{8}$$

where the amplitude A_q, the spinwave frequency ω_q, and the damping γ_q are parameters to be fit in the numerical simulation. If scaling holds, then the spinwave frequency and damping must satisfy some scaling relationships,

$$\omega_q(T) = \omega_s(T)\ \Omega(q\xi), \qquad \gamma_q(T) = \omega_s(T)\ \Gamma(q\xi), \tag{9}$$

where $\Omega(Q)$ and $\Gamma(Q)$ are temperature-independent scaling functions, and $\omega_s \equiv 2\pi/\tau_s$.

NUMERICAL SIMULATION AND RESULTS

For finite temperature dynamics,[7] a classical Monte Carlo simulation was used to generate initial states for an energy-conserving spin dynamics integration of the equations of motion resulting from Hamiltonian (1). The MC calculation was also used to obtain the correlation length, ξ. Dynamic quantities were averaged over five initial states to produce an ensemble average. The calculations were performed using a 100 x 100 square lattice with periodic boundary conditions. A fourth order Runge-Kutta MD integration scheme was used, out to times $t \approx 250\ \hbar/JS$.

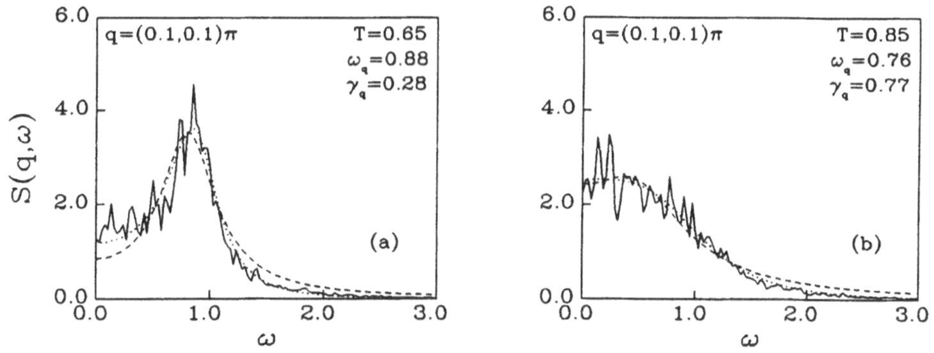

Fig. 1 Typical $S(q,\omega)$ data and curve fits for $q=(0.1,0.1)\pi/a$ at temperatures $T/JS^2=0.65$ and 0.85. The frequency is in units of $JS/\hbar$. The data were generated in the MCMD calculations described in the text. The dotted curve is a least-squares fit to a *product* of Lorentzians [Eq. 8], using spin-wave parameters shown. The dashed curve is a least-squares fit to a *sum* of Lorentzians for comparison (parameters not shown).

Some typical results for $S(q,\omega)$ are shown in Figure 1, together with the corresponding least square fits to Eq. (8). Similar fits to sums of Lorentzians were also made, but were not as good, with the fitted curves consistently being too low at low frequency and too high at high frequency. The damping, verses temperature, is shown in Fig. 2. At the lower temperatures, $T < 0.5$, the non-zero estimates of damping are due to the resolution of the simulation. The damping rate increases rapidly with q and T. The fitted spinwave frequency ω_q defined in Eq. (8) is shown verses wavevector in Fig. 3a. Softening with temperature is clear.

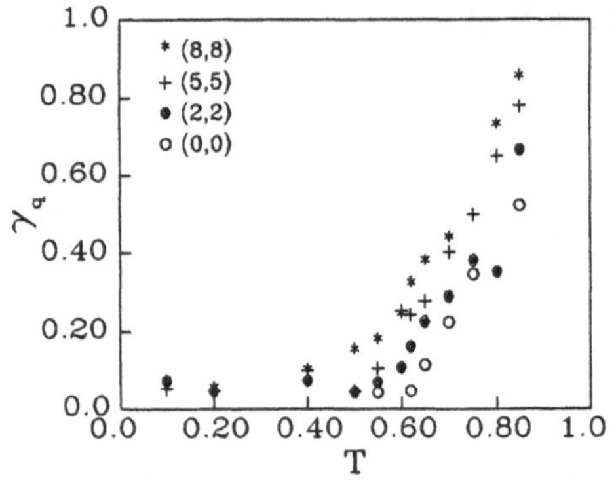

Fig. 2 The spin-wave linewidth γ_q vs. temperature for selected wave vectors q indicated in units of $\pi/50a$. The q=(0,0) data is used to give preliminary estimates of the scaling frequency ω_s.

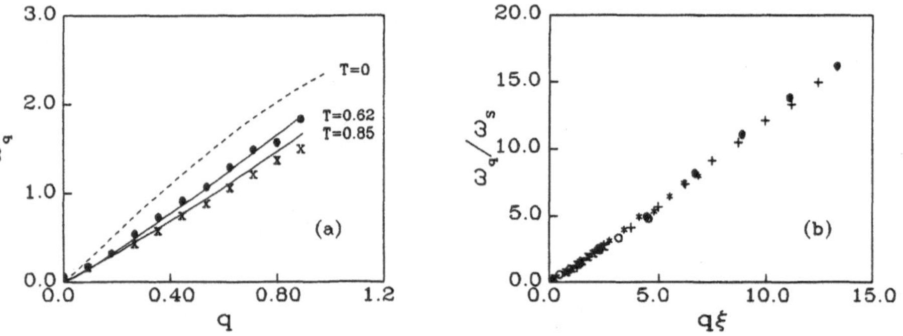

Fig. 3a Spin-wave frequencies ω_q vs. q, obtained from least-square fits as in Fig. 1. The MCMD data are compared with the zero-temperature dispersion, and the dispersion used in Ref. 6 [solid curves, with δ=2.5]. 3b) Scaled spin-wave frequencies ω_q/ω_s as a function of $q\xi$ using data derived at temperatures T/JS^2=0.62 (@), 0.65 (+), 0.70 (*), 0.75 (O), and 0.85 (×). Values of ω_s for each temperature (Table 1, Ref. 7) were first estimated from the q=0 linewidths in Fig. 2 and then corrected slightly such that the scaled frequencies from different temperatures fall on the same curve.

SCALING

The scaling frequency, ω_s, is proportional to the q=0 damping, $\gamma_{q=0}$, which can be read from Fig. 2 (open circles). Alternatively we can use the relationship to define the correlation time[7] τ_s,

$$\tau_s = \frac{\pi}{2} \frac{\int_{-\infty}^{\infty} dt\, S(q=0,t)}{S(q=0,t=0)} \tag{10}$$

For T > 0.6, ω_s increases approximately linearly with temperature. We have also found that the product, $\omega_s\xi$, first increases with T, but then decreases for T > 0.75, inconsistent with the CHN prediction, Eq.(7). This is probably because these temperatures are out of the range of validity of the CHN RG approach, which is an asymptotic theory ($T \rightarrow 0$).

The resulting scaled spinwave frequency, ω_q/ω_s, is shown in Fig. 3b, verses scaled wavevector, $Q = q\xi$, for a set of temperatures. The data at different T fall roughly along one curve, which defines the unknown scaling function, $\Omega(Q)$, and gives strong evidence for a scaling description. The damping rate verses Q determines $\Gamma(Q)$, but with less precision. In principle, even $\Phi(Q,\Omega)$ could be obtained by similar scaling methods but becomes technically difficult.

SUMMARY

The correlation length was estimated from the spatial decay of equal-time static correlations at long wavelengths ($q \rightarrow 0$ limit of Ornstein-Zernicke analysis). Correspondingly, the correlation time was estimated from the temporal decay of long wavelength ($q = 0$) correlations. The scaling theory is a long wavelength description but it should be kept in mind that damped propagating spinwaves make sense only when the wavelength is less than the correlation length ($q>\xi^{-1}$).

For the relatively high temperatures studied here, the dynamic structure function $S(q,\omega)$ is well-approximated by a product of symmetrically located Lorentzians. This implies that the spinwave correlations behave as a damped harmonic oscillator starting from rest with a finite initial displacement. The scaling function $\Phi(Q,\Omega)$ is similarly described by a product of Lorentzians. The spinwave frequency and damping were found to satisfy a scaling form. These results are consistent with earlier simulations of the classical rotor model.[6]

ACKNOWLEDGMENT

Extensive discussions with G. Reiter are gratefully acknowledged.

REFERENCES

1. S. Chakravarty, in "HTC: The Los Alamos Meeting," K. Bedell, D. Pines and J.R. Schrieffer, ed., Addison-Wesley, Redwood, CA (1990).
2. H.-Q. Ding and M.S. Makivic', Phys. Rev. Lett. 64:1449 (1990); J.D. Reger and A.P. Young, Phys. Rev. B37:5978 (1988).
3. S.R. White, D.J. Scalapino, R.L. Sugar and N.E. Bickers, Phys. Rev. Lett. 63:1523 (1989); M. Jarrell and O. Biham, Phys. Rev. Lett. 63:2504 (1989).
4. S. Chakravarty, B.I. Halperin and D. Nelson, Phys. Rev. Lett. 39:2344 (1989).
5. F.D.M. Haldane, Phys. Rev. Lett. 50:1153 (1983); Physics Letters 93A:464 (1983).
6. S. Tyc, B.I. Halperin and S. Chakravarty, Phys. Rev. Lett. 62:835 (1989).
7. G.M. Wysin and A.R. Bishop, Phys. Rev. B42:810 (1990).
8. This form was suggested by calculations of G. Reiter (private communication). A product of Lorentzians also describes $S(q,\omega)$ for the 1-D XY model, as in Eq. 3.2 of D.R. Nelson and D.S. Fisher, Phys. Rev. B16:4945 (1977).

CONTRIBUTORS

INDEX